# ECOLOGY OF EUROPEAN RIVERS

*Frontispiece:* The Rhine valley near Bacharach—a part of the river well known to tourists. The many physical and chemical changes which have taken place in the Rhine are described in Chapter 10.

# Ecology of European Rivers

EDITED BY

## B. A. WHITTON
MA, DSc
Department of Botany
University of Durham
England

BLACKWELL SCIENTIFIC PUBLICATIONS
OXFORD LONDON EDINBURGH
BOSTON PALO ALTO MELBOURNE

© 1984 by
Blackwell Scientific Publications
Editorial offices:
Osney Mead, Oxford, OX2 OEL
8 John Stret, London WC1N 2ES
9 Forrest Road, Edinburgh, EH1 2QH
52 Beacon Street, Boston, Massachusetts 02108,
    USA
706 Cowper Street, Palo Alto, California 94301,
    USA
99 Barry Street, Carlton, Victoria 3053,
    Australia

First published 1984

Set by Oxprint Ltd, Oxford

Printed and bound in Great Britain by
Butler & Tanner Ltd, Frome and London

DISTRIBUTORS

USA and Canada
    Blackwell Scientific Publications Inc
    PO Box 50009, Palo Alto
    California 94303

Australia
    Blackwell Scientific Book Distributors
    31 Advantage Road, Highett
    Victoria 3190

British Library
Cataloguing in Publication Data

Ecology of European rivers.
    1. Stream ecology—Europe
    I. Whitton, B.A.
    574.5′26323′094      QH135

    ISBN 0-632-00816-4

# Contents

Contributors

Preface

BELGIUM

1   Meuse, 1
    J-P. DESCY & A. EMPAIN

DENMARK

2   Gudenå, 25
    P. HEISE

ENGLAND & WALES

3   Wye, 51
    R. W. EDWARDS & M. BROOKER

4   Ebbw, 83
    R. W. EDWARDS, P. F. WILLIAMS & R. WILLIAMS

5   Mersey, 113
    D. G. HOLLAND & J. P. C. HARDING

6   Tees, 145
    B. A. WHITTON & D. T. CRISP

7   The Winterbourne Stream, 179
    A. D. BERRIE & J. F. WRIGHT

FRANCE

8   Lot, 207
    H. DÉCAMPS, J. CAPBLANQ & J. N. TOURENQ

9   Estaragne, 237
    P. LAVANDIER & H. DÉCAMPS

GERMANY

10  Rhine, 265
    G. FRIEDRICH & D. MÜLLER

11  Neckar, 317
    I. PINTÉR & D. BACKHAUS

IRELAND

12  Shannon, 345
K. HORKAN

13  Caragh, 363
H. HEUFF & K. HORKAN

14  Suir, 385
K. HORKAN

ITALY

15  Po, 401
G. CHIAUDANI & R. MARCHETTI

NETHERLANDS

16  Lower Rhine-Meuse, 437
G. VAN URK

NORWAY

17  Glåma, 469
O. M. SKULBERG & A. LILLEHAMMER

POLAND

18  Dunajec, 499
B. KAWECKA & B. SZCZĘSNY

SPAIN

19  Llobregat, 527
N. F. PRAT, M. A. PUIG, G. GONZALEZ, M. F. TORT & M. ESTRADA

SWEDEN

20  Rönneå, 553
A. ALMESTRAND & A. ALMESTRAND

Appendix A. Guide to Further Literature on European Rivers, 557

Appendix B. Authorities, Synonyms and English Names, 606

Organism Index, 611

Subject Index, 627

# Contributors

A. ALMESTRAND *Scandiaconsult AB, P.O. Box 17040, S-200 10 Malmo, Sweden*

A. ALMESTRAND *Department of Plant Physiology, University of Lund, P.O. Box 7007, S-220 07 Lund, Sweden*

D. BACKHAUS *Landesanstalt für Umweltschutz Baden-Württemberg, Institut für Wasser- und Abfallwirtschaft, Postfach 210752, 7500 Karlsruhe 21, Germany*

A. D. BERRIE *Freshwater Biological Association River Laboratory, East Stoke, Wareham, Dorset BH20 6BB, U.K.*

M. BROOKER *Welsh Water Authority, Directorate of Scientific Services, Nash Area Laboratory, West Nash, Newport, Gwent NP6 2YH, U.K.*

J. CAPBLANCQ *Laboratoire d'Hydrobiologie, Université Paul-Sabatier, 118 Route Narbonne, 31062 Toulouse Cedex, France*

G. CHIAUDANI *Dipartimento di Biologia (Sezione di Ecologia), Università degli Studi Milano, Via Celoria 26, 20133 Milano, Italy*

D. T. CRISP *Teesdale Unit, c/o Northumbrian Water Authority, Lartington Treatment Plant, Lartington, Barnard Castle, Co. Durham, DL12 9DW, U.K.*

H. DÉCAMPS *Centre d'Ecologie des Resources Renouvelables, 29 rue Jeanne-Marvig, 31055 Toulouse Cedex, France*

J-P. DESCY *Département de Botanique, Université de Liège, Sart Tilman, B 4000 Liège, Belgium*

R. W. EDWARDS *Department of Applied Biology, University of Wales Institute of Science and Technology, King Edward VII Avenue, Cardiff CF1 3NU, U.K.*

A. EMPAIN *Jardin Botanique de l'État, Domaine de Bouchout, B1860 Meise, Belgium*

M. ESTRADA *Instituto de Investigaciones Pesqueras, Paseo Nacional s/n, Barcelona 3, Spain*

G. FRIEDRICH *Landesamt für Wasser und Abfall Nordrhein-Westfalen, Auf dem Draap 25, 4000 Düsseldorf 1, Germany*

G. GONZALEZ *Department d'Ecologia, Facultat de Biologia, Universitat de Barcelona, Avda. Diagonal 645, Barcelona 28, Spain*

P. B. HEISE *The Gudenå Committee, Teknisk Forvaltning, Århus Amtskommune, Lyseng Allé 1, DK-8270 Højbjerg, Denmark*

H. HEUFF *Department of Botany, University College, Dublin, Ireland (but present address c/o Forest and Wildlife Service, 2 Sidmonton Place, Bray, Co. Wicklow, Ireland)*

D. G. HOLLAND *North West Water Authority, Rivers Division, Liverpool Road, Great Sankey, Warrington WA5 3LP, U.K.*

K. HORKAN *An Foras Forbartha Teoranta, Laboratories, Pottery Road, Kill of the Grange, Dun Laoghaire, Co. Dublin, Ireland*

J. P. C. HARDING *Wanlip Water Reclamation Works, Severn-Trent Water Authority, Soar Division, near Leicester, U.K. (article written during employment with North West Water Authority)*

B. KAWECKA *Laboratory of Water Biology, Polish Academy of Science, Sławkowska 17, Kraków, Poland*

# CONTRIBUTORS

P. LAVANDIER *Laboratoire d'Hydrobiologie, Université Paul-Sabatier, 118 Route Narbonne, 31062 Toulouse Cedex, France*

A. LILLEHAMMER *Zoological Museum, University of Oslo, Toyen, Norway*

R. MARCHETTI *Consiglio Nazionale della Ricerche, Istituto di Ricerca sulle Acque, 20047 Brugherio (Milano), Italy; (also at Departimento di Biologia (Sezione di Ecologia), Università degli Studi di Milano, Via Celoria 26, 20133 Milano, Italy)*

D. MÜLLER *Bundesanstalt für Gewässerkunde, Postfach 309, 5400 Koblenz, Germany*

I. PINTÉR *Landesanstalt für Umweltschutz Baden-Württemberg, Institut für Wasser- und Abfallwirtschaft, Postfach 210752, 7500 Karlsruhe 21, Germany*

N. PRAT *Department d'Ecologia, Facultat de Biologia, Universitat de Barcelona, Avda. Diagonal 645, Barcelona 28, Spain*

M. A. PUIG *Department d'Ecologia, Facultat de Biologia, Universitat de Barcelona, Avda. Diagonal 645, Barcelona 28, Spain*

B. SZCZĘSNY *Laboratory of Water Biology, Polish Academy of Science, Sławkowska 17, Kraków, Poland*

O. M. SKULBERG *Norwegian Institute for Water Research, P.O. Box 333, Blindern, Oslo 3, Norway*

M. F. TORT *Department d'Ecologia, Facultat de Biologia, Universitat de Barcelona, Avda. Diagonal 645, Barcelona 28, Spain*

J. N. TOURENQ *Laboratoire d'Hydrobiologie, Université Paul-Sabatier, 118 Route Narbonne, 31062 Toulouse Cedex, France*

G. VAN URK *Rijksinstituut voor Zuivering van Afvalwater, Postbus 17, Maerlant 6, 8200 AA Lelystad, The Netherlands*

B. A. WHITTON *Department of Botany, University of Durham, Co. Durham DH1 3LE, U.K.*

P. F. WILLIAMS *Nature Conservancy Council, 44 The Parade, Roath, Cardiff CF2 3AB, U.K.*

R. WILLIAMS *Department of Applied Biology, University of Wales Institute of Science and Technology, King Edward VII Avenue, Cardiff CF1 3NU, U.K.*

J. F. WRIGHT *Freshwater Biological Association River Laboratory, East Stoke, Wareham, Dorset BH20 6BB, U.K.*

# Preface

The aim of this book is to give an overall picture of the ecology of selected rivers in Europe. What are the key features of the environment? What generalizations can be made about their bacteria, plants and animals? What is known about the productivity in different parts of the river? How has man changed the river from its natural state and how is the river managed nowadays? The chapters have been written both to interest the student or researcher with a general interest in river ecology and also to provide more detailed information for those wanting to know about a particular river.

## An overview of the chapters

The twenty chapters include examples from twelve countries. The examples range in size from the Rhine to two tiny streams which have been the subject of intensive research and show features of particular interest. Most of the rivers might be described as medium-sized by European standards, running for distances of several hundred kilometres. No doubt they are small by world standards, but this has the advantage that in at least some cases it is possible to generalize about the whole river. Europe's largest river, the Danube, was not included because it received monographic treatment during the 1960s. (R. Liepolt, 1967, *Limnologie der Donau*, Schweizerbart, Stuttgart.)

Chapter 1 deals with the Meuse in Belgium, where the river is subject to a particularly wide variety of pollutants: heavy metals, radionuclides, organic and industrial wastes and thermal pollution. In response to this, researchers at Liège have developed a variety of biologically orientated methods for monitoring these pollutants. The chapter shows how these methods can provide general information of the ecology of the river. The Meuse is discussed again in Chapter 16.

The Danish, Norwegian and Swedish rivers (Gudenå, Chapter 2; Glåma, Chapter 17; Rönnea, Chapter 20) resemble each other in that they have relatively long lowland stretches with many lakes and reservoirs in the catchment which have a profound influence on their ecology. All three rivers are clearly very important for recreational purposes and subject to intensive management.

The inclusion of five rivers from the United Kingdom is a reflection of editorial bias rather than a lack of well-studied rivers elsewhere in Europe. It seemed at one time that the Wye (Chapter 3), which rises in Wales but flows into the Severn in England, would eventually be influenced by the building of a much increased reservoir at an upstream site. At the same time this river has been considered particularly important for conservation purposes. The possible conflict of interests has led to a wide range of studies on the Wye and it is possible to generalize in a way which is difficult for most rivers. The Ebbw—in Wales (Chapter 4)—provides a good example of a river with an upstream clean stretch and a downstream one heavily polluted by coal-mining and the effluent from a major steel-works. It is much more difficult to generalize on the Mersey system (Chapter 5), because of the variety of types of pollution. At one time this system included some of the most notorious stretches of river in Britain, but many have been improved greatly during the past two decades. That we know as much as we do about the ecology of the Mersey system, and the ecological changes which have taken place, is due to the exceptional enthusiasm of the biological staff of the local water authority.

The Tees (Chapter 6) is similar to the Wye in that it is fast-moving, subject to rapid changes in flow at all times of year and relatively unpolluted for the majority of its length. A chemical and biological survey made on this river in the 1930s was, at the time, the most detailed study which had been made on any British river and one of the most detailed for any river in Europe. It is therefore possible to make some comment on long-term changes.

A winterbourne is the name given to a stream rising on the chalk of southern England and which flows for only part of the year. The example chosen (Chapter 7), which is actually known as the Winterbourne Stream, dries out in the upper, but not the lower part of its course each summer. The Estaragne (Chapter 9) is another small stream, which drains a permanent field of consolidated snow in the High Pyrenees. There is again a marked seasonal pattern of flow changes, but in this case minimum flows are in winter, when the remaining water passes under the thick snow cover, which lasts for six to eight months near the top of the stream.

Movement of water in the Lot (Chapter 8), a major tributary of the Garonne in SW France, has been sufficiently influenced by on-stream reservoirs and weirs to permit dense phytoplankton populations to develop when summer flows are low. Although many features of this river still await study, some aspects of chemical and phytoplankton dynamics in the stretch below the junction of the Haut-Lot and the Truyère have been quantified to a much greater extent than on most rivers. The chapter summarizes this work, which includes modelling and behaviour of phytoplankton under different flow conditions.

The Rhine deservedly gets the longest treatment. Chapter 10 takes the river down to the German–Dutch border, while the downstream parts are included in Chapter 16. Some of the oldest accounts of river biology anywhere are for the Rhine and Chapter 10 refers to many older papers, as well as 1983 reports, in order to give a broad picture of changes which have taken place over the past century. Although the middle and lower stretches of the Rhine are still heavily polluted by wastes from the potash mines in Alsace, in general the authors' verdict on the state of pollution in the river is reasonably favourable. Improved conditions since the early 1970s are due largely to control measures, though also slightly to decreased economic activity. The pollution status of the Neckar (Chapter 11), one of the Rhine's main tributaries, has again much improved in recent years, but the condition of the effluents entering the river in the region of Stuttgart is still most unsatisfactory. Problems due to organic pollution on the Neckar are compounded by the heat load added by a series of power stations. The chapter has details of measures which have been adopted to minimize the effects of thermal pollution.

The three Irish rivers include Ireland's longest river, the Shannon (Chapter 12), one of its least polluted rivers, the Caragh (Chapter 13), and one of its rivers most subject to nutrient enrichment, the Suir (Chapter 14). The Caragh has no doubt changed considerably since pre-historic times because of the much reduced tree cover; nevertheless its soft water, low in inorganic nutrients but rich in humic materials, provides an interesting environment which deserves much more detailed study.

The Po (Chapter 15) has the second largest catchment of the rivers in this book. Many of its tributaries derive ultimately from mountain streams, but their influence is partly buffered by the presence of large lakes on the tributaries draining the Alps to the north. Much of the chapter is concerned with the middle and lower parts of the main river.

The Polish river, the Dunajec (Chapter 18), also rises in a high mountain region, but in this case the more upstream parts of the tributaries have received the most study. Its uppermost streams have some resemblance to the Estaragne, though with a much shorter period of snow cover. The river flows through some of the most beautiful scenery of any river described here, but some of the more downstream stretches receive heavy loads of poorly treated organic effluent. Two on-stream reservoirs also have a marked effect on the downstream river.

Chapter 16 (lower Meuse–Rhine) deals with the large river system flowing through the Netherlands. This is complicated not only because of the branching and coming together of different rivers, but also because of the huge changes which man has made to the behaviour of the rivers and the location of their channels. In particular a major dam, which was built to protect the south-western part of the country and completed in 1970, has had profound ecological effects. The chapter includes an account of the unique freshwater tidal area, which was largely destroyed by the dam, and of events since then.

Up to now there has been less research on rivers in Spain than in any other country covered in this book. The team of river ecologists at Barcelona are, however, clearly determined to change this situation and have assembled much information about the Llobregat within just a few years. This river (Chapter 19) arises on the other side of the Pyrenees to the Estaragne. It is a typical Mediterranean river in that flows are very high in spring at the time of snow melt, but low in late summer. The flow regime is somewhat modified by the presence of an on-stream reservoir on the main river and the chemistry of the water is changed downstream by the influence of various salt beds.

An understanding of the ecological processes operating in a river is essential if all the varying demands on the water are to be resolved, yet at the same time the river is to retain something of its original character. Progress can come partly from generalization based on observations of a wide variety of rivers and partly from intensive quantitative studies on a few selected ones. It is hoped that *Ecology of European Rivers* aids in both ways.

*Editorial comment*

With such a variety of countries and authors, it is not surprising that the original versions of the chapters had a great variety of names and terms for the same thing. I have gone some way towards standardizing these, but it seemed both impractical and unnecessary to bring about complete uniformity. Local names for towns and rivers have mostly been retained, except where the English equivalent is especially familiar, as with the Rhine. Terms and units for most physical and chemical variables are the same throughout the book. Among the exceptions are 'flow' and 'discharge', words which are in most cases used inter-changeably. Concentrations of ions such as nitrate and phosphate are almost always expressed as the element.

As far as possible the same Latin binomial has always been used for a particular species, though synonyms are sometimes added in brackets. No doubt I have overlooked some examples of synonymy, especially among the invertebrates. In the case of common flowering plants and fish, authors differ as to whether they use English, Latin or both names. A list of equivalents (Appendix B) should prevent any confusion.

It will be clear from the references how much information is available only in local journals and reports. An English translation of the title of a book or paper is added after each reference in Danish, Dutch, Italian, Norwegian, Polish, Spanish and Swedish, but not French or German. As this is based on the assumption that the last two languages are known more widely than the others, I hope it gives no offence to readers in other countries.

I thank all the contributors for their patience with the editor and his editing. Because of a delay in processing chapters which were submitted first, several chapters had to be modified considerably to bring them up-to-date at proof stage. The authors of Chapter 4 have now published a book on the Wye (*The Ecology of the Wye*, 1982, Junk), which gives a much fuller account of this river. I am also most grateful to the many people who have aided with advice, to Audrey Richardson for typing various early versions of chapters, to Margaret Creighton for typing the camera-ready version, with all the difficulties due to accents in the various languages, and to Penny Baker of Blackwell Scientific Publications for dealing with a range of practical problems.

Brian Whitton

# 1 · Meuse

J-P. DESCY & A. EMPAIN

1.1   Introduction                                          2

1.2   General data                                          3

1.3   Physico-chemical features of the rivers              6

    1.31   Natural typology                             8
    1.32   Pollution                                    8

        1.321   Mineral pollution                    8
        1.322   Organic pollution                   10
        1.323   Thermal pollution                   10
        1.324   Radioactive pollution               11

1.4   Ecology of aquatic communities                      12

    1.41   Algae                                       12
    1.42   Bryophytes                                  13
    1.43   Fish                                        14

1.5   Water quality assessment by communities             14

    1.51   Use of diatoms                              14
    1.52   Use of bryophytes                           15
    1.53   Use of macroinvertebrates                   17
    1.54   Use of fish                                 17

1.6   Use of bryophytes to monitor heavy metals and       18
      radionuclides

1.7   Discussion                                           20

References                                                 21

## 1.1   INTRODUCTION

This chapter deals with the 38.7% of the Meuse catchment which lies
within Belgium, while the more downstream part of the river is discussed
in Chapter 16.   The river basin is of considerable importance for the
south of Belgium, providing drinking water and also water for industry
and power stations; the river itself is also important for navigation
and, together with its tributaries, for recreation.   Although a wide
range of ecological studies have been made on the Meuse basin, few deal
in depth with fundamental ecological questions.   Nevertheless ecologists
have contributed by focusing their attention to the use of biological
methods for assessing the numerous pollutants which reach the rivers and
it is this aspect which the present chapters treat in most detail.   Part
of the work has come from our Laboratory of Hydrobiology at Liège
University; for the remainder we have included ecological results with a
special interest at the scale of the whole basin.

*Fig. 1.1*  Meuse at Liège
(200 m wide) at confluence
with R.Ourthe (right).   The
main industrial zone is
located upstream of this site.

    An indication of the importance of the Meuse basin may be provided
by some simple statistics.   It supplies most of the drinking water for
the southern part of the country and also Bruxelles and the great cities
of the north (Gent, Antwerpen); 80% of the water is collected underground and 20% by surface water treatment.   In addition artificial lakes
provide a total storage capacity of $58.3 \times 10^6$ m$^3$ per year.   Surface
waters are also needed for industry ($455 \times 10^6$ m$^3$ per year), agriculture
($235 \times 10^6$ m$^3$ per year) and navigation ($1800 \times 10^6$ per year)
(Vereerstraeten, 1971).   As the supply of underground water decreases,
the use of surface waters will rise, so the preservation of river water
quality is going to become more and more essential.


    *Fig. 1.2*  (opposite)   Meuse and its tributaries:  A) whole basin;
    Belgian part lies in shaded area; B) schematic geological features;
    C) Belgian part of basin, showing river system and main reservoirs.

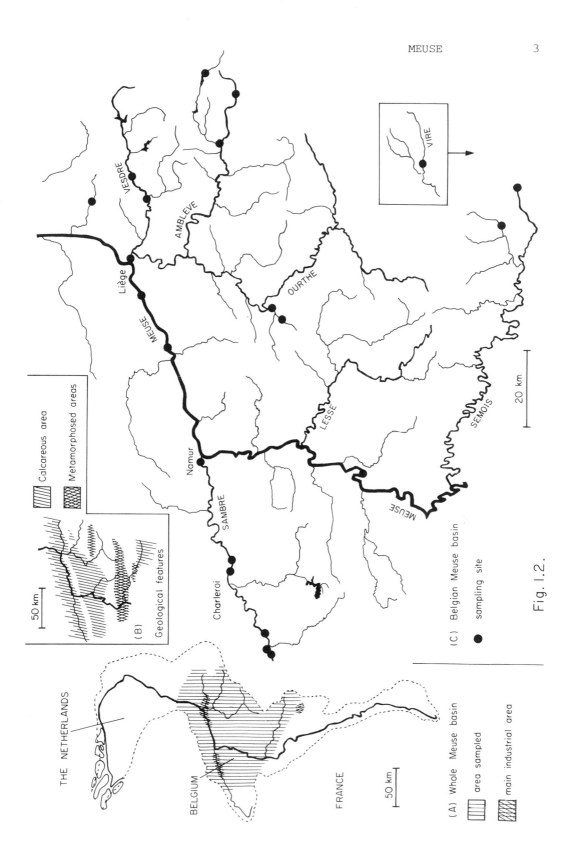

Fig. I.2.

(A) Whole Meuse basin
area sampled
main industrial area

(B) Geological features
Calcareous area
Metamorphosed areas

(C) Belgian Meuse basin
● sampling site

1.2   GENERAL DATA

DIMENSIONS

   Area of catchment           13489 $km^2$ in Belgium (38.7% of whole basin)

CLIMATE

   Precipitation               800-1400 mm $yr^{-1}$ (over 1000 mm $yr^{-1}$ in the
                               Ardenne)

   Temperature (mean annual)   7-9 $^{\circ}$C

GEOLOGY (see Fig. 1.2)

schistose and siliceous rocks in the east and south (Lower Devonian and
Cambro-Silurian)
limestone and calcareous rocks in remainder (middle and upper Devonian,
Carboniferous and secondary rocks)

LAND USE
   Ardennes                    forestry, agriculture, livestock
   Other rural areas           agriculture, livestock
   Industrial development
      iron-steel industry      mostly around Charleroi and Liège
      textile industries       Verviers
      power stations           include two nuclear plants, one in France,
                               near the border (Chooz, 270 MWe) and one
                               upstream of Liège (Tihange I, 879 MWe)

POPULATION

   Ardennes    10-50 people $km^{-2}$    Typical towns              3000 people $km^{-2}$
   Other rural                           Large cities along the    25000 people $km^{-2}$
      areas    50-300 people $km^{-2}$      "iron-coal" axis

PHYSICAL AND CHEMICAL FEATURES OF RIVER

(Figures for lengths given in brackets refer to the whole river; those for
sources in brackets refer to rivers which originate in France. Data for
precipitation and flow are based on 1954-1963.

|  | length (km) | source (m) | mean slope | mean rainfall | flow module | flow ($m^3$ $s^{-1}$) minimum monthly average | flow ($m^3$ $s^{-1}$) maximum monthly average |
|---|---|---|---|---|---|---|---|
| Meuse | 183 (890) | (402) | 0.25 | 898.8 | 244 | 70 | 436 |
| Semois | 177 | 413 | 1.54 | 1144.6 | 29 | 17.5 | 40 |
| Lesse | 83 | 405 | 3.7 | 947.7 | 24.3 | 11.6 | 39.9 |
| Sambre | 86 (190) | (212) | 0.25 | 832.8 | 27.5 | 7.6 | 59.7 |
| Ourthe | 175 | 507 | 2.5 | 976.8 | 61 | 10.3 | 32 |

Ranges for physico-chemical variables in the Meuse basin

| variable | unit | mean value for 'natural' stations | 'natural range' | extremes | |
|---|---|---|---|---|---|
| t | °C | - | 0 - 23 | 32 | Sambre et Marcinelle and Couillet |
| pH | - | 6.9 | 3.4 - 9.1 | min: 10.0 | Warche et Bullange |
| $O_2$ | % saturation | 98.2 | 74 - 130 | 0% | Sambre et Marcinelle Marchette at Rabosée |
| | | | | max:140% | Semois at Bohan |
| conductivity | $\mu S\ cm^{-1}$ at 25°C | 216 | 11 - 674 | 3375 | Sambre at Couillet |
| alkalinity | mg $l^{-1}$ $CaCO_3$ | 63 | 0.0 - 280 | 383 | Marchette at Rabosée |
| total organic C | mg $l^{-1}$ | 3.0 | <0.2 - 13.1 | 176 | Sambre at Couillet |
| Na | " | 6.5 | 1.0 - 18.0 | 288 | Sambre at Namur |
| K | " | 1.7 | <0.1 - 9.2 | 28.4 | Vesdre at Pepinster |
| Mg | " | 4.7 | 0.7 - 17.4 | 23.7 | Berwinne at Val Dieu |
| Ca | " | 28.6 | 1.1 - 112 | 423 | Sambre at Couillet |
| ammonia-N | " | 0.134 | <0.002 - 0.660 | 21.9 | Semois at Arlon |
| $NO_2$-N | " | 0.024 | <0.002 - 0.214 | 9.61 | Sambre at Marcinelle |
| $NO_3$-N | " | 2.2 | <0.1 - 6.3 | 10.1 | Vire at Latour |
| total reactive orthophosphate-P | " | 0.128 | <0.005 - 0.6 | 16.6 | Marchette at Rabosée |
| reactive-Si | " | 2.3 | 0.5 - 8.4 | 24 | Berwinne at Val Dieu |
| Cl | " | 14.5 | 2.3 - 43.8 | 840 | Sambre at Couillet |
| $SO_4$-S | " | 6.7 | 1.0 - 20.7 | 119 | Berwinne at Val Dieu |

The values for 'natural' stations are a selection of 531 results from 1040 (123 stations from 201): all obviously polluted stations are discarded from the pool, but 'natural' reference in Belgium is utopian. The 'natural' range includes the lowest value found, while the upper limit is defined by the largest type value (mean plus two sigma); due to the difficulty of defining true natural conditions, this upper side is often high (see ammonia, nitrite and phosphate).

1.3   PHYSICO-CHEMICAL FEATURES OF THE RIVERS

## 1.3   NATURAL TYPOLOGY

Geology is a key factor determining the ecological features of running
waters in the Belgian part of the Meuse basin.  Differences in litho-
logy lead to a great range in the chemical composition of the waters
and several attempts have been made to make a chemical classification
helpful in interpreting the ecology of rivers in the region.

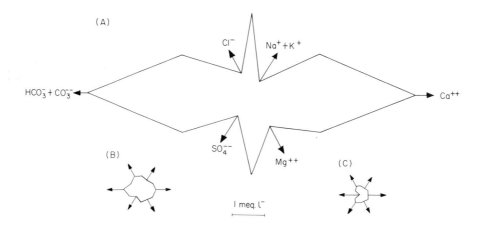

*Fig. 1.3*  Major elements dissolved in water, according to Symoens
(1957):  A, alkaline type;   B, neutral type;  C, acid type.


Symoens (1957) proposed a classification based on the mineral content of the water (Fig. 1.3).  As his terminology was derived from
local names, it is replaced here with acid, neutral and alkaline.

   Acid type ('type fagnard') acid water (pH 4 to 6, or lower) with
      low mineral content, especially $HCO_3^-$; usually rise in Ardennes
      peat-bogs and flow over Cambrian rocks; mountain and boreal
      algae present in flora (Fabri & Leclercq, 1979).
   Neutral type ('type ardennais') pH circumneutral and with a
      slightly higher mineral content than the acid type of water;
      usually flow over lower Devonian or Cambro-Silurian rocks.
   Alkaline type ('type condrusien') pH over 7.5 in natural
      conditions;  conductivity in range 250 to 600 µS cm$^{-1}$ at 25 °C
      and mineral  content dominated by $Ca^{2+}$ and $HCO_3^-$; flow over
      rocks rich in calcium carbonate.


*Fig. 1.4* (opposite)  Alkalinity in rivers of Belgian Meuse Basin.
(Mean for 1973-1980).

Concentration (mg l⁻¹ CaCO₃) proportional to radius

e.g.   ≤15   ≤30   ≤60   ≤120   ≤240

Fig. I.4. Alkalinity

In a synthesis we made for a general survey of the Meuse basin
(Descy & Empain, 1981), we attempted to redefine this chemical typology
statistically.  A comparison with the above types is given in Table 1.1.

*Table 1.1*  Types of water in the Meuse basin.  (Alkalinity range is
equal to mean ± 2 standard deviations, including about 95% of the
natural variation in each type).

| General type (Descy & Empain, 1981) | Symoens (1957) type | Alkalinity (mg $l^{-1}$ $CaCO_3$) |
|---|---|---|
| 1 | acid (fagnard) | 0 – 5 |
| 2 | neutral (ardennais) | 6 – 20 |
| 3 | | 16 – 50 |
| 4 | alkaline (condrusien) | 51 – 110 |
| 5 | | 130 – 265 |

Many rivers belong successively to different types, as they rise in the
Ardennes, often in peat-bogs, and then flow over Devonian or
Carboniferous limestone; such rivers rarely reach type 5, however;
these latter rise on limestone or chalk (Fig. 1.4).

This typology is important when considering the applied ecology of
the Meuse basin.  It provides a guide to natural productivity at the
different trophic levels and the composition and structure of the plant
communities (see below); it is also very useful when assessing
pollution, since the ecological effects of most pollutants appear to
depend on the natural type of the river concerned.

## 1.32  POLLUTION

### 1.321 *Mineral pollution*

Pollution leading to marked changes in the composition of major
elements occurs in the Meuse basin, mainly in industrial zones.  The
impact of such pollution depends on the natural chemistry of the water-
course.  For example, Na pollution at a particular level might be
acceptable for a calcareous river, but three times too high for a
stream of type 1.  There are also many sites where pollution by heavy
metals takes place.  The majority occur in the 'iron-coal' axis, along
the rivers Sambre and Meuse; others are associated with old mine
operations or ore processing (mainly Gueule and Vesdre) or with
particular industries like tanneries (Warche, Amblève, Viroin).

*Fig. 1.5* (opposite)   Reactive phosphate (as P) in Belgian Meuse
basin.    (Mean for 1973-1980).

Concentration ($\mu$g ($^{-1}$ P) proportional to radius

e.g. $\leq$125  $\leq$250  $\leq$500  $\leq$1000  $\leq$2000

Fig. I.5. Reactive phosphate P in Meuse basin

Particularly heavy contamination with Cu occurs in the upper Sambre and with chromium in the Warche ; (Mouvet, 1981) ; Zn, Cd and Pb pollution are associated with industrial areas and old mining areas. Waters are also enriched naturally by heavy metals in some areas ; typically these are associated with metamorphism and old geological faults.

## 1.322 Organic pollution

Heavy organic pollution occurs mainly where the population density is high and, in the large rivers, is often combined with industrial pollution.  In some other instances, extreme situations may be encountered where the sewage flow is much higher than that of the river itself ; two of the most striking examples are the Semois below Arlon, which is seriously polluted for about 30 km and the Magne, which receives slaughter-house effluent near the source of the river. The effects of these effluents are evident in the reduced levels of dissolved oxygen in many stretches.  Quite apart from these situations, most of the rivers are subject to marked eutrophication (Fig. 1.5), the only exceptions being the highest points of the basin, where the density of the population is low.  It will be clear that treatment of wastes is poorly developed in Belgium, which is probably one of the most backward countries in Europe in this respect ; for instance, only about 6% of the sewage is treated at present, as opposed to 80% in our nearest neighbour, the Grand-Duchy of Luxembourg.

## 1.323 Thermal pollution

Thermal pollution from the cooling systems of factories and power stations occurs at some sites in addition to other types of pollution. The amount of heat discharged is sometimes sufficient to bring about a marked rise in temperature.  For instance, a study carried out in 1973-74 in the Sambre below Charleroi showed that the mean temperature rise over a stretch of a few kilometres was 9.3°C, with a maximum recorded of 32°C (Descy & Empain, 1976).  At the Tihange I power plant on the Meuse, the heat discharge typically brings about a 2 to 3°C increase when mixing is complete (observed maximum 4.2°C).  This in itself is unlikely to be critical for many processes, but the total rise from a variety of sources may reach 10°C at Liège.  As these conditions occur with low flow (<50 $m^3$ $s^{-1}$) and consequently a high organic content in the water, the oxygen levels may drop below 50% saturation.  Mathematical models have been constructed (Smitz, 1976) to predict the thermal behaviour of the Meuse in relation to river flow. Unfortunately the results indicate that temperature rises resulting from pollution cannot be reduced sharply by using reservoir water to regulate flows at critical periods.  A system of monitoring has been developed to assess the ecological impact of the Tihange I power station ; this incorporates most of the techniques mentioned in the present chapter (Lambinon et al., 1977 ; Kirchmann et al., 1980).  A similar programme

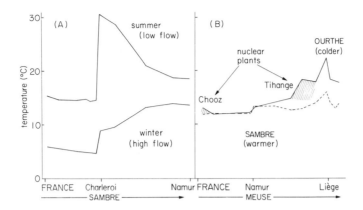

Fig. 1.6    Thermal profiles in Sambre and Meuse.
A)  Summer and winter profiles.  Summer has high warming, but good
    atmospherical cooling downstream due to low flow-rate.
B)  Typical profiles before and after commencement of nuclear plant
    at Tihange.


should be put into action now in the Upper Meuse in advance of further
nuclear power stations in order to provide ecological base-lines for
future comparison ('point zéro écologique').

   Problems associated with thermal pollution are also likely to arise
with the development of nuclear power stations at Chooz, just upstream
of the French-Belgian border.  Here a model of possible thermal changes
in the water has shown that it would be better to control the heat
discharge by using and releasing water at several (minimum of two) sites
along the main river.  Such a programme would however require the
construction of new dams and these themselves would undoubtedly cause
other types of damage to the environment.


*1.324 Radioactive pollution*

Two sites on the Meuse have been the subject of special checks on
radiocontamination of the aquatic ecosystem; these are the Belgian
stretch downstream of the Chooz nuclear plant (270 MWe) and the stretch
below the Tihange I nuclear plant (870 MWe).  Most of the ecological
monitoring is based on the measurement of activity in aquatic organisms,
particularly bryophytes (1.42) and fish (1.43).  The level of contamination is low under normal conditions, with activities in water and
suspended matter often below the limits of detection; some radio-
elements have however been found to reach detectable levels in various
organisms and in sediments (Kirchmann *et al.*, 1980).  Experimental
studies included in this programme deal with uptake and decontamination
kinetics, e.g. uptake of $^3$H by *Scenedesmus obliquus* in culture.

*Table 1.2*   Radionuclides in water, bryophytes and fish, downstream of the nuclear plant of Chooz (Kirchmann *et al.*, 1979); (pCi $1^{-1}$, pCi $kg^{-1}$ radionuclide wet weight; fish are eviscerated).

|  |  | 1971 | 1974 | 1977 |
|---|---|---|---|---|
| $^{54}$Mn | Water | 6.52 | 0.15 | 0.03 |
|  | Bryophytes | $10^4-10^5$ | $10^3-10^4$ | $10^3-10^4$ |
|  | Fish | $10^2-10^3$ | $<10^2-10^3$ | $<10^1-10^2$ |
| $^{60}$Co | Water | 2.19 | 0.16 | 0.05 |
|  | Bryophytes | $10^4-10^5$ | $10^3-10^4$ | $<10^2-10^4$ |
|  | Fish | $10^2-10^3$ | $<10^1-10^2$ | $<10^2$ |
| $^{134}$Cs | Water | 1.63 | 1.31 | 0.10 |
|  | Bryophytes | $10^3-10^4$ | $10^2-10^4$ | $10^2-10^3$ |
|  | Fish | $10^3-10^4$ | $10^2-10^3$ | $10^1-10^2$ |
| $^{137}$Cs | Water | 0.27 | 1.56 | 0.14 |
|  | Bryophytes | $10^3-10^4$ | $10^2-10^3$ | $10^1-10^3$ |
|  | Fish | $10^2-10^3$ | $10^2-10^3$ | $10^1-10^2$ |

## 1.4   ECOLOGY OF AQUATIC COMMUNITIES

### 1.41   ALGAE

The benthos has been studied in more detail than the plankton.  A complete study of the Belgian Meuse, Descy (1973, 1976) included both macroscopic and microscopic species.  Typically there is a *Cladophora glomerata* community, accompanied under particular conditions by other species such as *Lemanea fluviatilis*, *Vaucheria* sp.,  *Bangia atropurpurea* and blue-green algae.   If the basin as a whole is considered then the communities dominated by particular macro-algae can be seen to differ according to the natural water quality, as shown by Symoens (1957), who noted an ecological succession of *Microspora, Vaucheria* and *Cladophora* communities.  A more precise approach is now required, with statistical and multivariate analysis of quantitative samples taken from river sites over the whole basin.  Such an approach has already been adopted since 1976 for diatom communities (Descy, 1979, 1980).  This has led to a typology of communities closely related to the chemical types of the rivers.  The five 'biotypes', or ecological groups formed by dominant species, are also close to the associations described by Symoens (1957), but the use of quantitative data and data processing by factor analysis enabled the results to be analysed objectively.

Very few data are available on biomass and productivity of the
benthic algae.  Local measurements in the Meuse have indicated a
periphyton biomass ranging from 5 to 20 μg cm$^{-2}$ on artificial substrates.  The biomass measured in an Ardennes stream was at least five
times smaller (Fabri, 1977).

There have been no comprehensive studies of the phytoplankton of the
Meuse basin, although a number of papers do give records for the Meuse
(Pierre, 1974, for the upper and middle Meuse in France; Symoens, 1957
and Jansen, 1967, for the Belgian part).  The French and Belgian studies
of the Meuse both showed a similar flora, with the following dominants:
*Cyclotella meneghiniana, Fragilaria construens, Melosira varians,
M. italica, Stephanodiscus hantzschii, Scenedesmus quadricauda.*  Well
developed plankton communities probably only occur in the large
canalized stretches of the Meuse and Sambre.  Measurements on phytoplankton biomass upstream of Liège have showed chlorophyll contents
ranging from 3 to 40 mg chlorophyll a m$^{-3}$, with a maximum in summer
(Declercq-Versele & Kirchmann, 1982).

## 1.42  BRYOPHYTES

Bryophyte communities occur mainly on rocks and concrete surfaces in
fast flowing water, such as rapids and natural waterfalls or, in
downstream stretches of the Sambre and Meuse, the artificial falls
associated with fluvial traffic movement.  The biomass involved varies
greatly, ranging from well-defined strips limited to a few hundred
metres downstream of artificial falls to a relatively continuous cover
in rapids where the depth is not greater than 0.5 m and where there is
no disturbance of the bed by trampling.  The populations of the main
gated rivers, the Meuse and Sambre, are dominated by the genus
*Cinclidotus,* mainly *C. nigricans, C. fontinaloides* and *C. danubicus,*
with sometimes also *C. aquaticus* and *Dialytrichia mucronata* (= *C.
mucronatus*).

Fig. 1.7  Relationship between cover of bryophytes and alkalinity:
A, *Scapania undulata*; B, *Chiloscyphus polyanthus*; C, *Cinclidotus
danubicus*; D, *Platyhypnidium riparioides*; E, *Fontinalis antipyretica.*

The ecology of individual species has been studied in some detail
(Empain, 1974, 1978; Empain *et al.*, 1980), with data on the influence of
frequency of immersion, resistance to desiccation and solar heating,
photosynthesis and resistance to pollution.  Species distribution
within the basin shows a picture closely correlated with alkalinity
(Fig. 1.7).  Some species, such as *Scapania undulata, Chiloscyphus
polyanthus* and *Cinclidotus danubicus* are limited to a narrow range of
alkalinity concentrations; others like *Platyhypnidium riparioides
(= Rhynchostegium riparioides)* and *Fontinalis antipyretica* are found
over a wide alkalinity range.


## 1.43    FISH

According to the system proposed by Huet (1949), the changes in fish
communities on passing downstream depends under natural conditions
mainly on physical factors such as altitude, slope, width and water
temperature.  In the Meuse basin, the most frequent communities belong
to the grayling and barbel zones, the trout zone being limited mostly
to upland streams and stretches of the main rivers.  The Meuse itself
has communities dominated by still water cyprinids, but theoretically
it belongs to the lower barbel zone.  In unpolluted stretches, species
diversity is greatest in the barbel zone (up to 23 species) and least
in the trout zone (3-8 species); the more acid streams provide an
extreme example, with no fish present at all.

Measurements of fish biomass have been made by Huet and Timmermans
(1976) and Philippart (1979, 1980).  In natural conditions the biomass
may reach 500 - 700 kg ha$^{-1}$ in small rivers (e.g. upper Amblève,
520 kg ha$^{-1}$; Eau d'Heure in the Ourthe basin, 670 kg ha$^{-1}$; lower
Méhaigne, 470 kg ha$^{-1}$).  In the main rivers the mean biomass is lower,
with values of 250 - 350 kg ha$^{-1}$ being recorded.

A very detailed description of the features of walloon rivers has
been made in a guide for anglers in the south of Belgium (Balzat &
Dussart, 1978); even if the data are not always very precise, this book
gives much useful information on not only the fish fauna, but also
geography, geology, chemistry, vegetation and invertebrates.  Philippart
(1979) gives an account of the influence on population dynamics of
ecological and sociological factors and the economics of sport fishery;
it also describes the relationship between the growth rate of fish
species and physico-chemical factors.


## 1.5    WATER QUALITY ASSESSMENT BY COMMUNITIES

### 1.51    USE OF DIATOMS

A method for estimating water quality has been developed (Descy, 1979,
1980) which is of general applicability and which makes use of diatoms
and data about their response to pollution.  It involves a comparison
of communities found at a site with the 'normal' communities that might

be expected in the 'natural' type of the particular watercourse.
Briefly this 'diatom index' is based on quantitative data obtained by
counting (relative abundance of species in sample) and takes into
account the sensitivity to pollution of about 100 species, as well as
their 'indicator value' related to their ecological amplitude towards
pollution.

The computed diatom index ranges from 1 for heavily polluted sites
to 5 for clean water; in practice six classes of index values have been
adopted for interpretative purposes (Fig. 1.8). The method has been
applied successfully in very diverse situations within the Meuse basin;
it usually shows a very close relationship with chemical analysis and
has led in some cases to the detection of discontinuous pollution. Use
of the method for the Meuse basin survey led to the following conclusions:

  5% sampling sites are unpolluted; however 31% may be considered as
presenting a good water quality;

  44% are subjected to moderate eutrophication, mostly in the rivers
with lower mineral content (types 1,2,3) which are more sensitive to
pollution;

  9% show heavy eutrophication of medium pollution;

  16% present heavy pollution.


## 1.52  USE OF BRYOPHYTES

A somewhat similar approach to that with diatoms has been developed for
bryophytes. The first attempts (Empain 1973, 1978) were devoted mainly
to alkaline waters (type 5), where a high correlation was found with
the diatom index; the occasional differences may be explained by the
very different time taken for recovery - weeks for diatoms, years for
bryophytes. More recently a generalised approach (Descy & Empain,
1981) has been developed which takes account of the natural type of the
water. The key steps needed to compute the generalized bryophyte index
are the evaluation of sensitivity coefficient for each species and the
estimation of mean value of 'crude bryophyte index' for each type of
water. When pollution is too high, no bryophytes are found, such as
parts of the Sambre and Semois, and the crude bryophyte index equals 0.
The final steps are the calculation of crude bryophyte index for a
particular station and the generalization of the result by taking
account of the mean value of the water type of the sampling site. The
generalized bryophyte index gives results similar to those from the
diatom index. Among the 30 sites selected by the former as strongly
polluted, 27 were selected by the bryophyte index; the 3 discrepancies
are easily explained by local sampling constraints. Some other sites
were selected by the bryophyte index because of the discontinuous
nature of the pollution, such as rivers polluted by sugar-beet waste
(Méhaigne and Geer) or toxic wastes (Hoyoux). The main advantage of
the bryophyte index lies in the long-term integration of pollution
events, whereas the diatom index reflects more short-term events; it
is also more sensitive to eutrophication.

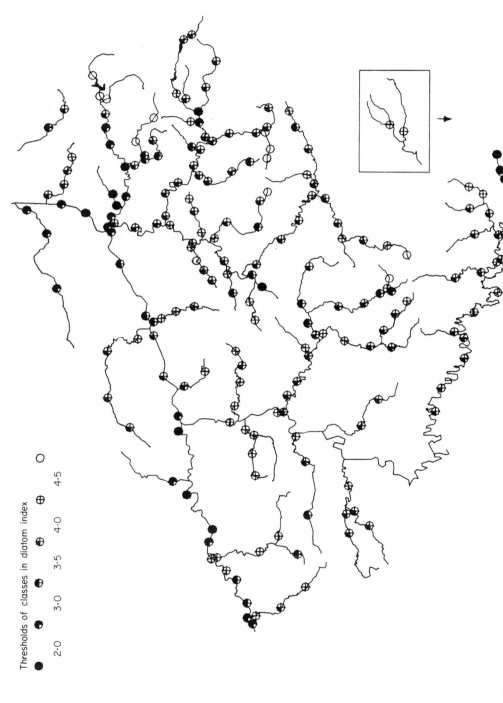

Thresholds of classes in diatom index

Fig. I.8. Diatom index for Meuse basin

1.53    USE OF MACROINVERTEBRATES

A survey of Belgian rivers based on the use of macroinvertebrates has
recently been published by the Ministry of Public Health (I.H.E., 1979),
following collaborative work between various institutes and laboratories.
The 'biotic index' method of Verneaux and Tuffery (1967) was employed,
which is itself derived from the Trent River Board index (Woodiwiss,
1964).  Unfortunately the results of this survey appear to give an
unrealistic picture of water quality in the Meuse basin.  As far as the
sites common with our own studies are concerned, there are some marked
differences.  47% of the sites in the I.H.E. survey were recorded as
having very good water quality (biotic index $\geqslant$ 9) versus 31% for the
diatom index.  In contrast, 11% of sites on the biotic index were
heavily polluted versus 19% on the diatom index.  However the total
number of sites of acceptable water quality were similar (73% v 75%).
We conclude that the differences lie mainly in the sampling procedure
used for macroinvertebrates, like the restriction to one or two periods
for sampling (autumn 1978, spring 1979), localization of the reaches
studied and difficulties in deep gates stretches.  Hawkes (1979) has
pointed out limitations in the use of macroinvertebrates, such as those
resulting from seasonality and how this should be taken into account
when planning sampling programmes.

    Our conclusion is that the I.H.E. pollution survey underestimates
some examples of heavy pollution, such as the Meuse between Liège and
Visé and the Vesdre downstream of Verviers.  On the other hand it
overestimates water quality in the less polluted parts of the catchment,
such as the Ourthe and Lesse.  This emphasizes the risks involved in
using biological methods in monitoring programmes without sufficient
care.  Biological and physico-chemical methods are complementary, as
they provide different sorts of information; both are needed for a full
assessment.


1.54    USE OF FISH

Many studies have been made by the Fish Research Unit of Liège University on the state of fish populations in the Meuse basin.  The programme
which started with a survey of the Ourthe basin (Philippart, 1979) is
now extended to the whole Meuse catchment.  A key aim is to provide
reference data for long-term comparison, but some estimates of change
are already possible; these have been made by comparison either with
any available earlier data or with reference stretches belonging to the
same type.  In the Ourthe basin, for example, the loss of biomass has
been estimated as 38.6 tonnes, mostly affecting *Salmo trutta, Chondrostoma nasus* and *Thymallus thymallus*.  In some cases (Fig. 1.9) the
change in biomass is minor, but the community structure is affected
greatly by the increase of tolerant species *(Leuciscus cephalus, L.
leuciscus, Gobio gobio, Phoxinus phoxinus)* without recovery of the

*Fig. 1.8* (opposite)  Diatom index for Belgian Meuse basin, 1973-1980
(Descy 1979, 1980).

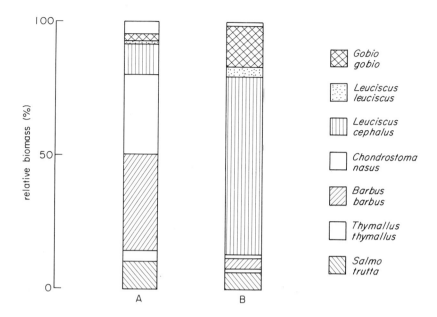

Fig. 1.9    Example of changes in a fish community following recovery
after pollution:  A, normal structure in the upper part of the grayling zone; B, unbalanced structure in the Amblève ten years after
heavy pollution.

normal population.  This type of study can provide information on long-
term changes long after the causes have disappeared.  Moreoever fish are
sensitive to alterations not often apparent with the other, monitoring
programmes, such as the destruction of biotopes by damming, canalization, dredging and water abstraction.


1.6   USE OF BRYOPHYTES TO MONITOR HEAVY METALS AND RADIONUCLIDES

As has already been mentioned, many sites on rivers in the Meuse basin
are influenced by heavy metal pollution.  The accumulation of these
metals in bryophytes has been monitored in the Meuse and Sambre since
1973 and in other rivers since 1977.  A semi-popular account of the
results has been given by Descy et al. (1981).  For monitoring purposes,
the main advantages are the ease of sampling, the huge increase in
levels to be analysed ($10^3$ to $10^5$ times the mean concentration in the
water) and the integration of large fluctuations in the water.  Metal
concentrations in the bryophytes are sometimes very high.  For instance
chromium has reached as much as 13 mg $g^{-1}$ in the Warche and Amblève
downstream of a tannery (Mouvet, 1981).  Reference to water type and

Fig. 1.10  (opposite)   Zinc in bryophytes (mg $kg^{-1}$ dry weight,
mean concentration) in Belgian Meuse basin, 1973-1980.

Concentration (μg g⁻¹) proportional to radius

e.g.  200  400  800  1600  3200

Fig. 1.10.  Zn in bryophytes of Meuse basin

the natural range of concentration makes it easy to discriminate between natural enrichment in particular areas and industrial discharges.  Copper is found at particularly high concentrations in mosses of the upper Belgian Sambre, downstream of industry producing electrical wires.  The mean values in the mosses vary between 5 and 20 times the natural limits, whereas the mean values for the water fit the natural range for this type of water (max. 40 µg l$^{-1}$); peak values in the water reach 2 to 6 times the natural upper limit, whereas those in the mosses 7 to 25 times their natural upper limit.

Bryophytes have also played an important role in monitoring radionuclides, since they are the most contaminated element of the river ecosystem (Fig. 1.11) (Kirchmann & Lambinon, 1973).  They have been used regularly downstream of the Tihange nuclear plant (Kirchmann *et al.*, 1979) and, together with other organisms, every three years on the upper Meuse (Lambinon *et al.*, 1976, 1977).

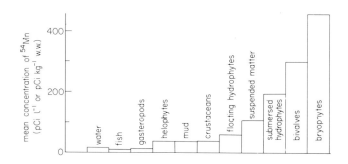

*Fig. 1.11* $^{54}$Mn in Meuse: mean concentration in water, mud and aquatic organisms downstream of the nuclear plant of Chooz (France, Ardennes).

## 1.7  DISCUSSION

Much of the knowledge of the ecology of the Meuse basin has been built up from the detailed studies of particular types of organism, often associated with programmes to assess water quality and pollution.  There are obvious gaps, such as the lack of detailed research on bacteria, macrophytes and plankton.  Among the physico-chemical studies there are very few systematic results for some elements and also pesticides and certain other compounds.  We hope that this chapter has made clear the need for integrated studies involving both abiotic parameters and community studies.  Reference to natural conditions is important for the assessment of water quality and the ecological impact of pollutants. A general approach involving all the important factors is complicated but needed if ecological information is to be used as an aid to optimal river management, which involves drinking water supply, the needs of industry, navigation, agriculture and angling and, finally, the maintenance of ecological equilibrium.

REFERENCES

Balzat N. H. & Dussart A. (1978)  *Guide de la pêche en Ardenne*, 384 pp.
Duculot, Paris - Gembloux.

Descy J.-P. (1973)  La végétation algale benthique de la Meuse belge
et ses relations avec la pollution des eaux. *Lejeunia* 66, 62 pp.

Descy J.-P. (1976)  Etude quantitative du peuplement algal benthique en
vue de l'établissement d'une méthodologie d'estimation biologique
de la qualité des eaux courantes : Application au cours belge de la
Meuse et de la Sambre.  In: *Recherche et Technique au service de
l'Environnement*.  382 pp., pp. 159-206.  Debedoc, Liège.

Descy J.-P. (1979)  A new approach to water quality estimation using
diatoms.  *Beih. Nova Hedw.* 64, 305-23.

Descy J.-P. (1980)  Utilisation des algues benthiques comme indicateurs
biologiques de la qualité des eaux courantes.  In: Pesson P., *La
pollution des eaux continentales.  Incidence sur les biocénoses
aquatiques*, 2me Edn. 345 pp., pp. 169-94.  Gauthier-Villars, Paris.

Descy J.-P. & Empain A. (1976)  Analyse physico-chimique de la Meuse
et de la Sambre belges, en 1973 et 1974.  In : *Recherche et Technique
au service de l'Environnement*.  382 pp., pp. 139-57.

Descy J.-P. & Empain A. (1981)  *Inventaire de la qualité des eaux
courantes en Wallonie (Bassin Wallon de la Meuse)*.  Rapport de
synthèse, 1, 87 pp., 2, 194 pp., 3, 37 pp.  Université de Liège
(limited circulation).

Descy J.-P, Empain A. & Lambinon J. (1981)  *La Qualite des Eaux
Courantes en Wallonie - Bassin de la Meuse*.  18 pp.  Secrétariat
d'etat à l'environnement, a l'aménagement du territoire et à l'eau
pour la Wallonie, Square de Meeus, 35, B-1040 Bruxelles.

Empain A. (1973)  La végétation bryophytique aquatique et subaquatique
de la Sambre belge, son déterminisme écologique et ses relations
avec la pollution des eaux. *Lejeunia* 63, 58 pp.

Empain A. (1974)  Relations quantitatives entre les bryophytes de la
Sambre belge et leur fréquence d'émersion : distribution verticale
et influence de la pollution. *Bull. Soc. r. Bot. Belg.* 107, 361-74.

Empain A. (1976)  Les bryophytes aquatiques utilisés comme traceurs de
la contamination en métaux lourds des eaux douces. *Mém. Soc. r. Bot.
Belg.* 2, 201-16..

Empain A. (1978)  Relations quantitatives entre les populations de
bryophytes aquatiques et la pollution des eaux courantes.
Définition d'un indice de qualité des eaux. *Hydrobiologia* 60, 49-74.

Empain A., Lambinon J., Mouvet C. & Kirchmann R. (1980)  Utilisation
des bryophytes aquatiques et subaquatiques comme indicateurs biologiques de la qualité des eaux courantes.  In: *La pollution des eaux
continentales.  Incidence sur les biocénoses aquatiques*, 2me Edn.
345 pp., pp, 195-223.  Gauthier Villars, Paris.

Fabri R. (1977)  Végétation, production primaire et caractéristiques
physico-chimiques d'une rivière de haute Ardenne (Belgique): la Warche
superiéure. *Lejeunia* 87, 43 pp.

Fabri R. & Leclercq L. (1979)  Flore et végétation algale des ruisseaux
de la Réserve naturelle domaniale des Hautes Fagnes (province de
Liège, Belgique).  Serv. Conserv. Nat., trav. n° 11, 48 pp.

Hawkes H. A. (1979)  Invertebrates as indicators of river water quality. In: James A. & Evison L. (Eds) *Biological Indicators of Water Quality*. pp. 2.1 - 2.45.  John Wiley & Sons, Chichester.

Huet M. (1949)  Aperçu des relations entre la pente et les populations piscicoles des eaux courantes. *Schweiz. Z. Hydrol.* 11, 332-51.

Huet M. & Timmermans J. A. (1976)  Influence sur les populations de poissons des aménagements hydrauliques de petits cours d'eau assez rapides. *Trav. Stat. Rech. Eaux et Forêts Groenendael,* D, 46, 27 pp.

I. H. E. (1979)  Carte de la qualité biologique des cours d'eau en Belgique. 61 pp. + 1 map, Institut d'Hygiène et d'Epidémiologie, Bruxelles.

Jansen P. (1967)  Een onderzoek naar de kwaliteit van het Maaswater in Frankrijk en België. *Nat. Hist. Maandbl.* 56, 74-84.

Kirchmann R. & Lambinon J. (1973)  Bioindicateurs végétaux de la contamination d'un cours d'eau par des effluents d'une Centrale Nucléaire à eau pressurisée.  Evaluation des rejets de la Centrale de la SENA (Chooz, Ardennes francaises) au moyen des végétaux aquatiques et ripicoles de la Meuse. *Bull. Soc. r. Bot. Belg.* 106, 187-201.

Kirchmann R., Lambinon J., Empain A. & Bonnijns-Van Gelder E. (1980)  Choix et performances de bioindicateurs pour la surveillance des sites d'implantation d'installations nucléaires.  In: *IIᵉ Symposium International de Radioécologie*. 751 pp., pp. 557-92.  CEA - EDF, Cadarache.

Lambinon J., Kirchmann R., Colard J. (et coll.) (1976)  Evolution récente de la contamination radioactive des écosystèmes aquatique et ripicole de la Meuse par les effluents de la Centrale Nucléaire de la SENA (Chooz, Ardennes françaises). *Mém. Soc. r. Bot. Belg.* 7, 157-75.

Lambinon J., Descy J.-P., Empain A., Kirchmann R. & Bonnijns-Van Gelder E. (1977)  La surveillance écologique des sites d'implantation des centrales nucléaires.  Effets des rejets d'effluents sur l'écosystème mosan: acquis et perspectives. *Ann. Ass. Belge Radiopro.* 2, 201-16.

Mouvet C. (1981)  Pollution de l'Amblève par les métaux lourds, en particulier le chrome : dosage dans les eaux et les bryophytes aquatiques. *Trib. Cebedeau* 445, 527-38.

Philippart J.-C. (1979)  Evaluation des ressources piscicoles et halieutiques dans les rivières du bassin de la Meuse.  Définition d'une méthode d'étude.  In: Calembert L. (Ed.) *Problématique et Gestion des Eaux Intérieures*. 967 pp., pp. 481-495.  Derouaux, Liège.

Philippart J.-C. (1980)  Incidences de la pollution organique et de l'eutrophisation sur la faune ichtyologique de la Semois. *Annls Limnol.* 16, 77-89.

Pierre J.-F. (1974)  Contribution à l'étude hydrobiologique des eaux superficielles du bassin Rhin-Meuse.  I. Evolution du phytoplancton des eaux du cours moyen et supérieur de la Meuse. *Bull. Acad. Soc. Lorr. Sciences* 13, 91-108.

Smitz, J. (1976)  Modèle thermique de rivière - température naturelle et rejets thermiques.  In: *Recherche et Technique au service de l'Environnement*. 382 pp., pp. 117-37, Cebedoc, Liège.

Symoens J-J. (1957)  Les eaux douces de l'Ardenne et des régions voisines: les milieux et leur végétation algale. *Bull. Soc. r. Bot. Belg.* 89, 111-314.

Vereerstraeten J. (1971)  Le bassin de la Meuse.  Etude de géographie
    hydrologique.  *Rev. belg. Géogr.* 94, 339 pp.
Verneaux J. & Tuffery G. (1967)  Une méthode zoologique pratique de
    détermination de la qualité biologique des eaux courantes.  Indices
    biotiques.  *Annals scient. Univ. Besançon, Zoologie* 3, 79-80.
Woodiwiss F. S. (1964)  The biological system of stream classification
    used by the Trent River Board.  *Chem. Ind.*, 443-447.

A shallow stretch of the Meuse below a weir near the France-Belgium
border, showing patches of *Scirpus lacustris*.  The red alga *Thorea
ramosissima* is frequent here in late summer.

# 2 · Gudenå

P. HEISE

2.1   Introduction and history                                    26

2.2   General data

2.3   Water quality                                               32

    2.31   Spring areas                                           32
    2.32   Pollution                                              32
    2.33   Nitrogen and phosphorus loadings                       35

2.4   Ecology                                                     37

    2.41   Macrophytes                                            37
    2.42   Macroinvertebrates and use of the Saprobien            39
           system
    2.43   Fish                                                   41

        2.431   Fauna                                          41
        2.432   Fish passes and the Gudenå salmon              42

2.5   River management                                            43

    2.51   Administrative relationships                           43
    2.52   Improvement of water quality in lakes                  43
    2.53   River modelling                                        44
    2.54   River management in the future                         47

References                                                        48

## 2.1  INTRODUCTION AND HISTORY

The Gudenå is Denmark's longest river.  Although its 146 km is short in comparison with many of the rivers in this book, nevertheless the river and its catchment have a long history of human use and the river itself has many features of ecological interest.  It also provides a good case study for considering the problems of river management in Denmark.

The river rises in springs at Tinnet in the hills of Midjutland close to the maximum extension of the ice during the last (Weichelian) glaciation.  Only 200 m from the Gudenå spring area, occur the springs of another major river, the Skjern Å.  The Gudenå discharges into the Randers Fjord, a typical Danish estuary, and further into the Kattegat; the combined catchments of the Gudenå and the Randers Fjord constitute about 8% of the area of Denmark.  The river runs through a series of lakes moulded during the last glaciation, but man-made impoundments have been added to these.  The Tange Lake, for example, was established in 1920 to provide hydro-electric power at Tange.  The Gudenå, its tributaries and the lakes, together with surrounding hills and woods, form the beautiful landscape of Midjutland known as 'The Danish Lake District'.

The Gudenå area has a long history of intensive use.  The Stone Age settlements on the banks of the river formed the so-called 'Gudenå-culture', while great monasteries were established close to the river and its lakes during the Middle Ages.  Hydro-power was used during the latter period for corn and paper mills and the waters also served the monasteries for fishing and transportation.  The Benedictines from the town of Vorre dug a canal - three kilometres long - and diverted a stretch of the Gudenå into the 'Klosterkanal' (Monastery Canal), which provided hydro-power for the 'Klostermølle' (Monastery Mill) at Mossø Lake.  (The mill is owned by Ministry of Environment and a water-mill museum is planned.)

Some transport of both people and goods took place on the Gudenå before 1850, but following the opening of the paper factory at Silkeborg in this year, the use of barges between the towns of Randers and Silkeborg was systematized.  From Randers to Silkeborg the barges were pulled upstream by horses or men ('hired men') from the shore; the flow of water was itself sufficient to carry the barges and their goods on their downstream journey.  At that time a great canalization project was fulfilled in the Silkeborg to Randers stretch, especially the part between Silkeborg and Tange, which was dredged to maintain a summer water level above one metre.

When the railroad connected the towns and the hydro-electric power station at Tange was built, the use of barges came to an end.  Today it is tourists who sail on the Gudenå and the lakes.  The tourist boats take some 120000 passengers a year, while canoes are hired by 25000 persons for tours of at least five days.  About 1500 motorboats are located on the lakes and the downstream stretch of the Gudenå.  There are 100000 anglers visiting the Gudenå and its tributaries each year. Tourists thus represent an important group of users demanding a high quality of water in the Gudenå system.

*Fig. 2.1* Upper Gudenå running through intensively used farm lands.

*Fig. 2.2* Middle Gudenå at Tvilum, 10 km upstream of Lake Tange: a place visited by many anglers and nature lovers.

*Fig. 2.3* The Gudenå drainage area: towns, lakes and water courses.

2.2   GENERAL DATA

DIMENSIONS

Area of catchment       2600 km$^2$ (3300 km$^2$ with Randers Fjord)
Length of river         146 km
Altitude at source      70 m

CLIMATE

Precipitation
mean:                   720 mm yr$^{-1}$
range:                  650 (in NE) to 750 mm yr$^{-1}$ (in SW)

LAND USE

agricultural areas      1700 km$^2$ (65%)
forests and moors        650 km$^2$ (25%)
towns, paved areas       150 km$^2$ ( 6%)
water surfaces*          100 km$^2$ ( 4%)

*In the Gudenå drainage area there are about 40 lakes or impoundments greater than 0.1 km$^2$, and the water surface of these 40 lakes amounts to 75 km$^2$.

The drainage area of the Gudenå covers three counties (amt):

Vejle amt               (10% of the drainage area)
Viborg amt              (20% of the drainage area)
Århus amt               (70% of the drainage area)
and 30 municipalities

POPULATION

Largest towns are Randers, Silkeborg, Skanderborg and Viborg.
Total = 190000 (= 73 people km$^{-2}$)

PHYSICAL FEATURES OF RIVER

Comparison of hydrological data for Gudenå with other Danish rivers
(Rate of flow given both as annual mean rate and median of annual minimum rate)

| river | length | catchment | flow | | precipitation |
|---|---|---|---|---|---|
| | | | annual mean | median minimum | |
| | (km) | (km$^2$) | (1 s$^{-1}$) | (1 s$^{-1}$) | (mm yr$^{-1}$) |
| Gudenå | 146 | 2600 | 31000 | 20000* | 720 |
| Storå | 100 | 1030 | 13900 | 6800 | 740 |
| Skjernå | 93 | 2210 | 32700 | 21700 | 750 |
| Suså | 81 | 815 | 6400 | 1000* | 650 |
| Karup Å | 77 | 820 | 9400 | 7000 | 690 |
| Odense Å | 60 | 606 | 5800 | 1750* | 650 |
| Kongeå | 60 | 460 | 6900 | 3100 | 770 |
| Mølleå | 40 | 116 | 340 | 50* | 620 |
| Nørreå† | 40 | 394 | 4500 | 2500* | 690 |
| Hadsten Lilleå† | 28 | 292 | 2600 | 1250 | 630 |
| Salten Å† | 22 | 124 | 1500 | 1200 | 730 |

*flow is influenced by lakes or impoundments with regulation
†important tributaries of Gudenå

## Topography of Gudenå

average slope (see Fig. 2.4)    0.047%, but there are 8 dams totalling 30 m
                                height. (Hydroelectric plants at Vestbirk
                                and Tange each have a height of fall of about
                                10 m)

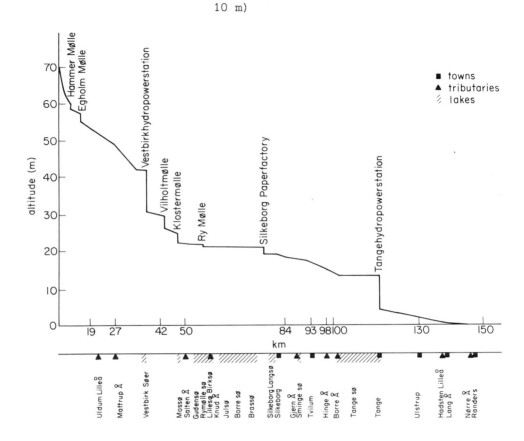

*Fig. 2.4*  Change in elevation in Gudenå on passing downstream

## Zonation

Three stretches can be recognized based on profile and shape:

| zone | length | limits | slope |
|---|---|---|---|
| upper | 46 km | Tinnet to Mossø | 0.03–0.40% |
| middle | 65 km | Mossø to Tange | |
| | | (including 40 km through lakes | 0.006–0.06% |
| lower | 35 km | Tange to Randers | 0.006–0.02% |

## Cross-section of river

| | width (m) | depth (m) | summer av. velocity (m s$^{-1}$) |
|---|---|---|---|
| km 10 (from source) | 6 | 0.4 (av.) | 0.5–2.0 |
| km 100–Randers Fjord | 25–40 | 2.5–4.0 | 0.1–0.5 |

Flow per unit area of catchment

average 12.5 1 s$^{-1}$ km$^{-2}$ (corresponds to 54% precipitation).
(Rate varies considerably according to part of catchment.
The tributaries and Salten Å, Funder Å and Nørre Å 'steal'
groundwater from outside the watershed and their average
flow is 16-20 1 s$^{-1}$ km$^{-1}$.)

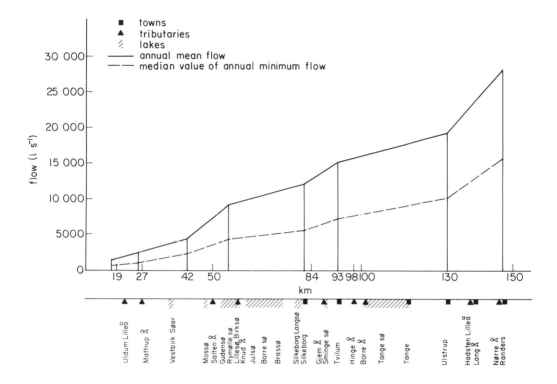

*Fig. 2.5* Change in flow in Gudenå on passing downstream. Results
from 'The Gudenå Investigation 1973-1975' (Water Quality Institute,
1976; Hydrological Survey of the Danish Land Development Service).

Sources

*c.* 550 springs and headwater streams

Main tributaries

| | | |
|---|---|---|
| Uldum Lilleå | Funder Å | Tange Å |
| Mattrup Å | Gjern Å | Hadsten Lilleå |
| Salten Å | Hinge Å | Nørre Å |
| Knud Å | | |

## 2.3  WATER QUALITY

### 2.31  SPRING AREAS

The natural quality of water in the Gudenå catchment is determined by the surface of the land and the geology of the area that is drained by the river.  However much of the water is influenced now by sewage discharges, effluents from industry and fish farms and pollution from rural activities, with consequent enrichment in organic matter, nitrogen and phosphorus.  Of about 550 spring areas which form the sources of the Gudenå and its tributaries, only some 15% are undisturbed today (F. Jensen, 1976;  Water Quality Institute, 1976).  The composition of water in some of these spring areas has been investigated intensively (e.g. Warncke, 1980) and typical concentrations for three examples are given in Table 2.1.  Estimates for the spring areas as a whole of average concentrations show:

$$\text{total dissolved N} \quad 0.38 \quad \text{mg l}^{-1}$$
$$\text{total dissolved P} \quad 0.06 \quad \text{mg l}^{-1}$$

*Table 2.1*  Composition (mg l$^{-1}$) of water in spring areas (see Warncke, 1980).

| variable | Tinnet (beginning of upper Gudenå) | Addit (in catchment of Salten Å) | Hald (beginning of River Nørreå) |
|---|---|---|---|
| Na$^+$ | 13.1 | 18.2 | 16.0 |
| K$^+$ | 0.85 | 1.4 | 1.0 |
| Mg$^{++}$ | 3.8 | 3.3 | 2.9 |
| Ca$^{++}$ | 44.1 | 21.0 | 36.2 |
| NH$_4$-N | 0.08 | 0.03 | 0.08 |
| NO$_3$-N | 0.01 | 0.01 | 0.006 |
| total-N | 0.27 | 0.08 | 0.37 |
| reactive PO$_4$-P | 0.05 | 0.014 | 0.07 |
| total filtrable P | - | 0.027 | 0.08 |
| total-P | - | 0.054 | 0.335 |
| Cl$^-$ | 16.8 | 21.0 | 22.5 |
| alkalinity (meq l$^{-1}$) | 2.10 | 0.75 | 1.81 |

### 2.32  POLLUTION

Effluents from domestic and industrial sewage are discharged into waters of the Gudenå system.  In addition to the 190000 people living in the area, discharges from industrial outfalls calculated on demand of water

amount to the equivalent of about 170000 persons.  Pollution from
sewage and industrial wastewater is widely reduced through biological
treatment and phosphorus removal.  A special group of 'water consumers'
has been formed by 60 fish farms, which together produce a pollution
load corresponding to 60000 persons (Table 2.2).

*Table 2.2*  Wastewater sources in Gudenå catchment.  (PE = person
equivalent)

| | |
|---|---|
| Towns, urbanized areas > 150 PE | 300000 PE |
| Industries with separate outlets | 15000 PE |
| Small villages and farms | 45000 PE |
| Fish farms | 60000 PE |
| Total | 420000 PE |

The water supply in the Gudenå catchment comes from groundwater wells
and approximately 1 $m^3$ $s^{-1}$ of drinking water is discharged as sewage
into the river; this value may be compared with the value of 31 $m^3$ $s^{-1}$
for the average flow where the river runs into the Randers Fjord.  The
sewage is of course not discharged uniformly, but mainly as point outlets
and often into small receiving waters.  In 1980 the sewage treatment
plants for about 100 towns with a pollution load above 150 PE used the
following methods:

| | |
|---|---|
| no treatment | 2% |
| mechanical treatment | 6% |
| biological treatment | 60% |
| biological-chemical treatment | 32% |

The pollution load reaching the river can be described by variables
such as biochemical oxygen demand over five days (BOD), total N, total
P, etc. (Table 2.3).

The range of concentrations shown for summer reflects the dilution
and decomposition of sewage effluents and in the primary production of
lakes and streams.  The high yearly average value for total nitrogen
results from the very high levels which occur in winter, with individual values sometimes as high as 12 mg $l^{-1}$ N.  The river is usually
relatively well oxygenated, with wintertime values mostly above 8 mg
$l^{-1}$ $O_2$ and summertime values seldom dropping below 5 mg $l^{-1}$ and much
higher by day due to photosynthesis by algae and other plants (Fig.
2.6).  The concentrations of heavy metals are low, with values at or
below the detection limits for all except zinc (as mg $l^{-1}$: Cr, <0.01;
Cu, 0.005; Zn, <0.09; Cd, 0.001; Hg, 0.00025; Pb, 0.025).

*Table 2.3* Composition (mg l$^{-1}$) of water in the Gudenå showing values for annual average and range in summer.

| variable | | Upper | Middle | Lower |
|---|---|---|---|---|
| BOD | annual | 2.5 | 3.5 | 2.8 |
| | summer | 1 - 7.5 | 2 - 7.5 | 3-4 |
| total | annual | 6 | 3.0 | 3.6 |
| N | summer | 1.5 - 2.3 | 1.5 - 1.9 | 0.8 - 1.0 |
| NH$_3$-N | annual | ~0.0 - 0.4 | 0.1 - 1.9 | 0.1 - 0.7 |
| | summer | 0.1 - 0.4 | 0.1 - 1.9 | 0.1 - 0.2 |
| NO$_3$-N | annual | 0.2 - 8.5 | ~0.0 - 4.1 | ~0.0 - 7.6 |
| | summer | 0.2 - 1.7 | ~0.0 - 0.4 | 0.1 - 0.25 |
| total | annual | 0.2 | 0.21 | 0.20 |
| P | summer | 0.16 - 0.25 | 0.10 - 0.29 | 0.15 - 3.8 |

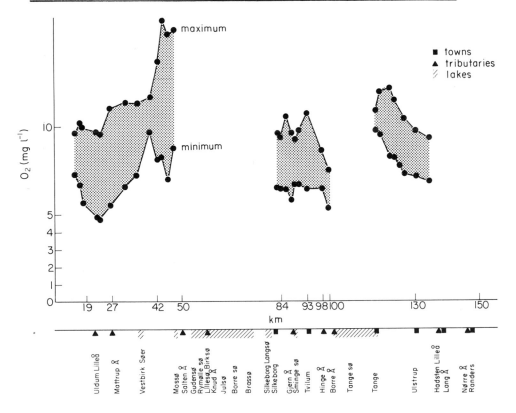

*Fig. 2.6* Diel range in dissolved O$_2$ in flowing water stretches of the Gudenå in summer.

## 2.33  NITROGEN AND PHOSPHORUS LOADINGS

Because of their fertilizing effect on the lakes of the Gudenå system, much emphasis has been placed on the investigation of the sources and importance of nitrogen and phosphorus.  Discharge per year within the catchment amounts to approximately 10000 tonnes of BOD, 4100 tonnes of nitrogen and 300 tonnes of phosphorus.  Domestic and industrial activities produce approximately 20% of the total nitrogen and 75% of the total phosphorus loadings.

*Table 2.4* Nitrogen and phosphorus loadings in run-off from non-point sources in the Gudenå catchment.  Voelbaek and Gjelbaek are two minor agricultural areas (each 12 km$^2$) in the drainage area of the middle Gudenå.

| | land-use | nitrogen (kg N ha$^{-1}$ yr$^{-1}$) | phosphorus (kg P ha$^{-1}$ yr$^{-1}$) |
|---|---|---|---|
| UPPER GUDENÅ | agricultural | | |
| reaches with strong slope | | 28 | 0.90 |
| reaches with flooded meadows in winter | | 22.5 | 0.02 - 0.05 |
| THE LAKE DISTRICT | wood | 7 - 10 | 0.15 - 0.20 |
| MIDDLE GUDENÅ | agricultural | 15 - 20 | 0.30 - 0.35 |
| LOWER GUDENÅ | woods & agricultural | 6 - 20 | 0.15 - 0.20 |
| Voelbaek (sandy) | agricultural | 14 | 0.17 |
| Gjelbaek (clay) | | 25 | 0.43 |
| storm sewers | urban | | |
| town of Viborg | | 14 | 1.0 |
| town of Skanderborg | | 8 | 2.9 |

The loadings from some non-point sources investigated in 1974-76 are shown in Table 2.4.  Detailed mass balances for two minor agricultural areas, Voelbaek and Gjelbaek, showed that about 20% of the nitrogen used as fertilizer was washed out, whereas less than 1% of the phosphorus was so.  Analyses of data for the daily runoff from the

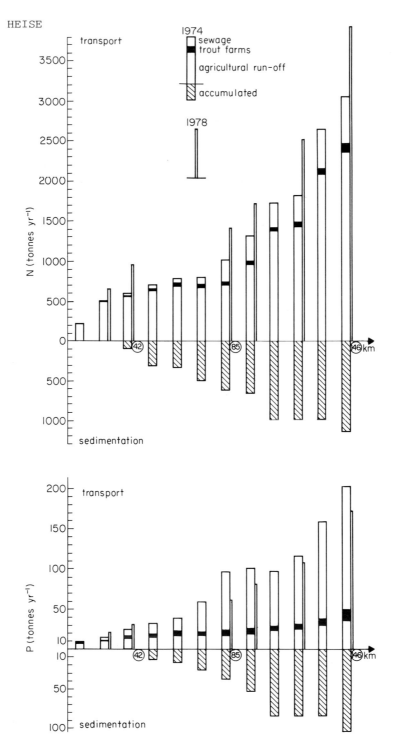

*Fig. 2.7* Transport and sedimentation of N and P on selected sites along the Gudenå in 1974, with some comparative data for transport in 1978.

agricultural areas showed a high correlation between water and the nitrogen flux (Strange Nielsen, 1977).  The correlation was less evident for phosphorus because of local disturbances caused by septic tank effluents and manure water.  The nitrogen loadings per unit area of urban runoff were similar to those of rural areas (Table 2.3), but the phosphorus loadings were higher.  As paved areas amount to only 3% of the Gudenå drainage area, urban stormwater runoff causes only local effects.

Annual loadings and transport of nitrogen and phosphorus have been investigated since 1974 at selected sampling sites throughout the whole river system.  About 30% of the nitrogen and 35% of the phosphorus are detained in the lakes and water courses because of sedimentation.  The amount of nitrogen transported in 1978 was higher than in 1974 (Fig. 2.7) due to a higher average water flow.  In contrast, the construction of new sewage treatment plant incorporating phosphorus precipitation during 1977-78 led to a reduction in the phosphorus flux downstream of Silkeborg (km 85).

## 2.4   ECOLOGY

### 2.41   MACROPHYTES

The combination of being low lying and having a high content of nitrogen and phosphorus means that the Gudenå and its tributaries have a high potential for producing macrophytes, although the extent of such production depends on local factors such as substratum, flow, water quality (especially turbidity) and shade.  Species in the upper Gudenå include the following:

| | | |
|---|---|---|
| Cladophora fracta | Butomus umbellatus | Batrachium (Ranunculus) |
| C. glomerata | Glyceria maxima | sp. |
| Fontinalis | Potamogeton alpinus | Callitriche sp. |
| antipyretica | P. crispus | Elodea canadensis |
| | P. pectinatus | Lemna trisulca |
| | P. perfoliatus | Nuphar lutea |
| | P. praelongus | Sium erectum |
| | P. pusillus | |
| | Sparganium erectum | |
| | S. simplex | |

Macrophyte diversity is less in the middle and lower Gudenå.  Submerged species include Butomus umbellatus, Glyceria fluitans, Potamogeton lucens, P. pectinatus, P. perfoliatus, P. praelongus, Sparganium simplex and Elodea canadensis; species at the sides of the river are Acorus calamus, Glyceria maxima, Phragmites australis, Sparganium erectum and Typha latifolia.

A study of macrophyte production in some tributaries during 1971-75 (Brøndum Jensen & Mathiesen, 1976) failed to establish a general

*Fig. 2.8* 'The Klosterkanal'. The river Gudenå still runs in the canal dug by the monks at Klostermølle (left-hand photo); there are important spawning beds for trout and grayling just upstream of the Klosterkanal.

picture of the extent to which dissolved nitrogen or phosphorus were limiting factors. However it was clear that massive growths of *Cladophora* can be expected if the average concentration of total phosphorus exceeds 0.05 mg $1^{-1}$ during the growing season. The authors recorded biomass values for aquatic macrophytes (excluding roots and rhizomes) as follows:

1-40 g m$^{-2}$ dry weight          reaches with shade effects from trees e.g. Spørring Å (Hadsten Lilleå), Mattrup Å; dominants were *Elodea* and *Callitriche*;

50-120 g m$^{-2}$ dry weight        reaches with some shade effects or effects from epiphytes e.g. Mattrup Å downstream of Lake Stigsholm; dominant was *Potamogeton perfoliatus*;

200-300 g m$^{-2}$ dry weight          reaches without shade effects e.g.
upper Gudenå and Hadsten Lilleå.
Dominants were *Potamogeton perfoliatus,*
*P. crispus, P. pectinatus, Elodea*
*canadensis* and *Sparganium simplex*;

500-700 g m$^{-2}$ dry weight          downstream of a fish farm in the lower
part of Mattrup Å, with *Sparganium*
*simplex* as the dominant (and only)
vegetation.

At their experimental sites in the tributaries, Granslev Å and Gjern Å,
Dawson and Kern-Hansen (1978) recorded maximum biomasses of 200-300 g
m$^{-2}$ and 480 g m$^{-2}$ dry weight, with the dominant species being *Elodea*
*canadensis* and *Sparganium simplex,* respectively.


## 2.42  MACROINVERTEBRATES AND USE OF THE SAPROBIEN SYSTEM

Most of the information on macroinvertebrates comes from the use of a
simplified Saprobien system as an aid to controlling the biological
quality of the watercourses.  The simplified Saprobien system is based
on the relative occurrence of a number of easily recognisable macroinvertebrates; water quality is distinguished in degrees from I to IV.
The system was issued in 1970 by the Ministry of Agriculture and taken
over by the National Agency of Environmental Protection in 1974.  The
following are the most important features:

I    (oligosaprobic)  characterized by the occurrence of clean water
     organisms, such as certain caddis fly larvae and stone fly
     nymphs.

II   (β-mesosaprobic)  characterized by abundance of species which
     are more or less pollution tolerant; dominance of *Gammarus pulex,*
     *Baetis rhodani* and various *Caddis* flies, but not *Chironomus* or
     *Asellus.*

III  (α-mesosaprobic)  characterized by occurrence of *Chironomus* and
     *Tubifex* together with some other pollution tolerant organisms
     such as *Asellus* and *Herpobdella.*

IV   (polysaprobic)  characterized by the exclusive occurrence of
     saprophilic organisms as *Eristalis, Tubifex* and *Chironomus,* the
     last two predominating under the most polluted conditions.

Organisms other than macroinvertebrates may also be valuable as
indicators e.g. *Beggiatoa* for pollution degree IV.  The sewage-fungus
community (dominated by *Sphaerotilus*) occurs most frequently in
pollution degree III.

The results of a detailed study of the Gudenå (J. Jensen *et al.,*

1976) are shown in Fig. 2.9. The upper Gudenå is not polluted, the pollution degree of the river being II or better. The clean water species include:

Plecoptera. *Perlodes microcephala, Chloroperla burmeisterii.*
*Ephemeroptera. Heptagenia sulphurea, Heptagenia fuscogrisea,*
*Ephemera danica.*

The more upstream part of the middle Gudenå and the lower Gudenå are influenced by the loadings of organic matter (phytoplankton) produced in the lakes, so evaluation using the Saprobien system has to be done with care. Comparison of the results obtained with the Danish system and with the Sládeček (1973) system (modified to take account of the Danish fauna) shows several differences. For instance, the former did not reflect the untreated sewage (corresponding to 30000 PE) which was discharged into the river from Bjerringbro at the time of the investigations, whereas the Sládeček system both indicated this and the self-cleaning of the river on passing downstream. The pollution index in this stretch is now about II as a result of biological treatment plants constructed at Bjerringbro in 1976 and Hadsten in 1978.

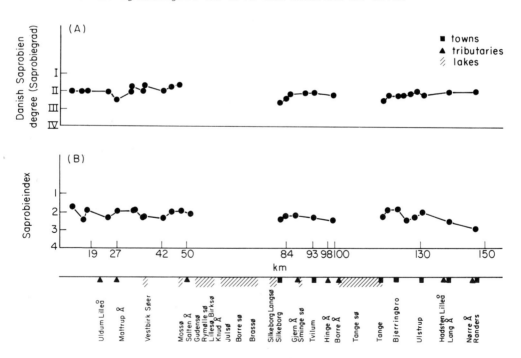

*Fig. 2.9* Use of macroinvertebrates to evaluate water quality of Gudenå according to two different Saprobien systems: A, Danish simplified system; B, Sládeček (1973) modified for Danish river fauna. (From J. Jensen *et al.*, 1976; the quality of lower Gudenå has since improved - see text)

2.43    FISH

*2.431  Fauna*

The Gudenå, its tributaries and the lakes have the reputation of being good waters for fishing.  Many foreign sports fishermen visit the area to compete in 'coarse fishing', while many individual anglers can be seen along the river banks.  The trout (*Salmo trutta*) is the dominant of the rivers.  Following at nearly 300 sites and involving about 11000 fish, Mortensen (1976) reported that trout was found at 68% of the sites and constituted 60% of the total catch (Fig. 2.10).

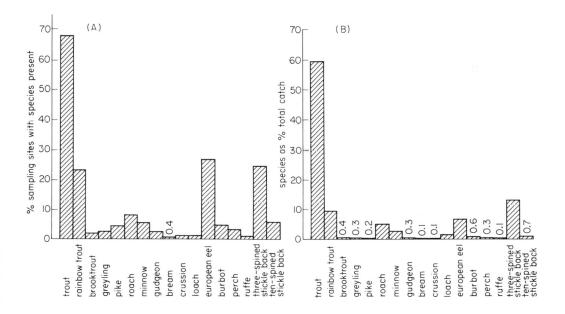

*Fig. 2.10* Occurrence of fish at 300 sites in Gudenå catchment during surveys in 1969, 1971 and 1974-75 involving 11000 individual fish:  A) Percentage of sample sites with species recorded; B) Percentage of total catch.  From Mortensen, 1976)

The diversity of the fish fauna was generally low.  About 30% of the sampling sites had only one species and another 50% only two species. Most of these were small streams.  In contrast five sites had six species and one site had seven species.  Investigations of fish productivity in small tributaries (Mortensen, 1977) showed a maximum value for trout of 22-27 g m$^{-2}$ yr$^{-1}$ in Brandstrup Baek, a stream entering the lower Gudenå at Ulstrup.

*2.432  Fish-passes and the Gudena salmon*

While many stretches of the Gudenå and its tributaries are suitable for nursing and growth of sea trout (*Salmo trutta*), obstructions in the form of dams in many cases prevent the migration of this and other salmonid fish.  The salmon (*S. salar*) was also common in the middle and lower Gudenå before the hydropower plant was built at Tange, but impoundments for the plant flooded key nursery streams by the middle Gudenå.  The fish-pass constructed in 1920 at the dam was never brought into function.

The evaluations which followed the Gudenå survey carried out in 1973-75 led to the Counties of Vejle, Viborg and Århus deciding to build a new fish-pass at Tange.  The type of pass chosen was based on principles developed in 1908 by a Belgian engineer, G. Denil; Lonnebjerg (1980) described the results of tests for the new fish-pass at Tange, while the pass was constructed in 1980.

*Fig. 2.11*  Fish-pass at Tange.  This consists of 7 resting pools and 8 flights, each 6.55 m long and with a gradient of 20%; the total rise is about 10 m; water flow through the pass is about 150 l s$^{-1}$.

In order to guide downstream migrants to the exit of the pass, a barrier of lights has been placed on the upstream side of the intake for the power station. Downstream, an electric screen has been placed across the river to prevent upstream migrants from entering the tail-race tunnels. The total cost of the pass amounted to U.S. $150000 (at 1980 prices).

The migration of fish in the pass at Tange is monitored intensively and, for example, in November-December 1980 one salmon, 23 sea trout and 15 brown trout were registered. Eel, roach and perch have also been found to migrate through the pass. In spite of the occasional salmon, however, salmon fishing in the Gudenå catchment is not expected to expand, because of the lost nursery streams and also the heavy netting which occurs in Randers Fjord. In contrast, there are great possibilities for expansion of sea trout fishing, particularly if other passes are established where there are obstructions at present e.g. Ry Mølle, Klostermølle, Vilholt Mølle and Vestbirk hydro-power station.

## 2.5   RIVER MANAGEMENT

### 2.51   ADMINISTRATIVE RELATIONSHIPS

The legal provisions for the use and protection of Danish watercourses, lakes, and other freshwater localities are mainly contained in the following four acts:

1. Watercourse Act (Vandløbsloven)
2. Environmental Protection Act (Miljøbeskyttelsesloven)
3. Freshwater Fisheries Act (Ferskvandsfiskeriloven)
4. Conservation of Nature Act (Naturfredningsloven)

According to these acts the Counties play a central administrative part to ensure that the protection of the environment has a high priority today and will have so in the future too.

The County Councils of Vejle, Viborg and Århus formed 'The Gudenå Committee' to co-ordinate and collaborate on the management of the Gudenå. Thus, the Gudenå Committee in 1978 agreed to set up guidelines to improve the water quality of the Gudenå and the associated lakes, based on the results of "The Gudenå investigation 1973-75" (Water Quality Institute, 1976) and the daily work of the Water Authorities i.e. the County Water Inspections. The County Councils of Vejle, Viborg and Århus agreed to the guidelines in 1979.

### 2.52   IMPROVEMENT OF WATER QUALITY IN LAKES

One example of the guidelines is the general emission limit for phosphorus in sewage discharged from towns upstream of Tange; this is needed in order to reduce the growth of phytoplankton in the lakes.

The emission limit is 1 mg P l$^{-1}$ and the treatment plants have to fulfil this condition, as follows:

   treatment plants > 5000 PE must establish phosphorus precipitation
      in 1981 or before;
   treatment plants > 2000 PE must establish phosphorus precipitation
      before 1985;
   treatment plants > 500 PE must establish phosphorus precipitation
      before 1990.

The investigation in 1973-75 showed that light and phosphorus limit the growth of phytoplankton, and for some selected lakes a mathematical model has been calibrated to simulate the effect on the water quality of altered nutrient loadings (Heise *et al.*, 1977; Nyholm, 1977). If the phosphorus from wastewater from towns greater than 500 persons was treated with chemical precipitation, growths of phytoplankton were predicted to decrease by 30-40%, at least up to the year 2000. The transparency (Secchi disc depth) was not influenced much more by inserting a 90% phosphorus reduction. The reason for this is the short detention time in the lakes (approximately 30 days) and the relatively high contribution of nutrients from non-point discharges and fish farm outlets, which maintain a high phytoplankton growth rate.


## 2.53   RIVER MODELLING

For the Gudenå and a number of the tributaries a mathematical model (e.g. Dahl-Madsen and Simonsen, 1974), simulating oxygen fluctuations was applied. The model used is based on the work of Streeter and Phelps (1925) describing 'The Oxygen Sag Curve for Flowing Waters'. In the case of the river Gudenå, however, not only have the decomposition of organic matter and reaeration been taken into account, but also other processes such as oxygen fluctuations due to respiration and photosynthesis by aquatic plants. The oxygen consumption due to nitrification is of minor importance, and so far it has not been taken into consideration.

   Table 2.5 gives the values calculated for certain parameters for the Gudenå and some tributaries (Simonsen *et al.*, 1976). Oxygen concentrations today are in fact satisfactory in the Gudenå system as a whole. The most severe pollution in 1973-75 was found in the tributary Hadsten Lilleå, where untreated sewage from 30000 PE (from the town of Hadsten) was discharged into the stream having a rate of flow of 300 l s$^{-1}$. Figs. 2.12 and 2.13 show the fluctuations of oxygen and BOD in the Hadsten Lilleå. Calibration of the model was made taking into consideration both the growth of plants and sediment respiration in the Hadsten Lilleå upstream of the sewage discharge. In order to decide the degree of biological treatment, the model was used to simulate the oxygen fluctuations in the stream as a function of the discharge from the town of Hadsten. Using the results of the model, it was decided to establish a biological treatment plant whose effluent BOD would not exceed 20 mg $l^{-1}$. The calculation predicted that the oxygen concentration under those circumstances would not be less than 2 mg $l^{-1}$ at any location along the river. The treatment plant now works well under dry weather conditions, and normally the oxygen concentration in the river in Hadsten Lilleå downstream Hadsten is higher than 4 mg $l^{-1}$, However, during storms the plant is hydraulically overloaded, and the pollution load (as BOD) from the plant rises by a factor of 5-10 times for some hours.

In summer 1980 some fish kills were recorded in a downstream fish farm at the same time as some heavy rain storms. As a consequence the treatment plant was accused of causing these kills. In order to see if this was the probable explanation, investigations were carried out during storm events in the summers of 1981 and 1982. The results showed that the oxygen content in the water body dropped only by about 1 mg $l^{-1}$ $O_2$ due to disposal of storm water and sewage from the treatment plant. No fish kills have been reported since the events of 1980. At present the most likely explanation for the kills would seem to have been the illegal disposal of manure effluents or other non-point pollution sources.

*Table 2.5* Range of values calculated for 20°C for the river parameters, reaeration coefficient, rate of oxygen consumption, total
daily production and total respiration (e.g. Simonsen, 1976)

| water-course | reaeration coefficient for oxygen $(K_2)$ $d^{-1}$ | rate of oxy-gen consump-tion due to BOD-decompo-sition $(K_1)$ $d^{-1}$ | total daily oxygen pro-duction (P) g $O_2$ $m^{-2}d^{-1}$ | total daily respiration (R) $O_2$ $m^{-2}d^{-1}$ |
|---|---|---|---|---|
| **UPPER GUDENÅ** | | | | |
| upper reaches | 2.0 - 6.1 | 0.5 - 6.0 | 3.9 - 17.6 | 1.7 - 29.9 |
| lower reaches | 2.4 - 2.8 | 0.5 - 0.75 | 11.8 - 27.7 | 7.9 - 19.2 |
| MIDDLE GUDENÅ | 0.4 - 3.4 | 0.25 - 2.0 | 3.8 - 22.5 | 3.9 - 26.4 |
| LOWER GUDENÅ | 0.7 - 1.5 | 0.25 - 1.75 | 2.0 - 25.9 | 3.0 - 29.6 |
| Mattrup Å | 1.0 - 9.7 | 0.5 - 2.0 | 1.0 - 17.8 | 9.6 - 46.7 |
| Hadsten Lilleå | 2.9 - 4.2 | 1.7 - 6.9 | 3.9 - 13.0 | 12.3 - 21.3 |
| Tange Å | 5.1 - 15.9 | 0.25 - 2.0 | 1.1 - 4.0 | 0.5 - 8.0 |
| Nørre Å | 0.4 - 3.8 | 0.75 - 1.7 | 0 - 8.2 | 4.0 - 17.1 |

The calibrated oxygen models have also been used as planning tools in Nørre Å and Tange Å. In Nørre Å the discharges from a fish farm and the biological treatment plant in Viborg were evaluated. Calculations showed that the fish farm discharge was the main cause of oxygen deficit in Nørre Å. In Tange Å the effect on the oxygen content when

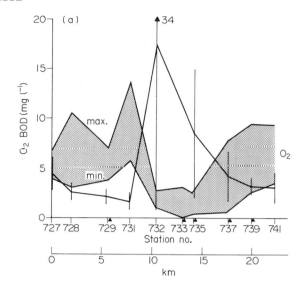

*Fig. 2.12* Relationship between dissolved oxygen and BOD in the tributary, Hadsten Lilleå. The shaded limits show the maximum and minimum values for $O_2$. BOD is plotted as the mean, with the vertical bar indicating maximum and minimum values.

Untreated sewage from the town of Hadsten was discharged between station no. 731-732, causing a rise in BOD concentration of the river water to 34 mg l$^{-1}$. The 'oxygen sag curve' is obvious.

*Fig. 2.13* Calculated 'duration curves for oxygen' as a function of different BOD discharges from the town of Hadsten at the 'critical' reach on the Hadsten Lilleå.

pumping river water for irrigation was simulated and evaluated.  No
critical oxygen concentrations became apparent; the reason for this is
because organic matter is removed with the irrigation water.

Modelling of processes in river water gives a good understanding of
the likely effects of various types of interference, though results of
the calculations must be interpreted with care.  Nevertheless, with
common sense and ecological experience, it is possible to set up guidelines for emission limits for discharges in order to improve the water
quality.


2.54    RIVER MANAGEMENT IN THE FUTURE

In order to fulfil the rules of the Environmental Protection Act the
County Councils have set up objectives for the use of the water systems
(e.g. bathing water, fishing water for migratory or for nursery purposes,
areas for scientific studies in the so-called receiving water plans
(recipientkvalitetsplaner).   The guidelines given by the Gudenå
Committee have now been incorporated into the planning worked out by the
Counties of Vejle, Viborg and Århus, and the Communities of the Gudenå
area.  It is expected that the sewage discharges from towns and industries will be treated biologically and chemically in the next decade to
improve the water quality.

Discharges from fish farms play a greater part today than they did
in 1973-75 and it is evident that the effluents from the fish farms
have to be treated too.  New sedimentation techniques, changes in
production and recycling of water are all processes that the fish
farmers have to implement, if the pollution from the fish farms is to
be reduced.

The non-point pollution from agricultural areas and from storm water
pollution has been drawing much attention.  Many farmers have been fined
for discharging manure water or silage joice into the receiving waters.
Because of the high price of fertilizers farmers are likely to be more
careful when spreading fertilizers on the fields, so this may reduce
the wash-out of nitrogen and phosphorus from the soil.

Pollution from storm sewers and overflows should be minimized.
This demands retention basins, grids or, in special situations, treatment facilities to handle urban rainwater running to the receiving
waters.  As one example, the County of Århus built in 1982 a filtration
plant (sedimentation and sand filtration) for storm water runoff from a
new highway to protect the upper reaches of the tributary Granslev Å.

Many attempts to improve or restore water quality can be expected in
the future.  However, the quality of the water alone is not enough to
raise the good quality of the watercourses: banks and substratum have
to be handled with care as well.  The 'Klosterkanal' provides an
example where this was not done.  The dredging of a 500 m long stretch,

where 500 m$^3$ sand was removed, caused the reduction of the macroinvertebrates from 60 to 24 species (F. Jensen, 1979). Weed cutting
may also be changed in the future. The Gudenå Committee is aware of the
problems when cutting aquatic macrophytes in the watercourses to improve
the run-off from rural areas. The investigations of Dawson & Kern-Hansen
(1978) have pointed out possible solutions, and in the County of Vejle
trees have been planted on an experimental basis along a stretch of two
kilometers on the banks of Gudenå to shade the aquatic macrophytes.
The Watercourse Act is being revised in 1983, and laws requiring
more gentle weed cutting and restoration of spoiled (canalized) watercourses can be expected and implemented in the Gudenå area.

Following the Conservation of Nature Act, the Counties of Vejle. Viborg,
Århus have been evaluating boating problems on the river Gudenå system.
Regulations for canoes in the upper Gudenå and many of the tributaries
were given in 1982.          Regulations of the speed of motorboats have
already been made for the middle Gudenå and the lakes and limitations
on the number of motorboats legally permitted are expected in 1987.

Conclusions: Nature around and in the river Gudenå and its tributaries
has many qualities that many people enjoy - thus the Gudenå has to be
used, but not abused.


*Acknowledgements*

I am most grateful to my colleagues Jens Møller Andersen, Jytte Heslop
Christensen, Jørn Jensen and Lis Stadil for their helpful comments.


REFERENCES

Brøndum Jensen U. & Mathiesen H. (1976) Botanisk-Økologiske under-
    søgelser over vandløbsvegetationen i udvalgte Gudenå vandløb.
    *Report to the Gudenå Committee* No. 23. (Danish = Botanical-ecolo-
    gical investigations of macrophytes in selected reaches of Gudenå
    and tributaries.
Dahl-Madsen K.I. & Simonsen J. (1974) Vandløbssimulering. *Research
    Report from the Water Quality Institute and the Laboratory of
    Sanitary Engineering, The Technical University of Denmark.*
    (Danish = Water Stream Simulation).
Dawson E.H. & Kern-Hansen U. (1978) Aquatic weed management in natural
    streams. The effect of shade by marginal vegetation. *Verh. int.
    Verein. theor. angew. Limnol.* 20, 1451-1456.
Heise P.B., Nyholm N. & Simonsen J.F. (1978) River Gudenå investi-
    gation 1973-75. *Verh. int. Verein. theor. angew. Limnol.* 20,
    1446-1450.
Hydrological Survey of the Danish Land Development Service (1978)
    *10. Beretning, afstrømningsmalinger i Danmark 1917-70.* (10. Report,
    Discharge data for Danish Streams 1917-70. Danish, with an English
    summary). Slagelse, Denmark.

Jensen C.F. & Jensen F. (1979)  Oprensningens virkning på den specielle
   rheophile rentvandsfauna ved Klostermøllebro, Gudenå.  *Natur-
   historisk Museum and the County of Vejle.*  (Danish = The effect of
   dredging in the Klostercanal to the Rheophile macroinvertebrates).
Jensen F (1976)  Kildeundersøgelser.  *Report to the Gudenå Committee*
   No. 33.  (Danish = Investigations of the springs).
Jensen J., Lastein, E. & Hansen I. (1976)  Vandløbsbiologi.  *Report to
   the Gudenå Committee* No. 13.  (Danish = The macroinvertebrates).
Lonnebjerg N. (1976)  Fiskepas af modstrømstypen.  *Meddelelser fra
   Ferskvandsfiskerilaboratoriet, Danmarks Fiskeri- og Havundersø-
   gelser* 1/80.  (Danish = Fishways of the Denil type).
Mortensen E. (1976)  Fiskeundersøgelser.  *Report to the Gudenå Committee*
   No. 18.  (Danish = Investigations of fish-fauna).
Mortensen E. (1977)  Fish production in small Danish streams.  In:
   *Folia limnol. Scand.* **17.**  (Danish = Limnology, Reviews and
   Perspectives).
Nyholm N. (1977)  A simulation model for phytoplankton growth and
   nutrient cycling in eutrophic, shallow lakes.  *Ecol. Modelling.*
Simonsen J., Strange Nielsen K. & Heise P. (1976)  Intensive vandløbs-
   undersøgelser.  *Report to the Gudenå Committee* No. 11.  (Danish =
   Intensive investigations according to river modelling).
Sládeček V. (1973)  System of water quality from biological point of
   view.  Ergebnisse der Limnologie, *Arch. Hydrobiol.* Beiheft 7,
   Stuttgart.
Strange Nielsen K. & Nyholm N. (1977)  The contribution of nutrient from
   diffuse sources.  *Prog. wat. Tech.* 8(4/5), 111-117.
Streeter H. W. & Phelps E.B. (1925)  A study of the pollution and
   natural purification of the Ohio River.  *Public Health Bulletin*
   No. 146, Treasury Department, U.S. Public Health Service.
Warncke E. (1980) Spring areas:  ecology, vegetation and comments on
   similarity coefficients applied to plant communities.  (Diss. Arhus).
   *Holarctic Ecology* vol. 3, no. 4.
Water Quality Institute (1976)  Kilder, søer, vandløb - samlerapport,
   Gudenåundersøgelsen 1973-75.  *Composite report to the Gudenå
   Committee.*  (Danish = Spring areas - lakes - watercourses, The
   Gudenå investigation 1973-75).

# 3 · Wye

R. W. EDWARDS & M. BROOKER

| | | |
|---|---|---:|
| 3.1 | Introduction | 52 |
| | 3.11 General | 52 |
| | 3.12 Geology and land-use | 52 |
| | 3.13 Hydrology | 54 |
| 3.2 | Chemistry | 56 |
| | 3.21 Introduction | 56 |
| | 3.22 Spatial distribution | 57 |
| | 3.23 Temporal variation | 58 |
| | 3.24 Catchment losses | 60 |
| | 3.25 Water quality model of the Wye | 60 |
| 3.3 | Plants | 61 |
| | 3.31 Algae | 61 |
| | 3.32 Bryophytes | 63 |
| | 3.33 Pteridophytes | 63 |
| | 3.34 Angiosperms | 63 |
| 3.4 | Macroinvertebrates | 66 |
| | 3.41 Introduction | 66 |
| | 3.42 Spatial distribution | 67 |
| | 3.43 Quantitative estimates | 69 |
| | 3.44 Drift | 70 |
| | 3.45 Autecological studies | 71 |
| 3.5 | Fish and Fisheries | 72 |
| | 3.51 General | 72 |
| | 3.52 Salmon stocks | 74 |
| | 3.53 Economic value of fishery | 76 |
| 3.6 | Future developments | 77 |
| References | | 80 |

## 3.1   INTRODUCTION

### 3.11   GENERAL

The scenic beauty of the Wye has been captured by writers, most notably Wordsworth and Shaw, by painters, particularly Turner, Cotman and Fielding, and by such addictive travellers as Borrow and the Reverend Gilpin.  During the late 18th century its attractions lured so many to its banks for the 'Wye Tour' that a holiday industry developed to accommodate visitors and carry them by boat from Ross-on-Wye to Chepstow - a journey which took about three days.  For many centuries before the arrival of the railways, around 1855, the lower reaches provided an important trade route in the valley, which contributed to the development of settlements, the largest being the City of Hereford with a current population of about 47000.

The river sustained large salmon stocks until the 19th century, when they were decimated both as juveniles (samlets) and as adults to titillate the palates of visitors and to sell elsewhere in the expanding markets made accessible by improved road and rail transport.  Strict control of the net fishery in the estuary, and its total prohibition in the river, led to a recovery of stocks in the early 20th century which facilitated an expansion of recreational rod fishing until it became the largest in England and Wales.

The waters of the Wye catchment, like its salmon, have been exported. At the end of the 19th century three reservoirs were constructed in the Elan catchment to supply water for Birmingham (see Fig. 3.8); since that period a further reservoir has been built (Claerwen in 1952) to match the increasing demand of that industrial conurbation.  The further development of the water resources of the catchment has been proposed (Water Resources Board, 1973) involving the enlargement of Craig Goch reservoir (see 3.6).

In addition the Wye is of considerable conservational significance. In *A Nature Conservation Review* (Ratcliffe, 1977) the River Wye was classified as a Grade 1 Site of Special Scientific Interest, the safeguarding of such sites being regarded as essential to the success of nature conservation in Britain.  Its importance stems largely from its virtual uniqueness as a large, relatively unpolluted river in southern Britain and from its catchment diversity, which gives rise to a biological richness.

### 3.12 GEOLOGY AND LAND-USE

The Wye, about 250 km long, rises at 677 m on Plynlimon in the northwest of the catchment and flows south-easterly to join the Severn Estuary at Chepstow (Fig. 3.1).  It has four major tributaries, the Ithon and Irfon in its upper reaches and the Lugg joining the main river downstream at Hereford and the Monnow at Monmouth (Fig. 3.1).

*Fig. 3.1* Catchment showing geological features, average annual flows at selected sites and seasonal flow pattern for Wye, Monnow and Lugg.

The Elan, before reservoir development in its catchment, was also a major contributor to the upper Wye but it now exports over 50% of its yield to the West Midlands.

The major geological division of the catchment is between the relatively impermeable lower Palaeozoic rocks (Ordovician and Silurian) in the north-west (41% of catchment), elevated above 200 m, and the Old Red Sandstone (55% of catchment), generally below 120 m.  There are outcrops of Carboniferous rocks, principally limestone, downstream of Ross-on-Wye and a small outcrop of Triassic strata between Chepstow and the estuary (Fig. 3.1).

Land-use broadly reflects the topographical and geological features. The elevated rough pastures in the north-west are grazed principally by sheep.  On somewhat lower land, intensive sheep rearing predominates with the inclusion of limited dairy farming.  On the richer lowlands there is a mixture of dairying, horticulture and arable crops.  In 1970 the principle agricultural uses in the catchment were permanent grass (37%), tillage (20%), rough grazing (15%), temporary grass (14%) and forestry (9%).

The population of the catchment, about 200000, represents an average density of 0.47 ha$^{-1}$ and in the upper 42% of the catchment, with no town having a population greater than 4000, the density is only 0.15 ha$^{-1}$.  Although the sewage from isolated dwellings is only treated in septic tanks, over 65% of the population now receives main drainage, sewage being given full biological treatment before discharge to streams and rivers within the catchment.


3.13 HYDROLOGY

The precipitation broadly reflects topography and ranges between 1800 and 2500 mm yr$^{-1}$ in the headwaters to less than 1000 mm yr$^{-1}$ in the lower elevations associated with Old Red Sandstone.  In the wetter areas there is a marked seasonal precipitation pattern, the period October-January being wettest and March-June being driest.

Except in the wetter elevated part of the catchment, about half the precipitation is lost in evapotranspiration, a process which also modifies the seasonal distribution of river flows.  Fig. 3.1 shows the average flow (discharge) at several recording stations on rivers within the catchment, seasonal rhythms being shown at three of these stations. Except during periods when the Elan valley reservoirs are full and overspill, flow in the Elan is maintained by 'compensation water', currently about 1.3 m$^3$ s$^{-1}$.  But average flows belie the wide variation which exists.  In Fig. 3.2 this variation is shown at Monmouth for only three years (1975-77), the summer of 1976 being the driest on record. Winter flows exceeding 200 m$^3$ s$^{-1}$ are not exceptional and, during the summer, flows are frequently around 10 m$^3$ s$^{-1}$, or 5% of this winter value.

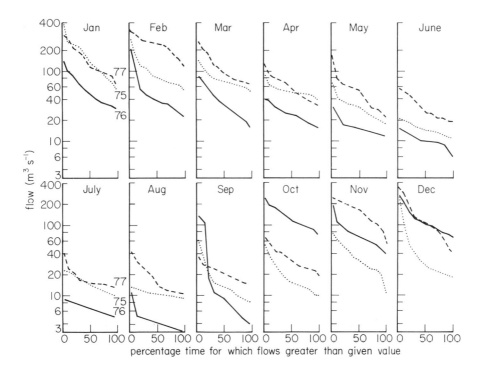

*Fig. 3.2* Distribution of daily flows for each month at Monmouth for 1975, 1976 and 1977.

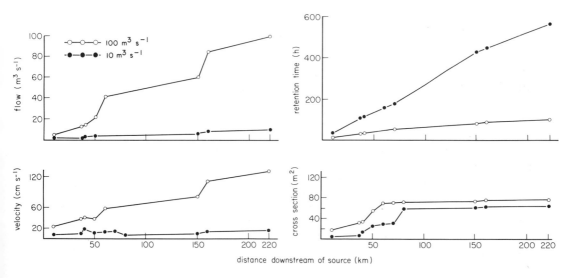

*Fig. 3.3* Flows, velocities, retention times and cross-sectional areas of the Wye when flows at Monmouth are 10 and 100 m³ s⁻¹, equivalent to dry summer and wet winter conditions.

Although important to water resource engineers, the flow *per se* has little ecological relevance except with respect to aspects of water quality (3.6). Changes in flow are reflected, however, in changes in velocity, which is reciprocally related to residence time, and in cross-sectional area, the latter finding tangible expression in depth and width: all these characteristics have considerable ecological relevance. Fig. 3.3 shows the effects of changing the equilibrium flows in the river, from 10 to 100 $m^3$ $s^{-1}$ at Monmouth, on velocity, residence time and cross-sectional area. It is based on calculations derived from iodine tracer studies at predominantly low flows and, whilst it may not be very accurate at high flows, it demonstrates the comparative response of parameters to flow - in the middle and lower catchment flow principally affects velocity whereas in the upper catchment both velocity and cross-sectional area are affected.

At low flows, velocities are generally insufficient to erode sediments but at flows more characteristic of the winter period particles of a wide size range may regularly be transported downstream. From turbidity data and studies of labelled bed material (Hey, 1980) it seems that at flows above 10 $m^3$ $s^{-1}$ at the Elan Valley confluence (equivalent to 100 $m^3$ $s^{-1}$ at Monmouth: Fig. 3.3) the movement of bed material (>10 cm diameter) will occur.

At low flows (10 $m^3$ $s^{-1}$, Monmouth), the total residence time of water in the main river is about three weeks, which provides ample opportunity for the development of planktonic algal blooms: in contrast, at high winter flows the residence time is three or four days. Whilst in the reaches downstream of the Lugg confluence, flow changes have comparatively little effect on cross-sectional area (Fig. 3.3), above that confluence, and particularly in the headwaters, substantial effects are apparent. In the lower reaches, with well-defined banks, such flow changes will result predominantly in depth changes, whereas in the upper reaches, with shallower bank profiles, fluctuations in width will also occur.


## 3.2   CHEMISTRY

### 3.21   INTRODUCTION

The waters of this predominantly rural catchment are free from major sources of sewage and industrial pollution and until relatively recently, when proposals to regulate the flow of the Wye and abstract considerable amounts from the lower reaches were considered, there seemed little need to undertake comprehensive monitoring programmes of water quality. Consequently, apart from a few simply-measured parameters of water quality such as ammonia, nitrate, orthophosphate, alkalinity and BOD, few measurements were made regularly and extensively.

The water quality is related largely to soil structure and the underlying geology of the catchment, which determines the potential source of most materials (other than sulphate and sodium, much of which are atmospheric in origin and phosphate which is derived from sewage) and to the pattern of rain-fall, infiltration and run-off which determines the leaching and dilution of such materials.  Man has influenced both the soils of the catchment by cultivation and the run-off processes by land-drainage (Jones, 1975) and the nature of land-use:  the upland pastures yield about 15% more run-off, with similar precipitation, than the coniferous plantations of the neighbouring Severn catchment (Newson, 1979).  Inputs are further modified by discharges of sewage effluent, although in the Wye catchment with a total population of only 200000, and 65% being connected to sewers, the effect of such effluent is small except with respect to phosphate concentrations.

Chemical quality may be modified within the river by chemical processes, such as the precipitation of calcium at high pH, and biological processes, such as silica utilization in diatom growth and denitrification:  such processes often prevent the construction of simple input-output models (Edwards *et al.*, 1978; Houston & Brooker, 1981).

3.22  SPATIAL DISTRIBUTION

As in many rivers, concentrations of most determinands generally increase steadily downstream in the Wye (Fig. 3.4).  The waters draining the mineral-poor impermeable mudstones in the upper catchment have low concentrations of minerals and nutrients and the more southerly part of the catchment, particularly that draining the arable lands of the Lugg basin, is characterized by much richer waters.  The major upland east-bank tributary, the Ithon, also carries higher concentrations than those in the upper Wye, reflecting drainage from the calcareous Wenlock beds.  In contrast, the upland west-bank tributaries, Elan and Irfon, are generally less rich than the Wye and the quality of the main river, below its confluence with the Elan, particularly during periods of natural low flow, is locally influenced by the acid, nutrient-poor water from the impoundments of this tributary.

One aspect of particular ecological significance relates to the relatively high concentrations of iron and manganese in the Elan, the mean concentrations of 500 µg $l^{-1}$ Fe and 100 µg $l^{-1}$ Mn being about ten times greater than those in the upper Wye.  About 75% of the total iron and 50% of the total manganese in the Elan are associated with particulate matter and almost half the load of these metals is derived from the effluent of a water-treatment works and not directly from the reservoirs.  Despite the generally high concentrations of iron and manganese in the deeper water of the Caban Coch reservoir, oxygen concentrations remain relatively high (>50% saturation) throughout the summer stratification period.

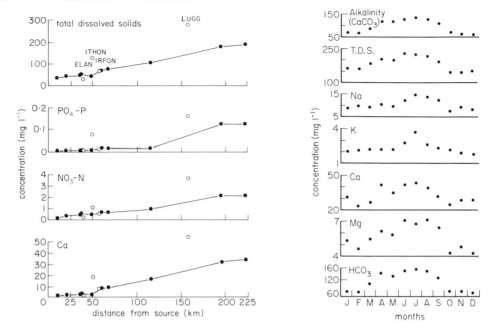

*Fig. 3.4* Average concentrations of total dissolved solids, $PO_4$-P, $NO_3$-N and Ca at sites down the Wye and in major tributaries.

*Fig. 3.5* Seasonal variation in monthly average concentrations of Ca, Mg, Na, K, alkalinity, total dissolved solids and $HCO_3$ for the period 1975-1977 at Monmouth.


## 3.23   TEMPORAL VARIATION

The water chemistry of the Wye is characterized by marked seasonal changes, particularly in the lower catchment, where concentrations are substantially higher than in the upper catchment (Fig. 3.5).   For example, alkalinity and the concentrations of total dissolved solids and the major ions, such as calcium, magnesium, sodium, potassium and bicarbonate, are generally lowest during the winter and highest during the summer:   these differences are predominantly related to the seasonal pattern of flows (Fig. 3.1) and simple concentration-flow regressions of these major ions, after logarithmic transformation, generally account for more than 40% of the variation although they rarely account for more than 70% in the lower reaches (Oborne *et al.*, 1980).

Throughout the catchment highest concentrations of phosphorus also occur principally during summer low-flow periods (Fig. 3.6) and, as expected, concentrations were particularly high during the drought of 1976.   Since the major source of phosphorus is relatively constant, being derived from sewage effluents, changes in concentration within the river principally reflect the dilution which such effluents receive.

*Fig. 3.6* Seasonal variation in concentrations of PO$_4$-P, NO$_3$-N and SiO$_2$ for the period 1973-1977 at Monmouth.

*Fig. 3.7* Flow-concentration relationships for dissolved silica (SiO$_2$) at sites down the Wye. The distance (km) from source of each site is shown.

Such behaviour contrasts with that of nitrate (the predominant form of combined nitrogen) which generally has highest concentrations during winter high-flow periods. The increase in concentration was particularly marked during the autumn of 1976 when the drought was terminated by heavy rainfall (Fig. 3.6). Substantial increases (0.1 to 5.0 mg l$^{-1}$ NO$_3$-N) were recorded in some lowland reaches; these probably resulted from flushing processes following the prolonged mineralisation of organic material in the soil during the long dry summer. In certain small lowland tributaries of the Lugg, such as the Frome, high concentrations of nitrate (7 mg l$^{-1}$ NO$_3$-N) occur during summer and are probably derived from nitrate-rich ground waters, which at this period of the year receive little dilution from other more superficial sources.

The concentration-flow relationships of dissolved silica illustrate the complex distribution of substances which are utilized and transformed biologically (Fig. 3.7). At the uppermost site (7 km from source), concentration and water flow were negatively related, behaviour being similar to that of several other major ions, but progressively down the catchment the relationship between concentration and flow became increasingly positive. This rotation seems to be caused by the utilization of silica by diatoms which are abundant for part of the year: at low flows, the appreciably longer residence times (see Fig. 3.3) provide an opportunity for substantial silica removal. During the winter, when diatoms are not growing appreciably, there is no down-

stream rotation of the concentration-flow relationship, but merely a
progressive increase in downstream concentration which reflects the
changing catchment topography and soil characteristics.  The importance
of diatoms in silica depletion is supported by the negative correlation
between concentrations of chlorophyll *a*, extracted from suspended
material, and dissolved silicate in the lower catchment (Edwards *et al.*,
1978).  It has been suggested by Edwards *et al.* (1978) that diatom
growth within rivers of the catchment annually utilizes about 3000
tonnes (as $SiO_2$) and, even on tributaries of short retention time (<4 d),
such as the Frome and Trothy, about 10% of the annual load is removed
(Houston & Brooker, 1981).


## 3.24   CATCHMENT LOSSES

Oborne *et al.* (1980) estimated catchment losses of several determinands
during two 12-month periods of contrasting hydrology - a dry period
from September 1975 to August 1976 and a much wetter period from September 1976 to August 1977 (Fig. 3.8).  There are substantial differences
in yield between different parts of the catchment related to geology
and land-use and between the wet and dry period.  Annual yields per
unit catchment of several substances, e.g. potassium, calcium, magnesium,
nitrate, phosphate, were greater in the lower Wye and Lugg than elsewhere and this difference was generally most marked during the wet
period.  In contrast, highest yields of sodium, probably derived from
rainfall, were in the upper reaches of the river which received the
highest annual precipitation.  The yield of phosphate could be account-
ed for solely from sewage effluents:  in contrast, less than 20% of
nitrate was derived from these sources, even during the dry period.
By relating sub-catchment yields to sewered-population densities it is
possible to calculate daily per capita outputs of phosphate and nitrate
- these were between 1.8 and 3.1 g $PO_4$-P and 6.8 g $NO_3$-N, generally
similar to values estimated from other studies (Oborne *et al.*, 1980).


## 3.25   WATER QUALITY MODEL OF THE WYE

Although flow-concentration regression equations may be used as simple
empirical models to predict the effects of changing flow in the whole
or parts of catchments, in the Wye these regressions rarely account
for more than 70% of the variation.  Water quality data for the lower
reaches have been used to test a more complex model based upon the
hydrological separation of river flow into surface run-off, base-flow
and effluent components, the quality of each hydrological component
being separately described (Oborne, 1981).  Models of the behaviour of
chloride, phosphorus, total hardness and alkalinity explained between
59 and 88% of test data after calibration (Fig. 3.9) and the performance of the model was considerably better than that of flow-concentration regressions derived from the same data (Sambrook, 1976).

| sub-catchment | NO$_3$-N D | W | PO$_4$-P D | W | SiO$_2$ D | W | Na D | W | Ca D | W |
|---|---|---|---|---|---|---|---|---|---|---|
| A | 2.8 | 2.8 | 0.07 | 0.08 | 28.8 | 33.8 | 70.0 | 80.1 | 27.4 | 32.0 |
| B | 4.1 | 6.5 | 0.04 | 0.06 | 15.5 | 25.3 | 37.9 | 61.0 | 25.8 | 40.5 |
| C* | 0.4 | 0.4 | 0.01 | 0.02 | 4.5 | 8.8 | 10.7 | 19.5 | 4.1 | 4.8 |
| D | 3.5 | 9.9 | 0.09 | 0.16 | 9.4 | 32.1 | 23.6 | 55.6 | 50.2 | 117.2 |
| E | 4.0 | 6.1 | 0.05 | 0.08 | 21.9 | 33.7 | 40.9 | 62.4 | 48.9 | 73.5 |
| F | 2.8 | 9.9 | 0.02 | 0.09 | 5.1 | 30.6 | 23.4 | 142.7 | 45.0 | 93.9 |
| G | 3.9 | 16.3 | 0.06 | 0.23 | 8.0 | 8.3 | 11.3 | 2.7 | 71.7 | 156.4 |
| H | 3.9 | 27.2 | 0.16 | 0.54 | 5.0 | 35.7 | 14.4 | 56.8 | 62.9 | 276.7 |
| J | 5.5 | 18.0 | 0.53 | 0.89 | -2.9 | 14.5 | 27.2 | 52.3 | 91.6 | 130.1 |
| total catchment | 4.1 | 16.7 | 0.21 | 0.44 | 5.3 | 23.3 | 21.6 | 52.9 | 64.3 | 154.6 |

*more than 50% of yields are exported and not shown here

*Fig. 3.8* Yields (kg ha$^{-1}$ yr$^{-1}$) of several substances in sub-catchments of the Wye (boundaries denoted by dotted lines) for dry (D) and wet (W) years. Numbers on map indicate sites used for invertebrate sampling (see Fig. 3.11). Insert shows reservoirs of Elan catchment with hatching giving the extent of proposed enlargement of Craig Goch.


## 3.3  PLANTS

### 3.31  ALGAE

The algae of the Wye have been described by Furet (1979) from water samples taken at several sites on the main river and at one site on each of the major tributaries. Two groups of species could be distinguished, those which grow planktonically, the 'bloom species' such as the coccoids (*Scenedesmus, Ankistrodesmus*) and discoid diatoms (*Cyclotella, Stephanodiscus, Thalassiosira*), and others which are adventitious, having been derived from benthic growth (*Synedra, Diatoma, Nitzschia*). On algal distribution, according to Furet, the river divides naturally into three zones:

*Fig. 3.9* Comparison of actual (–) and calculated (––––) concentrations of chloride, PO$_4$-P, total hardness and alkalinity during calibration and test periods using a flow separation model (Oborne, 1981).

    i)  upper zone, low in calcium (<10 mg l$^{-1}$), extends to km 60 from source and has less than 8 mg m$^{-3}$ chlorophyll *a* derived from algae suspended in the water, with no obvious seasonal peak.  Despite overall lack of seasonality, different algae are prevalent at different times: *Meridion circulare* occurs during the winter whereas *Scenedesmus* spp., *Ankistrodesmus falcatus*, *Rhoicosphenia curvata* and *Navicula cryptocephala* are dominant only during the summer.  Several benthic diatoms are abundant (*Synedra ulna*, *Diatoma vulgare*, *Nitzschia palea*, *N. amphibia*)

    ii)  middle zone, with calcium concentrations between 10 and 25 mg l$^{-1}$, extends from km 60 to 75, generally has chlorophyll *a* concentrations below 20 mg m$^{-3}$ with peaks in June-July and November.  As in the upper reach, *Meridion circulare* occurs during the winter and 'bloom species' (*Scenedesmus* spp., *Cyclotella meneghiniana*, *Thalassiosira fluviatilis*, *Carteria* spp.) during the summer.  Several benthic diatoms are abundant during the spring (*Melosira varians*, *Surirella ovata*, *Navicula viridula*, *Ceratoneis arcus*, *Diatoma vulgare*, *Pinnularia viridis*) and their growth coincides with a decrease in dissolved silica concentrations.

    iii)  lower zone, from km 75, with chlorophyll *a* peak concentrations between 40 (Ithon) and 100 (Lugg) mg m$^{-3}$, generally in June-July. Species composition is broadly similar to that upstream but *Monorhaphidium gracilis*, *M. tortile* and *Oocystis crassa* become abundant 'bloom species' during the summer.

Of the tributaries, the Elan is surprisingly poor in algae in view
of its impounded source; nevertheless *Tabellaria flocculosa* and *T.
fenestrata* are more abundant than elsewhere in the catchment.   In
contrast, the Ithon, which drains an area of relatively low relief, is
very rich in algae which are also abundant in the lower Wye, e.g.
*Diatoma vulgare, Cocconeis placentula*.   The Irfon, although generally
similar to the Upper Wye, is rich in the Chlorococcales, particularly
species of *Scenedesmus*.   The Lugg has the highest algal concentrations
in the Wye catchment and, like the Ithon, it drains an area of low
relief;   it is, however, much richer in plant nutrients, draining fertile farmland, much of it arable, in its lower course.

## 3.32   BRYOPHYTES

D. G. Merry (pers. comm.) surveyed the bryophytes at 49 riparian sites
down the length of the main river and, despite the difficulty of
separating aquatic from terrestrial species, of the 114 species he
recorded 44 might be regarded as generally associated with wet stream
margins or submerged habitats and only seven species as truly aquatic
(Table 3.1).   Of the aquatic species which were widely distributed,
*Nardia compressa, Scapania undulata* and *Fontinalis squamosa* were
generally restricted to upland sites and *F. antipyretica* to the lower
catchment.   Sites richest in bryophytes (>20 species per km) were all
within 60 km of the source and above an altitude of 130 m.

## 3.33   PTERIDOPHYTES

These are neither common nor diverse in the riparian habitat;   Merry
(pers. comm.) recorded only nine species and, of these, five were only
local - including *Lycopodium selago* at the highest site (600 m) and
*Phyllitis scolopendrium* near the Carboniferous Limestone gorge at
Symond's Yat.   The remaining four widely distributed species (*Equisetum
arvense, Pteridium aquilinum, Dryopteris filix-mas, Athyrium filix-
femina*) are common in Wales and not associated particularly with river
banks.

## 3.34   ANGIOSPERMS

Although several 19th and early 20th century botanists described the
aquatic and bank-side flora of limited sections of the Wye and, in more
recent times, Perring (1967) and Haslam (1978) have focused on particular sites or tributaries, the first extensive vegetation survey of the
main river was undertaken by D. G. Merry in 1976 and 1977.   During
Merry's survey, 380 species of angiosperms were recorded, but of these
only about 28 were truly aquatic and 88 riparian - the rest being
adventitious to the river corridor.   This summary of his study will be
confined principally to the aquatic species (Table 3.2).

*Table 3.1* Bryophyte species associated with stream margins or truly aquatic*. (Binomials for mosses according to Smith (1978))

---

mosses

*Sphagnum auriculatum*
*Atrichum crispum*
*Polytrichum commune*
*Fissidens crassipes*
*Campylopus atrovirens*
*Dichodontium pellucidum*
*Dicranella palustris*
*Cinclidotus fontinaloides*
*Racomitrium aciculare*
*R. aquaticum*
*Physocomitrium pyriforme*
*Bryum pseudotriquetrum*
*Rhizomnium punctatum*
*Philonotis fontana*
*Pohlia proligera*
*Hygrohypnum ochraceum*
*Hyocomium armoricum*
*Pohlia carnea*
*Climacium dendroides*
*Fontinalis antipyretica**
*F. squamosa**
*Thamnobryum alopecurum*
*Calliergon stramineum*

*Brachythecium plumosus*
*B. rivulare*
*B. rutabulum*
*Drepanocladus exannulatus*
*D. fluitans*
*D. uncinatus*
*Rhynchostegium riparioides**
*Amblystegium fluviatile**
*A. tenax**

liverworts

*Conocephalum conicum*
*Lunularia cruciata*
*Marchantia polymorpha*
*Pellia epiphylla*
*Riccardia pinguis*
*Barbilophozia floerkei*
*Chiloscyphus polyanthus*
*Diplophyllum albicans*
*Marsupella emarginata*
*Nardia compressa**
*Scapania undulata**
*Solenostoma triste*

---

*Table 3.2* Species of aquatic angiosperms found on the main river by Merry (1976-1977)

---

*Ranunculus flammula*
*R. omiophyllus*
*R. fluitans*
*R. penicillatus*
*R. peltatus*
*Rorippa nasturtium-aquaticum*
*Myriophyllum spicatum*
*M. verticillatum*
*Callitriche hamulata*
*Myosotis scorpioides*
*Veronica beccabunga*
*Mentha aquatica*
*Littorella uniflora*
*Polygonum amphibium*
*P. hydropiper*

*Elodea canadensis*
*Iris pseudacorus*
*Sparganium erectum*
*Lemna minor*
*Potamogeton crispus*
*P. lucens*
*P. pectinatus*
*P. perfoliatus*
*P. polygonifolius*
*P. salicifolius*
*Glyceria fluitans*
*G. plicata*
*Phalaris arundinacea*

Except for *Elodea canadensis* and most of the *Potamogeton* spp., the aquatic species showed some broad zonation (Fig. 3.10). At upland sites only *P. polygonifolius* and *Ranunculus omiophyllus* were found. Although not present at the uppermost sites (>350 m), *Callitriche hamulata, Glyceria fluitans* and *Myriophyllum verticillatum* occurred only in the upper 100 km of the river:  the presence of *M. verticillatum,* normally found in eutrophic lentic habitats, is somewhat unusual.  Only two species, *Veronica beccabbunga* and *Polygonum hydropiper,* were confined predominantly to the mid-reaches, the remainder being most abundant in lowland reaches. *Lemna minor* occurs only below the confluence of the Lugg and its occurrence in the main river depends upon production in back-waters of this tributary. Haslam (1978) found several aquatic and riparian species in the Lugg, including *Nuphar lutea, Schoenoplectus lacustris* and *Zannichellia palustris,* which have not been recorded in the main river.

distance from source (km)

| | 0 | 50 | 100 | 150 | 200 |
|---|---|---|---|---|---|

Ranunculus flammula
R. omiophyllus
R. penicillatus
Myriophyllum spicatum
M. verticillatum
Callitriche hamulata
Veronica beccabunga
Mentha aquatica
Polygonium amphibium
P. hydropiper
Sparganium erectum
Lemna minor
Potamogeton polygonifolius
Glyceria fluitans
Phalaris arundinacea

*Fig. 3.10* Distribution of aquatic plants down the Wye. Dotted line indicates regions where species are very patchily distributed.

In the lowland reaches and in the Lugg *Ranunculus* spp.* grow profusely during the early summer, with a peak biomass around 2 kg fresh wt m$^{-2}$ being recorded (Brooker *et al.,* 1978).  There is considerable inter-year variation in growth which appears to be related to river flows during the growth period (April-June).  The dense stands and their subsequent decay cause substantial changes in the oxygen economy

*In the lower Wye there is a mixture of *R. fluitans, R. penicillatus* and *R. peltatus* together with hybrids.

of the river, with pronounced daily oscillations in oxygen concentration
during the growth period.  During the drought of 1976 the low oxygen
concentrations at night (<1 mg $l^{-1}$), caused principally by plant respiration, coupled with high temperatures (>26 $^{o}$C) caused the mortality
of substantial numbers[*]  of adult salmon migrating upstream at the end
of June (Brooker et al., 1977).

Such dense stands of river-plants can also affect dramatically river
hydraulics by increasing frictional resistance to water-flow, with
consequent increases in the danger of flooding, as well as the need to
measure, with increasing frequency, velocity-depth relationships, where
these are used as a basis of flow-gauging.  Although plant stands
increased Manning's frictional coefficient roughly ten-fold at low
flows, from 0.02 to 0.25 on the Lugg and from 0.02 to 0.10 on the Wye
(Brooker et al., 1978), the river channels are sufficiently large to
accommodate all normal summer flows, particularly as frictional
effects of plants are reduced at high flows.

Of the 88 or so riparian species, two deserve particular mention:
Allium schoenopraesum has become more widespread over the past century
around Builth Wells, but it occurs at relatively few sites elsewhere
in Britain and Potentilla rupestre, found at only one site in the middle
reaches has only been recorded in Britain at two other sites within
Central Wales.


3.4  MACROINVERTEBRATES

3.41  INTRODUCTION

Most studies of riverine macroinvertebrates in the United Kingdom,
other than those in small streams, have concentrated upon polluted
river systems and rivers, like the Wye, which are generally of high
water quality, have been neglected.  Of course, many collectors have
visited the Wye over a considerable period and rare species, particularly mayflies, have been recorded from the river (Ratcliffe, 1977).
However, it was not until 1975 that a comprehensive systematic study
of the invertebrate fauna was undertaken.  This work had several aims,
including a qualitative assessment of the spatial distribution of
macroinvertebrates in the catchment, particularly in the main river,
using survey methods (Brooker & Morris, 1980a, b; Morris & Brooker,
1979; Lilley et al., 1979); a description of the temporal changes in
macroinvertebrate numbers, chiefly in the upper Wye (Brooker & Morris,
1978; Morris & Brooker, 1980) and an assessment of the drift of macroinvertebrates, particularly in relation to river flow (Brooker &
Hemsworth, 1978; Hemsworth & Brooker, 1979).

*About 1000 fish were collected from the river.

## 3.42  SPATIAL DISTRIBUTION

More than 230 different macroinvertebrates have been recorded from the Wye; these include, in the case of the midge (Diptera) larvae, species groups. In general the upper reaches are characterised by stonefly (Plecoptera) nymphs in spring, mayfly (Ephemeroptera) nymphs, caddisfly (Trichoptera) larvae and midge larvae being relatively more abundant during the summer months. In the lower reaches of the catchment the composition of the fauna is more variable, with oligochaetes, mayfly nymphs, caddisfly larvae, beetles and midge larvae all being relatively abundant at different locations and times.

Certain invertebrates (caddisfly larvae, beetles, mayfly nymphs) were found during the surveys to be sparse in the impounded tributary, the Elan, in comparison with the nearby naturally-flowing reaches of the Wye. Such differences probably resulted from the relatively constant compensation flow that the river received from the impoundment and the iron- and manganese-rich deposits which coat the bed of the river (Brooker & Morris, 1980b; Inverarity et al., 1983).

Total densities of invertebrates varied throughout the catchment from less than 1000 $m^{-2}$ to over 22000 $m^{-2}$, with highest densities in the lower reaches. A simple analysis of the number of types of macro-invertebrates and total density indicated that both showed a marked increase downstream of the confluence of the Ithon with the Wye, the contribution of minerals from the Ithon substantially increasing the total cation concentration in the Wye from about 330 to 440 µeq. $1^{-1}$; Egglishaw & Morton (1965) found similar effects of chemical changes on invertebrates in rivers in the Scottish Highlands.

A more complex assessment of the distribution of invertebrates using rank correlation coefficients and clustering techniques followed by nodal analysis (Brooker & Morris, 1980b) revealed that six groups of species or discrete taxa could be used to characterise five site groupings (Fig. 3.11), with the exception of one site each on the Elan and Irfon, these sites being on the main river. However, a full inter-pretation of those environmental factors responsible for such distri-butions is not yet possible, partly because of the close association between changes in physical and chemical factors throughout the catch-ment and partly because of limited knowledge of the ecology of many of the 'key' species.

However, on the basis of studies in other rivers (Boon, 1979; Hildrew & Edington, 1979), on  trichopterans belonging to the Hydro-psychidae, which are an important element of the stream fauna of the Wye, the characteristic grouping of species (Fig. 3.11) seems to be related to current velocity and water temperature. Although the general pattern of distribution of species was similar in the Wye and the nearby Usk (Hildrew & Edington, 1979) some minor differences were apparent. *Hydropsyche pellucidula* was found at all sites in the Wye but was not recorded from the upper reaches of the Usk whereas *H.*

Fig. 3.11 Distribution of the abundance of taxa groups of benthic
invertebrates in relation to site groups in the Wye catchment 1976-
1977 (position of sites is shown in Fig. 3.8). Groups are derived
from average linkage clustering of Spearman rank correlations.

*siltalai*, a species found throughout the Usk catchment, was not found in the most downstream 50 km of the Wye.  In both rivers *H. contubernalis* and *Cheumatopsyche lepida* were restricted to downstream reaches although *C. lepida* was found further upstream in the Wye.  *Diplectrona felix* and *Hydropsyche instabilis*, both species characteristic of headstreams, were not recorded from the Wye - probably because of the sampling locations - but were found in the Usk.  Only *H. siltalai*, *H. pellucidula* and *H. contubernalis* were recorded from the N. Tyne in Northern England (Boon, 1978).

The distribution of other invertebrates in the Wye catchment may be related to changes in chemical composition of the water.  For example, the snails *Sphaerium* and *Pisidium* were not found at mean concentrations of calcium below *c.* 8 mg $l^{-1}$ and the crustaceans *Gammarus* and *Asellus* have similar distributions.  In contrast the mollusc *Ancylus* was more widespread in the Wye catchment, being absent only from reaches with calcium concentrations less than *c.* 3 mg $l^{-1}$.  Maitland (1966) found similar distributions of these organisms in the R. Endrick, Scotland.

Another characteristic of the Wye catchment resulting from river chemistry was the restricted fauna of the Elan, which receives water from Caban Coch reservoir.  The stable flow and the iron- and manganese-rich (both soluble and particulate) water from the reservoir and from a water treatment works results in substantial deposits on the bed of the river and probably these have eliminated, or considerably reduced, the numbers of several species which might have been expected to inhabit these reaches.  These include *Hydropsyche siltalai, H. pellucidula, Rhithrogena semicolorata, Ephemerella ignita* and the beetles *Esolus parallelepipedus* and *Oulimnius tuberculatus*.  Such restrictions, particularly of beetles, are characteristic of pollution with ferric hydroxide (Scullion & Edwards, 1980) but have not been reported from other impounded rivers, such as the Tees in the U.K. (Armitage, 1978).

## 3.43  QUANTITATIVE ESTIMATES

Any assessment of changes in total numbers of benthic macroinvertebrates requires some knowledge of their distribution in the substrate with depth, since a substantial proportion of the population may not be collected by conventional surface sampling methods.  Experiments in the Wye indicated that, overall, 59%, 23% and 18% of benthic invertebrates were recorded in the top (0 - 11 cm), middle (12 - 22 cm) and bottom (23 - 33 cm) levels of special basket samplers buried in the river substratum (Morris & Brooker, 1979).  Some organisms were generally confined to the top level, e.g. mayfly nymphs, blackfly (Simuliidae) larvae, snails, and could be satisfactorily characterised by surface samples (e.g. Brooker & Morris, 1978) but others were either more uniformly distributed (worms, midge larvae) or increased with depth at certain times (e.g. *Sericostoma* - a trichopteran larva).

Quantitative estimates of invertebrate density derived from intensive sampling, were made at three sites in the upper catchment. Highest densities (>12000 m$^{-2}$) were found at the site furthest downstream (km 48), but at all sites peak estimates were recorded during the summer, particularly during the drought of 1976.


3.44  DRIFT

The drifting of macroinvertebrates is commonly characterised by seasonal and daily periodicity. In the upper Wye, daily variations of drift densities resulted principally from the activities of nymphs of mayflies and stoneflies and blackfly larvae and most often maximum drift density was recorded during the period of darkness. However, on some occasions maxima were recorded immediately before or after sunrise (Fig. 3.12) and during the long nights of winter two peaks in density were recorded. There were some organisms, e.g. *Hydra* and water mites, which had highest drift densities during daylight and, overall, most midge larvae drifted during daylight.

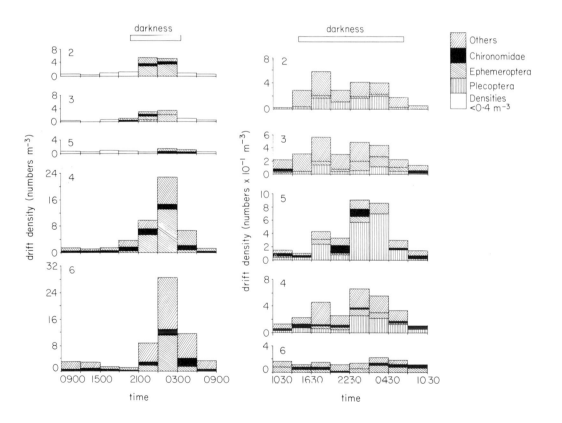

*Fig. 3.12* Densities of drifting invertebrates at sites down the Wye (position of sites shown in Fig. 3.8) over 24 h period in July and December.

From estimates of mean daily density, which ranged from about $6 \times 10^{-2}$ to $783 \times 10^{-2}$ m$^{-3}$ and with maxima generally recorded during July and August, it was estimated that downstream population displacement of selected insect species over the aquatic phase of the life cycle was not likely to be greater than 10 km (Hemsworth & Brooker, 1979) and the normal dispersive flight of adults provided adequate compensation for downstream losses.

The importance of changes in river flow to drifting organisms was illustrated when the flow was experimentally increased on the Wye from about 2 to 5 m$^3$ s$^{-1}$ by a release of water from the Elan Valley reservoirs during May–June. On the first day of the release the number of macroinvertebrates drifting at a site 20 km below the reservoirs was about seven times that of the preceding day: it declined on the second day but was still three times greater (Fig. 3.13). Some organisms responded immediately to the increased flow, such as larvae of the midge *Rheotanytarsus* whilst others, such as nymphs of the mayfly *Ephemerella*, increased some eight hours later during the night – their normal time of maximum abundance in the drift.

## 3.45  AUTECOLOGICAL STUDIES

Certain species in the Wye catchment have been studied because of their rarity or intrinsic interest:

a) *Potamanthus luteus*: This mayfly is widely distributed in Europe, but confirmed records in the United Kingdom are restricted to the Thames, Usk and Wye. In the Wye the nymphs were relatively widespread – km 70 and downstream – but estimated densities (<50 m$^{-2}$) were very low and no specimens were found in tributaries to the main river. Water quality conditions at sites supporting this species varied considerably (Brooker & Morris, 1980a).

Little is known of the life history of this mayfly in the U.K., but imagos and subimagos have been collected at Hereford during July and it seems likely from changes in the size of nymphs that the species is univoltine.

b) *Austropotamobius pallipes*: The oldest records of the crayfish in the Wye catchment date from 1805 when it was reported that stocking of the Irfon took place: the species introduced was referred to as *Astacus fluviatilis* but was probably *Austropotamobius pallipes*. Although not found in routine surveys, a special investigation at 103 sites on 57 streams in the catchment (Lilley et al., 1979) indicated its widespread distribution. It was not found, however, where the average calcium concentration was less than 6 mg l$^{-1}$. Although crayfish are known to have a high physiological requirement for calcium, other factors are clearly important, e.g. substrate, shading, current velocity, for it was rarely found at sites on the main river.

As in other studies, there were fewer males than females (0.6 : 1 in the Wye) and porcelain disease (*Thelohania contejeani*) was widespread, with at least 3.4% of the population affected.

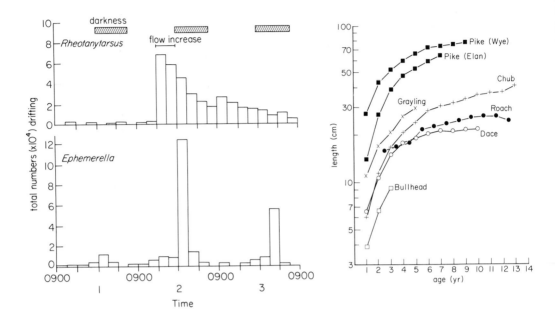

*Fig. 3.13* Total numbers per 3 h of larvae of the midge *Rheotany-tarsus* and nymphs of the mayfly *Ephemerella ignita* before and during an increase in water flow resulting from a release from impoundments in the Elan Valley.

*Fig. 3.14* Growth rates of several species of coarse fish in the Wye catchment. Pike growth both in the lower Wye and Elan is shown.

## 3.5  FISH AND FISHERIES

### 3.51  GENERAL

The Wye catchment contains many fish species (Table 3.3), some, such as the bleak and barbel being of very recent origin and probably introduced by anglers from neighbouring catchments. Of the non-salmonids, the chub, dace and pike are most profuse, judged by anglers' catches, parti-cularly in the lower reaches of the river and its major tributaries. Several coarse fish species (chub, dace, roach, perch, pike) are widely distributed, but none extends its range above Rhayader on the main river where water-falls limit access to the uppermost reaches which are probably too shallow and turbulent to sustain such species. Other species are comparatively local in distribution, the ruffe being limited to the R. Monnow, the rudd to the middle reaches and the carp to the

lower reaches of the Wye and the tench to the middle reaches of the Wye
and to parts of the Monnow and Lugg catchments.


*Table 3.3*  Fish species recently recorded in rivers of the Wye catchment

| | |
|---|---|
| *Lampetra planeri* | (brook lamprey) |
| *L. fluviatilis* | (river lamprey) |
| *Petromyzon marinus* | (sea lamprey) |
| *Alosa fallax* | (twaite shad) |
| *A. alosa* | (allis shad) |
| *Salmo salar* | (Atlantic salmon) |
| *S. trutta trutta* | (sea trout) |
| *S. trutta fario* | (brown trout) |
| *S. gairdneri* | (rainbow trout) |
| *Thymallus thymallus* | (grayling) |
| *Esox lucius* | (pike) |
| *Cyprinus carpio* | (carp) |
| *Barbus barbus* | (barbel) |
| *Gobio gobio* | (gudgeon) |
| *Tinca tinca* | (tench) |
| *Abramis brama* | (bream) |
| *Alburnus alburnus* | (bleak) |
| *Phoxinus phoxinus* | (minnow) |
| *Scardinius erythrophthalmus* | (rudd) |
| *Rutilus rutilus* | (roach) |
| *Leuciscus cephalus* | (chub) |
| *L. leuciscus* | (dace) |
| *Noemacheilus barbatulus* | (stoneloach) |
| *Anguilla anguilla* | (eel) |
| *Gasterosteus aculeatus* | (3-spined stickleback) |
| *Perca fluviatilis* | (perch) |
| *Gymnocephalus cernua* | (ruffe) |
| *Cottus gobio* | (bullhead) |
| *Platichthys flesus* | (flounder) |

The growth rates of some species of coarse fish (Fig. 3.14), including
the salmonid *Thymallus thymallus* (grayling), have been determined in
various tributaries and lengths of main river (A. S. Gee & N. J. Milner,
pers. comm.; Hellawell, 1969, 1971, 1972) and although the growth rate
of most species compares favourably with that in other catchments in
Southern Britain, that of grayling - which is widely distributed in the
Wye and its larger tributaries - appears slower than that in some other
European waters (Hellawell, 1969).  The growth of pike in the main river
near Monmouth has been compared with that in the Elan and whilst growth
rates in the main river are similar to those in the Stour (Mann, 1976),
Vistula (Backiel, 1971) and (lake) Windermere (Frost & Kipling, 1959),

growth rates in the Elan are appreciably lower.  This is probably
associated with the lower spring and summer temperatures of this tributary, which receives water only from the Caban Coch reservoir.

Of the anadromous non-salmonids, the Allis and Twaite shads, members
of the herring family and locally distributed in Britain, are of particular interest.  Both species migrate upstream as far as Builth Wells
to spawn during the early summer, after which survivors return to sea,
sometimes remaining in the Severn Estuary and Bristol Channel for
several weeks en route.  The recently hatched fry follow in the autumn,
but it seems from the work of Claridge & Gardner (1978), who caught shad
on the screens of power stations in the Severn Estuary, that some fry
overwinter in the river, migrating to sea in late-April when river
temperatures rise rapidly.

The sea- and river-lampreys, although not frequently caught, move
upstream as far as Rhayader, over 200 km from the sea, to spawn when
freshwater flows are favourable, the river-lamprey also spawning very
extensively in the Lugg catchment.

Of the catadromous species, the eel, although abundant and commonly
weighing up to 1 kg, is no longer caught commercially within the catchment as it is on the neighbouring Severn where there is also a thriving
elver industry.  Until recently an eel-trap was operated on the
Llynfi at the outlet to Llangorse Lake, a site where eel trapping had
been practised for many centuries.  The flounder is common in the lower
Wye, its distribution extending upstream to Ross-on-Wye, about 60 km
from the estuary.

Brown trout are extensively distributed in the catchment and their
productivity and mobility have been studied at several sites in the
headwaters (Milner *et al.*, 1978 and 1979).  In those tributaries where
2+ and 3+ fish remain and contribute substantially (66 - 88%) to
production, productivity is generally very high, exceeding 15 g m$^{-2}$
yr$^{-1}$.  In smaller tributaries from which larger trout have migrated
to deeper reaches, productivity is generally lower (<5 g m$^{-2}$ yr$^{-1}$).
Rainbow trout, which have been stocked in the catchment, are known to
have spawned successfully in localised areas, fry having been found
upstream of Builth Wells in routine surveys during the late 1970s.
Sea trout have never been common the Wye or its neighbours the Usk and
Severn: this is in marked contrast to the generally shorter rivers
further west.  Nevertheless, small numbers regularly ascend to the
upper Wye as far as Rhayader.


3.52  SALMON STOCKS

It is Atlantic salmon for which the Wye is justly famous although at
the end of the 19th century stocks were severely depleted by decades
of overfishing.  Not only were adults netted throughout the lower
reaches of the rivers in the estuary, but juveniles were caught through-

out the catchment and poaching was rampant in the headwaters. The
stocks became so depleted that the almost valueless netting rights were
purchased cheaply by the Wye Fisheries Association which recovered its
expenses by leasing rod fishing. With the prohibition of river netting
in 1901 and severe restrictions on estuary netting - which was eventually controlled by the River Board - stock recovery was rapid. Gee &
Milner (1980) have analysed the catch statistics since 1905 and demonstrated certain important changes:

    i) although there has been a decline in net catch this has been
compensated by an increase in rod catch, sustaining the mean annual
total catch at about 6800.

    ii) growth rate at sea has declined over the past 70 years for
fish of more than one sea-winter (1 SW)

    iii) the average weight of salmon in both net and rod fisheries has
declined appreciably since 1930, average net caught weights being consistently lower than rod caught weights over this period.

The decrease in average weight of salmon caught is not solely caused
by the declining growth-rate of salmon at sea (ii above), but is principally associated with a change in age composition of the catch.
Before 1945 3 SW was the dominant age class of the rod fishery and
formed 60 - 65% of the catch but since that time 1 SW and 2SW salmon
have increased to comprise 20 and 60% of the catch respectively. 4 SW
fish and previous spawners have also declined, the latter from 7.6 to
1.6% of the catch (Fig. 3.15).

Gee & Milner (1980) attribute the change in age composition principally to the markedly increased catch effort by the rods over the past
few decades, and the preferential exploitation of the older fish, these
being the first to enter in the early spring in contrast to most
grilse (1 SW) which do not enter until late August - towards the end of
the fishing season. If the number of seasonal licences may be used as
an index of rod exploitation, the latter has increased by 650% since
1945. It has been suggested that at current exploitation rates,
virtually all the 3 SW and 4 SW fish are caught, few remaining to spawn.

This explanation of the change in age composition of Wye salmon
stocks, which has considerable significance in fishery management,
hinges on there being a genetic basis to the duration of sea-absence as
claimed by Elson (1973). Although there is now evidence for this
genetic basis (Gardner, 1976) the relative importance of parent age-
composition and environmental factors is not yet established.

A computer simulation model has been developed for salmon stocks of
the Wye, which can explore the effects of fishery exploitation and
other mortality factors on stock abundance and composition (Gee &
Radford, in prep. ; Gee & Edwards, 1980). If reasonable assumptions
about exploitation rates of adult salmon by the rod fishery are made

(5, 50 and 90% for 1, 2 and 3 SW fish) and it is assumed that there is
a genetic basis to sea-absence but with spawners of one sea-age producing some progeny of different sea-age (Gardner, 1976) then the composition of the stock responds in the model similarly to that in the
river (Fig. 3.16) with a substantial increase in the proportion of
grilse.  The model also indicates that response of the stocks to
changing exploitation is rapid, a conclusion supported by evidence of
the quick recovery of stocks at the beginning of the 20th century when
netting was virtually banned throughout the Wye system.

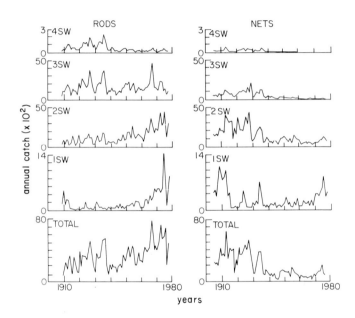

*Fig.3.15* Annual rod and net catches of salmon (over a 70-year period)
classified by years at sea.

## 3.53   ECONOMIC VALUE OF FISHERY

It was estimated from a survey carried out in the mid-1970s and based
on postal questionnaires (Randerson *et al.*, 1978) that there were about
750000 angler visits per year to the river catchment, each angler making
about 20 visits.  Of the total angler expenditure of £12.5 x 10^6
(adjusted to prices at September 1979) 88% may be attributed specifically to visits (entrance fees to sites, food, bait, travel), 11% to
basic angler expenditure (club membership, specialist literature,
replacement of tackle etc.)* and only 1% or £0.12 x 10^6 to licence

*Although each licence-holder who fished within the catchment spent on
 average £88.3 p.a. in this category, only 44% of angler visits were
 within the catchment so only £39 of this expenditure may be attributed
 to fishing within the area (total £1.38 x 10^6).

fees.  It was suggested (Gee & Edwards, 1980) that because of the high
proportion of expenditure attributable directly to visits, the large
number of coarse fish anglers could contribute up to 50% of this expenditure.  Expenditure of less than 30% on salmon fishing was considered
most unlikely and this would give each salmon caught a 'recreational
value' of £736 - contrasting with a 'commercial value' of less than £40
for fish caught in estuary nets.

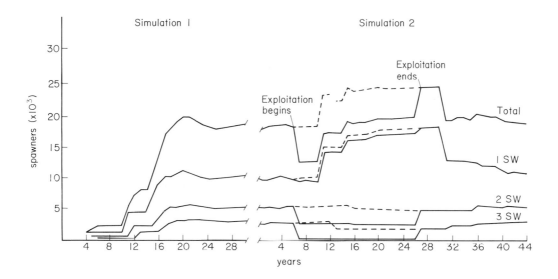

*Fig. 3.16* Simulated effect of differential exploitation of large
fish (see text) on population structure.  In Simulation 1 the equilibrium population of adult salmon, after recovery from a low number,
but with no exploitation in freshwater, is shown.  In Simulation 2
differential exploitation of large fish causes their reduced spawning
escapement and an increase in that of grilse.  The total salmon run
(dotted line) increases with exploitation at the particular cropping
level applied:  the total escapement, after a brief initial period,
is similar to that without exploitation.

### 3.6   FUTURE DEVELOPMENTS

In 1973 the Water Resources Board proposed the enlargement of one of
the existing reservoirs in the Elan Valley, Craig Goch, to a size
(volume, 534 x $10^6$ m$^3$; area, 1400 ha) that would make it one of the
largest in Europe, with sources, taken at times of high flow, from the
upper Wye and by transfers from the Severn, Ystwyth and Rheidol.  At
that time it was envisaged that it would provide water for the West
Midlands, South Wales, Avon and Thames Valley.  Recent reductions in
anticipated demand, coupled with alternative schemes for some demand
areas, have reduced the likely storage needs of Craig Goch reservoir

and it now seems probable that, with demand areas being limited to
South Wales and the West Midlands, enough water can be derived from the
immediate catchment of the reservoir and the adjacent upper Ystwyth
catchment.  South Wales would be supplied by regulation of the Wye,
with abstraction at Monmouth, and the West Midlands by regulation of
the Severn, with abstraction above Worcester.

The original scheme, in which water was being transferred to the Wye
from several catchments, posed problems of the transfer of both macro-
organisms such as fish species, not already present in the catchment,
and micro-organisms, particularly fish pathogens.  There was also concern over salmon stocks in view of the homing behaviour of salmon and
particularly the suggested role of race-specific (i.e. catchment-
specific) pheromones in homing behaviour (Solomon, 1973):  even with
recent modifications to the scheme, some water is transferred between
catchments, including an export from the Wye to the Severn.  Nevertheless, most effects of the Craig Goch scheme on the Wye will be associated with the supplementation of low summer flows ($<14$ m$^3$ s$^{-1}$ at
Monmouth) to match abstractions in the lower river.  These possible
effects derive principally from changes in velocity, depth and width of
the river associated with increased flow and from changes in water
quality consequent upon storage (Brooker, 1981; Edwards, 1984).

*Fig. 3.17*  Aerial view of Craig Goch Reservoir.

Effects in the lower catchment are also likely to be less discernible than those immediately downstream of the regulating discharge because of local disequilibrium effects produced by storage on factors such as oxygen and temperature, and because of the downstream dilution of regulating releases with water from natural sub-catchments.

Edwards (1984) attempts to predict the effects of regulation from Craig Goch reservoir on the Wye. Figs. 3.18 and 3.19 summarize local effects on the biological resources of the river and extensive effects on these resources and on water quality for abstraction. With respect to local effects, riparian and aquatic aspects of conservation are distinguished because in the upper catchment flow changes have a sub-stantial effect on river height and width (see Fig. 3.3). In view of the limited duration and degree of regulation in the summer months, extensive ecological effects are not likely to be pronounced but improvements in the salmon fishery, particularly resulting from an improved oxygen status, are foreseen. Local effects will probably be more diverse, but those resulting from reservoir stratification and hypolimnetic abstraction, such as low oxygen status and temperature together with high concentrations of iron and manganese, could be reduced or avoided by artificial destratification, now regarded as economically feasible for reservoirs of the size proposed.

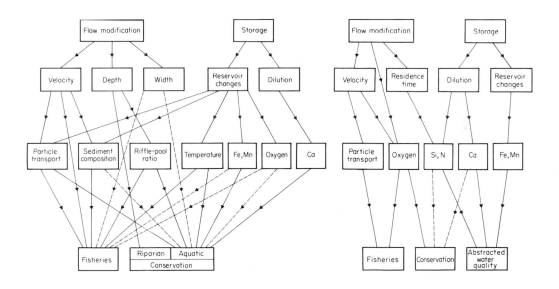

*Fig. 3.18* Pathways of possible local effects of river regulation in the upper Wye on conservation and fisheries. Dotted lines indicate where effects are likely to be least significant.

*Fig. 3.19* Pathways of possible effects of river regulation on con-servation and fisheries in the lower catchment and on the quality of abstracted water.

REFERENCES

Armitage P.D. (1978)  Downstream changes in the composition, numbers and biomass of bottom fauna in the Tees below Cow Green Reservoir and in an unregulated tribuary Maize Beck in the first five years after impoundment. *Hydrobiologia* 58, 145-56

Backiel T. (1971)  Production and food consumption of predatory fish in the Vistula River. *J. Fish. Biol.* 3, 369-405

Boon P.J. (1978)  The pre-impoundment distribution of certain Trichoptera larvae in the North Tyne river system (Northern England) with particularly reference to current speed. *Hydrobiologia* 57, 167-74

Brooker M.P. (1981)  The impact of impoundments on the downstream fisheries and general ecology of rivers. *Advances in Applied Biology*, Vol. VI.  Academic Press, pp. 91-152

Brooker M.P. & Hemsworth R.J. (1978)  The effect on the release of an artificial discharge of water on invertebrate drift in the River Wye, Wales. *Hydrobiologia* 59, 155-63

Brooker M.P. & Morris D.L. (1978)  Production of two species of Ephemeroptera (*Ephemerella ignita* and *Rhithrogena semicolorata*) in the upper reaches of the R. Wye, Wales. *Verh. int. Verein. theor. angew. Limnol.* 20, 2600-604

Brooker M.P. & Morris D.L. (1980a)  *Potamanthus luteus* (L.) (Ephemeroptera, Potamanthidae) in the R. Wye. *Entomol. Gaz.* 31, 247-51

Brooker M.P. & Morris D.L. (1980b)  A survey of the macroinvertebrate riffle fauna of the River Wye. *Freshwat. Biol.* 10, 437-58

Brooker M.P., Morris D.L. & Hemsworth R.J. (1977)  Mass mortalities of adult salmon (*Salmo salar*) in the R. Wye, 1976. *J. appl. Ecol.* 14, 409-17

Brooker M.P., Morris D.L. & Wilson C.J. (1978)  Plant-flow relationships in the R. Wye catchment.  In: *Proceedings 5th Symposium on Aquatic Weeds, 1978*, 63-70.  European Weed Research Council

Claridge P.N. & Gardner D.C. (1978)  Growth and movements of the twaite shad, *Alosa fallax* (Lacepede) in the Severn Estuary. *J. Fish. Biol.* 12, 203-11

Edwards R.W. (1984)  Predicting the environmental impact of a major reservoir development.  In: Roberts R.D. & Roberts R.M. (Eds) *Planning and Ecology.*  Chapman & Hall

Edwards R.W., Oborne A.C., Brooker M.P. & Sambrook H.T. (1978)  The behaviour and budgets of selected ions in the Wye catchment. *Verh. int. Verein. theor. angew. Limnol.* 20, 1418-422

Egglishaw H.J. & Morgan N.V. (1965)  A survey of the bottom fauna of streams in the Scottish Highlands.  Part II. The relationship of the fauna to the chemical and geological conditions. *Hydrobiologia* 30, 305-34

Elson P.F. (1973)  Genetic polymorphism in Northwest Miramachi salmon, in relation to season of river ascent and age at maturation and its implications for management of the stocks. *International Commission for Northwest Atlantic Fisheries.*  Research Document 73/76, 1-6

Frost W.E. & Kipling C. (1967)  The determination of the age and growth of pike (*Esox lucius* L.) from scales and opercular bones. *J. Cons. perm. int. Explor. Mer.* 24, 314-41

Furet J.E. (1978)  Algal studies on the River Wye system.  Ph.D. Thesis. University of Wales (Cardiff)

Gardner M.L.G. (1976)  A review of factors which may influence the sea-age and maturation of Atlantic salmon *Salmo salar* L.  *J. Fish. Biol.* 9, 289-327

Gee A.S. & Edwards R.W. (1980)  The recreational exploitation of the Atlantic salmon in the River Wye.  *International Symposium on Fisheries Resources Allocation (FAO)*, Vichy, France.  28 pp.

Gee A.S. & Milner N.J. (1980)  Analysis of rod and net catch statistics for Atlantic salmon, *Salmo salar*, in the R. Wye, 1905-1977, and implications for fishery management.  *J. appl. Ecol.* 17, 41-57

Haslam S.M. (1978)  River plants: the macrophytic vegetation of watercourses.  Cambridge Univ. Press

Hellawell J.M. (1969)  Age determination and growth of the grayling, *Thymallus thymallus* L. of the River Lugg, Herefordshire, tributary of the River Wye.  *J. Fish. Biol.* 1, 373-82

Hellawell J.M. (1971)  The autecology of the chub, *Squalius cephalus* (L) of the R. Lugg and Afon Lynfi.  1. Age determination, population, structure and growth.  *Freshwat. Biol.* 1, 29-60

Hellawell J.M. (1972)  The growth, reproduction and food of the roach of the R. Lugg, Herefordshire.  *J. Fish. Biol.* 4, 469-86

Hemsworth R.J. & Brooker M.P. (1979)  The rate of downstream displacement of macroinvertebrates in the upper Wye, Wales.  *Holarctic Ecology* 2, 130-36

Hey R.D. (1980)  Regime equations for gravel-bed rivers.  In: *Engineering Problems in the Management of Gravel-bed Rivers.* International Workshop: University of East Anglia

Hildrew A.G. & Edington J.M. (1979)  Factors facilitating the co-existence of hydropsychid caddis larvae (Trichoptera) in the same river system.  *J. Anim. Ecol.* 48, 557-76

Houston J. & Brooker M.P. (1981)  A comparison of nutrient sources and behaviour in two lowland subcatchments of the R. Wye.  *Wat. Res.* 15, 49-57

Inverarity R.J., Rosehill G.C. & Brooker M.P. (1983)  The effects of impoundment on the downstream macroinvertebrate riffle fauna of the River Elan, Mid-Wales.  *Environmental Pollution* (A) 32, 245-267

Jones A. (1975)  Rainfall, run-off and erosion in the upper Tywi catchment.  Unpublished Ph.D. Thesis. University of Wales

Lilley A.J., Brooker M.P. & Edwards R.W. (1979)  The distribution of the crayfish, *Austropotamobius pallipes* (Lereboullet) in the upper Wye catchment, Wales.  *Nature in Wales* 16, 195-200

Maitland P.S. (1966)  *Studies on Loch Lomond. 2. The Fauna of the River Endrick.*  Univ. Glasgow, Scotland

Mann R.H.K. (1976)  Observations on the age, growth, reproduction and food of the chub *Squalius cephalus* (L.) in the River Stour, Dorset.  *J. Fish. Biol.* 8, 265-88

Milner N.J., Gee A.S. & Hemsworth R.J. (1978)  The production of brown trout (*Salmo trutta* L.) in the tributaries of the upper Wye, Wales.  *J. Fish. Biol.* 13, 599-612

Milner N.J., Gee A.S. & Hemsworth R.J. (1979)  Recruitment and turnover of populations of brown trout, *Salmo trutta*, in the upper River Wye, Wales.  *J. Fish. Biol.* 15, 211-22

Morris D.L. & Brooker M.P. (1979)  The vertical distribution of macro-invertebrates in the substratum of the upper reaches of the River Wye, Wales. *Freshwat. Biol.* 9, 573-83

Morris D.L. & Brooker M.P. (1980)  An assessment of the importance of the Chironomidae (Diptera) in Biological Surveillance, 195-202. In:  Murray D.A. (Ed.) *Chironomidae, Ecology, Systematics, Cytology and Physiology.*  Pergamon Press

Newson M.D. (1979)  The results of ten years' experimental study on Plynlimon, mid-Wales, and their importance for the water industry. *J. Inst. Eng. Sci.* 33, 321-33

Oborne A.C. (1981)  The application of a water-quality model to the R. Wye, Wales. *J. Hydrol.* 52, 59-70

Oborne A.C., Brooker M.P. & Edwards R.W. (1980)  The chemistry of the River Wye. *J. Hydrol.* 45, 233-52

Perring F. (1957)  Llandrindod Wells.  August 4th-11th, 1956.  Field Meetings. *Proc. bot. Soc. British Isles* 11, 416-18

Randerson P.F., Edwards R.W. & Shaw K. (1978)  An economic evaluation of recreational fishing in areas of Wales, 203-218.  In: *Recreational Freshwater Fisheries: their Conservation, Management and Development.* Water Research Centre Conference, Oxford 1977

Ratcliffe D.A. (1977)  *A Nature Conservation Review*, 2 vols.  Cambridge U. P., Cambridge

Sambrook H.T. (1976)  Water quality: Wye.  Unpublished M.Sc. Thesis. University of Wales (UWIST)

Scullion J. & Edwards R.W. (1980)  The effects of coal industry pollutants on the  macroinvertebrate fauna of a small river in the South Wales coalfield. *Freshwat. Biol.* 10, 141-62

Smith A.J.E. (1978)  *The Moss Flora of Britain and Ireland.*  706 pp. Cambridge U. P., Cambridge

Solomon D.J. (1973)  Evidence for pheromone-influenced homing by migrating Atlantic salmon, *Salmo salar* (L.). *Nature, Lond.* 244, 231-32

Water Resources Board (1973)  *Water Resources in England and Wales.* H.M.S.O., London

# 4 · Ebbw

## R. W. EDWARDS, P. F. WILLIAMS & R. WILLIAMS

4.1   Introduction                                              84

4.2   Geology, hydrology and water quality                      88

      4.21   Geology                                            88
      4.22   Hydrology                                          88
      4.23   Effluent treatment                                 90
      4.24   Water quality                                      90

4.3   Plants and macroinvertebrates                             94

      4.31   Introduction                                       94
      4.32   Community diversity                                94
      4.33   Qualitative aspects of community structure         97

4.4   Fish                                                      100

      4.41   Distribution in catchment                          100
      4.42   Density, growth and condition                      103
      4.43   Trout movements and recruitment within
             the catchment                                      105
      4.44   Fish survival and water quality                    107

4.5   Conclusion                                                108

References                                                      109

### 4.1    INTRODUCTION

The River Ebbw Fawr, which discharges into the Usk Estuary at Newport just before its confluence with the Severn Estuary (Fig. 4.2) is fed by two major tributaries, the Ebbw Fach and the Sirhowy, and several smaller ones.  The river catchment, with a population of about 160000, is very small (226 km$^2$) near the eastern perimeter of the South Wales Coalfield.  The justification for its inclusion in this book is based on its dramatic biological response, recorded intensively during the 1970s, to the abatement of industrial pollution.

The decline of the river began two centuries earlier, in 1764, with the establishment of the first major iron-works in the catchment at Tredegar and the rapid development of other works such as those at Sirhowy, Nantyglo and Ebbw Valé (Fig. 4.1).  By the end of the 18th century there was a concentration of major iron-works along the northern rim of the coalfield at the heads of the valleys where there were supplies of iron-ore and of both limestone, used as a flux in the iron-making process, and coal, suitable for coking.  The general location of these early iron-works at the headwaters of the rivers of the coalfield was important in establishing the patterns of later development and in its consequences with respect to river quality.

*Fig. 4.1* An early 19th century engraving of industrialization on the banks of the R. Ebbw Fach at Nantyglo.

Iron-making had occurred within the catchment before the mid-18th century but these earlier furnaces were dependent on charcoal as a fuel and water power to operate bellows and this 'tyranny of wood and water' had produced only modest quantities of iron.  It seems that the technological opportunities resulting from Abraham Darby's use of coke as a fuel (1756) and Smeaton's improved blast engine (1765), coupled with the demands of the Seven Years War (1756-63) and American War of Independence (1775-83), were responsible for the growth of the iron-making industry in the late 18th century.

At Ebbw Vale the development of steel-making started in 1854 although the first system was soon replaced by a process invented by Henry Bessemer in 1856.  The other major iron-works within the catchment, not adapting to the new technology, declined and closed towards the end of the 19th century.  Ebbw Vale flourished, however, to become one of the major integrated producers of iron, sheet-steel and tin-plate in the U.K., making up to 1 million tonnes of steel annually until in 1977 it too reduced its activities, to tin-plating and galvanising.

Early coal mining in the catchment exploited outcrops, and soil and debris were frequently separated from the coal by scouring with water from local rivers.  During the period 1820-1840 the first shafts were introduced to exploit deeper seams, which extended over much of the catchment.  This early development of the coalfield was dependent on the needs of iron-making but with improved transport, firstly by canals and later by railways, the production of coal increased and by the beginning of the 20th century the industry in South Wales employed over 200000 miners and raised 50 million tonnes per year.  In the Ebbw catchment, the many small mines of the 19th century amalgamated, and the few resulting collieries continued to produce large quantities of coal.  During the 1970s, although only 7 deep mines remained (Fig. 4.2), there was an extensive legacy of spoil heaps or tips above ground and interconnecting shafts and galleries below ground which continued to influence the water quality and hydrology of the river.

Urbanization in the deeply incised valleys of the catchment was squeezed along the valley floor (Fig. 4.2) with consequences for river management.  Early industrial settlements had no sewerage systems or supplies of piped water.  Edmund Head in 1841 (Gray-Jones, 1970) wrote:

> "Very few or none of the houses have privies:  sewerage is
>   entirely unknown and the surface drainage is imperfect,
>   there is a great accumulation of filth around the houses .."

With such poor sanitary conditions it is not surprising that cholera epidemics occurred (1832, 1849, 1866), that of 1849 killing almost 2% of the inhabitants of the catchment.  In 1878 the first reservoir, Llangynidr, was built to provide piped water and within a decade a sewerage system was constructed which conveyed wastes to small ponds before their discharge into the river.  This system was not improved until 1912 when a trunk sewer was built to convey wastes to the coast.

*Fig. 4.2* Map of catchment (showing main land-use, geology, urban areas, tributaries, collieries).

This sewer, as in other valleys of the coalfield, was sited generally
adjacent to the river, with some lengths being along the river bed.
The capacity of such sewers was frequently exceeded by the intrusion of
flood-waters, inadequately separated from foul sewage, and the overflow
of sewage into rivers during storms was frequent.  Roads were similarly
located adjacent to rivers and were raised appreciably to avoid flooding
in an area of capricious and plentiful rain.  Tributaries, culverted
under such roads, were frequently inaccessible to fish migrating upstream from the main river, due to inadequacies in culvert design.

The major decline in water quality of the rivers of the Ebbw catchment took place in the early 19th century soon after the development of
iron-making.  Certainly at the end of the 18th century both the rural
character of much of the valley and the good quality of the fishery
seemed beyond dispute for according to Edmund Jones writing in 1778
(Gray-Jones, 1970) the river abounded with eels, trout and sewin (sea-
trout) and Coxe in his 'Tour throughout Monmouthshire, 1781', wrote:

> "I spent many a happy day
> On Sirhowy's banks so gay
> Gathering nuts that grew about
> And deceiving the old trout"

Nevertheless there were local pollution problems even then, for Edmund
Jones noted that fish were becoming less plentiful, the river water
'being often troubled with the pond waters scouring the coal works'.
Some decades later there is evidence that other rivers of the coalfield
had become polluted and their fish stocks decimated by the coal- and
iron-workers.  Of the Taff, some twenty kilometres to the west, Hansard
(1834) wrote:

> "The vicinity of Merthyr Tydfil has greatly contributed to
> render certain proportions of the Taf unworthy of anglers'
> attention.  The poisonous matter discharged into it from
> the iron-works, and the lawless practices of the forge-men,
> continually diminish the stock of fish.  In dry seasons these
> depredators assemble in bands, and, wading into the streams
> armed with sledge hammers, contrive, by violently striking
> the stones under which the trout are concealed, to destroy
> an incredible quantity of fish of all sizes".

By 1861, from the Report of the Commissioners appointed to enquire
into Salmon Fisheries (England and Wales), it was clear that the Ebbw
was grossly polluted and, although there were a few trout remaining,
these could not be eaten because of the taste imparted by tars and acids:

> "There are works up the river for making what is called
> pyroligneous acid.  The Ebbw receives all the sewage and
> all the filth from Bryn Mawr, Nant-y-Glo, Coalbrook Vale,
> Blaina, Cwm Celyn, Ebbw Vale, Beaufort and Victoria Works.
> Then at Abercarn there are the tar-works and at Risca there
> is another pyroligneous acid manufactory".

It was about a century later that the Usk River Authority proposed pollution control measures to improve river quality for the purpose of industrial abstraction and to restore life to the river by:

  i)   commissioning a full-scale treatment plant at the Ebbw Vale steel-works;

  ii)  constructing a large capacity trunk sewer to carry sewage to a coastal outfall;

  iii) improving the standard of waste treatment at five coal washeries remaining in the catchment.

The full treatment of the steel-works effluent, the most significant aspect of pollution abatement, was not accomplished until the autumn of 1973 although there were periods during the previous four years when partial treatment was achieved.  The replacement of the trunk sewer was started in 1971, construction being progressively upstream and reaching Cwm in 1980.  Only sections within the valley of the Ebbw Fach remain to be completed.


## 4.2   GEOLOGY, HYDROLOGY AND WATER QUALITY

### 4.21  GEOLOGY

The South Wales Coalfield is a complex syncline bounded by an escarpment of Old Red Sandstone (Fig. 4.2).  Below the Coal Measures - deposited in a brackish or freshwater environment and classified by George (1970) into Upper, Middle and Lower Coal Measures - are strata of Carboniferous Limestone and Millstone Grit.  The Middle and Lower Coal Measures, which over most of the coalfield can only be reached by sinking deep shafts, have the thickest and most important coal seams. Associated with these measures, in some areas, is iron ore, in the form of chalybite, occurring as nodules or bands.  The exposure of these measures, and the associated limestone and iron-ore, at the northern rim of the coalfield, were major factors in the early industrialisation of the area.  The Upper Coal Measures contain Pennant Sandstone and form the undulating plateau of the Ebbw catchment.

The Carboniferous rocks have been extensively mineralised, the sulphides of lead, zinc, copper-iron and nickel being common with those of cobalt and iron-nickel also occurring.  Although not commercially exploitable, extensive mining operations for iron ore and coal have exposed these sulphides to oxidation, allowing heavy metals to leach into local waters.


### 4.22  HYDROLOGY

Annual precipitation in the catchment is high, varying from 1650 mm in the northern uplands to about 1200 mm near the mouth of the river.  Fig. 4.3 shows the seasonal pattern of precipitation at Tredegar, 3 km from

Ebbw Vale, during the period 1969-1978. At this site, representative of the headwaters, precipitation during the winter (November-January) is approximately twice that of the period April-July.

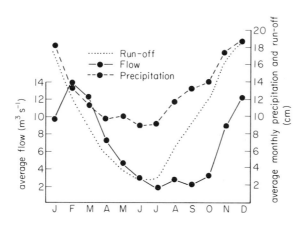

*Fig. 4.3* Monthly averages of daily flows for 1977-1979 near Risca and average monthly precipitation and run-off for the northern catchment during the period 1969-1978.

Of this precipitation about 70% flows down the river, evapotranspiration accounting for most of the difference. River flows have a pronounced seasonal rhythm (Fig. 4.3), mean river flows in the winter being about seven times those of July. Low summer flows reflect the enhanced evapotranspiration rates during this period as well as reduced precipitation. A reduced January flow probably reflects the storage of snow, particularly at the high altitudes (up to 500 m) in the northern part of the catchment. Whilst net precipitation increases in August, river flows do not increase appreciably until November, the delay being caused by soil and underground storage.

Of the average flow of the river at Rhiwderin, near Risca (Fig. 4.2), about 7 $m^3$ $s^{-1}$, minewaters contribute 0.5 $m^3$ $s^{-1}$ or 7%. During drought periods the river flow may decrease to values below 1 $m^3$ $s^{-1}$ and during such periods water pumped from underground mines is extremely important in maintaining flows, representing more than 50% of the river flow. Although there are outcrops of carboniferous limestone on the northern rim of the catchment (Fig. 4.2) these do not appear to provide a major component of storage (Parsons & Hunter, pers. comm.).

4.23  EFFLUENT TREATMENT

The major improvements in water quality during the 1970s resulted from
the introduction of improved methods of waste treatment at Ebbw Vale
steel-works, the treatment plant operating intermittently in 1971 and
becoming fully effective late in 1973.

   Effluents from steel-making and associated activities derive principally from pickling (sulphuric and hydrochloric acids, ferrous sulphate,
heavy metals), electro-plating (phenols derived from phenol-sulphonic
acid), tin-plating (chromium) and coke-making (ammonia, cyanide, phenols):
tars and oils are generally present and temperatures are high.

   Before 1970 treatment consisted of acid neutralization, initially
with powdered lime and subsequently with milk of lime, followed by a
brief period of settlement.  Neutralization procedures were inadequately
controlled and violent fluctuations in pH occurred.  Although some
metals were partially precipitated, other pollutants were discharged to
the river.  The ferrous sulphate became oxidized and hydrated leaving
the upper reaches of the river deficient in oxygen and rich in particulate and colloidal ferric compounds.  Improved treatment involved the
separation of strong and rinse pickling-liquors, followed by rapid
oxygenation, pH regulation with calcium carbonate, and then solids
removal.  Additionally, closure of the coke-ovens in 1972 reduced the
inputs of ammonia, cyanide and phenol.  The replacement of sulphuric
acid by hydrochloric acid for rinsing in 1975 led to further improvements, for calcium chloride formed in treatment, being more soluble
than calcium sulphate, does not coat lime particles and impede reaction.

   The coal washeries within the area have over the past twenty years
also introduced more effective techniques to remove particulate wastes
from discharges to the river.  There remains, however, the problem of
disposing of wastes and although at some sites some fine particles are
disposed of underground, most waste material is tipped and, unless the
tips are stabilized and re-vegetated, surface run-off carries such
particles to water-courses.

   During the 1970s, the trunk sewer, which carried wastes to Newport,
but which, being overloaded frequently, discharged sewage to the river
during rainy periods, was gradually replaced by one of larger capacity.


4.24  WATER QUALITY

Before the treatment plant of the Ebbw Vale steel-works became fully
operational in 1973 several aspects of water quality prevented the
establishment of a diverse flora and fauna (Fig. 4.4).  Dissolved
oxygen concentrations were occasionally negligible throughout the length
of the river despite the high reaeration capacity of the river which
was associated with its shallow and generally-fast current.  Although
intermittent discharges of sewage exacerbated the problem, the major

*Fig. 4.4* Aspects of water quality at sites down the Ebbw Fawr during the period of improvement. Three phases are shown:

1) up to 1973 the steel-works effluent was not fully operational;
2) 1973-1976 effluent quality improved dramatically;
3) 1976-1980 improved quality sustained.

Average values are shown by continuous lines and, where relevant, range limits are shown by hatched lines.

cause of deoxygenation was the considerable discharge of ferrous salts. Values of pH oscillated widely, the range at Cwm, 3 km downstream of Ebbw Vale, being from 3 to 12. Even 13 km downstream the pH range was between 4 and 9. Several soluble poisons were present in high concentrations, particularly ammonia, phenols and heavy metals (Cr, Ni, Cu, Zn, Cd, Pb) and the presence of these would have prevented the establishment of most organisms. The average suspended solids concentration, except in the headwaters, exceeded 400 mg $l^{-1}$, well above the range

regarded as acceptable for a viable fishery (E.I.F.A.C., 1964). Whilst
a high proportion of these solids was derived from coal wastes (pumped
mine-waters, washery discharges, tip run-off) there were high concentrations of total iron, averaging about 300 mg $l^{-1}$ below the steel-
works. These subsequently settled and coated the bed with heavy
deposits of ferric compounds.

After 1973 there were widespread improvements in water quality.
Dissolved oxygen concentrations, although occasionally low, were generally above 50% air saturation (Fig. 4.4) and pH fluctuations were
reduced to acceptable limits (pH 6 - 9) for many organisms (Scullion &
Edwards, 1980a). Total iron concentrations downstream of the steel-
works were reduced one hundredfold, contributing to the dramatic
decrease in total suspended solids concentrations but these, which are
now derived principally from the activities of the coal industry, have
since remained at about 200 mg $l^{-1}$, a value too high for the maintenance
of a successful fishery (Scullion & Edwards, 1980b) and for the establishment of a normal benthic fauna (Scullion & Edwards, 1980a). The
concentrations of soluble poisons also decreased, particularly of heavy
metals and phenols. Nevertheless occasionally high concentrations of
ammonia still occur in the lower catchment.

The toxicity of soluble poisons to fish may be summed when their
concentrations are expressed as fractions of their toxic threshold
concentrations (Brown, 1968). Fig. 4.5 shows that median and 95 percentile summed values for soluble poisons at sites downstream of the
steel-works during the period 1974-76 were less than 0.1 and 0.4
respectively and generally within a range considered tolerable to a
wide range of organisms (Alabaster et al., 1972).

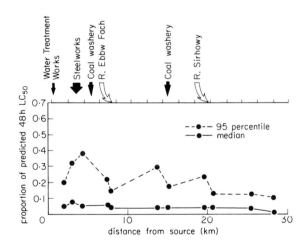

Fig. 4.5 Median and 95-percentile values of calculated toxicity to
rainbow trout of water at sites down the Ebbw Fawr during the period
1974-1976.

In most rivers the concentrations of many water quality determinands
are related to water flow, a log. concentration - log. flow regression
accounting for much of the variation in quality, but in the Ebbw, comparatively few, e.g. total hardness, are so related (Fig. 4.6a).  The
concentration of suspended solids is particularly flow-dependent in
many rivers, concentrations being considerably increased at high flows.
This is not so in the Ebbw (Fig. 4.6b) where coal washeries exert their
major effect on river quality at low flows and catchment run-off, particularly from spoil heaps, is most pronounced at high flows during periods
of rainfall.  Similarly, chloride concentrations are not markedly flow-
dependent (Fig. 4.6c).  In many catchments sewage effluent provides a
major constant source of chloride and consequently river concentrations
decrease at high flows.  In contrast, in the Ebbw catchment, with no
major treatment works, sewage enters principally at high flows as storm
sewage.  However, the annual contribution of chloride from sewage, even
if all were discharged to the river, is only about 1000 tonnes or 20%
of the total chloride yield of the catchment.  The steel-works effluent,
influencing river concentrations at low flows and mine drainage, making
its principal contribution after periods of heavy rainfall, are also
major sources of chloride.

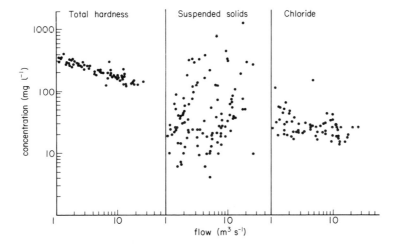

*Fig. 4.6* Relationship between flow and aspects of water quality
(water hardness, suspended solids and chloride) at Rhiwderin (gauging
station on Ebbw Fawr in Fig. 4.2) during the period 1976-1979.

4.3  PLANTS AND MACROINVERTEBRATES

## 4.31  INTRODUCTION

Sampling of the benthic flora of the Ebbw was carried out intermittently between 1958 and 1970, demonstrating that there was no change in the gross impoverishment of the river during this period (Benson-Evans & Williams, 1970), before improvements in water quality occurred following treatment of the steel-works effluent.  Between 1970 and 1975 regular, more-detailed surveys of both plants and macroinvertebrates were carried out at sites on the main river and its major tributaries, the principal objectives being to record the sequence of recolonization stages downstream of the steel-works effluent as water quality improved and to assess the influence of tributaries on such recolonization.

During each survey an estimate of cover of the major benthic plant communities was made at each site and samples taken for subsequent microscopic examination.  Macroinvertebrates were collected by kick-sampling, a procedure previously used in surveys of neighbouring catchments (Edwards *et al.*, 1972).

Until 1972 there was no visible plant growth in the main river downstream of Ebbw Vale.  Even sewage fungus was restricted to the middle and lower reaches where it covered the bed of the river for most of the year (Fig. 4.7).  In the spring of 1972 visible areas of green algal growth were noticed in downstream sites and a year later these had extended upstream to Cwm.  Nevertheless in those reaches immediately below the Ebbw Vale steel-works sewage fungus continued to dominate until 1975, the end of the period of detailed surveys:  in contrast its importance at downstream sites had declined by 1974.

Macroinvertebrates, particularly naidid and tubificid worms, chironomid midges and mayflies (principally *Baetis rhodani*) were present in small numbers at downstream sites in 1970, the first year of systematic observations (Fig. 4.8).  Some of these, particularly the mayflies, could well have drifted downstream from tributaries.  Recolonization at Cwm, 1 km below Ebbw Vale, was delayed until 1973 despite the high density of macroinvertebrates in the unpolluted headwaters.  Although there was a general increase in invertebrate abundance throughout the investigation at downstream sites, nowhere did abundance approach that of sites in the headwaters and major tributaries.  Nevertheless during 1974 and 1975 the density of macroinvertebrates in the lower reaches was similar to that in other rivers of the coalfield, such as the River Taff (Edwards *et al.*, 1972).

## 4.32  COMMUNITY DIVERSITY

In order to describe aspects of the change in structure of both plant and animal communities, several indices were computed.  Fig. 4.9 shows values for one of these, the Shannon Diversity Index, which reflects

*Fig. 4.7* Seasonal changes in percentage cover of groups of plants and bacteria at sites above and below Ebbw Vale during 1970-1975. In the period 1970-1973 surveys were carried out in spring, summer, autumn and winter; in 1974 and 1975 only spring and autumn surveys were conducted (see also Figs. 4.8 - 4.11).

not only the number of taxa but also the equitability of their abundance and has been widely used in surveys of rivers in South Wales (Edwards *et al.*, 1972; Williams *et al.*, 1974; Williams *et al.*, 1976).

Shannon diversity values for communities of plants and bacteria increased dramatically during the summer of 1972 with algae dominating except immediately below Ebbw Vale where fungi and bacteria contributed most to the diversity of the benthic communities. Plant diversity continued to increase during 1974 and 1975, with particularly high values during the summer, and was very similar to that recorded for polluted rivers elsewhere in the coalfield (Edwards *et al.*, 1972; Williams *et al.*, 1974). Indeed, diversity values in 1975 were very similar to those in unpolluted reaches upstream of Ebbw Vale and higher than those in certain tributaries, e.g. Ebbw Fach.

*Fig. 4.8* Seasonal changes in the abundance of macroinvertebrates at sites above and below Ebbw Vale during 1970-1975.

Values of macroinvertebrate diversity shown in Fig. 4.9 were lower than those of plants but this reflects, to some extent, the level of identification rather than a fundamental difference in 'community' structure. In the early years when only small numbers of animals were found, those drifting downstream could have been very significant. As with plants, there was a marked increase in diversity during the summer of 1972 in the middle and lower reaches to values comparable with unpolluted sites which was sustained until the end of the study period. Immediately downstream of Ebbw Vale comparable diversity values were not achieved until 1975.

The Trent Biotic Index, based on the variation in response of macroinvertebrates to predominantly-organic pollution and on a scale from 0 (grossly polluted) to 10 (clean), has been widely used to classify river quality (Woodiwiss, 1964). In Fig. 4.10 values based on the total sample are shown for the period 1970-1975 but it is possible that certain sample components have drifted downstream, particularly from clean tributaries, and influenced values of the index. The downstream

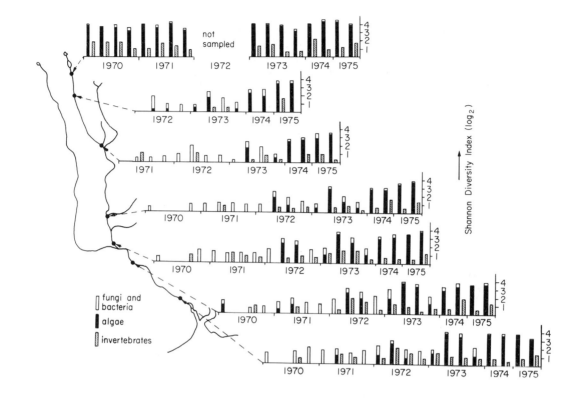

*Fig. 4.9* Seasonal changes in the Shannon Diversity Index for bacteria, plants and macroinvertebrates at sites above and below Ebbw
Vale during 1970-1975.

improvement, seen first in 1972, occurred progressively later with
nearness to Ebbw Vale. By 1975 values of the Trent Biotic Index were
still generally below 5 as compared with 6 or 7 upstream of the steel-
works effluent.

## 4.33 QUALITATIVE ASPECTS OF COMMUNITY STRUCTURE

Whilst there were no benthic plant communities in the 5 km reach below
Ebbw Vale in 1971 and 1972, water quality was such that a sewage fungus
association dominated by *Fusarium aqueductum* was able to grow in the
middle and lower reaches. With increasing distance downstream *Fusarium*
was mixed with bacterial zoogloea. The upstream dominance of *F.
aqueductum* is interesting for it is a major component in only about
0.6% of sewage-fungus outbreaks in rivers of the U.K. (Curtis & Curds,
1971). Several workers describe it as favouring acid waters (Paulsen,
1962) although Painter (1954) demonstrates its ability to grow well in

*Fig. 4.10* Seasonal changes in the Trent Biotic Index for macro-invertebrates at sites above and below Ebbw Vale during 1970-1975.

culture within the pH range 4 to 9 - a characteristic which might account for its dominance in the River Ebbw where pH ranged between 3 and 12 at upstream sites (Fig. 4.4) during the period 1970 to 1973. In 1972 *Sphaerotilus natans*, the usual dominant of the sewage-fungus association (Curtis & Curds, 1971), became more abundant at downstream sites as did *Stigeoclonium tenue*, a green alga often associated with polluted areas (Curtis, 1969), which, in the absence of *Cladophora glomerata*, is often an indication of combined high concentrations of heavy metals and nutrients (Whitton & Say, 1975; Whitton, 1979). Another alga which became widespread during these early recovery stages in the lower reaches was *Navicula avenacea*, where it was found in mucilage tubes rather than the normal free-living condition. Similar tubes were found in a neighbouring river, the Afon Lwyd, in 1962 when that river was recovering from pollution after the closure of a steel-works (Benson-Evans & Williams, 1970).

These initial stages of recovery were followed by an increase in algal diversity, particularly with the establishment of diatoms of the

genera *Navicula, Achnanthes, Nitzschia, Surirella, Gomphonema* and
*Diatoma*.   In the summer of 1973 *Cladophora* and two diatoms, *Cocconeis
placentula* and *Diatoma vulgare*, became widespread in the lower stretch.
*Cladophora* did not recolonize the stretch immediately below Ebbw Vale
during the study period.

Field observations in the initial recovery phase of the early 1970s
were supplemented by *in situ* exposure of *Cladophora* and the mosses
*Rhynchostegium riparioides* and *Fontinalis antipyretica* in perspex and
nylon cages at sites in the main river for periods up to 20 days during
the winter and at a control site of a clean tributary from which 'experimental' plant material was obtained.   A scale of damage from 0 (healthy)
to 4 (dead and decomposing) was used and Fig. 4.11 summarises the
results for *Cladophora*; mosses behaved very similarly.   At Cwm, 3 km
downstream of Ebbw Vale, no observations were undertaken until 1973 when
death of *Cladophora* was rapid; in 1974 there was a slow breakdown of the
plant material - behaviour consistent with the absence of *Cladophora*
from the site during this period.   Immediately above the Ebbw-Sirhowy
confluence, no observations were made after 1973, at which time, although
water quality was much improved, plant damage was still occurring.   At
the lowest site studied, Newport, little damage occurred in 1973 and
1974; this is consistent with the natural recolonization of the lower
reaches with *Cladophora* during this period.

After 1975 intermittent surveys have shown that diverse algal communities containing *C. glomerata* have extended upstream to within 5 km of
Ebbw Vale, with a mixed community of *Stigeoclonium tenue*, diatoms
(including *Navicula mutica*) and sewage-fungus occurring immediately
below Ebbw Vale.

Were the Ebbw Fawr unpolluted, bryophytes would almost certainly be
abundant, with liverworts such as *Scapania undulata, Solenostoma triste*
and *Nardia compressa* upstream and the mosses *Fontinalis squamosa, F.
antipyretica* and *Rhynchostegium riparioides* more downstream.   Aquatic
encrusting lichens (e.g. *Verrucaria* spp., *Dermatocarpon fluviatile*),
algae (e.g. *Phormidium, Lemanea, Hildenbrandia*) and several filamentous
and encrusting members of the Chlorophyta would probably be widespread
(N.T.H. Holmes, pers. comm.).

Most of the later surveys have, in contrast with earlier observations,
focussed on macroinvertebrates and have shown continued faunal recovery.
The Trent Biotic Index score had by 1977 improved to between 5 and 7
at all sites, except immediately downstream of Ebbw Vale where it
remained at two.   The fauna (Table 4.1) was regarded as typical of
moderately polluted rivers of the coalfield, containing molluscs, mayflies, caddis-flies, midge larvae, leeches and worms (Enchytraeidae,
Tubificidae and Naididae).   Although *Baetis rhodani*, the most tolerant
of the mayflies to pollution, was exceptionally abundant, others
(*Ephemerella ignita* and *Ecdyonurus dispar*) had begun to recolonize
the lower reaches.   Despite this marked improvement in species richness
the biotic score based on that of Chandler (1970) was, in 1977, only

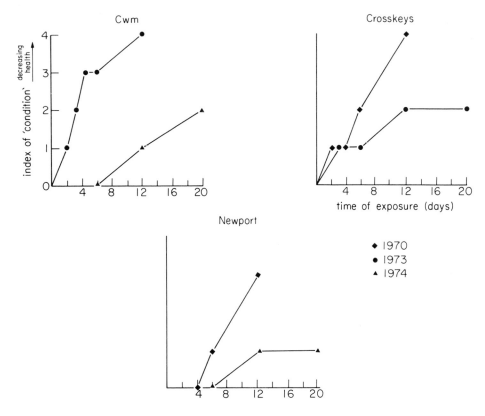

**Fig. 4.11** Condition of *Cladophora* placed in cages for up to 20 days at sites in the Ebbw Fawr. Condition is represented on scale of 0 (healthy) to 4 (dead and decomposing).

between 15 and 30% of that which might be expected for an unpolluted river of the region.

## 4.4  FISH

### 4.41  DISTRIBUTION IN CATCHMENT

Before the improved waste-treatment at Ebbw Vale steel-works in 1973, fish were confined to the headwaters of the Ebbw Fawr and to the tributaries. The two major tributaries, the Ebbw Fach and Sirhowy, were both intermittently polluted by washery effluents and material from colliery waste tips adjacent to the river. Despite these problems several species were present in both the Sirhowy (minnow, stickleback, bullhead, stoneloach, brown trout, eel) and Ebbw Fach (stickleback, stoneloach, brown trout) and fishing clubs, particularly on the lower

*Table 4.1*  Macroinvertebrates found at sites in the Ebbw Fawr downstream of Ebbw Vale in September 1979.  (Data from Welsh Water Authority)

Oligochaeta
   Enchytraeidae
   Tubificidae
    *Tubifex tubifex*
    *Limnodrilus hoffmeisteri*
    *Potamothrix hammoniensis*
    *Aulodrilus pluriseta*
   Naididae
    *Nais elinguis*
    *N. barbata*
    *N. bretscheri*
    *Stylaria lacustris*
   Lumbriculidae
    *Lumbriculus variegatus*

Hirudinea
    *Erpobdella octoculata*
    *Helobdella stagnalis*
    *Glossiphonia complanata*

Mollusca
    *Potamopyrgus jenkinsi*
    *Lymnaea peregra*
    *Ancylus fluviatilus*
    *Sphaerium* sp.
    *Planorbis* sp.
    *Physa fontinalis*

Crustacea
    *Gammarus pulex*
    *Asellus aquaticus*

Insecta
  Ephemeroptera
    *Baetis rhodani*
    *Ephemerella ignita*
  Trichoptera
    *Rhyacophila dorsalis*
    *Hydropsyche pellucidula*
    *Polycentropus flavomaculatus*
    *Hydroptila* sp.
  Diptera
   Chironomidae
   Tipulidae
   Empididae
   Tabanidae
   Muscidae

Hydracarina

reaches of the Sirhowy, had stocking programmes for brown trout.  A few
of these stocked fish entered the grossly-polluted Ebbw Fawr and were
subsequently found in the vicinity of small streams where there was
localised improvement in water quality.  The other cleaner tributaries,
draining upland moors and conifer plantations, were small and their
flow unreliable during dry summers;  of these the Carn and Gwyddon (Fig.
4.2) were most significant.  Generally, the small tributaries contained
bullheads and brown trout and a few held sticklebacks and stoneloach
(Fig. 4.12).  Roach and chub were caught in the lower Sirhowy in 1972
but it seems likely that these had escaped from the canal near the
Carn (Fig. 4.2) through a defective sluice.  At this time there were
also trout and stoneloach in the Ebbw Fawr at the Sirhowy confluence,
suggesting that water quality was suitable for fish in this local area,
at least for short periods.

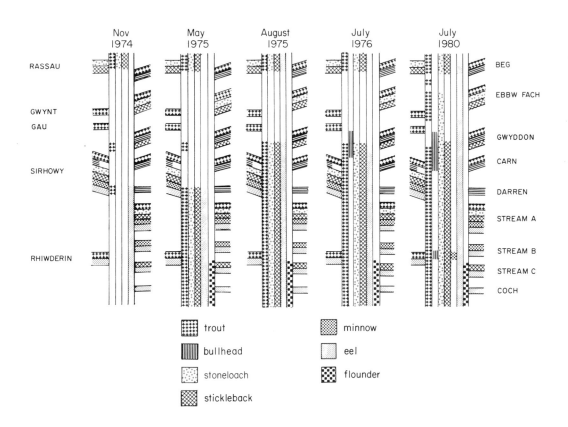

*Fig. 4.12*  Pattern of fish recolonization in the Ebbw Fawr since 1970.

In the years after 1973 there was a rapid recolonization of the main river by several species (Fig. 4.12). Late in 1974 eels extended upstream to the Sirhowy confluence and brown trout occurred at the mouths of both the Sirhowy and Gwyddon. In 1975 there were dramatic changes, leading to the recolonization of the Ebbw Fawr by brown trout, eels, stoneloach, stickleback and flounder, all but the flounder extending their range upstream to the Gwyddon, only about 16 km from Ebbw Vale. In 1976 the bullhead was recorded in a reach at the mouth of the Gwyddon.

By 1980 trout, stoneloach and eels had extended upstream from the sea to sites within a few kilometres of the steel-works effluent. Although the recolonization of stoneloach and eels was natural, that of trout was aided by a stocking programme carried out by the Welsh Water Authority in 1979 and 1980. Bullhead remained generally confined to areas adjacent to clean tributaries, such as the Carn and Gwyddon, but an isolated occurrence was recorded in the lower reaches at a site near Risca where minnows were also found. The slow recolonization of the river by minnow may be partly explained by its very limited distribution within the catchment, the only substantial population being in the upper reaches of the Sirhowy, and its apparent restricted mobility (Kennedy & Pitcher, 1975). A list of all fish species recorded recently within the catchment is given in Table 4.2.

## 4.42   DENSITY, GROWTH AND CONDITION

The densities of trout in tributaries (Table 4.3) are generally similar to those found elsewhere in the coalfield but most fish are small, few achieving a length greater than 20 cm, and consequently both biomass, densities and potential recruitment are low. Bullheads, although less widespread, reach high densities in the Beg, Gwyddon and Carn. Stoneloach and minnow are particularly abundant in the Sirhowy (the latter species being concentrated particularly in the headwaters) where they achieve densities of 144 and 107 per 100 $m^2$ respectively.

The largest populations of native trout in the Ebbw Fawr were found close to the mouths of the Carn, Gwyddon and Sirhowy. In 1976 at the mouth of the Sirhowy maximum densities ranged from 7 - 8 per 100 $m^2$ with biomasses between 810 and 1150 g per 100 $m^2$. In 1980, at the mouths of the Carn and Gwyddon similar maximum densities were recorded; these compare favourably with those found in other rivers of the coalfield with established but vulnerable trout populations (Learner et al., 1971; Scullion & Edwards, 1980b). Outside these very limited areas influenced by the dilution water of clean tributaries, trout densities were low.

Upstream migration of the flounder was probably prevented by Bassaleg Weir, 2 km above the tidal limit: in parts of this downstream reach, which were sampled, it was present at densities up to 13 per 100 $m^2$. Stickleback were found in backwaters and river margins downstream of tributaries with resident populations.

*Table 4.2* Fish species recorded recently in the Ebbw system.

| | | |
|---|---|---|
| Petromyzonidae | *Lampetra planeri* | brook lamprey |
| Salmonidae | *Salmo trutta* | brown trout |
| Cyprinidae | *Gobio gobio* | gudgeon |
| | *Phoxinus phoxinus* | minnow |
| | *Rutilus rutilus* | roach |
| | *Leuciscus cephalus* | chub |
| | *L. leuciscus* | dace |
| Cobitidae | *Noemacheilus barbatulus* | stoneloach |
| Anguillidae | *Anguilla anguilla* | eel |
| Gasterosteidae | *Gasterosteus aculeatus* | three-spined stickle-back |
| Percidae | *Perca fluviatilis* | perch |
| Cottidae | *Cottus gobio* | bullhead |
| Pleuronectidae | *Platichthys flcsus* | flounder |

*Table 4.3* Density and biomass (per 100 $m^2$) of trout and bullhead in tributaries of the Ebbw Fawr 1970-72.

| River | trout | | bullhead | |
|---|---|---|---|---|
| | density | biomass (g) | density | biomass (g) |
| Beg | 24 - 113 | 411 - 1282 | 6 - 117 | 29 - 267 |
| Gwynt | 31 - 71 | 866 - 1359 | - | - |
| Gwyddon | 61 - 167 | 828 - 2429 | 19 - 247 | 38 - 545 |
| Carn | 49 - 82 | 1122 - 1688 | 45 - 113 | 92 - 244 |
| Ystruth (off Ebbw Fach) | 9 - 91 | 173 - 648 | - | - |
| Sirhowy | 1 - 9 | 402 - 571 | - | - |
| Sirhowy tributaries | 3 - 125 | 220 - 1390 | - | - |

During 1975 and 1976 brown trout caught in the Ebbw Fawr, near the Sirhowy confluence, were tagged, and from subsequent recaptures in this reach their growth rate could be assessed.  The growth of many fish was remarkably fast, the average specific annual growth rate of four and five year-old fish being 36 - similar to those in unpolluted rivers of similar temperature regime in southern Britain (Williams & Harcup, 1974).  There is evidence from scale examination that slow-growing fish from small tributaries, such as the Gwyddon, grow much faster in the Ebbw Fawr but that the growth rate of those descending from the much larger Sirhowy does not change appreciably.

Average condition factors of trout captured in the Ebbw Fawr during 1975-76, derived from weight and length measurements, were about 1.35 and well above the value of unity often regarded as adequate for trout in reasonable condition (Frost & Brown, 1967).

## 4.43   TROUT MOVEMENTS AND RECRUITMENT WITHIN THE CATCHMENT

An extensive marking programme of native brown trout in the lower reaches of several small tributaries of the Ebbw Fawr and in the Sirhowy, together with the marking of brown trout stocked into the Sirhowy during the period 1971-78 (Williams & Harcup, 1974; Cresswell & Williams, 1979; Turnpenny, 1979), has facilitated an appreciation of the broad features of movement.

Native brown trout in the small tributaries, although relatively numerous, are not large and therefore egg production, or potential re-cruitment, is low.  It has been estimated that in the Gwyddon, Carn and Beg, the three more important unpolluted tributaries, the average density of eggs is only about 900 per 100 m$^2$.  The total egg production of these streams, principally because of their small size, is very low ($355 \times 10^3$ Gwyddon; $121 \times 10^3$ Carn; $15 \times 10^3$ Beg).  With the resulting low fry production there is little of the downstream emigration which occurs in many river systems soon after emergence from the gravel (Le Cren, 1973) and such emigration is further hindered by physical barriers (Williams, 1980).

In most river systems which support trout some fish move from main rivers into small tributaries to spawn.  In the Ebbw system, the two larger rivers containing trout, Sirhowy and Ebbw Fach, are intermit-tently polluted by coal wastes and sewage and contain low trout densities (Table 4.3 for Sirhowy).  These trout are prevented from moving into several clean tributaries for spawning by impassable cul-verts and other barriers and the gravels in the larger tributaries (Sirhowy and Ebbw Fach) are mostly unsuitable for spawning because of the low permeabilities principally associated with fine particles derived from colliery spoil heaps and coal washeries (Turnpenny & Williams, 1980).

As with fry, it seems that the downstream movement of older fish from the small, clean tributaries is very restricted for they move very little, most having a home range of less than 15 m over periods of several months (Harcup, 1978).

In consequence of this immobility of native trout caused by the topography, physical development and pollution of the catchment, the general recolonization of the Ebbw Fawr would be expected to be slow

following an improvement in water quality.  It seems that the relatively
fast but very localized recolonization during the period 1973-76 has
been assisted both directly and indirectly from stocking of brown trout
within the lower reaches of the Sirhowy where in the period 1974-76
about 4000 were introduced.  Turnpenny (1979) estimated that about 13%
of all trout caught in the Ebbw Fawr during these years were fish
stocked in the Sirhowy.  This downstream emigration is supported by
studies of Williams & Harcup (1974) who found that extensive downstream
movements occurred and that on one occasion up to 80% of stocked fish
were lost, presumably to the Ebbw Fawr, within a week of their introduction into the lower Sirhowy.  As stocking was with relatively large
fish (20 - 30 cm) it seems likely that stocking also led to the downstream displacement of native fish and after large fish were introduced
to the lower reaches of the Sirhowy substantial increases in the density
of native trout occurred in the Ebbw Fawr (Turnpenny, 1979).

The recolonization of the Ebbw Fawr with trout, by immigration, will
not necessarily lead to substantial improvements in recruitment within
the catchment for gravels in the river seem generally unsuitable for
spawning.  Turnpenny & Williams (1980) showed that most ova planted in
substrates at sites in the Ebbw Fawr died, whereas at sites on clean
tributaries most survived:  in the Sirhowy, polluted by coal wastes,
intermediate results were obtained.  Both embryo survival and alevin
length were related to substrate permeability which influenced oxygen
supply (Fig. 4.13).

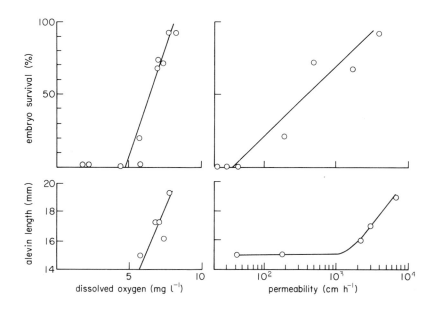

*Fig. 4.13* Relation between success of embryo development (embryo
survival and alevin length) and substrate quality (permeability and
dissolved oxygen content of interstitial water).

The concentrations of some metals in sediments of the Ebbw Fahr, when compared with other rivers of the catchment, were very high - particularly Cr, Fe, Zn and Pb (Table 4.4). In the case of certain metals, e.g. Co, Cd, Mn, the most contaminated site was in the headwaters of the Ebbw Fawr, upstream of steel-works discharges. Contamination probably results from the leaching of metalliferous wastes adjacent to the river derived from earlier mining activities. Despite these high metal concentrations, their effect on the survival of fish embryos seems minimal because of their generally low solubilities at normal pH values.

Table 4.4  Mean concentrations (mg kg$^{-1}$) of heavy metals in sediments of the Ebbw Fawr, Sirhowy and the Gwyddon in 1975.

| site | Cr | Mn | Fe (x $10^3$) | Co | Cu | Zn | Cd | Pb |
|------|----|----|------|----|----|----|----|----|
| Gwyddon (baseline) | 36.8 | 495 | 15.9 | 25.5 | 39.0 | 117 | <0.2 | 27.0 |
| Sirhowy | 11.5 | 987 | 18.6 | 31.1 | 53.1 | 114 | <0.2 | 30.3 |
| Ebbw Fawr (headwaters) | 6.4 | 5229 | | 52 | 65.4 | 20665 | 78.2 | 565 |
| Ebbw Fawr (Cwm) | 136 | 1347 | | 65 | 146 | 384 | 3.6 | 224 |
| Ebbw Fawr (Crosskeys) | 39.7 | 784 | | 48 | 80.3 | 225 | <0.2 | 143 |

## 4.44  FISH SURVIVAL AND WATER QUALITY

That the improvement in water quality in the Ebbw since 1973 has been adequate to support fish life, has been demonstrated by their capture within the river. But the capture of fish does not indicate the length of time they might survive in the system and so independent assessments of the adequacy of water quality to support them have been made. In other industrial river systems it has been demonstrated that if other water quality characteristics are satisfactory, natural fish populations will be sustained where the median concentrations of soluble poisons, expressed as fractions of their lethal threshold concentration to rainbow trout, do not exceed about 0.25 (Alabaster et al., 1972). Turnpenny & Williams (1980) have shown that during the period 1974-76 at all sites down the river, this median fraction of the 48 h LC$_{50}$ did not exceed 0.1 (Fig. 4.5). Curiously, the calculated toxicity of sites in the headwaters was similar to that downstream of Ebbw Vale: this toxicity is derived largely from metals which seem to emanate from ground and tip drainage.

Further evidence of the adequacy of water quality has been acquired
from experiments in which trout (Fig. 4.14) and several other fish
species have been retained in cages at sites within the river.  Such
fish are inevitably stressed and the adequacy of feeding, important in
long-term tests, presents a problem.  Nevertheless, in the Ebbw the
median period of survival of trout at sites only 5 km downstream of the
Ebbw Vale steel-works increased from about 0.5 h in 1970 to several
days after 1974.  More importantly, at downstream sites fish generally
survived the test period, about two months for rainbow trout and up to
three weeks for other species (bullhead, minnow, stickleback, stoneloach, brown trout).

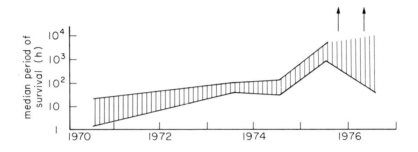

*Fig. 4.14* Improvement in survival times of caged trout in the Ebbw
Fawr, 1970-76.

## 4.5  CONCLUSION

The efforts made by the Usk River Authority during the 1960s to improve
the water quality of the grossly-polluted Ebbw Fawr, which led to major
pollution abatement programmes during the early 1970s, has provided an
excellent opportunity to study the recovery of a river, formerly almost
devoid of living organisms over most of its length.  Furthermore the
fact that the major effluent discharge and the one which has improved
most dramatically, the effluent from the Ebbw Vale steel-works, was
located in the headwaters, has meant that the recovery sequence could
be followed down the whole river, a distance greater than 30 km.
Chemical and biological surveillance has demonstrated an improvement
in water quality leading to the establishment of associations of
organisms of increasing diversity.  Surveillance has been supported by
*in situ* survival tests which have demonstrated the increasing ability
of the river to support both plants and animals.  In the case of fish,
and particularly brown trout, it has been possible to compare survival
of captive fish in the river with that expected from prevailing water
quality and laboratory toxicity data.

It is perhaps the work on the establishment of fish populations in the river that demonstrates most clearly the dramatic change which has taken place in the past decade.  Seven species are now found in the main river, some distributions extending downstream from the steel-works to the estuary.  Although siltation of the river bed, in part by coal wastes prevents natural spawning, a 'put and take' fishery of brown trout has been established with the aid of regular stocking.

Despite these recent major improvements, the Ebbw Fawr, along with other rivers in the industrial and highly-urbanised valleys of the South Wales Coalfield, will continue to be affected by pollution problems.  Such factors as the linear urban settlements mostly near the river, mining subsidence leading to the periodic fracture of the sewers, which are often routed along river-banks, high rainfall and shallow soils causing unsteady river-flows and the intermittent discharge of storm-sewage, historical accumulation of mining spoil which is easily washed into water-courses, are all significant constraints on further improvement.

The capital costs of improving the effluent of the steel-works (£8 million*), of constructing a new trunk sewer (£13 million) and of modifications to coal washeries and colliery spoil disposal arrangements (£1 million), must be coupled with the considerable on-going costs of effluent treatment plant at the Ebbw Vale steel-works.  Apart from the benefits to amenity and recreation, particularly fishing, the river now provides water of acceptable quality for abstraction for a variety of industrial uses in the lower reaches, thus saving 'imported' water of higher quality for potable and agricultural supply.

*Acknowledgements*

The authors wish to thank Dr K. Benson-Evans (University College, Cardiff), Mr J. Bartlett (UWIST) and the many staff of the Welsh Water Authority, particularly Mr M. Henderson, Mr M. Hunter, Mr F. Jones and Mr R. Streeter, who provided information and commented upon the text, and Mrs M. Thomas and Mrs S. Fish for the preparation of typescript and figures.

REFERENCES

Alabaster J.S., Garland J.H.N., Hart I.C. & Solbé J.F. de L.G. (1972) An approach to the problem of pollution and fisheries. *Symp. Zool. Soc. Lond.* 29, 77-114
Benson-Evans K. & Williams P.F. (1970)  Biological review of 1960-70 position in some Usk River Authority streams. *Usk River Authority Annual Report* 5, 75-85

*all costs adjusted to 1980 values

Brown V.M. (1968)  The calculation of the acute toxicity of mixtures of poisons to rainbow trout. *Wat. Res.* 2, 723-33

Chandler C.M. (1970)  A biological approach to water quality management. *Wat. Pollut. Control* 66, 415-21

Cresswell R.C. & Williams R. (1979)  Studies on trout stocking in an industrial river and their management implications. *Proc. 1st British Freshwater Fisheries Conference*, Univ. Liverpool, pp. 285-95

Curtis E.J.C. (1969)  Sewage fungus: its nature and effects. *Wat. Res.* 3, 289-311

Curtis E.J.C. & Curds C.R. (1971)  Sewage fungus in rivers in the United Kingdom: the slime community and its constituent organisms. *Wat. Res.* 5, 1147-59

Edwards R.W., Benson-Evans K., Learner M.A., Williams P.F. & Williams, R. (1972)  A biological survey of the River Taff. *J. Wat. Pollut. Control* 71, 144-60

E.I.F.A.C. (European Inland Fisheries Advisory Commission) (1964)  Water quality criteria for European freshwater fish. *Report on Finely Divided Solids and Inland Fisheries EIFAC/T1.*  FAO Rome

Frost W.E. & Brown M.E. (1967)  *The Trout,* 316 pp.  Collins, London

Gray-Jones A. (1970)  *A History of Ebbw Vale,* 271 pp.  Gray-Jones

Harcup M.F. (1978)  *Fish Movement in some Streams of the Ebbw Catchment.*  M.Sc. Thesis, University of Wales (UWIST)

Kennedy G.J.A. & Pitcher T.J. (1975)  Experiments on homing in shoals of the European minnow, *Phoxinus phoxinus* L. *Trans. Am. Fish. Soc.* 104, 454-57

Le Cren E.D. (1973)  The population dynamics of young trout (*Salmo trutta*) in relation to density and territorial behaviour. *Rapports et Procès-verbaux des Réunions,* Conseil International pour l'Exploration de la Mer 164, 241-47

Learner M.A., Williams R., Harcup M.F. & Hughes B.D. (1971)  A survey of the macro-fauna of the River Cynon, a polluted tributary of the River Taff (South Wales). *Freshwat. Biol.* 1, 339-67

Painter H.A. (1954)  Factors affecting the growth of some fungi associated with sewage purification. *J. gen. Microbiol.* 10, 177-90

Paulsen B.B. (1962)  Pollution of the lower reaches of the Otteraa. *Vattenhygien* 18, 115-27

Scullion J. & Edwards R.W. (1980a)  The effects of coal industry pollutants on the macro-invertebrate fauna of a small river in the South Wales coalfield. *Freshwat. Biol.* 10, 141-62

Scullion J. & Edwards R.W. (1980b)  The effects of pollutants from the coal industry on the fish fauna of a small river in the South Wales coalfield. *Environ. Pollut. (Series A)* 21, 141-53

Turnpenny A.W.H. (1979)  *Aspects of the Recovery of Fish Populations in the River Ebbw Fawr.*  Ph.D. Thesis, University of Wales (UWIST)

Turnpenny A.W.H. & Williams R. (1980)  Effects of sedimentation on the gravels of an industrial river system. *J. Fish. Biol.* 17, 681-93

Whitton B.A. (1979)  Plants as indicators of river water quality. In: James A. & Evieson L. (Eds.) *Biological Indicators of Water Quality,* pp. 5.1-5.34.  Wiley, Chichester

Whitton B.A. & Say P.J. (1975)  Heavy Metals. In: Whitton B.A. (Ed.) *River Ecology,* 725 pp., pp. 286-311.  Blackwell, Oxford

Williams P.F., Benson-Evans K. & Jones A.K. (1974)  Further analysis of data from biological surveys of the River Taff. *J. Wat. Pollut. Control* 75, 576-83

Williams R. (1980)  *The Fish of the Ebbw System.*  Ph.D. Thesis, University of Wales (UWIST)

Williams R. & Harcup M.F. (1974)  The fish populations of an industrial river in South Wales. *J. Fish Biol.* 6, 395-414

Williams R., Williams P.F., Benson-Evans K., Hunter M.D. & Harcup M.F. (1976)  Chemical and biological studies associated with the recovery of the River Ebbw Fawr (1970-1974). *J. Wat. Pollut. Control* 75, 428-44

Woodiwiss F.S. (1964)  The biological system of stream classification used by the Trent River Board. *Chemy Ind.*, 443-47

# 5 · Mersey

D. G. HOLLAND & J. P. C. HARDING

5.1   Introduction                                          114

5.2   General data                                          115

5.3   The Mersey catchment                                  117

5.4   Historical changes                                    119

    5.41   The Industrial Revolution                    119
    5.42   The twentieth century                        121

5.5   The Mersey system today                               125

    5.51   Introduction                                 125
    5.52   Mersey                                       126
    5.53   Goyt                                         128
    5.54   Tame                                         128

5.6   Life in the river                                     129

    5.61   Introduction                                 129
    5.62   Plants                                       129

        5.621   General comments                     129
        5.622   Mersey                               131
        5.623   Goyt                                 132
        5.624   Tame                                 133
        5.625   Changing distribution                135

    5.63   Invertebrates                                135

        5.631   General comments                     135
        5.632   Mersey                               136
        5.633   Goyt                                 138
        5.634   Tame                                 138
        5.635   Changing distribution                139

    5.64   Fish                                         141

5.7   Discussion                                            143

References                                                  144

## 5.1   INTRODUCTION

Ever since the start of the Industrial Revolution in the 18th century,
many rivers in Britain have been subjected to a wide range and
excessive volume of man's wastes.  The river described here, the Mersey,
has a long history of particularly severe pollution and the local
author, Charles Kingsley, who knew the river well, probably had it in
mind when he wrote:

> "Dank and foul, dank and foul.
>
> By the smoky town with its murky cowl;
>
> Foul and dank, foul and dank,
>
> By wharf and river and slimy bank;
>
> Darker and darker the farther I go,
>
> Baser and baser the richer I grow."

Today, despite many years of pollution control measures, that in-
heritance is still evident for all to see.  An indication of this
severe pollution is the fact that the Mersey, with several of its main
tributaries, forms a major component of the region's dirty rivers,
which themselves make up 26% of the lowest water quality rivers of
England and Wales (National Water Council, 1981).  It naturally follows
that this poor water quality has an adverse effect on the biology of
the river system, which is typified largely by pollution-tolerant flora
and fauna and limited species diversity.

The description of the biology of the Mersey catchment presented in
this chapter has been compiled by two Water Authority biologists who
are involved in practical day-to-day studies.  Although the catchment
has received little detailed study in the past, the authors are fortu-
nate in having access to a large body of data collected during routine
work.  In order to describe the present situation, it is necessary to
give first an historical account of the area, which suggests how the
biology of the river may have been affected by the changing conditions
induced by man.  The past history is then linked to the relationship
between present-day water quality and plant and animal communities.
Since the theme of change will be evident throughout the account, some
attempt will also be made to forecast future trends.

## 5.2  GENERAL DATA

DIMENSIONS
  Area of catchments

|                         |              |                            |
|-------------------------|--------------|----------------------------|
| Etherow                 | 155.7 km$^2$ |                            |
| Goyt (excl. Etherow)    | 216.4 km$^2$ | Sources of Etherow, Goyt and |
| Tame                    | 145.3 km$^2$ | Tame all in range 460-520 m |
| Mersey (excl. E/G/T     | 161.2 km$^2$ |                            |
| total Mersey            | 678.6 km$^2$ |                            |

  Highest point in catchment 631 m (Kinder Scout)

CLIMATE
Precipitation
  top of Etherow     2400 mm yr$^{-1}$
  Tame-Etherow-Goyt  900-2400 mm yr$^{-1}$
  Mersey             800-900 mm yr$^{-1}$

Temperature
(Data for whole area of the former: Mersey & Weaver River Authority,
First Periodical Survey, 1969)
  Annual data:                              Bright sunshine:
  mean daily minimum  mostly 6-7 $^{o}$C    mean daily near coast  4-4.5 h
  mean daily maximum      12-13 $^{o}$C     mean daily on high ground
  mean                 c. 9.5 $^{o}$C                              3-3.5 h

GEOLOGY
  (i) Carboniferous outcrop of both Namurian (Millstone grit) and West-
phalian (coal measure) ages on the high ground of the upper catchment.
  (ii) Glacial sand and gravel, and alluvium in narrow bands following
the courses of the Etherow, Goyt and Tame and then broadening out into
the Mersey flood plain.
  (iii) Boulder and laminated clays on the low ground between rivers.

LAND USE
  hill farming                46%        urban     34%
  heathland and poor pasture  27%        arable     3%

POPULATION
  Total in catchment (1971 Census)    828000 people

PHYSICAL FEATURES OF RIVER

On-stream reservoirs and major weirs
  total number of reservoirs including  39 (totalling 22700 m$^3$ capacity
  4 service
  total number of major weirs           23 (as per Reservoir Act limit)

Discharge at selected points
  data given for modal flows, i.e. most frequentiy occurring

|                     | natural | artificial | total           |
|---------------------|---------|------------|-----------------|
| Etherow above Goyt  | 0.3     | 0.54       | 0.93 m$^3$ s$^{-1}$ |
| Goyt above Tame     | 1.45    | 0.93       | 2.37            |
| Tame above Goyt     | 0.66    | 0.81       | 1.47            |
| Mersey (Irlam Weir) | 2.78    | 2.66       | 5.44            |

  Artificial = compensation from reservoirs + sewage discharge +
industrial discharge

Seasonal changes in flow (snow melt etc)
    Effect of limited snow melt mostly absorbed by reservoirs

Influence of regulation
    Artificial flow = Reservoir compensation + industrial and sewage
    effluents

| | | | | |
|---|---|---|---|---|
| Etherow | 0.54 | = | >83% | + | <17% |
| Goyt | 0.93 | = | >27% | + | <73% |
| Tame | 0.81 | = | >27% | + | <73% |

Important abstraction points
    Etherow consented)        $0.12 \ m^3 \ s^{-1}$ sewer
    Goyt (consented)          $0.49 \ m^3 \ s^{-1}$ river and sewer

Substratum
    Mostly stable stone and gravel beds in Etherow, Goyt and Tame—giving
way to silt in the slower Mersey.

Temperature

|  | mean annual values ($^{o}$C) |
|---|---|
| Etherow above Goyt | 10-11 |
| Goyt above Tame | 10-12 |
| Tame above Goyt | 11-13 |
| Mersey (Irlam Weir) | 11-13 |

MANAGEMENT
    Water quality management within the North West Water Authority is
undertaken principally by the Scientist's Department of the Rivers
Division, whose duties cover the entire region.  Close cooperation with
Head Office staff enables this function to be related to the work of the
three dual purpose Divisions on the management of water supply and waste
water treatment.   The Rivers Division Scientist is supported by:
  (i) a team of River Inspectors who liaise with individuals and
      organizations instrumental in causing river pollution;
 (ii) a technical support team providing expertise in the biology and
      chemistry of water quality control.

## 5.3   THE MERSEY CATCHMENT

The Mersey catchment above the tidal limit covers 203000 hectares (Fig. 5.1), an area only half that of some of the major British rivers, such as the Thames, the Severn and the Trent.  It is characterised by extensive urbanisation.  Most of the upper river valleys are well developed and the Mersey and its tributaries pass from town to town almost without break.

The upper Mersey catchment consists of three similar sized tributaries, the Tame, Etherow and Goyt, but the defined point at which the Mersey actually begins has been the subject of debate for many years. Although the river named on present-day maps as the Mersey starts at Stockport, where the Goyt and Tame join, Palmer (1950) claims that 13th century documents give the Etherow, rising on the Pennine Hills 40 km to the east, as the source of the Mersey.  For ease of description, the Mersey will be taken from this source in the Pennines as the Mersey-Etherow.  Below the confluence with the Goyt, the river will be referred to as the Mersey-Goyt, which then joins with the Tame to become the Mersey proper at Stockport.

The Mersey-Etherow and these two major tributaries rise at similar altitudes (460 - 520 m) on the high peaty plateaus of a section of the Pennine range from Saddleworth Moor in the north to High Peak in the south (Fig. 5.2).  Similar gradients (10 m km$^{-1}$ or more) result in fast currents and high aeration rates, so that even in stretches subject to heavy organic pollution, concentrations of dissolved oxygen are generally much higher than they would be in lowland rivers of comparable size. Downstream from the confluence with the Tame the gradient of the Mersey decreases markedly as it passes through a narrow gorge and onto a broad flood plain to the south of Manchester.  32 km from Stockport the Mersey drops over Irlam Weir to unite with the Manchester Ship Canal.

The Mersey catchment above Irlam Weir has a high proportion of urban development (24%) and there is a rapid run-off to the river from all the impermeable surfaces.  Pearson (1939) quotes water level rises after heavy rain of up to 5 m in 3 to 4 hours and, because of the flat gradient of 1 m km$^{-1}$ between Stockport and Irlam, flooding has been a severe problem for several centuries.  The earliest attempts to control flooding can be traced back to the end of the 18th century when local landowners constructed a variety of floodbanks, many of which were ill-designed and subject to frequent breaching.  The present-day flood control measures of high channel embankments and flood relief basins have largely overcome the threat of widescale inundation.

In medieval times the lower Mersey meandered across a broad boggy floodplain before broadening out into the Mersey Estuary west of Warrington.  Today, however, the situation is more complex as a result of extensive alterations made to the river to enable merchant ships to pass between the Irish Sea and docks in the heart of Manchester.  Downstream of Irlam Weir, the Mersey joins the Manchester Ship Canal, which

*Fig. 5.1* Mersey system, showing sampling sites.  (u/s, upstream; d/s, downstream)

was constructed in the late 19th century on the course of the Irwell,
a major tributary flowing south-west from Manchester.  The combined
natural flow of the Mersey and Irwell continues in the Ship Canal for
14 km, and picks up two smaller tributaries before being diverted away
from the canal at Rixton.  After a further 10 km the Mersey discharges
over Howley Weir into the tidal reach of the estuary at Warrington.
Downstream of Rixton the Ship Canal skirts the course of the Mersey to
connect with the estuary at Eastham Locks.

*Fig. 5.2* Examples of altitude, river distance and gradient in the
Mersey catchment.

*Fig. 5.3* River quality classification of Mersey catchment in 1980.

## 5.4   HISTORICAL CHANGES

### 5.41  THE INDUSTRIAL REVOLUTION

Many small settlements were present on the Mersey catchment in Roman
times, but intensive urban development of the area did not begin until
the  start of the Industrial Revolution.  Cotton shipped into Liverpool
formed the foundation of a massive industry which took advantage of the
abundant supplies of clean soft river water and established many spin-
ning and weaving mills in the valleys over a wide area of South
Lancashire, North Cheshire and the area now termed Greater Manchester.
The availability of water was improved further by the construction of
a large number of reservoirs between 1840 and 1870 to supply the valley

factory towns and a large part of Manchester.

The development of communications, first of canals, then later of railways and the Manchester Ship Canal, made the import of raw materials and the dispersal of finished goods much easier.  At the same time the South Lancashire coal industry supplied the area's fuel needs.  During the 1800s industrial interests diversified into paper and machine-making, dyeing and bleaching and textile printing.  In the present century the manufacture of man-made fibres has gradually taken the place of the cotton industry, but this industry in turn is now going into a period of recession.  In more recent years industry has diversified further to include engineering, chemicals and many other interests.

With the expansion of Manchester and with additional townships springing up in the surrounding valleys, refuse, sewage and industrial wastes found their way into the river system and water quality deteriorated.  However, Bracegirdle (1973) records that even as late as 1830 drinking water was still being taken from the Irwell in Manchester - undoubtedly in ignorance of health risks.  By the mid 1800s it was becoming clear that sewage-contaminated drinking water was an excellent vehicle for the transmission of diseases such as cholera and typhoid, which by then had become endemic in the area.

Apart from the adverse effect on the quality of human life, declining water quality in the 19th century had a severe impact on fish stocks in the river.  The only available detailed documentary evidence is that by Corbett (1907), who described the demise of salmon and coarse fisheries in the Irwell through to the Mersey in the period between 1820 and 1850. As industry continued to expand, water quality deteriorated further so that by 1870 even the commercial fishery downstream in the Mersey Estuary had become affected and was eventually destroyed completely.

Sadly there is an apparent shortage of documentary evidence about the 19th century conditions in the catchment upstream of the confluence with the Irwell, but because of the similarities of industrial development throughout the Manchester area in this period, Corbett's recollections of the general condition of the Irwell can be taken as representative of all the rivers.  In particular, Corbett recorded a visible deterioration in the Irwell from about 1855 to the mid 1880s, by which time water-weeds had disappeared from the river, and the banks to a height of nearly 1 m above summer level were also devoid of plant growth.

During this period local concern was being expressed about what was happening to the river.  The first of a series of Acts of Parliament aimed at controlling river pollution (The Mersey Etc. Protection Act, 1862) was passed and the first river inspector ever appointed commenced duties in 1867.

## 5.42   THE TWENTIETH CENTURY

Documentary evidence becomes easier to find after 1892 when the Mersey
and Irwell Joint Committee was formed as the first statutory body
specifically charged with the responsibility of controlling pollution
to the river system.  In 1906 the Committee reported that 444 manufacturers discharged wastes to the Mersey and Irwell catchments, and,
with the addition of no less than 419 sewage works effluents, there was
clearly considerable scope for improvement.

   Despite the introduction of statutory pollution control powers in
the latter half of the 19th century, the appalling river water quality
conditions at the beginning of the 20th century showed very little
improvement over the next fifty years.  Indeed it was not until the
Rivers (Prevention of Pollution) Act 1951 that effective powers of
persuasion were made available to the controlling authority of the time
(Mersey River Board).  Further tightening of legislation has enabled
considerable progress to be made over the last thirty years and this
latter period has seen an impressive reduction in the number of sewage
and industrial waste effluents discharged to river.  Chemical data from
1953 onwards demonstrate clearly the changes which have taken place
(Fig. 5.4).  The 5-day Biochemical Oxygen Demand (B.O.D.) values can be
interpreted as representing the amount of organic material in the river,
whilst percentage saturation values of dissolved oxygen express the
effect of the B.O.D. on the river.  During 1962-64 there was a particularly marked improvement all over the catchment demonstrated most
strikingly by the increase in dissolved oxygen.  Several important
changes in the polluting discharges account for this.

   One clear-cut example of the kinds of pollution and improvement that
have occurred throughout the catchment is found in the Mersey-Etherow.
Ever since the beginning of the 20th century one of two paper manufacturing companies in Britain using the sulphite process for digesting.
wood had been discharging highly polluting acidic organic wastes to
Glossop Brook, a tributary of the Mersey-Etherow.  The company ceased
to operate in 1963 after having failed throughout its entire period of
operation to provide adequate treatment facilities for its wastes.  At
the same time extensions to Glossop Sewage Works brought to an end
twenty years of exceptionally poor sewage treatment.  Prior to these
two events the pollution produced by the two effluents, in combination
with that of a textile printing works, had made the water quality of
the Mersey-Etherow amongst the worst in the Manchester area, equalled
only by the Irwell and Tame.

   Immediately after 1963 the improvement in the Mersey-Etherow could
be traced downstream chemically through the Mersey-Goyt to below
Stockport.  In the 20 years since then, water quality conditions
throughout the upper Mersey catchment have, on the whole, continued to
improve at a more gradual pace.

Total effluent discharges:  sewage    38        27    27    25    25
                            industrial  81        31    21    12    10

*Fig. 5.4* Changes in 5-day B.O.D. and dissolved oxygen (as % saturated) at key sites in Mersey catchment over period 1953-1982, together with number of effluent discharges in whole catchment.  Chemical data based on mean of monthly readings.

*Fig. 5.5* The Tame at Stockport upstream of the Mersey.

In addition to organic pollution from treated and untreated sewage and industrial waste, the Mersey and its two tributaries have been subject to other forms of pollution, such as heavy metals, cyanide, detergents (Fig. 5.6), phenol and chlorine. Industrial development of the river valleys included the establishment of several factories that made use of these toxic substances in a variety of processes. In some cases elevated concentrations provided significant contributions to the industrial wastes being discharged to the rivers. However, it is only in the last few years that the importance of the heavy metals has become apparent.

*Fig. 5.6* Mersey at Woolston Weir, Warrington, showing flocs of detergent foam.

A well documented and clear-cut example of the kind of problem which can occur is again provided by the Mersey-Etherow. Factory closures and improvements to sewerage and sewage treatment had brought about such a marked decrease in organic loads by the mid 1970s, that hopes were high for the establishment of a mixed coarse/trout fishery. However, attempts to stock the river met with no success and it soon became apparent that most of the Mersey-Etherow was toxic to fish. The use of caged fish and laboratory fish toxicity tests established the existence of a previously unsuspected source of toxic waste to the river and chemical analysis of water and the moss *Fontinalis anti-pyretica* enabled the poison to be identified as zinc from a specialised paper treatment process (Fig. 5.7). Following this discovery the firm responsible for the input began to reduce the amount of zinc entering

A. <u>Chromium</u>. Concentrations increase markedly downstream of the effluent from Dukinfield S.T.W., which treats tannery waste. High levels were found also in the Etherow, providing evidence for short-term releases from industrial premises.

B. <u>Zinc</u>. The high concentrations in the Etherow were due to continuous input of metal from an industrial source reaching the river via old culverts.

C. <u>Lead</u>. A marked increase occurred on the Tame downstream of the effluent from Denton S.T.W. The high concentrations probably result from battery manufacture.

*Fig. 5.7*  Cr, Zn and Pb in samples of shoot tips (2-cm long) of *Fontinalis antipyretica* collected between October 1979 and January 1980.

the river, but progress was slow because the metal reached the river via
an extensive underground system of derelict culverts and contaminated
aquifers.  However, by late 1980 zinc leaching from the culverts had
decreased markedly (Say *et al.*, 1981), and following the closure of the
works in 1982 it seems likely that trout and coarse fish will be able
to spread up the river in the near future.  The history of the zinc
and organic pollution and its impact on plant and animal life in the
Mersey-Etherow has been summarised by Harding *et al.* (1981).

In seeking to improve river water quality, one of the approaches of
the controlling bodies has been to arrange for industrial wastes to be
diverted into the sewers, so that they may be treated more effectively
in admixture with sewage.  The second approach has been to improve
industrial waste treatment plants and sewage works.  Increased capital
investment and closure of small inefficient sewage works, leading to
new regional treatment facilities, have all contributed to improved
river quality, and the factory closures that have occurred have also
helped to lessen the impact of industrial discharges.  Despite the
reduction in numbers of discharges, the total volume of all effluent
to river has changed little over the years, as the remaining sewage and
industrial works have increased their output and new discharges have
come into being.


## 5.5  THE MERSEY SYSTEM TODAY

### 5.51  INTRODUCTION

It is clear from the previous section that from the beginning of the
Industrial Revolution the Mersey has been subjected to a long history
of severe pollution, with marked and sometimes dramatic changes taking
place from time to time.  Whilst the trend in recent years has been
towards substantial improvement, it is unlikely that a stable situation
has been reached.  A description of present conditions can apply only
to a single point in time, but is needed in order to relate the plant
(5.62) and animal communities (5.63) to their environment.

Most industrial concerns pass their waste effluents to sewage works
for treatment and only the Goyt still receives the impact of a significant number of direct industrial discharges.  As a result, the largest
contribution today to the still relatively poor water quality of the
Mersey catchment comes from the many sewage works and also from untreated sewage overflowing from inadequately sewered areas.

An important feature of the Manchester conurbation, in common with
most long-established urban areas, is the existence of a large number
of overflows on the sewerage system (at least 220 upstream of Irlam
Weir).  The frequency of operation of these overflows is influenced by
the condition and capacity of the sewers they serve and since many of
these sewers are decrepit and overloaded they tend to surcharge to

river at irregular and frequent intervals with a minimum of stimulus
from blockage or rainfall.  Because of their irregular behaviour their
effect on the river is difficult to document and assess.

The water quality of the catchment can be summarised conveniently in
the terms of the classification scheme in current use by water authorities (National Water Council, 1981).  The main chemical criteria
employed are dissolved oxygen, B.O.D. and ammonia, with modifications
where necessary, to allow for biological data, toxic substances and
general water usage.  The classification is used by water authorities
for setting water quality management policies and five classes of river
are recognised:

| | | |
|---|---|---|
| Class 1A | Good quality | Water of high quality suitable for potable supply abstractions ; game or other high class fisheries ; high amenity value. |
| Class 1B | Good quality | Water of less high quality than Class 1A but usable for substantially the same purposes. |
| Class 2 | Fair quality | Waters suitable for potable supply after advanced treatment; supporting reasonably good coarse fisheries ; moderate amenity value. |
| Class 3 | Poor quality | Waters which are polluted to an extent that fish are absent or only sporadically present ; may be used for low grade indus-trial abstraction purposes; considerable potential for further use if cleaned up. |
| Class 4 | Bad quality | Waters which are grossly polluted and are likely to cause nuisance. |

The classification of the Mersey catchment using this scheme (Fig.
5.3:  see p. 119) can be compared with the chemical data presented for
five key river sampling points (Table 5.1).  A description of the more
important influences on the various parts of the river system will help
to explain why so much is of poor or bad water quality.


5.52  MERSEY

In the Mersey-Etherow all that remains of the severe pollution described
in 5.4 is the input from Glossop Sewage Works, which is now operated to
a reasonable standard, and the steadily declining concentrations of
zinc.  The river is also subject to intermittent contamination by other
toxic metals, notably chromium, copper and cadmium, but these probably
rarely reach concentrations high enough to endanger the biota.  Unlike
the situation with zinc, inputs of these metals usually take the form
of short-term contributions from spillages and contaminated surface
run-off to the Glossop Brook.  These incidents are very difficult to

*Table 5.1*  Summary of data on water chemistry at five sites during 1979-1980.  (Total alkalinity and elements other than $O_2$ as mg $l^{-1}$)

| | | Tame u/s of Goyt | Etherow u/s of Goyt | Goyt u/s of Etherow | Goyt u/s of Tame | Mersey u/s of Manchester Ship Canal |
|---|---|---|---|---|---|---|
| discharge | n | 70 | 16 | 2 | 86 | 78 |
| ($m^3 s^{-1}$) | $\bar{x}$ | 4.55 | 3.20 | 2.48 | 5.48 | 13.9 |
| | s.d. | 4.35 | 2.31 | | 4.35 | 12.4 |
| suspended | | 93 | 15 | 2 | 104 | 91 |
| solids | | 34.2 | 16.6 | 25.5 | 18.6 | 46.7 |
| | | 38.04 | 9.15 | | 11.0 | 55.8 |
| conductivity | | 93 | 15 | 2 | 104 | 91 |
| ($\mu S cm^{-1}$) | | 617 | 347 | 415 | 395 | 544 |
| | | 186 | 118 | | 92 | 152 |
| $O_2$ (%) | | 84 | 15 | 1 | 96 | 77 |
| | | 84.0 | 91.3 | 94.3 | 87.1 | 74.8 |
| | | 9.1 | 7.1 | | 8.9 | 13.4 |
| B.O.D. | | 93 | 15 | 2 | 104 | 90 |
| | | 9.26 | 3.95 | 4.3 | 5.10 | 7.3 |
| | | 4.56 | 2.22 | | 3.01 | 4.42 |
| C.O.D. | | 93 | 15 | 2 | 104 | 88 |
| | | 60.4 | 25.7 | 42.5 | 32.6 | 48.5 |
| | | 23.1 | 9.41 | - | 10.8 | 17.3 |
| pH | | 93 | 15 | 2 | 104 | 91 |
| | | 7.24 | 7.16 | 7.45 | 7.23 | 7.18 |
| | | 0.21 | 0.21 | | 0.22 | 0.16 |
| total alk. | | 92 | 15 | 2 | 104 | 90 |
| ($CaCO_3$) | | 91.4 | 47.3 | 72.5 | 64.7 | 93.4 |
| | | 29.2 | 17.7 | | 24.4 | 32.9 |
| Cr | | 59 | 5 | | 48 | 70 |
| | | 0.015 | 0.013 | | 0.017 | 0.015 |
| | | 0.005 | - | | 0.012 | 0.004 |
| Zn | | 81 | 9 | | 53 | 80 |
| | | 0.10 | 0.16 | | 0.07 | 0.09 |
| | | 0.08 | 0.08 | | 0.04 | 0.05 |
| Cd | | 59 | 1 | | 38 | 59 |
| | | 0.006 | 0.007 | | 0.006 | 0.006 |
| | | 0.003 | - | | 0.002 | 0.002 |
| Pb | | 61 | 3 | | 43 | 62 |
| | | 0.06 | 0.03 | | 0.04 | 0.05 |
| | | 0.05 | - | | 0.02 | 0.05 |
| Cl | | 93 | 15 | 2 | 104 | 90 |
| | | 65.6 | 37.6 | 35.5 | 39.0 | 57.8 |
| | | 30.0 | 17.9 | | 20.7 | 24.6 |
| $NH_3$-N | | 93 | 15 | 2 | 104 | 91 |
| | | 1.88 | 0.54 | 0.85 | 0.78 | 2.21 |
| | | 1.29 | 0.37 | | 0.61 | 1.37 |
| $NO_2$-N | | 93 | 15 | 2 | 104 | 91 |
| | | 0.31 | 0.11 | 0.15 | 0.16 | 0.22 |
| | | 0.24 | 0.11 | - | 0.10 | 0.16 |
| $NO_3$-N | | 93 | 15 | 2 | 104 | 91 |
| | | 4.22 | 2.48 | 2.12 | 2.82 | 3.45 |
| | | 1.83 | 1.16 | | 0.78 | 1.15 |
| $PO_4$-P | | 91 | 15 | 2 | 103 | 99 |
| | | 1.19 | 0.34 | 0.27 | 0.48 | 0.88 |
| | | 0.99 | 0.32 | | 0.33 | 0.65 |

detect by analysis of water samples alone, but evidence of their occurrence  may be gained by the chemical analysis of submerged river plants,
which accumulate metals from the water passing over them (Say *et al.*,
1981).

The next section of the river, the Mersey-Goyt, receives additional
organic loading from the Goyt, one large sewage works and a textile
works.  Unfortunately, the benefit of natural river self-purification
is counter-balanced by each polluting input so that water quality
through this reach shows little downstream change.

The same comments apply to the first lowland section of the Mersey
from Stockport to the Manchester Ship Canal.  Contributions of heavy
organic pollution from the Tame, the large sewage works serving Stockport, two smaller sewage works and untreated sewage from a poorly
sewered part of Manchester all combine to depress water quality.  The
situation deteriorates even further where the Mersey joins with the
Manchester Ship Canal, which is itself affected by several badly
polluted Irwell tributaries and imposes an additional organic burden.
A slight recovery takes place before the Mersey finally discharges to
the Estuary at Warrington.


5.53   GOYT

The Goyt is characterised by having more direct industrial discharges
than any other river on the Mersey catchment.  Four textile processing
firms, a paper mill, a manufacturer of brake linings and two sewage
works together provide a heavy organic load to the river.  The largest
industrial contributor discharges kier wash liquor and a significant
amount of chlorine from bleaching processes to the head of the Goyt.
Fortunately, the steep gradient immediately downstream provides excellent aeration in the river and a high rate of self-purification reduces
the effect of the total load of organic material, whilst the chlorine
is rapidly lost by volatilization.  Dilution from a clean tributary
(Sett) helps self-purification and the overall result is a downstream
improvement.  However, the final output of the Goyt to the Mersey is
a poor quality river in need of further improvement; hopefully, present
work on extending the sewage treatment facilities in the valley will
bring this about.


5.54   TAME

In spite of dramatic improvements that have taken place in the last
twenty years (Fig. 5.4), the Tame gives the best impression of what
most of the Mersey catchment would have been like at the end of the
19th century.  From source to the Mersey at Stockport the Tame Valley
is completely urbanised and the river is bounded on both banks by
housing and industrial development.  The situation today is dominated
by the effluents from six sewage works, all treating a wide variety of

industrial wastes and by the frequent operation of several sewer overflows.  Two of the works are grossly overloaded and contribute substantial sub-standard effluents to the river.

The upper stretch, which supports a fishery, is relatively unpolluted,
but a paper works and then the six sewage works discharge to the river
and the water quality deteriorates progressively downstream.  In the
lower stretch untreated sewage and continuous growths of sewage-fungus
dominate the scene.

Although less well documented than the Mersey-Etherow, the Tame also
provides examples of pollution by metals from industrial sources.  The
treated effluents from Dukinfield and Denton Sewage Works contain
relatively high concentrations of chromium and lead, respectively, due
to the wastes from tanning processes and battery manufacture.  The
effects of these two inputs on the metal composition of *Fontinalis* can
be seen clearly in Fig. 5.7, and their impact can be traced downstream
into the Mersey below Stockport despite dilution from the Mersey-Goyt.

## 5.6  LIFE IN THE RIVER

### 5.61  INTRODUCTION

Most of the data used for the following descriptions are taken from
surveys conducted since 1970 by the North West Water Authority and
previously by the Mersey and Weaver River Authority.  Analysis of invertebrate communities has proved an invaluable tool in the day-to-day
management of water quality, but it is only with the passage of time
that the value of a long-term review has become apparent.  The authors,
as two members of a large team of scientists, have found that the
exercise of describing the current situation has greatly enhanced their
understanding of the biology of the Mersey - a benefit which has been
passed straight back into the operation of routine river surveillance.

In the most recent years of this short history of biological investigation further techniques have been developed to expand the Water
Authority's ability to monitor and understand the changes taking place.
Analysis of heavy metals in river plants and surveys of plant distribution have now become part of the routine monitoring programme and
together with fish surveys give a comprehensive picture.

### 5.62  PLANTS

#### 5.621 General comments

As is the case with many British rivers the Mersey has until recently
received very little attention from local botanists, and historical
documentation of plants in the catchment is non-existent.  In pre-

historic times the whole Mersey valley was densely wooded and growths
of photosynthetic plants in the river itself must have been severely
restricted by dense shading.  However, as man began to clear the
forests for farming and timber, increased illumination, accompanied by
increased nutrient run-off from the cleared land, must have led to a
marked increase in submerged plant growth throughout the Mersey system.
A few clues to the nature of these early plant communities can be seen
in the present vegetation, especially in the upper reaches of the
Mersey-Etherow and Goyt, which have never been subject to the kind of
pollution affecting the rivers further downstream.  In both cases the
bed is rocky and the vegetation is dominated by mosses and liverworts,
with four species, *Hygrohypnum ochraceum*, *Fontinalis squamosa*,
*Racomitrium aciculare* and *Scapania undulata*, being especially characteristic.  Since occasional specimens of these species have been
recorded recently in fast-flowing riffles further downstream in both
the Mersey-Etherow and Goyt, it does seem probable that these four
bryophytes were abundant in both rivers prior to the Industrial Revolution.

It is much more difficult to obtain evidence concerning the possible
early distribution of rooted aquatic angiosperms in the catchment.
Although the middle and lower Mersey has probably always contained areas
of gravel, sand and mud suitable for the establishment of rooted plants,
none have remained unaffected by pollution or artificial channel management.  Only one angiosperm, *Ranunculus fluitans*, has a present-day
distribution that suggests it may have been present in the catchment
before the onset of severe pollution.  Surveys of the distribution of
*R. fluitans* throughout the N.W.W.A. area (Harding, 1981) indicate that
although the species is unable to form large beds in rivers affected by
severe organic pollution, occasional plants may survive for long periods
in such situations even in competition with dense growths of pollution-
tolerant plants.  Isolated plants of *R. fluitans* occur on gravel substrata in the middle stretches of the Goyt, Mersey-Goyt and Mersey, and
it is tempting to speculate that these plants represent the last
survivors from beds that may have flourished in the Mersey system from
the time the forests were first cleared to the beginning of the Industrial Revolution.

As pollution intensified during the 19th century there seems little
doubt that many submerged plants were eliminated rapidly from the
system, or became severely restricted in their occurrence.  Corbett's
(1907) brief description of the plantless state of the Irwell in
Manchester in the mid 1880's can be used to infer that similar conditions prevailed in the Mersey at that time, and his obvious delight at
the appearance of the pollution-tolerant alga *Vaucheria* in the Irwell
in 1905 can be taken as evidence that even the most resilient photosynthetic plants were absent during the latter half of the 19th century.
In the face of continuous organic pollution heterotrophs must have
played an important part in the ecology of the river at that time, and
even today visually obvious growths of sewage-fungus occur regularly
in the lower Tame and intermittently elsewhere.  Stretches of river

affected by such growths often also support the tolerant filamentous alga *Stigeoclonium*, but there is no documentary evidence to suggest that this alga was present when pollution was more severe. Similarly, the pollution-tolerant angiosperm *Potamogeton pectinatus*, which has dominated the vegetation of the lower Tame, Mersey-Goyt and Mersey for at least ten years, has probably increased in abundance as organic pollution has decreased.

Apart from pockets of woodland on the lower Mersey-Etherow and Goyt, the catchment today is well illuminated, and with the exception of a small area of the Goyt affected by chlorine pollution, photosynthetic plants occur throughout the system. However, chemical conditions are still changing quite rapidly, and it seems likely that only the most upstream parts of the Mersey-Etherow, Tame and Goyt support plant communities that have reached stable equilibria with their environment. Despite this the system today contains a wide variety of plant communities influenced to varying degrees by the different types of pollution that still occur. Several workers (e.g. Haslam, 1978; Westlake, 1975) have stressed the importance of physical factors such as bedrock geology and channel topography on plant communities and the following brief description of the vegetation of the Mersey attempts to distinguish between these various influences.

*5.622 Mersey*

Thirty-three species of macrophyte were recorded from the Mersey-Etherow during 1978 (Harding, 1979). Submerged or emergent angiosperms were present at all sites, but the vegetation was in all cases dominated by attached filamentous algae (*Cladophora, Vaucheria*) or by submerged bryophytes. This is probably due to the largely rocky nature of the river bed and the relatively high current speeds at most sites, but it is also possible that rooted aquatic macrophytes have not yet realised their full potential range of distribution within the river. Improvements in the water quality of the Mersey-Etherow have taken place much more quickly in recent years than in the Mersey-Goyt or Mersey downstream, and whilst communities of algae and bryophytes probably change quickly in response to these improvements, some of the angiosperms recorded from the lower part of the Mersey-Etherow have been established for only a relatively short period (Harding *et al.*, 1981). Conditions at several sites appeared suitable for the growth of certain angiosperms (e.g. *Potamogeton crispus, P. pectinatus, P. natans, Ranunculus fluitans*) that were recorded elsewhere in the Tame and Goyt, and plants such as these might be expected to colonise the river in the future.

A downstream zonation of macrophytes was apparent in the Mersey-Etherow in 1978, and several species showed clearly defined ranges of occurrence. This zonation is illustrated most clearly by the seven submerged bryophytes. Two, *Scapania undulata* and *Racomitrium aciculare*, were confined almost exclusively to the top site, upstream

of any inputs of treated sewage.  The relative abundances of both
*Fontinalis squamosa* and *Hygrohypnum ochraceum* were also greater at this
site than further downstream.  In contrast, *Rhynchostegium riparioides*
was absent here, but appeared as abundant growths immediately downstream of a minor sewage effluent and continued as the dominant bryophyte at all other sites in the Mersey-Etherow.  *Fontinalis antipyretica*
was recorded only at sites downstream of Glossop Brook, the tributary
receiving intermittent pollution from the Glossop area.  Another moss,
*Amblystegium riparium* is particularly tolerant of organic pollution in
rivers and appears to be a reliable indicator of marked nutrient enrichment within the area of the North West Water Authority.  Its presence
in the Mersey-Etherow was restricted to sites downstream from Glossop
Sewage Works.

   This zonation of bryophyte species appeared to be largely a response
to downstream increases in nutrient concentrations, and none of the
observed changes occurred in the first part of the Mersey-Etherow
affected by high concentrations of zinc.  In fact, the only obvious
sign of the toxic effect of zinc on plants was the absence of the alga
*Cladophora* immediately downstream of the zinc input, in a section which
carries levels of nutrients that should be high enough to ensure its
success.  The marked sensitivity of *Cladophora* to zinc (Whitton, 1979)
seems the only plausible explanation for its absence from the shallow,
well-illuminated riffles that occur in this part of the river.  It seems
very likely, therefore, that *Cladophora* will spread upstream as further
reductions in zinc pollution take place.  *Cladophora* was also absent
from the lower reaches of Glossop Brook, which although not carrying
high concentrations of zinc, does receive intermittent chromium
pollution (Say *et al.*, 1981).

   Despite the numerous polluting discharges and sewer overflows affecting the Mersey and the catchment upstream of Stockport, the eight
sites surveyed on the Mersey-Goyt and Mersey downstream of Stockport
supported a fairly wide variety of macrophytes in 1978 (26 species
recorded).  A high proportion of the species found in the lower Goyt
and lower Tame were recorded also in the Mersey-Goyt and the Mersey,
and it seems probable that small populations of even quite sensitive
species are maintained in the main river by continuous downstream
inoculation (Fig. 5.6).  *Potamogeton pectinatus* (inoculated from the
Tame) and *Cladophora* dominated the vegetation at all sites on the
Mersey and the other species were mostly present at a few sites only.


*5.623 Goyt*

Forty species of macrophyte were recorded from the Goyt.  The uppermost
site surveyed, upstream of Whaley Bridge, supported a very diverse and
dense vegetation dominated by mosses (especially *Rhynchostegium
riparioides* and *Fontinalis squamosa*), reflecting the excellent water
quality in this part of the river and the stable nature of the rocky
bed.

At Whaley Bridge, where the river receives the severely polluting effluent from the textile works containing organic waste and free chlorine from bleaching processes, true aquatic plants were absent and only bankside macrophytes (e.g. *Agrostis stolonifera* and *Phalaris arundinacea*) were recorded in the water. Even these had an unhealthy bleached appearance. Progressively downstream the vegetation showed a steady increase in diversity and cover, with first the pollution-tolerant *Amblystegium* appearing and then the slightly less-tolerant *Cladophora* and *Rhynchostegium*. Further improvement by dilution from a small clean tributary, the Sett, was indicated by an increase in the abundance of *Rhynchostegium* relative to that of *Amblystegium*, but high nutrient concentrations throughout the Goyt resulted in an extensive cover of *Cladophora* from the Goyt to the Mersey at Stockport. Plants of *Ranunculus fluitans* were recorded at sites downstream of the Sett and into the Mersey-Goyt, although nowhere did this species reach high cover values.

*5.624 Tame*

Upstream of any sources of pollution the Tame had a vegetation that was fairly typical of a small upland stream flowing over millstone grit; that is, it was dominated by bryophytes (e.g. *Fontinalis antipyretica, Rhynchostegium*) and with relatively few rooted angiosperms. There was some indication, however, that even these upper sites were subject to nutrient enrichment, with quite large growths of *Cladophora* in places. Further enrichment was indicated downstream as, successively, the abundance of the *Cladophora* increased, *Scapania undulata* disappeared, *Amblystegium* appeared and *Rhynchostegium* started to decrease in abundance. Downstream of the entry of Chew Brook, which at the time of the survey was heavily polluted by organic waste from a paper mill, there was a particularly marked decrease in the abundance of *Rhynchostegium*, and a corresponding increase in the cover of the more tolerant *Amblystegium*.

Several rooted macrophytes (e.g. *Potamogeton crispus, P. natans, Sagittaria sagittifolia, Sparganium emersum, S. erectum*) were recorded in the mid-section of the Tame in artificially created slow-flowing stretches with soft muddy beds. However, turbulent riffles alternated with these much slower stretches downstream to the confluence with the Mersey-Goyt, and *Amblystegium, Fontinalis antipyretica* and *Cladophora* continued to occur. In the lower Tame, beds of *Potamogeton pectinatus* were common and most sites were dominated by either this or *Cladophora*. As in the Mersey-Etherow, there was a stretch in the lower Tame below Dukinfield Sewage Works in which *Cladophora* was absent from an entire 0.5 km length (Fig. 5.8). Suitable substrata for *Cladophora* were abundant at the site, and the alga reappeared at the next site downstream. It seems probable that the high chromium concentrations revealed by analysis of *Fontinalis antipyretica* collected from the site in 1979 may have been responsible for the discontinuous distribution of *Cladophora* in 1978. Surveys of other rivers in N-W England (Harding,

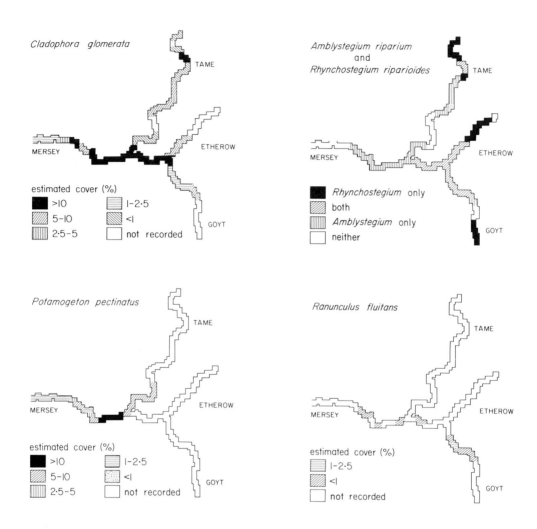

*Fig. 5.8* Distribution and estimated cover of selected macrophytes in the Mersey catchment, as shown by a survey of 0.5 km lengths of river in summer 1978.

1979) have shown that *Cladophora* may be absent from rivers carrying
very high loads of organic material with low levels of metals, but it
seems unlikely that the absence of the alga from this particular site
is due to organic pollution, since it reappeared further downstream
despite inputs from several other large sewage works.

### 5.625 *Changing distribution*

It will be clear from the above descriptions of vegetation in the
catchment that plant communities have undergone rapid change in the
system and that it is unlikely that any species has attained a pattern
of distribution within the system that will remain stable in the near
future.  This picture of instability is reinforced by examination of the
distribution of individual species throughout the catchment.  For
example, it can be seen in Fig. 5.8 that five species (*Cladophora
glomerata, Amblystegium riparium, Rhynchostegium riparioides, Potamogeton pectinatus, Ranunculus fluitans*) showed highly discontinuous patterns of distribution and abundance.  These are likely to change
markedly in response to pollution control measures in the near future,
as well as by inoculation with other rooted species from outside the
catchment or between the tributaries of the Mersey.

In addition to the effects of changing water quality, plant growth
in the Mersey catchment as a whole has obviously been influenced by the
introduction of alien species in the past.  For example, the banks of
the Mersey through Stockport are coloured pink in the summer by *Impatiens glandulifera* which first spread in England in the 19th century.
During 1980 dense beds of the exotic species *Elodea nuttallii* were
recorded for the first time in the Mersey at Warrington.  This macrophyte is at present invading rivers and standing waters in N-W. England,
and it will be interesting to observe its future success at colonising
rivers of the Mersey catchment.

## 5.63   INVERTEBRATES

### 5.631 *General comments*

Of all the facets of plant and animal life in the Mersey the distribution of invertebrates perhaps gives the best demonstration of the
effect of man's activities on the catchment.  The Upper Goyt Valley
is the most unspoiled part of the system and can be used to suggest
the conditions of former times.  The wide range of invertebrate
species present today is usually dominated by *Gammarus pulex* and
representatives of the Plecoptera, Ephemeroptera, Trichoptera and
Chironomidae and is characteristic of many clean upland rivers in other
parts of Britain.  Prior to the Industrial Revolution it is likely that
such a fauna would have been common throughout much of the upper catchment, at least downstream to the head of the Mersey lowland plain where
Stockport now stands.  Further downstream, slower-water communities

would probably have been more prominent, but there are no clues to
their possible composition because of the absence today of comparable
lowland unaffected by pollution.

From the beginning of biological survey work in 1970, it was known that
a wide range of water quality conditions prevailed and that many stretches were also subject to short-term fluctuations.  The Trent Biotic
Index (T.B.I.) (Woodiwiss, 1964), which depends on a sequential loss of
key invertebrate groups with increasing organic pollution, was quickly
established as a very useful means of transmitting the results of biological surveys to non-biological water quality control staff.  Throughout the period all biological samples have been examined in the field in
order to provide a quick appraisal of river conditions.  Changes in
T.B.I. values since 1970 (Fig. 5.9) suggest that water quality changes
have taken place.

Fig. 5.9  Trent Biotic Index values from 1969 to
1980 for selected invertebrate sampling sites.

## 5.632 Mersey

The headwaters of the Mersey-Etherow flow from peaty moorland and are
generally somewhat unproductive because of the soft, slightly acidic
water (pH 5.5 - 7.0).  The invertebrate community present upstream of
all known sources of pollution usually includes Nemouridae, *Leuctra*
spp., Polycentropidae, *Rhyacophila* spp., *Baetis* spp., Dytiscidae and
Chironomidae.  An increase of faunal variety, from the seven taxa in
1971 to 25 taxa in 1980 and a gradual rise of T.B.I. values to 7-9

(Fig. 5.9) suggest that conditions for invertebrates have improved, but without any readily identifiable cause.

During the period of review the river upstream of Glossop Brook has shown marked improvement from a community once dominated by tubificid worms and chironomid larvae and often supporting dense growths of sewage-fungus. In 1971 there was a reduction of organic inputs upstream, but continuing minor pollution still produces evidence in the fauna of organic enrichment. Several of the upstream groups, including Plecoptera and Dytiscidae, are suppressed, leaving *Baetis, Rhyacophila,* Polycentropidae and Chironomidae dominant, together with the Tubificidae which appear as a response to the organic inputs. The effects of zinc superimposed on organic pollution are described on p.139. Whilst accidental industrial pollution and the operation of sewer overflows on the Glossop Brook have an adverse effect on the Mersey-Etherow, most organic pollution comes from Glossop Sewage Works further downstream. The fauna below the works is usually dominated by *Baetis, Asellus,* Chironomidae and Tubificidae. There has been a positive benefit to the river from extensions completed in 1973 and various cleaner-water organisms have been recorded since then, with T.B.I. values showing an upward trend throughout the period to the present range of 5 to 8.

Steep gradients downstream to the Goyt confluence produce high self-purification rates with improved water quality, resulting in Trichoptera and Simuliidae being part of the dominant fauna. Up to the early part of the 1970s the downstream improvement had been reversed by the addition of the poorer quality Goyt, but a gradual restoration of water quality of this tributary has had a beneficial effect on the Mersey-Goyt downstream, pushing T.B.I. values up to 6-8. Whilst the invertebrate community continues to be dominated by the more pollution-tolerant taxa such as *Baetis* spp., *Asellus,* Chironomidae and Tubificidae, clean-water indicators such as Trichoptera and Ephemeroptera are usually present also. These latter organisms are eliminated lower downstream by the effluent from Hazel Grove Sewage Works, but even here a long-term improvement is still evident.

The next influence comes from the badly polluted Tame and below Stockport the Mersey fails to make any recovery as the beneficial effects of self-purification are cancelled out by additional organic loadings from three sewage works and a number of frequently or continuously operating sewer overflows. The biological picture downstream to the Manchester Ship Canal is thus fairly uniform with T.B.I. values at all survey sites varying between 4 and 6 (Fig. 5.9). The community of pollution-tolerant organisms consists of *Baetis* spp., *Asellus,* *Erpobdella* spp., *Lymnaea,* Chironomidae and Tubificidae.

Although the present condition of the Mersey is unsatisfactory it was in a much worse state in the early 1970s. Downstream from Stockport Sewage Works the fauna was limited to Chironomidae and Tubificidae, with *Asellus* appearing only occasionally. Improvements to the works were completed in 1973 and produced an immediate response in the river.

*Baetis* spp. appeared in 1974 for the first time and marked the start of the upward trend evident in Fig. 5.9; the total number of invertebrate species recorded has been steadily increasing ever since.

Little is known about the biology of the Mersey from the confluence with the Manchester Ship Canal, but occasional samples from just above the tidal limit at Warrington have only ever produced chironomids and tubificid worms, thus confirming the poor water quality indicated by the chemical data.

### 5.633 *Goyt*

Downstream from the pollution-free stretch of the upper Goyt the effluent from the textile works at Whaley Bridge has a particularly severe impact on the fauna. The deterioration to a Tubificidae/ Chironomidae community makes the biggest contrast of any on the Mersey catchment with T.B.I. values dropping from 8-10 upstream to 1-3 downstream. Its severity may be due in part to the free chlorine discharged.

Despite a succession of textile, papermill and sewage effluents adding to the organic load to the river, changes in the faunal community demonstrate a downstream recovery which is due to the excellent aeration provided by the steep gradient. The first stages are seen in a limited appearance of *Baetis* spp. and *Asellus*, but conditions improve with the diluting effects of the Sett. *Baetis, Asellus* and Simuliidae become firmly established, although high numbers of Chironomidae and Tubificidae continue as a direct response to the polluted river upstream. In recent years the faunal variety below the Sett has been steadily increasing with the tributary providing an inoculation of cleaner-water invertebrates, such as Plecoptera, Heptageniidae, *Ephemerella ignita* and *Rhyacophila* spp. so that T.B.I. values have risen to 7-9. Although a pollution-tolerant fauna still dominates this stretch, the increasing T.B.I. values over the last few years (Fig. 5.9) provide a convincing picture of steady improvement. This is due probably to the gradually decreasing volumes of industrial effluent combined with tighter quality control.

### 5.634 *Tame*

In the absence of any major discharges the type of invertebrate community of the headwaters of the Tame down to the Chew Brook has changed little over the last ten years. The water quality indicated by T.B.I. values of 5-10 and a dominant fauna of *Baetis, Ephemerella, Rhyacophila* and Chironomidae falls well short of that of both the Mersey-Etherow and the Goyt headwaters. *Asellus* appears in the lower part of the stretch confirming the slight downstream deterioration shown by the sequence of plant species and probably results from contaminated surface water run-off. The immediate response to the effluents from

Saddleworth Sewage Works and the paper mill on Chew Brook is the loss
of *Ephemerella* and *Rhyacophila*. *Baetis* spp. and *Asellus* have become
established only since the completion of extensions to the sewage works
in 1977.

From this point downstream a succession of sewage effluents, sewer
overflows and industrially polluted tributaries causes a gradual deterioration. T.B.I. values generally remain within the range 2 - 5, due
to a pollution-tolerant fauna of *Baetis* spp., *Asellus, Erpobdella* spp.,
*Limnaea peregra,* Chironomidae and Tubificidae. The worst conditions
prevail near the bottom end of the Tame where the effluent from Hyde
Sewage Works eliminates all but Chironomidae and Tubificidae, and
extensive growths of sewage-fungus are recorded throughout the year.
The reappearance of *Asellus* upstream of the confluence with the Goyt
marks a final slight recovery, though its distribution throughout the
lower Tame serves to emphasise the polluted state of this tributary.


*5.635 Changing distribution*

It will be clear that the water quality of the Mersey has undergone
many changes over a long period of time and that both plant and invertebrate communities reflect these changes. The distribution of individual species throughout the catchment provides additional information
for interpreting survey data and forecasting future trends.

As an example, Harding *et al.* (1981), looking at the Mersey-Etherow
in isolation, failed to detect any abnormalities of downstream distribution associated with the zinc pollution problem, but Fig. 5.10
highlights some quite striking anomalies in distribution through the
catchment as a whole. In view of the relatively high organic content
of the river the absence or scarcity of *Ancylus fluviatilis, Limnaea
peregra, Erpobdella* spp., *Asellus* and Simulidae is somewhat surprising.
The absence of *Asellus* downstream from Glossop Brook is particularly
notable, considering it has been present in the tributary on at least
50% of all sampling occasions since 1970. Studies of zinc and lead
polluted rivers in West Wales have shown molluscs, leeches, worms,
crustaceans and fish to be particularly susceptible to these heavy
metals (Hynes, 1960). The range of zinc concentrations (0.25 - 0.5 mg
$l^{-1}$) determined in the Mersey-Etherow in 1979 and 1980 (Say *et al.*,
1981) has also been found to limit the distribution of molluscs in
West Wales (Jones, 1940; Newton, 1944). It seems highly probable,
therefore, that the limited distributions mentioned for the Mersey-
Etherow are caused by zinc. If this is so, then as the forecasted
decrease in concentration takes place these animals should extend their
distribution upstream. *Gammarus pulex*, which is remarkably scarce
throughout the rivers of the Mersey catchment (Holland, 1976), appeared
in the Mersey-Etherow for the first time in 1979 and may be included in
this expected upstream invasion.

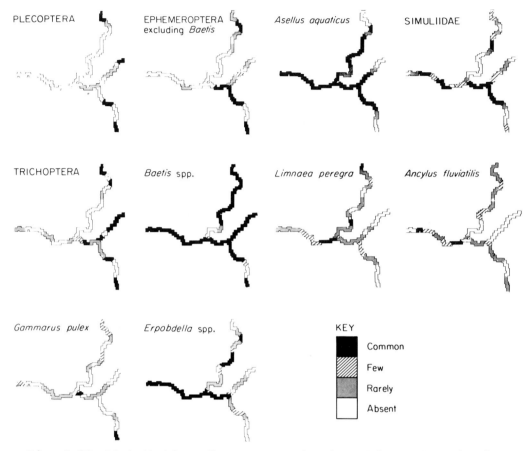

*Fig. 5.10* Distribution of representative invertebrate taxa in the Mersey catchment, 1978-1980. (For details, see text)

Of the various invertebrates in the river upstream of the zinc input, *Baetis vernus, Rhyacophila* and Polycentropidae appear to be resistant to the zinc pollution downstream. Since *Baetis* spp. and the Trichoptera have been recorded frequently since 1971, it is obvious they have a resistance to the even higher zinc concentrations which must have occurred in earlier years before control measures started to take effect. The decreased abundance of Plecoptera is probably solely a response to the presence of organic material, as Jones (1940) has shown them to have a high tolerance to zinc.

The lower Tame appears to provide a parallel to the Mersey-Etherow in that *Erpodella, Lymnaea* and *Ancylus* are unexpectedly absent or scarce (Fig. 5.10). It seems possible that a similar explanation, the influence of heavy metals, applies here too; the metals in this case are chromium and lead from tannery and battery manufacturing processes.

5.64   FISH

Apart from Corbett's (1907) description of fish life in the Irwell and
the Mersey below Salford in the early years of the Industrial Revolution, there is no information on the distribution of fish on the Mersey
before pollution took its toll.  The physical types of river suggest
that the entire catchment upstream of Stockport would have supported a
salmonid fishery, with the probable inclusion of fast-water coarse
fish such as grayling, chub and dace.  In the slower waters of the lowland Mersey downstream from Stockport conditions would have been more
suitable for coarse fish, but would still have been available for
migrating salmon and sea trout.

Fish were, of course, the first noticeable casualties of the
declining water quality in the Industrial Revolution and over a short
time interval in the mid 1800s fish populations were destroyed in all
but the top reaches of the upper catchment.  Ever since those years
poor water quality has continued to prohibit the re-establishment of
viable fisheries and it is only in the last ten years or so that any
recovery has taken place.  Part of the biological investigations which
have been in progress on the Mersey since 1970 have been directed
towards the fishless state of the catchment.  Between 1973 and 1978
trout in cages placed at strategic sites were found in some instances
to survive only a few days or even only hours (Fig. 5.11).  These
short survival times emphasised how wide the gap was between the
existing water quality and that required for supporting fish life.
As part of the continuing investigations, samples of industrial and
sewage wastes were subjected to static laboratory fish toxicity tests.
The results showed that few sewage effluents were even slightly toxic
and those that were had no toxic significance after dilution in the
river.  However, several industrial effluents were found to contain
highly toxic constituents and these provided explanations for the
fishless state of the upper Goyt (textile works) and the upper Tame
(paper mill).  The toxic concentrations of zinc in the Mersey-Etherow
became known only as a result of the moss analyses.  Elsewhere, it
was concluded from the chemical data that low dissolved oxygen concentrations caused by sewer overflows and poor quality sewage effluent
were responsible for the lethal conditions.

The extent of present pollution damage to the fisheries is shown in
Fig. 5.12.  River sections regularly fished by anglers are limited to
the headwaters of the Mersey-Etherow, Goyt and Tame, with short
stretches lower downstream in the Mersey-Etherow and Goyt offering
newly established fisheries.  Upstream of the zinc pollution on the
Mersey-Etherow natural perch populations and artificially maintained
stocks of brown and rainbow trout form the basis of a restricted
fishery.  Another limited but natural brown trout fishery with bullheads and stone loaches occupies a short reach of the Goyt upstream of
Whaley Bridge.  A similar fishery exists in the top stretch of the Tame
upstream of the Chew Brook with several coarse fish (roach, perch,
grayling, chub, carp) being recorded in small numbers from time to time,

*Fig. 5.11*  Results of fish survival tests in cages during surveys from 1974-1978 (Mersey-Etherow, 11 surveys; Mersey, 3 surveys; Tame, 17 surveys).

*Fig. 5.12*  Distribution of fish in 1980, based both on electro-fishing surveys and information provided by anglers.

probably as a short-lived result of the stocking activities of local angling clubs.

The few changes which have taken place over the last ten years were preceded by several unsuccessful attempts to establish fisheries in the Mersey-Etherow at the Goyt confluence and the Mersey below Stockport under unsuitable water quality conditions.  Since 1975, however, the declining zinc concentrations in the Mersey-Etherow have permitted the establishment of a fishery dominated by gudgeon and accompanied by small numbers of other species (perch, brown trout, roach, tench) near the Goyt confluence.

After the closure of a paper mill on the Sett in 1970, trout and grayling became established in the tributary through to the confluence with the Goyt.  Over the last five years, as conditions in the Goyt have gradually improved, the fish population of the Sett has extended into the main river for a short distance downstream of the confluence, and has been permanently resident there since 1980.

In the near future it is anticipated that fish in the Mersey-Etherow will move upstream as zinc concentrations decrease further.  It is also hoped that the water quality of the Goyt will continue to improve and thus allow the fish to extend their distribution.  The presence of sticklebacks in the Mersey-Goyt holds out good prospects for this

part of the river system.  Whilst sticklebacks in the mid reaches of
the Tame point to the near suitability of the river for coarse fish,
there are, in fact, no immediate prospects for improvement in water
quality.

## 5.7  DISCUSSION

A 200-year history of development of industry and housing on a large
part of the Mersey catchment has left a legacy of pollution problems
which makes the Mersey one of the worst systems in Britain today.
Man's activities have produced an example of a river system heavily
used in both upland and lowland stretches.  The various rivers are
subject to wide and rapid fluctuations in flow and the pollution
created by the large volume of sewage and industrial wastes has a
particularly severe impact on flora and fauna at times of low flow.
The present account (5.6) of plant and animal life illustrates both
the marked differences in water quality across the catchment and also
the many changes which have taken place with time.  Indeed, from the
onset of the Industrial Revolution stable conditions have probably
never been maintained anywhere for long, and it is also unlikely that
a state of equilibrium will be achieved in the near future in view of
further expected improvements.

The main thrust of biological investigation has from the outset in
1970 centred on examination of invertebrate communities.  The response
of animal life to different kinds of pollution has provided an
invaluable tool for monitoring water quality.  As other lines of
investigation have been introduced, understanding of the life processes
in the river has increased.  As an example, the macrophyte survey
work initiated in the area in 1978 has shown that the interplay of
distribution of the two bryophytes, *Amblystegium riparium* and
*Rhynchostegium riparioides*, can be used in a similar fashion to describe water quality.  Increasing organic pollution is suggested by
the sequence:  neither present $\longrightarrow$ *Rhynchostegium* only $\longrightarrow$ both
present $\longrightarrow$ *Amblystegium* only $\longrightarrow$ neither present (Fig. 5.8).  The
high tolerance of these bryophytes to heavy metals removes what would
otherwise be a complicating factor.  The alga *Cladophora* on the other
hand, is sensitive to heavy metal contamination and gaps in its distribution provide a useful pointer for further investigation.

Although each biological survey technique has been developed singly,
the results can be brought together to produce an integrated picture
of the biology of the river system.  The invertebrates have been seen
to respond most quickly to changes in water quality, whilst the macrophyte communities change more slowly and present a long-term summary
of conditions.  Fish populations are the last to recover in response
to improving water quality.  Secure fisheries will be a most worthy
end-product of several decades of intensive effort to restore the
river system to an acceptable quality.

*Acknowledgements*

The authors are grateful to the North West Water Authority for permission to publish the data used in this paper. They also acknowledge their debt to the many colleagues who have contributed physical help and constructive discussion to what is essentially a team effort.

REFERENCES

Bracegirdle C. (1973) *The Dark River*, 175 pp. John Sherratt & Son, Altrincham.
Corbett J. (1907) *The River Irwell*, 168 pp. Abel Heywood & Son, Manchester.
Haslam S.M. (1978) *River Plants*, 396 pp. Cambridge U.P., Cambridge.
Harding J.P.C. (1979) *River Macrophytes of the Mersey and Ribble Basins Summer 1978*. North West Water Authority Rivers Division, Ref. No. TS-BS-79-1. British Lending Library loan collection.
Harding J.P.C. (1981) *Macrophytes as Monitors of River Quality in the Southern N.W.W.A. Area*. North West Water Authority Rivers Division, Ref. No. TS-BS-81-2. British Lending Library loan collection.
Harding J.P.C., Say P.J. & Whitton B.A. (1981) River Etherow: plants and animals of a river recovering from pollution. *Naturalist, Hull* 106, 15-31.
Holland D.G. (1976) The distribution of the freshwater Malacostraca in the area of the Mersey and Weaver River Authority. *Freshwat. Biol.* 6, 265-76.
Hynes H.B.N. (1960) *The Biology of Polluted Waters*, 202 pp. Liverpool U.P., Liverpool.
Jones J.R.E. (1940) A study of the zinc-polluted River Ystwyth in north Cardiganshire, Wales. *Ann. appl. Biol.* 27, 368-78.
National Water Council (1981) *River Quality: the 1980 Survey and Future Outlook*, 39 pp. National Water Council.
Newton L. (1944) Pollution of the rivers of West Wales by lead and zinc mine effluent. *Ann. appl. Biol.* 31, 1-11.
Palmer W. (1950) *The River Mersey*, revised edition, 250 pp. Robert Hale, London.
Pearson S. (1939) The Mersey and Irwell Basin. *The Royal Engineers Journal*. March, 1-12.
Say P.J., Harding J.P.C. & Whitton B.A. (1981) Aquatic mosses as monitors of heavy metal contamination in the River Etherow, Great Britain. *Environ. Pollut. Ser.B.* 2, 295-307.
Westlake D.F. (1975) Macrophytes. In: Whitton B.A. (Ed.) *River Ecology*, 725 pp., pp. 106-28. Blackwell, Oxford.
Whitton B.A. (1979) Plants as indicators of river water quality. In: James A. & Evison L. (Eds) *Biological Indicators of Water Quality*, pp. 5/1-5/34. Wiley, Chichester.
Woodiwiss F.S. (1964) The biological system of stream classification used by the Trent River Board. *Chemy Ind.* 11, 443-47.

# 6 · Tees

## B. A. WHITTON & D. T. CRISP

| | | |
|---|---|---|
| 6.1 | Introduction | 146 |
| 6.2 | General data | 148 |
| 6.3 | Geography and geology | 150 |
| 6.4 | Hydrology | 151 |
| | 6.41  Discharge | 151 |
| | 6.42  Transport of eroded peat | 151 |
| 6.5 | Physico-chemical features | 153 |
| 6.6 | Plants | 155 |
| | 6.61  Sources of information | 155 |
| | 6.62  Phytobenthos and macrophytes | 155 |
| | 6.63  Phytoplankton | 161 |
| 6.7 | Invertebrates | 162 |
| | 6.71  Sources of information | 162 |
| | 6.72  The fauna and its distribution | 163 |
| | 6.73  Drift at upstream sites | 163 |
| | 6.74  Influence of impoundment and regulation downstream of Cow Green Reservoir | 164 |
| | 6.75  Comments on individual species | 165 |
| 6.8 | Fish | 166 |
| | 6.81  Sources of information | 166 |
| | 6.82  Species distribution | 166 |
| | 6.83  Population, density, growth and production | 167 |
| | 6.84  The Tees fishery and pollution | 169 |
| | 6.85  Effects of regulation | 171 |
| 6.9 | Concluding comments | 171 |
| | References | 173 |

## 6.1  INTRODUCTION

The Tees is one of a number of rivers draining the eastern slopes of the Pennines in northern England.  Three in particular, the Tyne, the Wear and the Tees, have been the subject of a wide range of studies (Horne, 1977).  All three are similar in that the rivers are predominantly fast-flowing and relatively unpolluted for much of their length; they do however become highly polluted towards the estuaries, where much of the population and industry is located.

Just over half a century ago, the Tees, the southern-most of these three rivers, received the first detailed survey of any British river. The authors (Butcher *et al*, 1937) showed that the inflow of a downstream tributary, the Skerne (Fig. 6.1), with very hard and grossly polluted water, led to major changes in the main river.  This study has been widely quoted in the literature since then, although the condition of the Skerne has long been improved.

The upper part of the Tees catchment became a focus for research in the late 1960s and early 1970s.  The construction of a new reservoir on the main axis of the river at Cow Green threatened an area of arctic-alpine vegetation famous for its flowering plant 'rarities'.  Largely as the result of a gift of £100000 from Imperial Chemical Industries Ltd., many studies were made in Upper Teesdale, with the aims of establishing just what was so special about the area and advising on conservation of the remaining parts of high biological importance.  Some studies on streams were included.  In spite of the many interesting papers (see Clapham, 1978), no clear explanation for the rarities emerged, though the post-glacial history of the area, severe winter climate, exposed surfaces of highly metamorphosed limestone and perhaps also the presence of old mine workings are all important factors.  The changes brought about by the Cow Green reservoir on the downstream Tees are one of the topics discussed in the present chapter.

A third burst of research took place in the mid-1970s, this time on the main river.  A major water transfer scheme was planned to bring water from the Tyne to both the Wear and the Tees (Burston & Coats, 1975) and completed in 1983.  This involved the construction of the largest reservoir in Britain (Kielder Water) on the N. Tyne, abstraction of water from the Tyne and its transfer southwards by means of a tunnel.  Apart from initial tests, water has so far been transferred to the Tees only once.

In addition to research stimulated by special problems, fundamental studies on hydrology, chemistry and biology have been made by the Freshwater Biological Association, the University of Durham and other organizations.  A number of these studies have dealt with streams in the Moor House National Nature Reserve, located in the upper part of the catchment.  Hydrological records and routine chemical surveys have been made by the responsible water management body for many years.

Useful summaries are given in the Annual Reports of the Wear and Tees River Board (1951-1965) and of the Northumbrian River Authority (1967-1972).

In spite of the fact that there have probably been more scientific publications on the Tees and its tributaries than on any other British river, there has never been an integrated study of the ecology of the river since the survey by Butcher *et al.* (carried out in 1929-1932). The available information is patchy and the reader of this chapter will no doubt be aware of many obvious questions which it is impossible to answer.

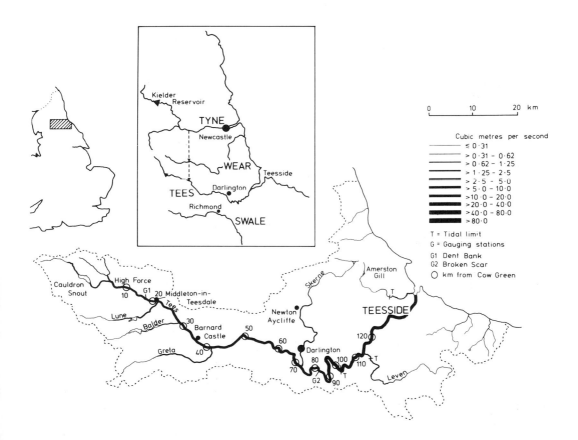

*Fig. 6.1* Catchment of Tees and adjacent coastal streams, showing tidal limits, mean flows and distances downstream from Cow Green dam. Insert shows route of Tyne - Tees water transfer scheme. (Map is redrawn from River Pollution Survey, England and Wales 1975 - published by Department of the Environment and Welsh Office, 1978- by permission of HMSO)

## 6.2 GENERAL DATA

The Tees was separated into upper and lower stretches in 1970 by the completion of the Cow Green Reservoir.   Km points used here refer to distance downstream of the dam, whether or not reference is to an event subsequent to construction of the dam.  (Km values are those used by Durham University.)

### PHYSICAL FEATURES OF CATCHMENT
area (including streams entering estuary, but not coastal streams) 1805.9 km

| | |
|---|---|
| highest point | 762 m |
| height below Cow Green dam | 463 m |
| length of river: | |
| dam - tidal limits (High Worsall) | 103.0 km |
| dam - North Sea | 141 km |

### CLIMATE
Precipitation for period 1916-50 ranged from 2285 mm $yr^{-1}$ just south of highest point to 635 mm $yr^{-1}$ near mouth of river.

### ABSTRACTION
On 1 September 1976, towards end of period of dry weather.  (Data from J. Brady of Northumbrian Water Authority)

| | distance from Cow Green dam (km) | abstraction ($m^3 s^{-1}$) | estimated natural discharge ($m^3 s^{-1}$) |
|---|---|---|---|
| Broken Scar abstraction | 64.4 | 2.31 | |
| Blackwell abstraction | 69.6 | 0.30 | |
| Low Worsall abstraction | 106.5 | 1.00 | |
| discharge to sea | | 1.39 | 1.78 |

### WATER TEMPERATURE
(Data of Crisp, 1977; see also Smith, 1968 and Crisp *et al.* 1975) monthly water temperatures during 1971-75 at:

| | mean | min. | max. |
|---|---|---|---|
| Maize Beck | 7.2 °C | 1.8 °C (Feb.) | 14.4 °C (July) |
| regulated Tees (75 m downstream of dam) | 6.8 °C | 1.6 °C (Feb.) | 13.7 °C (Aug.) |

### IMPOUNDMENTS

| reservoir | stream | area (ha) | capacity ($m^3$ x $10^6$) | year filled | compensation discharge ($m^3 s^{-1}$) |
|---|---|---|---|---|---|
| Cow Green | Tees | 312 | 40.9 | 1972 | 0.45 |
| Selset | Lune | 111 | 15.3 | 1960 | |
| Grassholme | Lune | 57 | 6.1 | 1915 | 0.34 |
| Balderhead | Balder | 117 | 19.7 | 1965 | |
| Blackton | Balder) | | | | |
| | ) | 78 | 6.0 | 1894/6 | 0.18 |
| Hury | Balder) | | | | |
| Hurworth | Skerne | 23 | 0.73 | c1870 | |
| Crookfoot | Amerston Gill (to Greatham Creek) | | | | |

### MANAGEMENT
Water management in catchment is the concern of the Northumbrian Water Authority, except for Hurworth and Crookfoot reservoirs owned by the Hartlepools Water Co.

*Fig. 6.2*   Tees, showing tributaries and reservoirs.

## 6.3  GEOGRAPHY AND GEOLOGY

In spite of its relatively small size, there are marked variations in climate and land use within the catchment (Smith, 1970).  Annual precipitation varies from over 2000 mm in the headstream area to 640 mm or less in the lowlands.  The upland area has in general a greater amount of cloud cover and temperatures are lower.  Minima can fall below freezing point in all months at Moor House (556 m).  Snow can persist at higher altitudes for some weeks.

The highest parts of the catchment are mostly covered in peat or are heather moors.  Most of the middle and much of the lower altitudes are agricultural, with pasture tending to give way to crops the nearer the location is to the coast.  There are many villages and a number of small towns, but Darlington (population 97200) is the only major town whose effluent reaches the Tees system above the tidal limit.  The upland parts of the Tees and its tributaries mostly lack a tree cover on their banks, but some stretches of the middle river and extensive stretches of the lower river have a dense cover on one or both banks (Fig. 6.3).

The geology of the region is described by Dunham (1948).  A number of streams in the Cow Green region receive springs draining the metamorphosed limestone and flow over alternating bands of limestone and other types of rock.  The coarsely crystalline marble ('sugar limestone') here has a low magnesium content, in marked contrast to the Magnesian Limestone of parts of the Skerne catchment.  There are a number of old lead or barytes mines in the upper part of the catchment, but at present only one barytes mine is still operational.  The upstream part of the Skerne lies within the Durham coalfield.

The possibility cannot be ruled out that upland lead and barytes mining had an obvious impact on the river in the 18th and 19th centuries, but supporting evidence, which is available for the Wear, appears to be lacking for the Tees.  Present-day ecological effects of the old mine workings are probably only very local, though the possibility of more widespread effects at times of sudden flood needs critical study.

Until the closure of the Cow Green dam (in 1970), the Tees was unusual in having an upland, slow-flowing stretch (the Weel) with well-developed stands of macrophytes (6.62).  Downstream of the dam there are two natural barriers to upstream fish migration, Cauldron Snout (km 0.5) and High Force (km 10).  The latter, at least, has almost certainly been a complete barrier to migratory fish, since there is a vertical drop of 21 m.

## 6.4  HYDROLOGY

### 6.41  DISCHARGE

The Tees, which rises at an altitude of 762 m, has the steepest average channel slope (2.7 m km$^{-1}$) and steepest headwater slope (4.3 m km$^{-1}$) of any of ten U.K. rivers listed by Newson (1981).  By the time the river has been joined by Trout Beck (550 m), its mean discharge is about 1.3 m$^3$ s$^{-1}$. Between there and Cow Green (490 m), the mean discharge is augmented to about 2.5 m$^3$ s$^{-1}$.  The Tees is unusual among British rivers in being of comparatively large size at a relatively high altitude.

The mean discharge at Broken Scar, the main gauging station (catchment 817 km$^2$) is 16.6 m$^3$ s$^{-1}$.  The minimum natural discharge at Broken Scar is about 0.8 m$^3$ s$^{-1}$, but the present regulation of the river raises the minimum summer discharge to about 5 m$^3$ s$^{-1}$.

The Tees is a notoriously 'flashy' river and Butcher *et al.* noted that at times the rise in water level is so rapid that the flood water passes down the river 'like a tidal wave' and the first waters of a spate are very turbid.  The Base Flow Index (Institute of Hydrology, 1978) can be regarded as an indication of flashyness.  Streams with a B.F.I. approaching 1.0 have relatively even flow regimes, while streams with low values are relatively flashy.  All except one of the upland tributaries (down to and including the Greta) have B.F.I. values (Table 6.1) less than 0.25, while all the lowland ones have values of 0.25 or more.  These values may be compared with 0.31 and 0.29 for the upper reaches of the Wye and Severn respectively (M. D. Newson, pers. comm.).

Because snow can last for some weeks at higher altitudes, the eventual onset of mild weather can lead to spates, as documented for a small tributary by Crisp & Gledhill (1970).  Snow-melt sometimes leads to floods in the main river (Smith, 1971), but there is no regular period for such floods each spring, as there is for some European rivers.  Study of old records for the region suggests that major floods have occurred at all times of year.

During recent decades the construction of reservoirs and programmes for regulation and abstraction have favoured a reduction of flow extremes. Most other changes brought about by man's activities will probably have had the opposite influence, though their quantitative importance is not clear.  95% or more of the catchment was probably once covered by forest. Subsequently much of this was destroyed and the upland area became covered by peat or peaty soils.  Within historical times much of the peat was drained, the extent of drainage depending on the economics of upland farming and other types of land use.  Within the past five years there has been another surge of upland drainage, largely as a result of government subsidies.

One quantitative study is that of Conway & Miller (1960), who examined four small (4 - 8 ha) catchments on blanket peat in the Upper Tees catch-

*Table 6.1* Values of Base Flow Index, calculated from hydrographs, for a selection of tributaries of the Tees and for two stations on the main river. (Information on Eggleshope Burn and Carl Beck from P. A. Carling and for other stations from Institute of Hydrology)

|  | National Grid Reference | Base Flow Index |
|---|---|---|
| **Tributaries** | | |
| Trout Beck | NY 759336 | 0.15 |
| Langdon Beck | NY 852309 | 0.21 |
| Harwood Beck | NY 850309 | 0.20 |
| Eggleshope Burn | NY 984288 | 0.34 |
| Carl Beck | NY 946229 | 0.23 |
| R. Balder | NY 928183 | 0.18 |
| R. Greta | NZ 034100 | 0.21 |
| Dale Beck | NZ 260156 | 0.25 |
| R. Skerne (Preston-Le-Skerne) | NZ 292238 | 0.53 |
| R. Skerne (Bradbury) | NZ 318285 | 0.64 |
| Rushyford Beck | NZ 280290 | 0.78 |
| Mardon Stell | NZ 323274 | 0.46 |
| Clow Beck | NZ 282101 | 0.54 |
| R. Leven (Leven Bridge) | NZ 445122 | 0.45 |
| R. Leven (Easby) | NZ 585087 | 0.63 |
| Billingham Beck | NZ 408227 | 0.29 |
| **Main River** | | |
| R. Tees (Dent Bank) | NY 932260 | 0.24 |
| R. Tees (Low Moor) | NZ 364105 | 0.39 |

ment. The least managed of these had an extensive *Sphagnum* cover and only one drainage channel. At the other extreme was a catchment which had been severely burnt and had an extensive pattern of drainage channels. Relative to the natural catchment, the burnt and drained catchment had a far more sensitive response to rainfall, with earlier and higher peak discharges, but gave very little discharge in dry weather. Of the other two catchments, the hydrological performance of the burnt one was not so different from that of the natural catchment, whereas performance of the drained one was quite similar to that of the one which had been both drained and burned. Although this particular study suggests that drainage has a much greater effect than simply the removal of the peat, discussion on the management of the Tees catchment has often assumed that peat plays an important buffering role on stream discharge. Jansson (1982) concluded from a statistical survey of 45 catchments in northern Sweden that bogs do not act as buffering reservoirs, so there is clearly a need for further study within the Tees catchment.

## 6.42  TRANSPORT OF ERODED PEAT

Several studies have quantified the transport of peat by streams in the catchment.  Crisp (1966) found that in a small headstream catchment (83 ha), about 77% of the input rainfall flowed down the stream, while the rest was lost as evaporation.  About 30% of the stream discharge occurred during about 3% of the time and during these same spates (>0.2 $m^3$ $s^{-1}$) the stream transported about 75 t dry weight of eroded peat, some 80% of the total transported each year.  Crisp & Robson (1979) found that there was a quantitative relationship between the concentration of peat in suspension and the quantity transported per unit time; the peat concentration reached a peak before the pea, discharge. Peat concentrations of up to 500 g $m^{-3}$ water occurred during spates of up to 1 $m^3$ $s^{-1}$ and, during one exceptionally large spate ($\underline{c}$ 1.5 $m^3$ $s^{-1}$), reached 2000 g $m^{-3}$.  These quantities and patterns are probably typical of headwater tributaries of the Tees arising on blanket peat, where peat is eroding at a relatively steady rate.

A less frequent and more violent form of erosion also occurs from time to time.  During a thunderstorm on the evening of 6 July 1963, about 6000 $m^3$ of peat slid from the face of Meldon Hill and passed down Lodgegill Sike into the Tees (Crisp et al., 1964).  Some of the peat blocks were deposited on the banks of the river and subsequently broke down slowly.  However, sufficient of this material remained in the river and became finely divided to give a peat concentration of 2200 g $m^{-3}$ at the Broken Scar gauging station the next morning.  A similar phenomenon happened in the catchment of Langdon Beck on 17 July 1983.

It is evident that considerable quantities of peat enter the Tees from its upper tributaries.  Much of this peat may be deposited in the lower stretches of the river or washed out to sea, but a certain amount settles within the tributaries and the upper river during the recession of each spate.

## 6.5  PHYSICO-CHEMICAL  FEATURES

Small 3rd and 4th-order streams encounter the widest temperature range. The most upstream ones sometimes freeze almost entirely in severe winter weather, although there is probably always some moving water if part of the flow is spring-fed.  In contrast, summer stream temperatures sometimes exceed 20º for short periods.  Crisp (1977) reported that the mean annual temperature of Maize Beck was 7.2º and of the regulated Tees just downstream of the dam, 6.8º.  The mean temperature of the river increases by about 2º by the time it has reached the tidal limit. Crisp & Howson (1982) showed in a study of six streams that there is an approximately linear relationship between mean air temperature and mean water temperature (over 5- or 7-day periods), except for periods when the mean air temperature is below 0º.  Thermal stratification in the Cow Green reservoir is rare and generally of short duration

(Crisp, 1977), so has little influence on downstream temperature. The reservoir has, however, modified the temperature immediately downstream of the dam in several ways. The annual temperature fluctuation is reduced by 1 - 2° and there is a marked reduction in diel fluctuations. The spring rise in water temperature is delayed by 20 - 50 days and the autumn fall by 0 - 20 days.

Some indication of changes in water chemistry on passing down the river is given in Table 6.3. The composition of the water at most sites does, however, vary markedly according to flow conditions, with water which is much softer, rich in humic materials and with lower conductivity and pH values when discharge is high.

The chemistry of the Skerne is sufficiently different from that of the main river to bring marked changes in the concentrations of several elements between Tees km 67.5 and km 78.8. Increases in Mg and Ca reflect the influence of the Magnesian Limestone and pumped mine water; increases in Na, P and Cl are due to effluents entering the Skerne from Darlington and other sewage treatment plants. The Skerne is also polluted by phenols and cyanide from a coke-works.

*Table 6.3* Mean values (mg $l^{-1}$) for selected variables in 0.2 μm filtrable water sampled at four sites in the Tees system during a one-year study (Holmes & Whitton, 1981a). Data of Carling (1983) for rainfall in the Great Eggleshope Beck basin are included for comparison.

| | Sand Sike | Tees (km from dam) | | | rainfall |
| | | 0.1 | 67.5 | 78.8 | |
|---|---|---|---|---|---|
| Na | 3.6 | 2.6 | 7.8 | 30.4 | 3.16 |
| K | 0.71 | 1.31 | 2.16 | 3.71 | 1.30 |
| Mg | 1.42 | 0.86 | 5.27 | 12.7 | 0.51 |
| Ca | 40.0 | 8.47 | 32.4 | 51.3 | 0.47 |
| Fe | 0.14 | 0.32 | 0.21 | 0.27 | |
| Zn | 0.069 | 0.044 | 0.029 | 0.038 | |
| $NH_4$-N | 0.017 | 0.044 | 0.026 | 0.280 | |
| $NO_3$-N | 0.166 | 0.195 | 0.889 | 1.05 | |
| $PO_4$-$P_i$ | 0.025 | 0.028 | 0.126 | 1.27 | |
| Cl | 6.6 | 5.0 | 15.7 | 39.1 | |

## 6.6  PLANTS

### 6.61   SOURCES OF INFORMATION

The 1929-32 survey (Butcher *et al.*, 1937) paid particular attention to
the photosynthetic organisms, especially the microalgae of the main river
and the macrophytes below the entry of the Skerne.  The survey led to
generalizations on observations at 24 sites and also descriptions of
several new algal species, which have since been shown to be widespread
in rivers of many countries.  The (now drowned) Weel was studied in some
detail during 1967-68 by Proctor (1971, 1978), his account also including
casual observations as far back as 1937.  Selected sites in the lower
stretches of the Tees and Skerne were surveyed during 1964-65 (Whitton
& Dalpra, 1968) and the Tees from the entry of the Skerne to km 100 in
June-July 1970; the emphasis in both studies was on comparison with the
earlier survey.  A much more detailed survey was made in 1975 (Holmes &
Whitton, 1977), six years after the closure of Cow Green dam.  Records
were made for both 'river' and 'bank' in each 0.5 km length downstream
of the dam.  The records included subjective estimates (expressed on
1 - 5 scale) for both the relative biomass of each species and the
percentage of the area covered by the species.  The occurrence of
*Grimmia agassizii* on rocks subject to very high current speeds (Holmes,
1976) provided the second record for this species in the British Isles.

The phytobenthos of six sites down the river system was studied at
monthly intervals over one year during 1976/77 (Holmes & Whitton,
1981a).  Besides floristic records, estimates were made of the relative
proportions both of various life-forms and individual species.  Phytoplankton was sampled at the same time as the phytobenthos at the two
most downstream sites and also on three different occasions at other
sites (Holmes & Whitton, 1981b); comparisons were made with the phytoplankton of adjacent rivers at the same time.

The only upland streams studied are those subject to the influence of
the metamorphosed limestone (Carter, 1972; Hughes & Whitton, 1972;
Livingstone & Whitton, 1984).  The blue-green alga *Calothrix parietina*,
an important species in Sand Sike, has been the subject of a number of
experimental studies and its behaviour in the laboratory compared with
that in the stream (see Livingstone & Whitton, 1984a, for references).
The fixation of $CO_2$ and $N_2$ by *Rivularia* in Red Sike is described by
Livingstone & Whitton (1984b).

### 6.62   PHYTOBENTHOS AND MACROPHYTES

The vegetation of the upland streams is usually dominated by algae,
though bryophytes (e.g. *Acrocladium cuspidatum, Philonotis fontana*) are
often conspicuous immediately below the source of calcareous springs,
and several rooted macrophytes can occur in slow-flowing stretches.
Rooted macrophytes were well represented in the Weel (see below).  Nowadays submerged and floating-leaved species occur only in the mid- and

D

E

*Fig. 6.3* Views of Tees system. A, Harwood Beck just above gauge; B, Cow Green dam; C, view downstream from B; D, Eggleston (km 25.6), showing extensive area of cobbles exposed under low flow; E, Hurworth (km 72.8) downstream from Croft Bridge, showing stretch with dense growths in summer of *Ranunculus fluitans* (in photo), *Cladophora* & *Fontinalis*.

lower stretches.  Bryophyte crops are highest immediately below the Cow
Green dam and in general become less important on passing downstream.
Below the entry of the Skerne, only one bryophyte (*Fontinalis antipyretica*) remains abundant, in contrast to flowering plants which show
a marked increase in abundance.  Other abundant or frequent species
below the Skerne are *Ranunculus fluitans*, *Potamogeton pectinatus*,
*Cladophora glomerata*, *Vaucheria sessilis* and, slightly further downstream, *Myriophyllum spicatum*.  Only non-vascular species occur in 'river'
habitats downstream of km 113.0.

*Rivularia* is abundant in streams on the metamorphosed limestone,
though it appears to be restricted to those parts which receive both
peat and limestone drainage, possibly due to the importance of organic
phosphate in such waters (Livingstone & Whitton, 1984a).  These streams
usually show conspicuous growths of diatoms in spring, mostly species
of *Cymbella*, several of which form small hemispherical colonies.  The
development of obvious growths of other species (especially filamentous
Conjugales) appears to be determined at least as much by recent flow
history as season.  Filamentous green algae are often abundant in the
larger, soft-water streams in late winter, these usually being species
of *Ulothrix*.  Such growths develop during periods of low flow and
bright sunshine and, if there is no scouring flood, may be succeeded by
a mass development of the large hemispherical colonies of *Didymosphenia
geminata*.

In spite of the presence of old mine workings in the area, it is doubtful if the slight elevation in the concentrations of heavy metals in the
water is sufficient to have much influence on phytobenthos biology.  An
isolate of *Calothrix parietina* from Sand Sike did however prove to be
much more resistant to zinc than might be expected from ambient zinc
levels (Shehata & Whitton, 1981).  This alga grows in microhabitats
alternatively wetted and dried and is perhaps at times subject to zinc
enrichment in this zone.

Proctor (1971) suggested that differences between his own observations
in the Weel and those of R. W. Butcher might have been due in part to
the complicated history and effects of the Cow Green barytes mine.
Wastes from the mine appear to have favoured the growth of aquatic
vegetation.  One way this might have been brought about was due to the
destructive effects of geese.  Some 60 - 70 geese were summered on the
Weel until 1937 when many died, apparently due to ingesting barytes
sludge.  The species found by Proctor in 1967-68 were, in decreasing
order of abundance: *Potamogeton alpinus*, *Myriophyllum alterniflorum*,
*Equisetum fluviatile*, *E. palustre*, *Callitriche platycarpa*, *C. intermedia*,
*Potamogeton natans*, *Ranunculus aquatilis*.

During the study of six sites down the river system (Holmes & Whitton,
1981a), a spring 'burst' of diatoms occurred at four sites, the two most
upstream and the two most downstream.  The latter occurred a month later
than the former, in spite of higher downstream temperatures.  It seems
likely that the higher irradiation received at the surface of the sub-

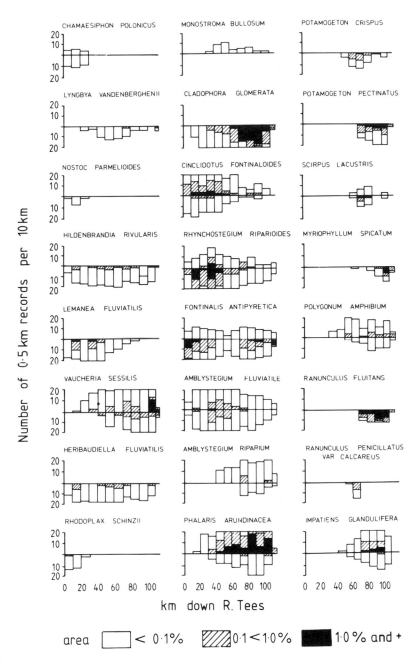

*Fig. 6.4* Distribution of selected macrophytes in Tees. The histograms show the number of 0.5 km lengths from which each species was recorded in the 20 lengths of each successive 10 km of river; plotted above the line are records from bank habitats, below the line, those from river habitats. In addition to giving presence/absence data, the histograms also show the percentage cover of a species in each 0.5 km length. (Data of Holmes & Whitton, 1977)

stratum in the shallow upstream waters was an important factor.  The
lack of a diatom burst below the Cow Green dam and at km 25.4 may have
been due to the lower silica concentrations in the water there, this
almost certainly resulting from the removal of silica in the reservoir.
Apart from the lack of a spring diatom burst and the greater cover of
bryophytes, the site below the Cow Green dam differed in other ways.
Filamentous green algae of many species were abundant in late summer
and early autumn and there was a conspicuous autumn cover of *Phormidium*.
General observations (the authors and P. D. Armitage, pers. comm.) have
confirmed that bryophyte crops are consistently higher here than at the
same site prior to regulation or elsewhere in the river.

Another apparent effect of regulation is evident much further downstream.  Apart from the populations in the Weel, submerged angiosperms
were restricted in 1929-32 to the river below the entry of the Skerne.
This was again the situation in 1965.  However in 1975 four species
were established upstream of the Skerne, although none were abundant.
*Potamogeton crispus*, *Zannichellia palustris* and *Myriophyllum spicatum*
had spread 25, 9.5 and 4.5 km, respectively, upstream from their
previous known most upstream localities.  *Ranunculus penicillatus* var.
*calcareus* (Fig. 6.4) was an apparent invader to the river and crops of
this species above the Skerne have increased still further since 1975.
It seems probable that the four submerged angiosperms which became
established above the Skerne did so during the period after the full
regulatory effects of Cow Green Reservoir started in 1971.  Although
Butcher *et al*. described the Tees above the Skerne as 'virtually
unhabitable by higher plants', this does not apply now.

Three macrophyte species appear to have been lost between 1929-32
and 1975.  Of these, *Nitella opaca* disappeared before 1964-65, while
*Potamogeton perfoliatus* and *P. pusillus* have disappeared since then.
None have been found in tributaries.  *Elodea canadensis*, present in
1965, was absent in 1970, but present again in 1975.  It seems quite
likely that the main river population was washed out by an exceptional
flood in March 1968 and that the species was re-inoculated from tributaries where it is known to be present.

The macrophytic green alga *Enteromorpha flexuosa* was absent from the
main river in 1929-32 and 1975, yet was widespread during 1964-65 and
1970 in the Skerne and below its inflow.  Its absence from the main
river in 1975 is especially striking in view of the fact that it was
present in Clow Beck, which presumably provided an inoculum to the
Tees at km 71.6; it was also abundant elsewhere in north-east England
in 1975.

Holmes & Whitton (1981a) reported that surveys of photosynthetic
plants of the Tees and tributaries have shown about 600 species.  In
view of the large number of streams to be studied and their marked
differences in water chemistry, it seems probable that the complete
flora of flowing waters in the Tees catchment is not far short of
1000 species.

## 6.63    PHYTOPLANKTON

The survey of phytoplankton in 1975 showed that the most important groups were, in decreasing order of abundance, diatoms, green algae and cryptomonads.  Most taxa were ones which have been found also in the benthos.  Of the sixteen taxa found only in the plankton, only two (*Rhodomonas lacustris, Actinastrum hantzschii*) were ever frequent and many could easily have been overlooked during surveys of the benthos. Nine of the species found only in the plankton were motile: *Lepocinclis ovum, Lobomonas* sp., *Gymnodinium* sp., *Ochromonas* sp., *Gonium pectorale, G. sociale, Lobomonas* sp., *Pteromonas angulosa, Pyramimonas* sp.

Among the diatoms, the density of empty frustules was usually of a similar order to that of live cells, so counts made only on cells cleared in acid would have given a false impression.  The density of all types of live cell ranged from 100 to 5000 cells (or units) $ml^{-1}$. These values are quite similar to those in the Tyne, Wear and Swale, but about one order of magnitude less than in the Neckar (p. 333).  The three studies made on the same day of the sites on the Tyne and Tees associated with the present water transfer scheme showed that plankton densities were only slightly higher in the Tyne.  So far no studies have been made of the fate of plankton on passing through the tunnel.

The plankton flora immediately downstream of the Cow Green dam probably reflected the composition of plankton in the reservoir, the most frequent species being *Melosira italica*, *Achnanthes* spp., *Asterionella formosa*, *Diatoma elongatum* and *Synedra* spp.  Live cells of all these species were found further downstream, but overall their densities tended to decrease on passing downstream.  It is uncertain what effect passage over High Force has on the viability of cells, as the downstream sample site at km 25.7 may receive inocula from the Lunedale reservoirs as well as Cow Green.  While *Melosira italica* became infrequent by km 67.5, another centric diatom, *Cyclotella meneghiniana,* which was absent below the dam, became frequent in the downstream river, expecially in August when it was the most abundant species at km 78.8.

## 6.7   INVERTEBRATES

### 6.71   SOURCES OF INFORMATION

The most detailed accounts are for the Upper Tees and the tributaries
within and around the Moor House National Nature Reserve.  These include
the Simuliidae (Davies & Smith, 1958; Phillipson, 1957; Wotton, 1976,
1977, 1978), Plecoptera (Brown *et al.*, 1964), Ephemeroptera (Crisp &
Nelson, 1965) and Chironomidae (Birkett, 1976).  In addition, papers by
Coulson (1959) and Nelson (1971) have records of aquatic Tipulidae and
other groups.  There are also accounts of the benthic invertebrates of
four headstream sites within the Moor House Reserve (Armitage *et al.*,
1975) and in tributaries within the Cow Green area (Armitage *et al.*,
1974).  Whitehead (1929) and Brown & Whitehead (1938) list several species,
but all ones included by Butcher *et al.*

The 1929-33 study (Butcher *et al.*, 1937) included a survey at sites
down the whole of the non-tidal part of the main river.  Macan (1957)
added two species of Ephemeroptera to the faunistic list.  A general
survey of the fauna was made also in April 1976 (by P. D. Armitage;
data held by Freshwater Biological Association).  The effects of impoundment and regulation upon the invertebrate fauna just downstream of Cow
Green have received considerable attention (see 6.74).

*Table 6.3* Percentage composition of fauna from two collections of
samples taken at different altitudes.  Collection 1 is a composite of
kick samples taken in May, August and October  over several years
from four stations in headstreams on Moor House N.N.R. (Armitage *et
al.*, 1975); collection 2 is a similar composite for the main river,
one major tributary and several minor tributaries within Cow Green
area (Armitage *et al.*, 1974).  (Representatives of some minor groups
were excluded from this analysis)

|              | composition (%) | |
|              | collection 1 | collection 2 |
|--------------|:------------:|:------------:|
| Plecoptera    | 34.7 | 21.0 |
| Ephemeroptera | 26.2 | 44.4 |
| Diptera       | 11.3 |  8.2 |
| Trichoptera   | 10.3 |  5.0 |
| Coleoptera    |  8.3 |  4.1 |
| Crustacea     |  4.5 | 12.1 |
| Annelida      |  2.8 |  1.6 |
| Hydrachnellae |  1.6 |  0.6 |
| Mollusca      |  0.3 |  3.0 |
|              | 100.0 | 100.0 |

6.72   THE FAUNA AND ITS DISTRIBUTION

A total of 135 taxa have been recorded for the Tees system in the various
papers listed above.  (Species limits and nomenclature used here follow
those of Maitland, 1977.)  The records include 76% of known British
Plecoptera, but only 19% of gastropod molluscs.

Some indication of the relative proportions of various taxa among
upstream invertebrate populations is given in Table 6.3.  The table also
illustrates the decrease in importance of Plecoptera and increase of
Ephemeroptera with decreasing altitude, though the Ephemeroptera
decrease again in the downstream stretch of the main river.  The differences shown in the table for Crustacea and Mollusca are largely a
reflection of local conditions.

The 1976 survey of the main river showed that overall the most common
invertebrates were *Caenis rivulorum, *Naididae, Tubificidae, Baetis
rhodani, Hydropsyche siltalai, Orthocladiinae, *Limnius volckmari,
*Elmis aenea and *Rhithrogena semicolorata.  Some of these (indicated
by *) were apparently more abundant in 1976 than in the 1929-32 survey.
The most important changes are probably immediately below the Cow Green
dam and in the stretch of the river below the entry of the Skerne,
though the records are insufficient for a detailed comparison to be
made for the latter.


6.73   DRIFT AT UPSTREAM SITES

Drift of benthic organisms has been followed at two upstream sites
(Table 6.4).  In the small headstream, about $4 \times 10^6$ individual animals
(c 1.6 kg dry weight) passed a point in the stream annually.  Within
individual spates the drift concentration  has been found (Crisp &
Robson, 1979) to reach its peak before that of the hydrograph, a result
similar to that found for peat fragments (6.42).  For most organisms
concentrations were low during the spate peak and increased again as

Table 6.4  Comparison of estimates of mean discharge, total annual
drift past a point and of mean concentration in terms of numbers and
weight for a small headstream (Rough Sike: Crisp & Gledhill, 1970)
and an unregulated major tributary (Maize Beck: Armitage, 1977).
Data from a southern chalk stream are included for comparison (East
Stoke Mill stream: Crisp & Gledhill, 1970).

|  | Rough Sike | Maize Beck | East Stoke Mill Stream (Crisp & Gledhill, 1970) |
|---|---|---|---|
| mean discharge ($m^3 s^{-1}$) | 0.04 | 1.48 | 0.71 |
| total drift (kg $yr^{-1}$) | 1.4 | 15.6 | 28.0 |
| mean conc. (number $m^{-3}$) | 2.92 | 4.09 | 0.912 |
| mean conc. (mg $m^{-3}$) | 1.0 | 0.3 | 1.2 |

the spate receded.  In contrast, terrestrial casualties peaked with the
spate.  As terrestrial casualties formed 10 - 15% of annual drift in
terms of numbers and had a greater mean weight per individual than the
aquatic component, the terrestrial contribution was appreciable.  Despite
these within-spate patterns, on a round-the-year basis the concentration
(number or weight per unit volume of water) and transport (drift per unit
time in terms of weight) showed a highly significant positive correlation
with stream discharge.  The relationships between discharge and numbers
were fitted best by the equation:

$$C_n = 1.5 \, Q^{0.72} \quad \text{and} \quad T_n = 40744 \, Q^{1.71}$$

where $C_n$ is individuals $m^{-3}$, $T_n$ is individuals $h^{-1}$ and $Q$ is discharge
(expressed as $m^3 \, s^{-1}$).

The observation that drift concentration increased with discharge
agrees with that found for the Wye (see pp. 70-71) by Brooker & Hemsworth
(1978) during regulation releases.


6.74   INFLUENCE OF IMPOUNDMENT AND REGULATION DOWSTREAM OF COW GREEN
       RESERVOIR

Information on the pre-regulation benthos of the main river some 500 m
downstream of the present dam and in the lower stretch of the unregulated
Maize Beck is given by Armitage et al. (1974) and accounts of both after
regulation by Armitage (1976, 1978a, b).  The following is a simplified
summary of the main findings.

Both population density and biomass increased substantially in the
Tees following regulation; Maize Beck, whose faunal composition and
abundance was similar to that of the Tees prior to regulation, showed
no obvious change over the same period.  Species changes on the Tees
included both losses (mainly Plecoptera) and gains, though all were
species which occurred elsewhere in the area.  Probably of more importance were changes in the relative abundance of groups.  The abundance
of Gammarus, Chironomidae and Mollusca has increased in the Tees and
the abundance of Ecdyonuridae nymphs has decreased.  In addition, there
has been the appearance of large numbers (but small biomass) of Hydra
and Nais.

Following closure of the Cow Green dam in June 1970 there was a
rapid build-up of zooplankton in the reservoir and of Hydra in the
reservoir and the river downstream of the dam.  Armitage & Capper
(1976) estimated that 150 kg (dry weight) of this material was discharged from the reservoir annually, comprising about 136 kg of Cladocera,
13 kg of Copepoda and 1 kg of Hydra.  About 98% of this material was
discharged during July-October and for much of this period the concentrations in the water column were about 1300 Cladocera and 200
Copepoda $m^{-3}$.  Densities had however dropped to 2% or less of these
values by km 6.5.

In contrast to this drift of material from the reservoir, drift of benthos from the river bed was changed only slightly as a result of regulation. There was an increase in *Nais* in the drift, reflecting the greatly increased importance in the benthos. Comparison with Maize Beck also suggested that nocturnal peaks of *Baetis* and *Simulium* were less pronounced in the regulated river.

The various observations on seston, benthos and drift suggest that impoundment has also had a marked effect on trophic relationships within at least the first 0.5 km. Organic sediment, whose importance as a primary or secondary food source for detritivores and as case-building material for some Chironomidae is becoming apparent (McLachlan *et al.*, 1979; Walentowicz & McLachlan, 1980), has almost certainly decreased markedly downstream of the dam. Extrapolation from results for Rough Sike (6.42) to the Tees at Cow Green suggests that about $4 \times 10^6$ km of peat enters Cow Green reservoir annually, while less than $0.2 \times 10^6$ kg of organic sediment is discharged downstream. On the other hand, the 150 kg of small animals from the reservoir is a rich source of direct or indirect food for some species and is taken up rapidly by the benthic system. During August and September the rate of uptake averages about $0.15$ g m$^{-2}$ d$^{-1}$, though it may reach four times as much (Armitage & Capper, 1976).

It is not yet possible to assess the importance for the ecosystem of the input of reservoir animal material and loss of detrital material. However, the former is only 4% of the latter and is concentrated within a four-month period. Although no studies have yet been made, the qualitative composition of the peat fraction discharged from the reservoir is also likely to be changed by its passage through the reservoir, with a smaller average particle size but perhaps an increased bacterial content.

The post-regulation density and biomass of benthos is substantially higher in the Tees than in the Maize Beck (Armitage, 1978b). This may represent an increase in available food for fish. Trout, but apparently not bullheads, consume the Cladocera and Copepoda from the reservoir (Crisp *et al.*, 1978). Trout also feed avidly on terrestrial casualties such as earthworms, slugs and larval Tipuliidae, which one would expect to be reduced considerably downstream of the impoundment.

## 6.75 COMMENTS ON INDIVIDUAL SPECIES

The large mayfly *Ephemera danica* was described by Butcher *et al.* as uncommon and nymphs were recorded only at Eryholme (km 83.9). As the nymphs burrow in sand and fine gravel, it would not be expected to be widespread or abundant in the Tees, but it appears more widespread than indicated. Nymphs have been recorded at Dubby Sike (Cow Green) during electrofishing (Armitage *et al.*, 1974) and large numbers of imagines were observed in the late 1970s in the Cotherstone area

(A. Clarke, pers. comm.) and a single nymph at Red Sike in 1983.  Either
the earlier workers overlooked the species at some sites or its status
has changed.

The lack of records for the crayfish *Austropotamobius pallipes* by
Butcher *et al.* is almost certainly due to its being overlooked.  The
species is in fact widespread both in the main river in the Barnard
Castle area and in the lower parts of some tributaries.  It is also
found at High Coniscliffe (c̲ km 58) and the R. Leven.

## 6.8  FISH

### 6.81  SOURCES OF INFORMATION

Published information on fish in the main rivers includes:  Longstaffe
(1954), Jackson (1930), Pentelow (1932), Pentelow *et al.* (1933a, b),
Southgate *et al.* (1932), Brown *et al.* (1970), Butcher *et al.* (1937),
Crisp *et al.* (1974, 1975), Hancock (1977).  The Tees fishery for migratory salmonids is summarized in the annual reports of the Northumbrian
Water Authority and its predecessors.  Butcher *et al.* dealt mainly
with the distribution of species in the main river.  In contrast, Crisp
*et al.* covered distribution, population density, growth, production and
reproduction, but only in the river system above the confluence with
Maize Beck.  Hancock surveyed the main river from Cauldron Snout to
Hurworth and described species distribution, but gave growth and population data only for the brown trout.

### 6.82  SPECIES DISTRIBUTION

At least twenty-two species are present in the Tees system (Table 6.5).

*Table 6.5*  Fish species present in Tees system.  B, listed by
Butcher *et al.* (1937); H, additions by Hancock (1977); N, further
records of Northumbrian Water Authority.

| | | | | | |
|---|---|---|---|---|---|
| N | brook lamprey | N | bream | N | 10-spined |
| B | salmon | B | minnow | | stickleback |
| B | sea and brown trout | H | rudd | B | perch |
| N | rainbow trout | B | roach | B | bullhead |
| B | grayling | B | dace | B | flounder |
| B | pike | B | chub | | |
| N | carp | B | stone loach | | |
| N | barbel | B | eel | | |
| B | gudgeon | B | 3-spined stickleback | | |

Two others mentioned by Longstaffe, the sturgeon and the burbot, have
long since disappeared, the former already being a rarity in 1854.
Butcher *et al.* listed only fifteen species, but several were probably
overlooked, while others have been introduced since then.

The rainbow trout is present as a stocked fish in some of the Lunedale and Baldersdale reservoirs and, not only can downstream escapes
occur, but there is evidence of breeding in Carl Beck.  The bream was
introduced by the Water Authority at Croft in 1974/75.  A carp of about
230 g was taken at Over Dinsdale in 1979 and small barbel were found at
Croft in 1978.  Butcher *et al.* recorded the three-spined stickleback
as far up the estuary as Yarm and also in the Skerne.  Nowadays, at
least, the species is certainly more widespread.

Butcher *et al.* found that dace, chub and minnows were the most frequent fish in the lower river, with gudgeon and roach less frequent
and pike and perch rare.  Of these, all except the minnow decreased in
numbers above Croft (km 72.8) and the most upstream site for dace,
roach and chub was at Piercebridge (km 56.3).  Hancock's (1977) observations are consistent with these earlier ones, but they do extend the
ranges of some species, with bullhead and minnow at Cauldron Snout
(km 0.5) and stone loach at Middleton-in-Teesdale (km 19).  If the
Tees system as a whole is considered, there are records for minnows
at 503 m altitude and of bullhead at 628 m (Crisp *et al.*, 1974, 1975).

Eels were described by Butcher *et al.* as common throughout the Tees,
but despite intensive sampling for over a decade, they have not been
recorded at or above Cauldron Snout (Crisp *et al.*, 1974, 1975) and it
is doubtful whether they occur upstream of High Force.

Among Salmonidae, the brown trout occurs throughout almost the
length of the river.  According to Butcher *et al.* the grayling was
common in the middle stretch of the river, with its upper limit at
about km 48.  It now occurs at least as far upstream as Dent Bank (A.
Clarke, pers. comm.).  Butcher *et al.* mentioned a local tradition that
the grayling was introduced about 1890.  It would appear, therefore,
that since its original introduction it has slowly spread upstream.
The changing status of the salmon and sea trout in the Tees is covered
in 6.84.


6.83  POPULATION DENSITY, GROWTH AND PRODUCTION

The larger, older brown trout tend to occur at low density and relatively low biomass in the main river (Table 6.6).  These fish spawn in
the tributaries, which act as nursery streams and contain mostly
smaller, younger fish at higher density and biomass.  Recruitment in
the main river is largely by downstream dispersal of fish from the
tributaries and headwaters.  There is considerable variation in the
annual production of different tributaries (Table 6.7).  The annual
survival rates of both brown trout and bullhead are high, as can be

*Table 6.6* Population density and biomass of brown trout and bull-head based on data of Crisp *et al.* (1974, 1975, unpublished) and Hancock (1977).   O, species absent; -, not estimated.

| | brown trout | | bullhead | |
| --- | --- | --- | --- | --- |
| | numbers m$^{-2}$ | g m$^{-2}$ | numbers m$^{-2}$ | g m$^{-2}$ |
| **Tributaries** | | | | |
| Trout Beck | 0.1 -0.2 | 2.0 - 4.3 | 0.04 | - |
| Weelhead Sike | 0.25-1.37 | 1.6 -16.6 | 0 | 0 |
| Dubby Sike | 0.21-0.92 | 1.7 - 6.9 | 0 | 0 |
| Lodgegill Sike | 0.03-0.13 | 0.3 - 5.3 | 0 | 0 |
| Mattergill Sike | 0.04-0.19 | 1.0 -11.8 | 0 | 0 |
| Gt. Dodgen Pot Sike | 0.14-0.30 | 3.0 - 6.8 | 0 | 0 |
| Moss Burn | 0.02-0.15 | 0.8 - 1.6 | 0 | 0 |
| **Tees** | | | | |
| Tees Bridge | - | - | 0.18-0.25 | 0.48-0.87 |
| Tees above Weel | 0   -0.2 | 0.01- 0.37 | 0.33-0.96 | 0.41-0.74 |
| Tees below Cauldron Snout | 0.02-0.06 | 1.1 - 3.7 | 0.06-0.62 | 0.38-1.38 |
| Middleton-in-Teesdale (km 19) | 0.07 | - | - | - |
| Egglestone Abbey (km 25.5) | 0.03 | - | - | - |

*Table 6.7* Estimates of fresh weight production (g m$^{-2}$ year$^{-1}$).   O, species absent; -, not estimated; *, based on tray samples.

| | production (g m$^{-2}$ yr$^{-1}$) | |
| --- | --- | --- |
| | brown trout | bullhead |
| **Tributaries** | | |
| Trout Beck | 2.2 | - |
| Weelhead Sike | 8.6 | 0 |
| Dubby Sike | 6.0 | 0 |
| Mattergill Sike | 5.9 | 0 |
| Gt. Dodgen Pot Sike | 3.5 | 0 |
| Moss Burn | 2.2 | 0 |
| Cow Green tributaries | 4.4 | 0 |
| **Tees** | | |
| Tees Bridge | - | *7.4 |
| Tees in reservoir basin (Cow Green) | 0.7 | 1.1 |
| Tees below Cauldron Snout (c km 0.6) | - | 1.1 |

seen by comparing results summarized in Table 6.8 with the values of
0.095 and 0.047, respectively, found for chalk streams in southern England by Mann (1971).

Table 6.9 gives details of the age of fish going to sea. About 95%
of Tees salmon were found to go to sea at age 2 or 2+ by Pentelow *et al.*
(1933a) and this percentage is similar to those for the Tyne, Tweed,
Forth, Wye and Findhorn, but different from those for the Spey and Don.
Size is an important factor governing the time of seaward migration of
both salmon and sea trout. Thus those salmon which went to sea at age
2 were the larger specimens of their year class.

*Table 6.8* Annual survival rate (fraction of fish present at time t
which survive to time t + 1 years)

| site(s) | annual survival rate (but confounded by movements) | |
| --- | --- | --- |
| | brown trout | bullhead |
| Trout Beck | 0.52 | – |
| Tees Bridge | – | 0.30 |
| Tees & tributaries at Cow Green | 0.57 | – |
| Cow Green tributaries | 0.40 | – |
| Tees at Cow Green | 0.75 | 0.33 |
| Tees below Cauldron Snout (c km 0.6) | 0.49 | 0.52 |

*Table 6.9* Percentages of salmon and sea trout smolts going to sea
at each age based on combined data for 1930 and 1931 (Pentelow *et
al.*, 1933a)

| age (years) at migration | salmon | sea trout |
| --- | --- | --- |
| 2 | 56.6 | 44.6 |
| 2+ | 38.4 | 29.6 |
| 3 | 5.0 | 24.5 |
| 3+ | – | 1.3 |

## 6.84 THE TEES FISHERY AND POLLUTION

Coarse fishing occurs chiefly in the lower stretches of the river,
where some stocking is done. The brown trout, which occurs throughout
the river, is however the main quarry of anglers. Stocking with hat-
chery trout is a procedure of long-standing in the Tees. 64750 trout
were distributed in the Tees between 1904 and 1915 and further stocking

took place in seven different years up to the time of the survey by
Butcher *et al*. Hancock (1977) referred to the stocking of 10420 trout
in the Tees below Middleton-in-Teesdale in 1974, 8460 in 1975 and 5900
in 1976. It is instructive to examine this in more detail. If the
trout population density between Middleton-in-Teesdale and Hurworth is
taken as 0.03 fish $m^{-2}$ (probably a low estimate), then there are about
65000 trout in this stretch and Hancock's data suggest that 30 - 40%
of these are of takeable size. At the time of stocking, the stock fish
of takeable size are equivalent to 3.7% of the indigenous fish of similar size. The remaining stock fish are undersized at the time of
stocking and, given growth and survival rates similar to those of the
indigenous fish, will reach takeable size two years after stocking and
be equivalent to some 10 - 30% of the number of indigenous fish reaching takeable size at the same time. This suggests that stocking may
have an appreciable impact on the Tees fishery, though this rests on
the assumption that survival and growth of stocked fish are similar to
that of indigenous fish.

During the late 19th and 20th centuries the Tees had a substantial
fishery for migratory Salmonidae. The Skerne apparently also had at
one time a reputation as an important salmon river, but lost this at an
early stage of the Industrial Revolution. The Tees fishery consisted
of two elements, a rod and line fishery within the river and a net
fishery within the estuary, the latter being the more important. There
has been a drastic decline of both salmon and sea trout in the main
river within the past century (Table 6.10), with a major change taking
place shortly after the First World War. Butcher *et al*. state that the
first official recognition that there was a serious pollution problem
in the Tees was in 1920 when the Tees Fishery Board reported substantial pollution mortality of smolts in the estuary. If 1920-24 is
accepted as the start of serious pollution problems in the Tees, then
the virtual destruction of this fishery took only a decade.

*Table 6.10* Summary of Tees rod catches of salmon and sea trout
between 1886 and 1978, based on reports by Northumbrian Water Authority and its predecessors.

| years | mean catch (no. of fish $yr^{-1}$) ($\pm$ 95% confidence limits) | | % sea trout |
|-------|---------|-----------|-------------|
|       | salmon | sea trout | |
| 1886–1912 | 502.0 $\pm$ 11.07 | 28.5 $\pm$ 30.3 | 5.4 |
| 1914–1930 | 144.5 $\pm$ 57.3 | 15.6 $\pm$ 9.54 | 9.8 |
| 1931–1939 | 0.7 $\pm$ 0.94 | 0.8 $\pm$ 0.92 | 53.8 |
| 1951–1978 | 0 | 1.4 $\pm$ 1.79 | 100.0 |

Recently there have been signs that rehabilitation of the fishery is possible, with an adult sea trout found in a small tributary of the Lune in 1976 and a salmon near Barnard Castle in 1980.  By 1983 it was clear that salmon were spawning successfully within the Tees system. The most likely explanation for this change is the depressed state of the economy of Teesside during the late 1970s and early 1980s, which led to a reduced pollutant load in the estuary.

In the autumn of 1983 there was a severe pollution of the Tees by "flux oil".  The pollutant entered the river near High Force and there was a substantial fish kill.  The casualties included several hundred adult salmon and sea trout.  This gives some indication of the recovery of stocks up to that time.  The effects of this pollution upon the recovery will not be clear for some years.


6.85  EFFECTS OF REGULATION

An account of the effects of the Cow Green regulation scheme upon the populations of fish within 0.6 km of the dam was given by Crisp et al. (1983) (Table 6.11), of regulation upon stomach contents by Crisp et al. (1978) and of the effects of changed temperature regime upon the growth and metabolism of brown trout (Crisp, 1977).  The main changes observed among trout were an increase in mean population density (by c 40%) and a small, but statistically significant, increase in growth rate.  There was a threefold increase in the mean population density and biomass of the bullhead and this was accompanied by changes in reproductive performance.  This improvement in conditions for trout and bullhead may reflect an improved food supply.  However, as growth of individual fish is limited by temperature rather than food supply, this improvement has been exploited through an increase in population density, rather than by improved growth of individuals.


6.9   CONCLUDING  COMMENTS

In spite of the many changes to the Tees brought about by man's activities, most of the river upstream of the estuary is in good condition. The changes which have taken place can be grouped broadly as either chemical or hydrological.

The main sites for pollution in the Tees system are the R. Skerne and the estuary.  The Skerne probably underwent a progressive deterioration during the 19th century and was at its worst in the middle of this century.  The predicted cessation of a coke-works effluent should lead to a marked improvement in the condition of this tributary, making it suitable for an increased range of organisms, including a number of fish species.  Provided this river is subject to careful management, it could develop into an important amenity.

*Table 6.11*  Simplified summary of changes in the population characteristics of brown trout and bullhead in the Tees below Cauldron
Snout as a result of regulation of the Tees.  Values observed before
regulation are followed by values observed after regulation.  (Derived from Crisp *et al.*, 1983)
-, inadequate data for analysis; +, no significant change;
    *, P <0.05;  **, P <0.01.

|  | brown trout | bullhead |
|---|---|---|
| population density (no. 100 m$^{-2}$) | 3.5 to 4.9* | 9.9 to 33.4* |
| biomass (g m$^{-2}$) | - | 0.5 to 1.4** |
| fecundity (eggs per female | + | 148 to 116 (for 7 cm female)* |
| fecundity (eggs m$^{-2}$ yr$^{-1}$) | + | 2.3 to 8.4 |
| growth rate | small increase** | + (small increase in age group 0 only) |
| age at first sexual maturity | - | 50% to 84% (% mature in age group II) |
| instantaneous rate of "loss" | + | 0.7 to 1.3* |
| production (g m$^{-2}$ yr$^{-1}$) | 0.3 to 0.4 | 0.5 to 3.6 |

In contrast to the Skerne, pollution in the estuary became a serious
problem only in this century, with a sharp drop in the condition of the
river in the 1920s continuing until the 1970s.  The recent passage of
migratory salmon and trout appears to be due in part to improvements
in the quality of industrial effluents, but probably even more to an
overall reduction in pollution resulting from economic decline in the
region.  It is to be hoped that any future resurgence of industrial
activity is matched by still further improvements in effluent quality.

Superficially the hydrological changes in the Tees system appear
more profound than those due to pollution, since they are unlikely
ever to be reversed.  Construction of the Cow Green reservoir destroyed
the Weel and has led to some changes in chemistry and in plant and
animal populations in the downstream river.  However, with the exception of the stretch immediately below the reservoir, these changes
appear relatively slight.  The fact that the dam acts as an upstream
barrier to migration is of minor importance, because High Force, the
waterfall 10 km downstream of the dam, has always acted as another
such barrier.  The Tyne-Tees water transfer scheme, whose ecological
significance at one time caused considerable apprehension, also seems

likely to have little impact on the Tees, both because of the infrequency with which the system is used and because the chemistry and ecology of the donor river are quite similar. Overall, therefore, the hydrological management of the river has probably had much less influence on its ecology than a superficial glance at the map would suggest. In view of this and the decreasing influence of pollution, the Tees should certainly be considered among those British rivers where nature conservation is an important criterion in making future decisions on management.

## Acknowledgements

Many people have supplied data or helpful conversation. These include: P.D. Armitage, P.A. Carling, A. Clarke and E.M. Ottaway, from the Freshwater Biological Association; M.D. Newson from the Institute of Hydrology; N.T.H. Holmes from the Nature Conservancy Council; J.A. Brady, J. Cave, A.S. Champion, J.W. Hargreaves, P. Johnson, W.O. Ord, J.R. Pomfret and W.J. Walker, from the Northumbrian Water Authority.

## REFERENCES

Armitage P.D. & Capper M.H. (1976) The numbers, biomass and transport downstream of microcrustaceans and *Hydra* from the Cow Green reservoir (Upper Teesdale). *Freshwat. Biol.* 6, 425-32.

Armitage P.D., MacHale A.M. & Crisp D.T. (1974) A survey of invertebrates in the Cow Green basin (Upper Teesdale) before inundation. *Freshwat. Biol.* 4, 369-98.

Armitage P.D., MacHale A.M. & Crisp D.T. (1975) A survey of the invertebrates of four streams in the Moor House National Nature Reserve in Northern England. *Freshwat. Biol.* 5, 479-95.

Armitage P.D. (1976) A quantitative study of the invertebrate fauna of the River Tees below Cow Green reservoir. *Freshwat. Biol.* 6, 229-40.

Armitage P.D. (1977a) Invertebrate drift in the regulated River Tees, and an unregulated tributary Maize Beck, below Cow Green dam. *Freshwat. Biol.* 7, 167-83.

Armitage P.D. (1977b) Development of the macroinvertebrate fauna of Cow Green Reservoir (Upper Teesdale) in the first five years of its existence. *Freshwat. Biol.* 7, 441-54.

Armitage P.D. (1978a) The impact of Cow Green reservoir on invertebrate populations in the River Tees. *Freshwat. Biol. Ass. U.K., Annual Reports* 46, 47-56.

Armitage P.D. (1978b) Downstream changes in the composition, numbers and biomass of bottom fauna in the Tees below Cow Green reservoir and in an unregulated tributary Maize Beck, in the first five years after impoundment. *Hydrobiologia* 58, 145-56.

Birkett N.L. (1976) Chironomidae (Diptera) trapped in a Pennine stream, including two species new to Britain. *Ent. Gaz.* 27, 161-70.

Brooker M.P. & Hemsworth R.J. (1978)  The effect of the release of an artificial discharge of water on invertebrate drift in the R. Wye, Wales. *Hydrobiologia* 59, 155-63.

Brown J.M. & Whitehead H. (1938)  The Trichoptera or caddisflies of Yorkshire. *Naturalist, Hull,* 260-65.

Brown V.M., Cragg J.B. & Crisp D.T. (1964)  The Plecoptera of the Moor House National Nature Reserve, Westmorland. *Trans. Soc. Br. Ent.* 16, 123-34.

Brown V.M., Shurben D.G. & Shaw D. (1970)  Studies on water quality and the absence of fish from some polluted English rivers. *Water Research* 4, 363-82.

Burston U.T. & Coats D.J. (1975)  Water resources in Northumbria with particular reference to the Kielder Water Scheme. *J. Instn Wat. Engrs* 29, 226-51.

Bussell R.B. (1979)  Changes of river regime resulting from regulation which may affect ecology : a preliminary approach to the problem. *Central Water Planning Unit, U.K. 1979.* pp. 48.

Butcher R.W., Longwell J. & Pentelow F.T.K. (1937)  Survey of the River Tees. Part III. The non-tidal reaches - chemical and biological. *Wat. Pollut. Res. Techn. Pap.* 6, 1-189 (HMSO London).

Butcher R.W. (1933)  Studies on the ecology of rivers 1. On the distribution of macrophytic vegetation in the rivers of Britain. *J. Ecol.* 21, 58-91.

Carling P.A. & Reader N.A. (1982)  Structure, composition and bulk properties of upland stream gravels. *Earth Surface Processes and Landforms* 7, 349-65.

Carling P.A. (1983)  Particulate dynamics, dissolved and total load, in two small basins, northern Pennines, U.K. *Hydrological Sciences - Journal* 28, 355-75.

Carter J.R. (1972)  The diatoms of Slapestone Sike, Upper Teesdale. *Vasculum* 57, 35-41.

Clapham A.R. (Ed.) (1978)  *Upper Teesdale. The Area and its Natural History.* 238 pp. Collins, London.

Conway V.M. & Millar A. (1960)  The hydrology of some small peat-covered catchments in the northern Pennines. *J. Inst. Wat. Engrs* 14, 415-24.

Coulson J.C. (1959)  Observations on the Tipuliidae (Diptera) of the Moor House National Nature Reserve, Westmorland. *Trans. R. Ent. Soc. Lond.* 3, 157-74.

Crisp D.T. & Gledhill T. (1970)  A quantitative description of the recovery of the bottom fauna in a muddy reach of a mill stream in Southern England after draining and dredging. *Arch. Hydrobiol.* 67, 502-41.

Crisp D.T. & Howson G. (1982)  Effect of air temperature upon mean water temperature in streams in the north Pennines and English Lake District. *Freshwat. Biol.* 12, 359-67.

Crisp D.T., Mann R.H.K. & Cubby P.R. (1983)  Effects of regulation of the River Tees upon fish populations below Cow Green reservoir. *J. appl. Ecol.* 20, 371-86.

Crisp D.T., Mann R.H.K. & McCormack J.C. (1974)  The populations of fish
    at Cow Green, upper Teesdale before impoundment.  *J. appl. Ecol.* 11,
    969-96.
Crisp D.T., Mann R.H.K. & McCormack J.E. (1975)  The populations of fish
    in the River Tees system on the Moor House Nature Reserve, Westmor-
    land.  *J. Fish. Biol.* 7, 573,93.
Crisp D.T., Mann R.H.K. & McCormack Jean C. (1978)  The effects of im-
    poundment and regulation upon the stomach contents of fish at Cow
    Green, Upper Teesdale.  *J. Fish. Biol.* 12, 287-301.
Crisp D.T. & Nelson J.M. (1965)  The Ephemeroptera of the Moor House
    National Nature Reserve, Westmorland.  *Trans. Soc. Brit. Ent.* 16,
    181-87.
Crisp D.T., Rawes M. & Welch D. (1964)  A Pennine peat slide.  *Geog. J.*
    130, 519-24.
Crisp D.T. & Robson S. (1979)  Some effects of discharge upon the trans-
    port of animals and peat in a north Pennine headstream.  *J. appl.*
    *Ecol.* 16, 721-36.
Crisp D.T. & Robson S. (1982)  Analysis of fishery records from Cow
    Green Reservoir, Upper Teesdale, 1971-80. *Fish Mangmt* 13, 65-78.
Crisp D.T. (1966)  Input and output of minerals for an area of Pennine
    moorland : the importance of precipitation, drainage,  peat erosion
    and animals.  *J. appl. Ecol.* 3, 327-48.
Crisp D.T. (1977)  Some physical  and chemical effects of the Cow Green
    (Upper Teesdale) impoundment.  *Freshwat. Biol.* 7, 109-20.
Davies L. & Smith C.D. (1958)  The distribution and growth of *Prosi-*
    *mulium* larvae (Diptera: Simuliidae) in hill streams in northern
    England.  *J. Anim. Ecol.* 271, 335-48.
Dunham K.C. (1948)  *Geology of the Northern Pennine Orefield. Vol. 1.*
    282 pp.  H.M.S.O., London.
Edwards R.W., Densen H.W. & Russell P.A. (1979)  An assessment of the
    importance of temperature as a factor controlling the growth rate of
    brown trout in streams.  *J. Anim. Ecol.* 48, 501-7.
Elliott J.M. (1975a)  The growth of brown trout, *Salmo trutta* L., fed
    on maximum rations.  *J. Anim. Ecol.* 44, 805-21.
Elliott J.M. (1975b)  The growth rate of brown trout (*Salmo trutta* L.)
    fed on reduced rations.  *J. Anim. Ecol.* 44, 823-42.
Elliott J.M. (1975c)  Weight of food and time required to satiate brown
    trout (*Salmo trutta*).  *Freshwat. Biol.* 5, 51-64.
Gorham E. (1956)  On the chemical composition of some waters from the
    Moor House National Nature Reserve.  *J. Ecol.* 44, 375-82.
Hancock R.S. (1977)  The impact of water transfers and associated regu-
    lation reservoirs on the fish populations of the Tyne, Tees and
    Swale.  *Proc. Eighth British Coarse Fish Conference*, 137-58.
Holmes N.T.H., Lloyd E.J.H., Potts M. & Whitton B.A. (1972)  Plants of
    the River Tyne and future water transfer scheme.  *Vasculum* 57,
    56-78.
Holmes N.T.H. & Whitton B.A. (1977)  The macrophytic vegetation of the
    River Tees in 1975 : observed and predicted changes.  *Freshwat.*
    *Biol.* 7, 43-60.
Holmes N.T.H. & Whitton B.A. (1981)  Phytoplankton of four rivers, the
    Tyne, Wear, Tees and Swale.  *Hydrobiologia* 80, 111-27.

Holmes N.T.H. & Whitton B.A. (1981)  Phytobenthos of the River Tees
   and its tributaries.  *Freshwat. Biol.* 11, 139-63.
Holmes N.T.H. (1976)  The occurrence and ecology of *Grimmia agassizii*
   (Sull. & Lesq.) Jaeg. in Teesdale.  *J. Bryol.* 9, 275-78.
Horne J.E.M. (1977)  *A Bibliography of Rivers Tyne, Wear, Tees and
   Swale.*  54 pp.  Occas. Publs. No. 3, Freshwater Biological Assoc,
   U.K.
Hughes M.K. & Whitton B.A. (1972)  Algae of Slapestone Sike, Upper
   Teesdale.  *Vasculum* 57, 30-35.
Institute of Hydrology (1978)  *Low Flow Study.*  *Rep. No. 1.*  61 pp.
   Institute of Hydrology (U.K.), Wallingford.
Jackson W.H. (1930)  Pre-pollution days in the Tees.  *Salmon Trout
   Mag.* 58, 87-89.
Jansson B.-O. (1982)  Myrat, svampar och statistik.  Bogs, sponges and
   statistics.  *Vatten* 38, 50-58.  (In Swedish, English summary)
Lavis M.E. & Smith K. (1972)  Reservoir storage and the thermal regime
   of rivers, with special reference to the River Lune, Yorkshire.
   *Sci. total Environ.* 1, 81-90.
Livingstone D. & Whitton B.A. (1984a)  Water chemistry and phosphatase
   activity of the blue-green alga *Rivularia* in Upper Teesdale streams.
   *J. Ecol.* (in press).
Livingstone D. & Whitton B.A. (1984b)  Diel variations in nitrogen and
   carbon dioxide fixation by the blue-green alga *Rivularia* in an up-
   land stream.  *Phycologia* (in press).
Longstaffe W.H.D. (1954)  *The History and Antiquities of the Parish of
   Darlington, in the Bishoprick.*  374 pp. + appendices.  Proprietors
   of Darlington & Stockton Times, Darlington.
Macan T.T. (1957)  The Ephemeroptera of a stoney stream.  *J. Anim.
   Ecol.* 26, 317-42.
Macan T.T. (1979)  Key to British freshwater Crustacea - Malacostraca.
   *Sci. Publs Freshwat. Biol. Ass. U.K.* 32, 1-72.
Maitland P.S. (1977)  *A Coded Checklist of Animals occurring in Fresh
   Water in the British Isles.*  76 pp.  Inst. of Terrestrial Ecology,
   Edinburgh.
McLachlan A.J., Pearce L.J. & Smith J.A. (1979)  Feeding interactions
   and cycling of peat in a bog lake.  *J. Anim. Ecol.* 48, 851-61.
Mann R.H.K. (1971)  The populations, growth and production of fish in
   four small streams in southern England.  *J. Anim. Ecol.* 40,
   155-90.
National Water Council (1981)  *The 1980 Survey and the Future Outlook.*
   National Water Council, 1 Queen Anne's Gate, London SW1H 9BT.
   (Name now changed to Water Authorities Association).
Nelson J.M. (1971)  The invertebrates of an area of Pennine moorland
   within the Moor House National Nature Reserve in northern England.
   *Trans. Soc. Br. Ent.* 19, 173-235.
Newson M.D. (1981)  Mountain streams.  In: Lewin J. (Ed.) *British
   Rivers.*  Allen & Unwin.
Northumbrian River Authority (1967-1972)  *Ann. Reps. for Years Ending
   31st March 1966-31st March 1971.*  Northumbrian River Authority,
   Newcastle-on-Tyne.

Northumbrian River Authority (1973)  *Report on Survey of Water Re-
    sources.*  191 pp.  Northumbrian River Authority, Newcastle-on-Tyne.
Northumbrian Water Authority (1975-    )  *Annual Water Reports 1974/75-*
           Northumbrian Water Authority, Gosforth.  (Published
    annually; analytical data not included in this series of reports.)
Northumbrian Water Authority (1976)  *Report on Experimental Releases
    from Cow Green Reservoir on the R. Tees (23rd-28th June 1976).*
    Directorate of Planning and Scientific Services, Northumbrian Water
    Authority, Gosforth.
Pentelow F.T.K., Southgate B.A. & Bassindale R. (1933a)  I. The rela-
    tion between the size, age and time of migration of salmon and sea
    trout smolts in the River Tees. *Fishery Invest., London.*  Ser. 1,
    3, No. 4, 1-10.
Pentelow F.T.K., Southgate B.A. & Bassindale R. (1933b)  II. The pro-
    portion of the sexes and the food of smolts of salmon (*Salmo salar*
    L.) and sea trout (*Salmo trutta* L.) in the Tees Estuary.  *Fishery
    Investigations, London.*  Ser. 1, 3, No. 4, 11-14.
Pentelow F.T.K. (1932)  The food of brown trout (*Salmo trutta* L.).
    *J. Anim. Ecol.* 1, 101-7.
Phillipson J. (1957)  The effect of current speed on the distribution
    of the larvae of the black-flies: *Simulium variegatum* (Mg.) and
    *S. monticola* Fried. (Diptera).  *Bull. Ent. Res.* 48, 811-19.
Proctor H.G. (1971)  Aquatic macrophytes in the Wheel of Tees.
    *Vasculum* 56, 59-66.
Proctor H.G. (1978)  Changes in the large aquatic flora of the Cow
    Green basin.  In: Clapham A.R. (Ed.) *Upper Teesdale. The Area and
    and its Natural History.*  238 pp., pp. 191-94.  Collins, London.
Shehata F.H.A. & Whitton B.A. (1981)  Field and laboratory studies on
    blue-green algae from aquatic sites with high levels of zinc.
    *Verh. int. Verein. theor. angew. Limnol.* 21, 1466-71.
Smith K. (1968)  Some thermal characteristics of two rivers in the
    Pennine area of northern England.  *J. Hydrol.* 6, 405-16.
Smith K. (1969)  The baseflow contribution to runoff in two upland
    catchments.  *Water & Water Engineering,* Jan. 1969, 18-20.
Smith K. (1970)  Weather and climate.  In: Dewdney J.C. (Ed.) *Durham
    County and City with Teesside.*  522 pp., pp. 58-74.  British
    Association, Durham.
Southgate B.A., Pentelow F.T.K. & Bassindale R. (1932)  An investi-
    gation into the causes of death of salmon and sea trout smolts in
    the estuary of the River Tees. *Biochem. J.* 26, 273-84.
Walentowicz A.T. & McLachlan A.J. (1980)  Chironomids and Particles :
    a field experiment with peat in an upland stream.  In: Murray D.A.
    (Ed.) *Chironomidae Ecology, Systematics, Cytology and Physiology.*
    Pergamon.
Wear & Tees River Board (1951-1965)  *Annual Reports for Year Ending
    31st March 1951-31st March 1965.*  Wear and Tees River Board,
    Darlington.
Whitehead H. (1929)  Plecoptera (stone-flies) of Yorkshire. *Natura-
    list, Hull* 875, 403-7.
Whitton B.A. & Dalpra M. (1968)  Floristic changes in the River Tees.
    *Hydrobiologia* 32, 545-50.

Wotton R.S. (1976)  The distribution of blackfly larvae (Diptera: Simuliidae) in Upper Teesdale, Northern England.  *Hydrobiologia* 51, 259-63.
Wotton R.S. (1977)  Sampling moorland stream blackfly larvae (Diptera: Simuliidae).  *Arch. Hydrobiol.* 79, 404-12.
Wotton R.S. (1978)  Life-histories and production of blackflies (Diptera: Simuliidae) in moorland streams in Upper Teesdale, Northern England.  *Arch. Hydrobiol.* 83, 232-50.

# 7 · The Winterbourne Stream

A. D. BERRIE & J. F. WRIGHT

7.1 Introduction                                    180

7.2 General data                                    183

7.3 Physico-chemical background                     185

7.4 Study sites                                     188

7.5 Macrophytes                                     188

7.6 Invertebrates                                   191

    7.61 Fauna in 1972                              191
    7.62 Classification of sites                    192
    7.63 Faunal groupings                           194
    7.64 The 1976 drought                           199

7.7 Fish                                            199

    7.71 Bullheads                                  199
    7.72 Trout                                      201

7.8 Conclusions                                     203

References                                          204

## 7.1     INTRODUCTION

The Winterbourne Stream is the only tributary of the River Lambourn, a chalk stream which drains the Berkshire downs in southern England. With a total length of 8 km, of which only the lower 2 km have permanent flow, it might be regarded as an unusual choice for detailed investigation.  The reason for the attention now being given to a number of similar streams in England will become apparent later, but first it is important to describe the features which characterize all winterbourne streams.

Chalk outcrops over much of southern and eastern England, and its porous structure gives rise to streams with a high groundwater component to their discharge.  The headwaters and tributaries of these chalk streams are of an intermittent nature, their flow largely being determined by the level of water in underground storage.  In winter, when the aquifer is replenished, water is discharged into the stream through springs and the stream bed itself.  The water-table drops as the aquifer is depleted through the summer due to lower rainfall and consequently the upper limit of the running water moves down the valley.  The stream thus has an intermittent section where flow is seasonal and a perennial section downstream, where flow is permanent.  This intermittent section is commonly known as a winterbourne, i.e. a winter stream, and this name is given to the stream which forms the subject of this chapter, despite the fact that it is perennial in its lower reaches.

Our understanding of the structure and functioning of chalk streams has increased considerably in the last 20 years (Westlake *et al.*, 1972; Le Cren & Lowe-McConnell, 1980), but until recently winterbournes received very little attention.  In an attempt to rectify this, Casey & Newton (1972) and Casey & Ladle (1976) published accounts of the water chemistry, discharge regime, macrophytes and invertebrate communities at a series of sites along the South Winterbourne stream, a tributary of the River Frome in Dorset.  During this survey, Ladle & Bass (1975) discovered a blackfly, *Metacnephia amphora*, new to science.  They also examined a small chalk stream in Dorset, the Waterston stream (Ladle & Bass, 1981), which dried up for a period of four months during a minor drought in 1973.

Perhaps the greatest stimulus to the investigation of winterbournes resulted from the apparent need to develop groundwater pumping schemes to meet the predicted increase in consumption of water (Water Resources Board, 1973).  One of the largest demands was for London and a scheme involving the Lambourn and Winterbourne valleys was planned and commissioned to provide an additional supply in dry periods (Brettell, 1971).  The water is pumped from boreholes located some distance from the rivers and piped to outfalls in the perennial sections from where it flows downstream to be abstracted again from the R. Thames near London.  This avoids the need for expensive pipelines and provides additional water in the rivers at times when the flows would otherwise be low.  Suitable facilities already exist for treating the river water

before passing it to the supply system.  During the development of the
scheme a team from the University of Reading carried out ecological
studies on the R. Lambourn and the Winterbourne Stream to obtain base-
line information prior to any ecological effects of the scheme.  The
research programme for the Winterbourne Stream involved an extensive
survey of the macrophytes and macroinvertebrates in 1972.  This was
followed by more limited observations in 1976/77 when extreme drought
in summer 1976 led to operation of the groundwater scheme between
August and November.  High rainfall during the autumn and winter of
1976 and the return of water to many of the study sites on the Winterbourne Stream provided the opportunity to make some observations on
recolonization by the macroinvertebrate fauna in the aftermath of the
drought.  Observations on the fish were made from 1971 to 1976.

Ecological studies have also been undertaken in connection with at
least four other groundwater schemes in the chalk regions of southern
and eastern England.  These involve pumping groundwater into the Candover
Brook in Hampshire (Southern Water Authority, 1979) and into the Rivers
Thet, Rhee and Little Ouse in eastern England (Klee, 1974; 1976);
lining the bed of the Gussage stream in Dorset and altering its flow
regime (Furse, 1977a & b); and pumping groundwater direct to supply
from the upper catchment of the R. Wylye in Wiltshire with effects on
flow in the headwaters and in tributaries (Avon & Dorset River Authority, 1973).

In reviewing the conservation value of sites in Great Britain,
Ratcliffe (1977) highlighted the special character of chalk streams and
listed the R. Lambourn as a good example of an unspoilt chalk stream.
Winterbournes appear to support some scarce species, they may provide
nursery areas for the trout populations of chalk streams and they are
under threat from a variety of sources.  These are some of the practical reasons why winterbournes have received attention in the past few
years.  They also have much to tell us about the response of organisms
to drought conditions, methods and rates of recolonization and the
importance of physical factors in determining the composition of
biological communities.  Whilst these topics have received considerable
emphasis in relation to standing waters (Wiggin *et al.*, 1980) only now
is attention being focused on temporary streams (Williams & Hynes,
1976;  1977).

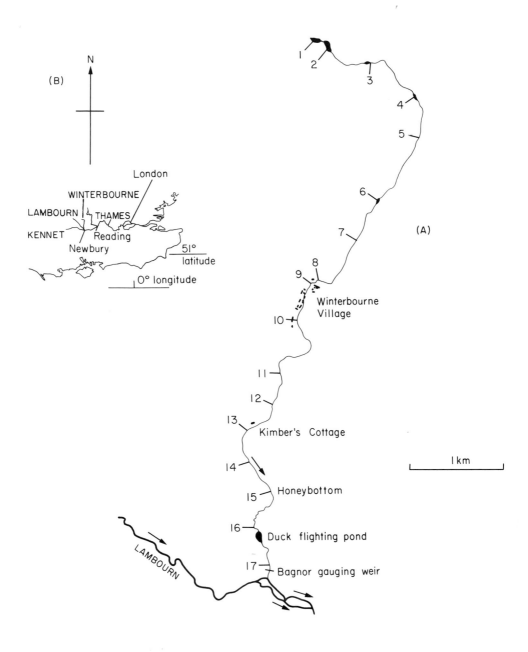

*Fig. 7.1* A) Winterbourne Stream, showing the 17 macroinvertebrate sampling sites and additional locations mentioned in the text. B) Location of the Winterbourne Stream in S-E. England.

## 7.2  GENERAL DATA

DIMENSIONS
  Area                         49 km$^2$
  Length                        8 km (total length, including inter-
                                     mittent section)
  Highest point in catchment   230 m
  Drop in height                32 m (from 113 m to 81 m)

LOCATION
  51.5° N  1.5° W

CLIMATE
Precipitation  705 mm yr$^{-1}$ for 1941-1970
  driest month = April (44 mm)
  wettest month = November (73 mm)
  120-140 wet days per year (>1 mm rain)

Temperature
  mean January air    3.5-4 °C
  mean July air       16.5-17 °C

GEOLOGY
Upper chalk is exposed over much of the catchment, but along the valley
floor there are river and valley gravels with some alluvium.  On higher
ground in the northern part of the catchment the upper chalk is covered
by clay-with-flints, whilst further south the higher ground is capped
by Eocene sands and clays overlain by Plateau gravel.

LAND USE
Arable farming and pasture land in chalk upland areas; trees are restricted to occasional copses.  Woods and heathland occur on the higher
ground in the southern part of the catchment, where the chalk is overlain by sands, clays and finally gravelly drift.

POPULATION
Approximately 500 people in small villages and isolated farms.

PHYSICAL FEATURES OF RIVER

On-stream reservoirs and major weirs
No reservoirs or major weirs.  A gauging weir and a duck-flighting pond
(Fig. 7.1) may influence the movement of some fish and invertebrates.

Flow at selected points
See Fig. 7.2 for mean monthly discharge (1972-1977) at gauging weir.

Seasonal changes in flow
In a typical year peak discharge occurs in March (~0.25 m$^3$ s$^{-1}$) followed
by a steadily decreasing hydrograph to early winter (<0.1 m$^3$ s$^{-1}$) when
discharge starts to rise again (see Fig. 7.2 and text for further details)

Influence of regulation
Between August and November 1976, water from boreholes high in the
catchment was used to augment the Winterbourne stream above the gauging
weir (Fig. 7.1).  Some water was also added at the head of the perennial
section to maintain the amenity value of the stream.  Pumping delayed

the return of water to the intermittent section of the stream through a
lowering of the water table.

## Important abstraction points

None on the Winterbourne itself, but Chieveley pumping station (SU 476751)
takes on average 148.5 $m^3$ $day^{-1}$ from a borehole located near the edge of
the catchment.

## Substratum

Terrestrial vegetation or emergent macrophytes normally dominate the
intermittent section of the stream, but submerged and emergent macrophytes,
gravel and silt are all important components of the substratum in the
perennial section (see Table 7.2).

## Temperature

A thermograph located upstream of the gauging weir indicates a very regular
water temperature regime under normal meteorological conditions in which
the winter minimum is $c$ 4 $^{\circ}$C and the summer maximum is $c$ 17.5 $^{\circ}$C.   (See
Fig. 7.2 and text.)

### 7.3   PHYSICO-CHEMICAL BACKGROUND

The course of the stream is shown in Fig. 7.1 together with 17 sites
chosen for a survey in 1972.  The perennial head is normally taken to
be between sites 12 and 13 just above Kimber's Cottage.  However, the
Winterbourne Stream does not dry out to the perennial head every year
and in fact parts of the intermittent section remained wet in 1972.

Fig. 7.2 presents mean monthly discharge at the gauging weir on the
Winterbourne Stream before its confluence with the R. Lambourn.  Since
the major outfall of the pumping scheme is just upstream of the weir,
it includes both natural and pumped discharge for July 1976 (test pumping) and for August to November 1976 (operational pumping).  In a
typical year, the peak discharge occurs in March followed by a steady
decline through the summer and autumn when at least part of the intermittent section of the stream becomes dry.  Winter rains then lead to
an increase in the height of the water-table in the catchment, breaking
of the springs and inundation of the entire intermittent section.  This
is reflected by the rising hydrograph at the weir.  The very stable
flow regime observed at the weir is a consequence of the high proportion
of the discharge derived from groundwater and the low proportion of
surface runoff.  In 1973 (Fig. 7.2) a minor drought was experienced
whilst in 1975 a peak spring discharge in excess of 400 l $s^{-1}$ occurred.
This was followed by a prolonged period of low rainfall which resulted
in a failure of the springs in 1976 and a major drought.

Fig. 7.2 also includes the monthly maximum and minimum stream
temperatures recorded at a site just upstream of the gauging weir and
groundwater outfall.  The natural groundwater enters the stream through
the river bed and springs at a temperature of approximately 11 $^{\circ}$C
before being influenced by the seasonally varying air temperature.
Superimposed upon the very regular annual temperature regime is evidence of higher summer maxima in years of drought (1973, 1976) and
warmer winter minima when discharge is exceptionally high (1975).

The chemical composition of the water is similar to a number of
other chalk streams previously investigated in the region (Westlake *et
al.*, 1972).  Monthly water samples (taken by the Thames Water Authority)
near the gauging weir show that many chemical determinands maintain
relatively stable values with time, despite seasonal changes in discharge.  This feature has also been observed in a chalk spring (Casey,
1969) and in the South Winterbourne stream in Dorset (Casey & Newton,
1972; Casey & Ladle, 1976), but is less pronounced in larger chalk
streams and rivers (Casey, 1969; Casey & Newton, 1973).

Between 1972 and 1977, pH was recorded over the range 7.2 - 8.3
(n = 90) and levels of dissolved oxygen were invariably high (8.0 -
14.3 mg $l^{-1}$, n = 89) at the time of sampling.  Suspended solids were
almost always low (range 0.5 - 21.7 mg $l^{-1}$, n = 75) resulting in very
clear waters for most of the year.

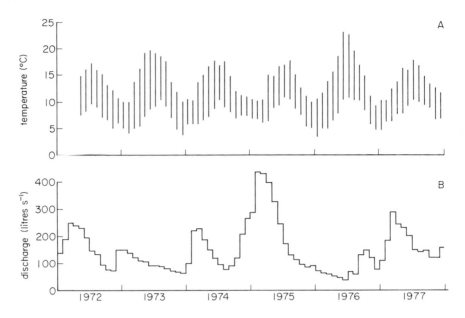

*Fig. 7.2* Records of temperature and discharge. A. Monthly maximum and minimum temperatures from a thermograph just upstream of both site 17 and groundwater outfall. B. Mean monthly discharge at Bagnor gauging weir. (Discharge between July and November 1976 includes pumped discharge.)

Table 7.1 presents further water quality data in the form of annual means and ranges. Total oxidized nitrogen which is due almost entirely to nitrate-N, failed to demonstrate a significant positive correlation with discharge as observed for nitrate by Casey & Newton (1972) in the South Winterbourne stream; the drought years of 1973 and 1976 did, however, have the lowest yearly means. Particularly low values were noted during both test pumping (1.4 mg $l^{-1}$) and operational pumping (4.0 mg $l^{-1}$) of the groundwater scheme in 1976.

Orthophosphate never exceeded 100 µg $l^{-1}$ P over the period, with the exception of the first four months of 1972 when values up to 261 µg $l^{-1}$ P were recorded. Alkalinity normally maintained relatively stable values and there was no inverse relationship between discharge and alkalinity, as observed by Casey (1969) in the R. Frome. Results for sodium, potassium, magnesium and calcium also showed remarkable stability with time and Casey (1969) and Casey & Newton (1972) present similar findings for a chalk spring and the South Winterbourne stream respectively. A high value for sodium of 23.4 mg $l^{-1}$ in the Winterbourne Stream in November 1975 may have been a result of road salting.

*Table 7.1* Water quality data for the Winterbourne Stream at the Bagnor gauging weir. Yearly mean values and ranges expressed as mg $l^{-1}$ and based on 10 or more analyses. (Data provided by Thames Water Authority)

| determinand | Year | | | | | |
|---|---|---|---|---|---|---|
| | 1972 | 1973 | 1974 | 1975 | 1976 | 1977 |
| alkalinity as $CaCO_3$ | 252 (230–260) | 252 (244–264) | 251 (229–257) | 241 (135–259) | 248 (238–257) | 243 (194–254) |
| sodium | 7.6 (6.3–8.6) | 8.1 (6.1–10.8) | 9.6 (7.1–11.8) | 12.5 (9.2–23.4) | 8.6 (6.5–12.3) | 9.5 (8.6–11.9) |
| potassium | 1.5 (1.2–1.8) | 1.3 (1.0–2.8) | 1.6 (1.1–2.4) | 1.5 (0.9–1.9) | 1.4 (1.2–1.7) | 1.5 (0.9–2.2) |
| magnesium | 1.7 (1.4–2.0) | 1.7 (1.5–1.8) | 1.7 (1.4–2.4) | 1.7 (1.6–1.8) | 1.8 (1.5–2.1) | 1.7 (1.3–2.0) |
| calcium | 108 (95–116) | 107 (87–115) | 113 (104–120) | 111 (108–114) | 110 (98–121) | 118 (109–142) |
| total oxidized nitrogen as N | 6.7 (5.5–7.8) | 5.5 (4.9–6.3) | 6.0 (5.0–8.9) | 6.5 (5.5–8.5) | 5.3 (1.4–6.1) | 6.3 (5.5–7.3) |
| orthophosphate as P | 0.08 (0.01–0.26) | 0.02 (0.01–0.06) | 0.05 (0.03–0.10) | 0.06 (<0.03–0.10) | 0.02 (0–0.06) | 0.03 (<0.03–0.10) |

## 7.4   STUDY SITES

In 1972 a sampling programme was established to investigate the macrophyte and invertebrate communities of the stream.  Sampling sites for
the invertebrate programme were chosen along the entire length of the
stream (Fig. 7.1) to give comprehensive coverage of both the intermittent (sites 1-12) and permanent stream (sites 13-17).  Sites 1-4, 6
and 8 were ponds in the stream channel but all other sites were 25 m
sections of stream.

   In February 1972 the invertebrate survey included each of the 17
sites, after which attention was focused on sites 8-17 in a bimonthly
programme which continued to the end of the year.  Several factors led
to the decision to exclude the upper sites from further detailed study.
Between sites 1 and 7 the stream has very little amenity value and for
much of its length was channelled in a ditch.  In contrast, downstream
of the pond at site 8, the Winterbourne Stream had considerable amenity
value as it flowed first through Winterbourne village and then continued
alongside a minor road towards Bagnor.  Its lower reaches also supported
a trout population.  Finally, since sites 8-17 encompassed both
intermittent and perennial sections of the stream, the scientific
interest of comparing the fauna of the contrasted sections was still
retained.

## 7.5   MACROPHYTES

Each of the 25 m long sites from 9-17 was mapped bimonthly to record
the occurrence of major macrophytes and substrata.  Information was
also collated on mean widths and shading (Table 7.2). Note in particular
the increase in mean width downstream and the impact of total and
partial shading by trees and bushes on the mean cover of macrophytes
during the year.

   *Rorippa nasturtium-aquaticum* (= *Nasturtium officinale*) occurred at
all sites except the heavily shaded site 13, but its pattern of growth
differed between intermittent and perennial sites.  At intermittent
sites 9-11 it was unimportant between February and June but showed
spectacular growth by August.  In contrast, at sites 15-17 the area of
*Rorippa* in February was not markedly different from August.  Since
*Rorippa* is killed by frost, the effect of cold may be greater on the
shallow exposed reaches of sites 9-11, thereby reducing the area during
winter.  At sites 15-17 the high discharge from March to June (Fig. 7.2)
probably removed much of the *Rorippa* present in February, causing a
reduction in area before the onset of summer growth.  The depth and
current speed of the water at the downstream sites probably prevented
*Rorippa* from extending far from the banks and thus attaining the high
percentage cover recorded at sites 9-11.

   Another species, *Apium nodiflorum*, was important only in the intermittent zone (sites 9-12) and, although it dominated a number of sites

*Table 7.2* Features of sites 9-17 in 1972. O = unshaded; I = partially shaded; II = total shading by trees. A = *Apium nodiflorum*; C = *Callitriche* sp.; E = emergent vegetation including grasses; M = *Mentha aquatica*; N = *Rorippa nasturtium-aquaticum*; R = *Ranunculus* spp.; V = *Veronica beccabunga*

| site number | 9 | 10 | 11 | 12 | 13 | 14 | 15 | 16 | 17 |
|---|---|---|---|---|---|---|---|---|---|
| shading | O | O | O | O | II | I | I | O | O |
| mean width (m) | 1.0 | 2.6 | 3.3 | 3.5 | 4.5 | 4.2 | 4.3 | 4.3 | 6.4 |
| mean area in 1972 as % — macrophytes | 61 | 71 | 72 | 66 | 2 | 40 | 28 | 47 | 43 |
| silt | 9 | 15 | 9 | 8 | 73 | 19 | 56 | 33 | 47 |
| gravel | 30 | 14 | 19 | 26 | 25 | 41 | 16 | 20 | 10 |
| area of macrophytes each month as %, with dominant — February | 54-A | 68-C | 51-A | 46-A | 0 | 63-C | 22-N | 34-N | 55-N |
| April | 33-A | 75-E | 57-E | 42-R | 1-M | 37-V | 18-N | 22-N | 35-N |
| June | 68-A | 83-E | 75-E | 67-N | 0 | 36-C | 21-N | 34-R | 35-N |
| August | 67-N | 73-C/N | 93-N | 96-N | 5-M | 31-E | 39-N | 66-N | 45-N |
| October | DRY | 59-N | 96-N | 84-N | 2-M | 28-C | 43-N | 70-N | 46-N |
| December | 82-E | 68-E | 59-E | 59-N | 3-M | 45-M | 28-N | 55-N | 45-N |

*Fig.* 7.3 (above)    The intermittent section (site 11).  In spring, emergent grasses and gravel predominate before discharge decreases and *Rorippa* increases in area during late summer.

*Fig.* 7.4 (below)    The perennial ⌐ection (site 15).  A substratum of of gravel and silt supports *Ranunculus* in mid-stream and *Rorippa* along the left-hand margin, whilst further silt has accumulated under trees on the right-hand bank.

in February, it tended to be overgrown by other plants, particularly
*Rorippa* in summer.  Casey & Ladle (1976) found *Apium* well represented
near the source of the South Winterbourne stream and Furse (1977b) also
noted that *Apium* dominated winterbourne sites on the Gussage.

*Ranunculus penicillatus* var. *calcareus* and *R. peltatus* were both
present in the Winterbourne Stream and the maximum area was recorded at
perennial sites around August, as in the Lambourn at Bagnor (Ham *et al.*,
1981).  *Callitriche* sp. occurred in small quantities in both intermittent and perennial sites but only *Mentha aquatica* appeared able to
tolerate the low light intensities to be found at the heavily shaded
site (13) near the perennial head.


## 7.6   INVERTEBRATES

### 7.61   FAUNA IN 1972

The primary aim of the 1972 survey was to obtain comprehensive species
lists for study sites 8-17.  Each sample was a 5-minute collection with
a pond-net, obtained by both kicking and sweeping all available biotopes
at the study site in proportion to their occurrence.  At sites 1-7,
limited sampling took place in February and June 1972 but sites 8-17
were sampled (4 samples per site) every second month from February to
December except for site 9 which was dry in October.  Hence, the data
set for the major study sites (8-17) consists of 4 samples x 6 months
= 24 qualitative samples for each site (20 only for site 9).

   Table 7.3 presents the number of taxa in each major invertebrate
group for study sites 8-17 and also indicates the number of additional
taxa confined to sites 1-7.  217 taxa were recorded at sites 8-17 in
1972 but only 16 additional taxa were found at sites 1-7.  Of the 16,
just  four were regarded as typical of cold or intermittent waters, the
remaining species being characteristic of ponds, ditches and small
trickles (Wright *et al.*, 1984).

   Within the 10 major study sites (8-17) there was a richer fauna
downstream from site 14 to 17 as the Winterbourne Stream approached its
confluence with the Lambourn.  Attempts to interpret this as being the
result of greater habitat variability as the stream increased in size
or through colonization from the Lambourn would be speculative.  However,
there were a number of clearly defined patterns of distribution between
sites 8 and 17 for several of the major invertebrate groups.  The
Gastropoda, Ephemeroptera and Trichoptera all increased in species
richness downstream.  In contrast, site 8 was a pond situated in the
course of the stream and this allowed the Hemiptera and Coleoptera to
develop their most diverse fauna at this site.  A number of community
changes were associated with the perennial head between sites 12 and 13.
The Hemiptera, Coleoptera and Chironomidae, which were well represented
in the pond and intermittent section, decreased in species richness at
the perennial head.  But it could be argued that this was a response to

*Table 7.3* Number of taxa recorded during 1972 survey confined to sites 1 - 7, and number at each of sites 8 - 17. Groups marked with an asterisk were not necessarily identified to species; in the Oligochaeta, the families Naididae and Enchytraeidae were not identified to species and in the Chironomidae, many specimens were taken to genus only.

| | taxa confined to sites 1-7 | 8 | 9 | 10 | 11 | 12 | 13 | 14 | 15 | 16 | 17 | total taxa at sites 8-17 |
|---|---|---|---|---|---|---|---|---|---|---|---|---|
| Tricladida | 1 | 2 | 2 | 2 | - | - | 2 | 3 | 3 | 2 | 4 | 4 |
| Acanthocephala* | - | 1 | 1 | 1 | 1 | 1 | 1 | 1 | 1 | 1 | 1 | 1 |
| Nematoda* | - | 1 | 1 | 1 | 1 | 1 | 1 | 1 | 1 | 1 | 1 | 1 |
| Gastropoda | - | 4 | 4 | 5 | 4 | 4 | 4 | 4 | 7 | 7 | 10 | 11 |
| Bivalvia | - | - | - | 2 | 3 | 4 | 4 | 4 | 3 | 3 | 2 | 5 |
| Oligochaeta(*) | - | 6 | 6 | 6 | 5 | 4 | 8 | 8 | 8 | 8 | 9 | 11 |
| Hirudinea | - | - | 1 | - | 2 | 5 | 5 | 5 | 6 | 5 | 5 | 6 |
| Hydracarina* | - | 1 | 1 | 1 | 1 | 1 | 1 | 1 | 1 | 1 | 1 | 1 |
| Cladocera* | - | 1 | - | - | - | - | - | 1 | - | 1 | - | 2 |
| Ostracoda* | - | 1 | 1 | 1 | 1 | 1 | 1 | - | 1 | 1 | 1 | 1 |
| Copepoda* | - | 1 | 1 | - | - | - | - | 1 | 1 | - | 1 | 1 |
| Malacostraca | - | 2 | 2 | 1 | 1 | 3 | 3 | 2 | 3 | 3 | 3 | 4 |
| Ephemeroptera | 1 | 2 | 3 | 3 | 3 | 3 | 3 | 5 | 6 | 5 | 8 | 10 |
| Plecoptera | - | - | - | 1 | - | - | - | 1 | 1 | 1 | 1 | 2 |
| Odonata | - | 1 | 1 | - | - | - | - | - | - | - | - | 1 |
| Hemiptera | - | 8 | 2 | 2 | 4 | 4 | - | 1 | 1 | 3 | 5 | 13 |
| Coleoptera | 6 | 28 | 12 | 15 | 13 | 12 | 7 | 12 | 12 | 17 | 18 | 46 |
| Megaloptera | - | - | - | - | - | - | 1 | 1 | 1 | 1 | 1 | 1 |
| Trichoptera | 2 | 5 | 8 | 10 | 13 | 14 | 15 | 18 | 19 | 22 | 24 | 31 |
| Chironomidae* | 5 | 18 | 16 | 18 | 19 | 18 | 11 | 15 | 23 | 25 | 35 | 48 |
| other Diptera* | 1 | 7 | 13 | 11 | 11 | 11 | 12 | 11 | 13 | 13 | 13 | 17 |
| number of taxa | 16 | 89 | 75 | 80 | 82 | 86 | 79 | 95 | 111 | 120 | 143 | 217 |

the heavy shading at site 13 which, lacking abundant macrophyte growths (Table 7.2), may not offer suitable living conditions. On the other hand, the Tricladida, Oligochaeta and Megaloptera increased in species richness between sites 12 and 13 and are known to be more characteristic of permanent water. The fact that the stream did not dry down to the perennial head in 1972 may be another factor which allowed a number of invertebrates, including *Gammarus pulex*, to colonize the intermittent section, thereby blurring the faunal distinction between the perennial and intermittent sites.


7.62   CLASSIFICATION OF SITES

Species lists have been compiled for the study sites (Wright *et al.*,

1984; Wright, 1984 and, by the use of cluster analysis, this information
has been used to produce a classification of the sites. Since 24
samples were taken at each site (except site 9), each species was given
a frequency of occurrence from 0 - 24. (The frequencies for site 9
were standardised to the same range by multiplying by 24/20.) Two
dissimilarity coefficients were used on the site x species frequency
matrix followed by group average clustering (UPGMA of Sneath & Sokal,
1973) in each case (Fig. 7.5).

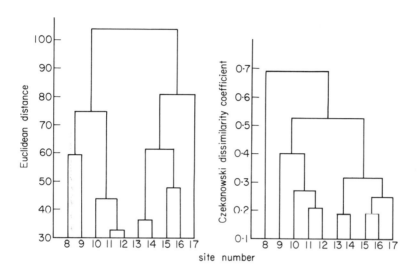

*Fig. 7.5* Classification of sites 8-17 based on the invertebrate
communities using dissimilarity coefficients (Euclidean distance and
Czekanowski coefficient) followed by group average clustering.


The dendrogram based on Euclidean distance followed by group average
clustering provided a clear indication of the importance of the discharge regime, since intermittent sites (8-12) formed one group and
perennial sites (13-17) formed a second group. This feature was particularly notable in view of the enhanced opportunities for migration in
1972 when the stream did not dry out to the perennial head. Within the
group of perennial sites, site 17 was the most distinct, probably
reflecting the species richness at this site. The dendrogram based on
Czekanowski dissimilarity coefficient followed by group average
clustering, provided a slightly different perspective. The pond (site
8) was first delimited from the remaining sites before the next major
division successfully differentiated the intermittent (9-12) from the
perennial sites (13-17). The physical environment of the pond
contrasted strongly with the stream sites (9-17) and this was reflected
in the many taxa only recorded from the pond. In addition, a number of
typical stream  invertebrates failed to colonize the pond.

    Ham *et al.* (in prep.), using similar clustering techniques, also demonstrated a major division between the invertebrate communities of intermittent and perennial sites on the Lambourn, as did Furse (1977a) for the Gussage, using a combination of ordination and clustering procedures.


## 7.63   FAUNAL GROUPINGS

Table 7.4 lists the taxonomic groups which were normally identified to genus or species.  They have been divided into three major groups as follows: species present at all sites; species restricted to the intermittent sites (8-12); and species restricted to the perennial sites (13-17).  Within the intermittent sites, pond species restricted to site 8 are distinguished with an asterisk, as are species in the perennial stream which were restricted to site 17, adjacent to the Lambourn.

In autumn 1972, when discharge remained too high for the stream to dry to the perennial head and the distinction between intermittent and perennial sites was not reinforced, some species normally restricted to one zone may have moved downstream or upstream.  Species which retained their distinctive distributions are likely to be those most character-istic of either intermittent or perennial sites.

The occurrence of *Gammarus pulex* at all ten sites throughout 1972 implies an ability to survive in damp gravel during periods of drought or to migrate upstream from the perennial zone when water returns. The latter explanation is favoured since only large specimens of *G. pulex* were found in reinundated sections of the Lambourn, which suggested upstream migration of large individuals (Ham *et al.*, in prep.). Clifford (1966) noted that only small amphipods (*Crangonyx*) survived summer and autumnal dry periods by burrowing deep into the damp substratum.

Of the Ephemeroptera, only *Baetis vernus* occurred at all 10 sites, but *B. rhodani* and *Ephemerella ignita* were found at 9 and 8 sites, respectively.  These three species were also successful at colonizing the intermittent section of the Lambourn (Ham *et al.*, in prep.).

The ability of Trichoptera of the family Limnephilidae to exploit intermittent sites is already well documented (Clifford, 1966; Williams & Williams, 1975; Wiggins *et al.*, 1980) and three species were recorded at each of the 10 sites.  Furse (1977a) also recorded one of these, *Limnephilus lunatus*, as an early colonizer of winter-bourne reaches of the R. Piddle in Dorset.

Since there are approximately 500 species of Chironomidae known to occur in the British Isles, it was not surprising that they also contributed the most taxa to colonize the 10 sites.

Finally, four species of oligochaetes also occurred at 10 sites and all appear to have some resistance to drought. *Eiseniella tetraedra* can aestivate when pools dry out (Reynolds, 1977) and several authors record its ability to survive drought (Hynes, 1958; Legier & Talin, 1973; Casey & Ladle, 1976; Furse, 1977a). In general, tubificids are largely confined to permanent waters, but Williams & Hynes (1976) observed that some tubificids, probably *Tubifex tubifex*, secreted mucous envelopes around themselves in drought conditions and Casey & Ladle (1976) and Furse (1977a) produced evidence of *T. tubifex* colonizing intermittent sites. Of the two lumbriculids, *Lumbriculus variegatus* is known to be capable of resisting drought conditions by encysting (Cook, 1969). Although *L. variegatus* and *Stylodrilus heringianus* were both found in the Winterbourne Stream (Table 7.4) and also the South Winterbourne (Casey & Ladle, 1976), Ham *et al.* (in prep.) did not find either in the intermittent section of the Lambourn following the severe drought of 1976.

There are a number of additional taxonomic groups with species restricted to intermittent sites, compared to the wide-ranging species in the previous column of Table 7.4 although no oligochaetes are confined to the intermittent zone. *Niphargus aquilex* is widely distributed in groundwater in southern England and Wales (Gledhill *et al.*, 1976) and is liable to occur where this reaches the surface. It has also been recorded at intermittent sites on the Lambourn (Ham *et al.*, in prep.) and the Gussage (Furse, 1977a).

The group with the greatest number of species colonizing the intermittent sites was the Coleoptera and, although nine species were restricted to the pond and some were facultative colonizers of streams, a few were typical of intermittent streams. Legier & Talin (1973) reported that Coleoptera were better represented in intermittent than perennial streams and Hynes (1958) found that after an unprecedented drought, additional species of dytiscids colonized the Afon Hirnant in Wales for a limited period of time. Nevertheless, it is clear that the Winterbourne Stream has an unusually rich fauna. Without further investigation we can only speculate on the possible causes of this. The pond at site 8 has been shown to attract Coleoptera, but farms alongside the stream and houses in Winterbourne village may also have small water bodies from which species can colonize the stream. The lower part of the catchment supports heathland and copses which may offer suitable refuges for species not usually associated with the few bodies of open water found on chalk downland. The lack of drying out of this part of the Winterbourne Stream during and immediately prior to 1972 also means that colonization was possible over an unusually long period of time.

During an extensive survey of 98 chalk stream sites including sites 1-17 on the Winterbourne Stream, Cooling (1981) recorded *Hydroporus marginatus, H. pubescens, Agabus bipustulatus* and *Colymbetes fuscus* from intermittent sites only. In contrast, he recorded *Agabus biguttatus* and *Hydrobius fuscipes* from both intermittent and perennial

Table 7.4   A list of taxa recorded at every major study site (8-17), together with those restricted either to intermittent (8-12) or perennial sites (13-17). All taxa with some records from both intermittent and perennial sites are excluded unless they were recorded at every site.

| taxon | present at 10 sites | restricted to intermittent sites (8-12); *species restricted to site 8 (pond) | restricted to perennial sites (13-17); *species restricted to site 17, adjacent to R. Lambourn |
|---|---|---|---|
| Malacostraca | Gammarus pulex | Niphargus aquilex | |
| Ephemeroptera | Baetis vernus | Cloeon dipterum*; Caenis horaria | Centroptilum pennulatum*; Paraleptophlebia submarginata*; Centroptilum luteolum; Ephemera danica; Caenis rivulorum. |
| Coleoptera | Helophorus spp. | Laccophilus minutus*; Guignotus pusillus*; Coelambus confluens*; Graptodytes pictus*; Hydroporus incognitus*; Agabus sturmi*; A. nebulosus*; Acilius sulcatus*; Laccobius bipunctatus*; L. striatulus*; Coelambus impressopunctatus; Hydroporus marginatus; H. memnonius; H. nigrita; H. pubescens; Agabus biguttatus; A. didymus; Colymbetes fuscus; Dytiscus marginatus; Hydrobius fuscipes; Anacaena limbata | Potomonectes depressus ssp. elegans*; Brychius elevatus; Oreodytes sanmarki; Platambus maculatus; Agabus chalconatus; Hydraena gracilis; Limnebius truncatellus; Oulimnius tuberculatus; Riolus subviolaceus |
| Trichoptera | Limnephilus lunatus; Halesus sp. Stenophylax sp. | Limnephilus vittatus; Glyphotaelius pellucidus | Chaetopteryx villosa*; Melampophylax mucoreus*; Bereodes minutus*; Lype reducta; Hydroptila sp. Ithytrichia sp.; Oxyethira sp; Beraea maurus; Odontocerum albicorne; Anthripsodes albifrons; Adicella reducta; Brachycentrus subnubilus; Sericostoma personatum |

| | | | |
|---|---|---|---|
| Chironomidae | Macropelopia sp.; Conchapelopia gp.; Brillia modesta; Prodiamesa olivacea; Micropsectra sp. A | Psectrotanypus various*; Chironomus sp. B*; Micropsectra sp. C*; Natarsia sp.; Diamesa insignipes; Limnophyes sp. | Eukiefferiella sp. B*; Orthocladius sp.*; Epoicocladius flavens*; Polypedilum sp. A*; Larsia sp.; Cricotopus sp. D; Heleniella ornaticollis; Phaenopsectra sp.; Polypedilum sp. B; Polypedilum sp. C; Polypedilum sp. D; Rheotanytarsus sp.; Stempellina bausei |
| Oligochaeta | Tubifex tubifex; Lumbriculus variegatus; Stylodrilus heringianus; Eiseniella tetraedra | | Psammoryctides barbatus; Peloscolex ferox; Aulodrilus pluriseta |
| Odonata | Agrion splendens | | |
| Bivalvia | Pisidium personatum | | Pisidium milium |
| Hemiptera | | Sigara nigrolineata*; S. lateralis*; Notonecta maculata; Corixa punctata; Sigara concinna | Gerris sp. nymph*; Corixa panzeri* |
| Plecoptera | Nemoura cinerea | | Nemurella picteti |
| Tricladida | | | Polucelis felina |
| Gastropoda | | | Valvata piscinalis*; Physa fontinalis |
| Hirudinea | | | Theromyzon tessulatum |
| Megaloptera | | | Sialis lutaria |

sites in other streams.  These two species were restricted to intermittent sites in the Winterbourne Stream in 1972 but not in 1976
(Cooling *et al.*, in prep.).  Foster (1980) has examined the records of
the two 'subterranean' species *Hydroporus marginatus* and *Agabus
biguttatus* in the U.K. and although the former does appear to be
closely associated with hard water, the latter has a much wider distribution and its occurrence in hard water is  probably because this
provides suitable subterranean habitats.

The Ephemeroptera, Trichoptera and Odonata which were limited to
the intermittent section (Table 7.4), were still or slow-flowing water
forms which occurred in the pond at site 8 or in nearby sites (9, 10)
downstream.

Chironomidae were also well represented, with both still water
(*Psectrotanypus varius*) and cold water (*Diamesa insignipes*) forms plus
a member of the Metriocnemini (*Limnophyes* sp.) which are mainly terrestrial or semi-aquatic.  All the Hemiptera except *Sigara concinna* (site
11 only) occurred in the pond (site 8) but some also penetrated further
downstream.

*Pisidium personatum* was the only bivalve limited to the intermittent
stream but *P. casertanum* also occurred predominantly in this zone and
both species are known to be capable of tolerating drought conditions
(Boycott, 1936).

The plecopteran *Nemoura cinerea,* confined to the intermittent stream,
has been widely reported in these conditions (Legier & Talin, 1973;
Casey & Ladle, 1976; Ham *et al.*, in prep.).  It has also been recorded
in the Afon Hirnant in Wales in the aftermath of a drought (Hynes,
1961); the species had not been recorded previously in this stream and
did not reappear after passing through a single generation.  Khoo
(1964) found that *N. cinerea* had a long egg diapause and also showed
that hatching from a single batch of eggs occurred over a long period.
Since the eggs have a gelatinous coat which protects them from drying
out amongst the stones of a dry river-bed, the species has an effective method of coping with the intermittent nature of the flow regime,
provided that nymphs can grow and emerge before the stream dries in
summer.

Taxa restricted to perennial sites need less detailed consideration
than those of the intermittent section.  The major feature which
distinguishes the two lists is the greater species richness of the
Ephemeroptera, Trichoptera, Chironomidae and Oligochaeta in the
perennial zone and the occurrence of a number of species in other
taxonomic groups (Tricladida, Gastropoda, Hirudinea, Megaloptera) which
were absent from the intermittent section.  In contrast, it was notable
that fewer Coleoptera were restricted to the perennial zone than to the
intermittent zone.  The majority of the species recorded exclusively
from the perennial zone also occurred in the Lambourn and relied on
continuous discharge throughout the year for their survival.

7.64   THE 1976 DROUGHT

When a drought appeared imminent early in 1976, a new sampling
programme was initiated to record changes during and after the drought
(Cooling *et al.*, in prep.).  In April 1976 the upstream limit of water was
the perennial head at site 13 and by August 1976 the upstream limit was
site 15.  Only sites 15-17 remained wet until December, when flow
returned to 13 and 14 and by February 1977 to all of sites 8-17.  No
macrophytes were eliminated by the drought but several showed poor
growth and sites which remained wet in 1976 became heavily silted.

Although the number of taxa at the lowest two perennial sites (16,
17) remained fairly stable between April 1976 and August 1977, the
fauna at site 15, which became the upstream limit of water in 1976,
suffered a substantial decrease in the number of taxa, but recovered
during the study period.  Sites upstream of 15, which dried out, showed
a rapid recolonization upon the return of water and this was still
continuing in August 1977.


7.7    FISH

The brown trout *Salmo trutta* and the bullhead *Cottus gobio* are the most
common fish species in the Winterbourne Stream, as they are in the
Lambourn.  Three-spined stickleback *Gasterosteus aculeatus*, ten-spined
stickleback *Pungitius pungitius* and brook lamprey *Lampetra planeri* occur
locally, particularly in beds of *Rorippa* in the lower reaches of the
perennial section.  The grayling *Thymallus thymallus* is common in the
Lambourn but was never caught in the Winterbourne Stream although it
has been reported in the lower part.  The Winterbourne Stream has been
used successfully as a nursery stream for young salmon (*Salmo salar*).
These were introduced by Thames Water Authority as part of their
programme to re-establish the species in the catchment of the Thames
(M. Bulleid, pers. comm.).


7.71   BULLHEADS

The bullhead population of the Winterbourne Stream was studied in
detail by Green (1975) who obtained samples by electric fishing short
sections of the stream.  He found that bullheads survived for up to
three years but that fish in their first year were always the majority
of the population.  Numbers increased during the summer with the
recruitment of fry and then fell through the rest of the year.  Some
fish spawned when one year old but most did not do so until the end
of their second year and very few survived beyond that age.

A survey of bullheads at 17 sites was carried out from 27 March to
18 April 1973.  Population densities ranged from 0.2 - 1.2 m$^{-2}$ with a
mean of 0.69 m$^{-2}$ at the 15 sites in the perennial section. No bullheads

were found at the two sites sampled in the intermittent section but this may be due to a small weir at Kimber's Cottage which may obstruct upstream movement of bullheads. Green found that the highest population densities occurred at sites with a high cover of macrophytes and with coarse gravel or loose stones. Areas of exposed, compact gravel or of silt were less-well populated.

The uppermost study site in the perennial section, at Kimber's Cottage, was first sampled in March 1972. At that time no O group bullheads were found and the population consisted of I group fish only. This suggests that no fry had hatched there in 1971. The stream dried as far down as Honeybottom for a short period in the autumn of 1969 as a result of trial pumping operations. The site at Kimber's Cottage may not have been recolonized by bullheads within the following 15 months, but a further year had allowed recolonization to occur. Fry were found at the site after the 1972 breeding season. This indicates that bullheads do move upstream to colonize available areas but that the rate of movement is probably slow. Recolonization of the Gussage by non-salmonid species was also slow (Furse, 1977a).

During 1972-73 Green estimated the biomass and production of bullheads at three sites on the Winterbourne Stream (Table 7.5). Biomass varied greatly at different times and both biomass and production were highest at the uppermost site; the missing age-class (mentioned above) at this site is apparent in Table 7.5. The results are similar to those obtained for sites on the Lambourn (Green, 1975) and on the Bere Stream and the R. Tarrant (Mann, 1971).

Table 7.5 Estimates of biomass and production of bullheads in 1972-73 (from Green, 1975).

| site | biomass ($g\ m^{-2}$) | | production ($g\ m^{-2}\ yr^{-1}$) | | | | |
|---|---|---|---|---|---|---|---|
| | min | max | egg | 0 | I | II | total |
| A (near 13 in Fig. 7.1) | 0.98 | 10.84 | 0.6 | 15.9 | 0.0 | 2.8 | 19.3 |
| B (between 15 and 16 in Fig. 7.1) | 0.11 | 2.89 | 0.3 | 2.2 | 0.6 | 0.4 | 3.5 |
| C ( " " " " " | 0.09 | 8.94 | 0.2 | 2.9 | 0.5 | 0.5 | 4.1 |

Green found that bullheads fed entirely on benthic invertebrates. The fry ate ostracods, cladocerans and chironomid larvae while larger fish fed mainly on Gammarus but also took larval stages of various aquatic insects. The observations are similar to those from other chalk streams (Mann & Orr, 1969; Abel, 1973) and the differences that occur are probably due to the local abundance of particular invertebrates.

## 7.72   TROUT

The trout population was sampled by electric fishing at least once each
year from 1972 to 1976.  This always covered the 1.5 km from the
flighting pond to site 13 and sometimes included the 230 m between the
pond and the gauging weir or a stretch above site 13.  On most occasions two fishings were carried out and all fish caught on the first
fishing were marked by injection of a spot of dye so that they could
be identified on the second fishing.  A different batch mark was used
on each occasion.  The results have not been fully examined and only a
preliminary account of some aspects is presented here.

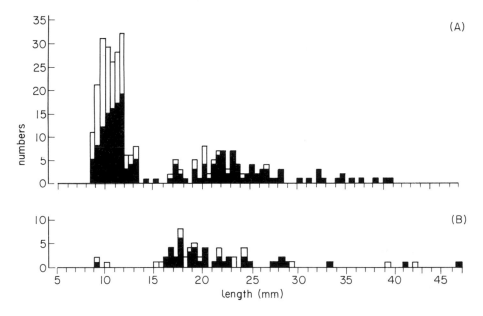

*Fig. 7.6* Length frequency distributions of trout from above the
flighting pond in (a) January 1972, and (b) December 1972.  The
black areas indicate marked fish that were recaptured.

The size structure of the trout population above the flighting pond
in January and December 1972 is shown in Fig. 7.6.  These were second
fishings and recaptured fish are indicated in black.  The January
histogram corresponds conveniently with three age groups which can be
deduced from experience of ageing Winterbourne trout from their scales.
The large group of fish under 15 cm in length are young trout which
hatched in the spring of 1971.  The group between 15 and 30 cm in
length includes trout that are one and two years old while the fourteen
individuals over 30 cm in length are probably three and four years old.
By comparison the December histogram shows that very few young trout
hatched in spring 1972 and the numbers of older fish are considerably
lower than in January.  These impressions are confirmed when the recap-

ture data are used to obtain estimates of the trout population at both times (Table 7.6). The calculation for the young fish must be treated with caution since the efficiency of catching them is lower than for the larger fish and they are more liable to die as a result of handling during the marking operation. Similar changes in the trout population and especially in the recruitment of young fish were noted in 1971 and 1972 in the Chitterne Brook, a winterbourne tributary of the R. Wylye in Wiltshire (Avon & Dorset River Authority, 1973).

*Table 7.6* Population estimates for trout above site 16 (in Fig. 7.1) derived from numbers marked and recaptured.

| date | age class | number marked | number re-captured with marks | total re-capture | popu-lation estimate |
|------|-----------|---------------|-------------------------------|------------------|----------------------|
| January 1972 | 0 | 254 | 105 | 199 | 481 |
|  | >0 | 130 | 82 | 97 | 154 |
| December 1972 | 0 | 4 | 1 | 3 | 12 |
|  | >0 | 55 | 43 | 61 | 78 |

The next fishing during the mild drought in August 1973, showed that the number of larger fish had fallen to an estimated 17 and indicated a better hatch of young trout that year. During 1974 and 1975 the numbers of older trout rose to around the December 1972 levels but the high population of January 1972 was not equalled. In three other small chalk streams Mann (1971) has shown that the biomass and production of trout are generally less than those of bullheads and that 0 group trout contribute most to total production.

In August 1972 electric fishing was carried out for about 1 km above Kimber's Cottage and 20 trout were caught. Four of these had been marked below Kimber's Cottage in December 1971 and the group included no 0 group trout. When this section was fished again in December 1972 no trout were caught. Between 1971 and 1973 a substantial number of trout were marked in the perennial section of the Winterbourne Stream and in the part of the Lambourn downstream from the confluence. Good recoveries were obtained for fish in their home streams but there were only a few cases in which trout were found to have moved between the two streams. In contrast, results from a trap at the lower end of the Chitterne Brook showed that in 1972 mature trout migrated into it from the R. Wylye after flow began in the winter and that adult trout and large numbers of young trout left the brook before it dried in the summer (Avon & Dorset River Authority, 1973). However, in other years substantial numbers of fish had been found stranded when the brook dried. A study on the Candover Brook (Solomon & Temple-

ton, 1976) showed that there was an annual pattern of local movements
in the trout population but that less than 1% made extensive movements.
Spawning fish moved as far as 1.5 km above the perennial head when the
winterbourne section was flowing but most failed to return downstream
before the section dried again.  It is still not clear whether winterbournes act as nursery streams which supplement recruitment of trout
in the main river.

Young trout fed mainly on *Gammarus* and the larval stages of aquatic
insects but older trout relied less on these and took substantial
quantities of surface food and fish.  One third of all large trout
examined from the Winterbourne Stream had eaten fish, mainly bullheads
and sticklebacks.  Thus, young trout utilized the same food as bullheads
but older trout extended their range of food.


## 7.8  CONCLUSIONS

The Winterbourne Stream has substantial growths of macrophytes, a rich
invertebrate fauna and viable populations of fish.  Clear differences
exist between the communities of plants and invertebrates in the
perennial and intermittent parts of the stream despite the fact that
the stream does not always dry to the perennial head each year.  The
data on fish are more limited but indicate that neither bullheads nor
trout are successful in exploiting the intermittent zone.  Williams &
Hynes (1977) have suggested that the invertebrates found in temporary
streams may be divided into three main groups.  First, stream species
capable of surviving drought for limited periods; second, facultative
colonizers capable of living in both lotic and lentic waters; and
third, species highly adapted and often restricted to temporary waters.
These groups can be recognised in the Winterbourne Stream.  The intermittent zone contains a number of invertebrates that have not been
recorded at many other sites in Great Britain whereas the perennial
zone is essentially similar to other chalk streams.

Several small chalk streams in the south of England are already
affected by water engineering schemes and others are likely to be
affected in the future.  If these schemes produce substantial changes
in the flow conditions of winterbournes they may have detrimental
effects on species whose survival depends on an intermittent flow
regime.  Information is still lacking on the precise requirements of
these species and on their adaptations for surviving and exploiting
the alternating wet and dry conditions.


*Acknowledgements*

This study was financed by the Thames Conservancy and the Water
Resources Board and by their successors, Thames Water Authority and the
Central Water Planning Unit, as part of an ecological study of chalk
streams.  We are grateful to both organizations for their support and

to successive Heads of the Department of Zoology in the University of Reading for making facilities available for the work. Special thanks are due to the landowners in the Winterbourne valley who gave us access to the stream and to colleagues who took part in the field surveys and in processing the samples, particularly Dr P.D. Hiley, Dr G.P. Green, Mrs A.C. Cameron, Mrs P.R. McLeish and Mrs M.E. Wigham. Thames Water Authority provided data on discharge and water chemistry together with some of the information for Section 7.2. We are also grateful to Mr D.A. Cooling who developed the cluster package; Mrs A.M. Matthews who drew the figures; and Mr M.T. Furse who provided helpful comments on the manuscript.

## REFERENCES

Abel R. (1973) The trophic ecology of *Cottus gobio* L.. Unpublished Thesis, University of Oxford, England

Avon and Dorset River Authority (1973) Upper Wylye investigation. *A report on the Effects of Abstraction from a Chalk/Upper Greensand Aquifer*, 120 pp.

Boycott A.E. (1936) The habitats of freshwater Mollusca in Britain. *J. Anim. Ecol.* 5, 116–86

Brettell E.J. (1971) *Report on the Lambourn Valley Pilot Scheme, 1967–1969*, 170 pp. Thames Conservancy.

Casey H. (1969) The chemical composition of some southern English chalk streams and its relation to discharge. *River Auth. Ass. Yb. 1969*, 100–13

Casey H. & Newton P.V.R. (1972) The chemical composition and flow of the South Winterbourne in Dorset. *Freshwat. Biol.* 2, 229–34

Casey H. & Newton P.V.R. (1973) The chemical composition and flow of the River Frome and its main tributaries. *Freshwat. Biol.* 3, 317–33

Casey H. & Ladle M. (1976) Chemistry and biology of the South Winterbourne, Dorset, England. *Freshwat. Biol.* 6, 1–12

Clifford H.F. (1966) The ecology of invertebrates in an intermittent stream. *Invest. Indiana Lakes & Streams. VII*, 2, 57–98

Cook D.G. (1969) Observations on the life history and ecology of some Lumbriculidae (Annelida, Oligochaeta). *Hydrobiologia* 34, 561–74

Cooling D.A. (1981) Records of aquatic Coleoptera from rivers in southern England. *Entomologist's Gaz.* 32, 103–13

Cooling D.A., Wright J.F., Ham S.F. & Berrie A.D. (in prep.) The effects of the 1976 drought on the invertebrate fauna of a small chalk stream in Berkshire, England

Foster G.N. (1980) Subterranean statistics II. *The Balfour-Browne Club Newsletter* 15, 3–7

Furse M.T. (1977a) *An ecological study of the Gussage, a lined winterbourne 1973–1977*. A report to the Water Research Centre, 363 pp.

Furse M.T. (1977b) The ecology of the Gussage, a lined winterbourne. *Annual Report, Freshwat. biol. Ass. U.K.* 45, 30–6

Gledhill T., Sutcliffe D.W. & Williams W.D. (1976)  A Revised Key to
    the British species of Crustacea:  Malacostraca occurring in Fresh
    Water with notes on their Ecology and Distribution, 72 pp.  *Sci.
    Publs Freshwat. biol. Ass., U.K.* 32

Green G.P. (1975)  The food and production of the bullhead, *Cottus
    gobio* L. in the River Lambourn.  Unpublished thesis, University of
    Reading, England

Ham S.F., Cooling D.A. & Berrie A.D. (in prep.)  The distribution of
    macroinvertebrates in the River Lambourn, Berkshire, England

Ham S.F., Wright J.F. & Berrie A.D. (1981)  Growth and recession of
    aquatic macrophytes on an unshaded section of the River Lambourn,
    England, from 1971 to 1976.  *Freshwat. Biol.* 11, 381-90

Hynes H.B.N. (1958)  The effect of drought on the fauna of a small
    mountain stream in Wales.  *Verh. int. Verein. theor. angew. Limnol.*
    13, 326-33

Hynes H.B.N. (1961)  The invertebrate fauna of a Welsh mountain stream.
    *Arch. Hydrobiol.* 53, 344-88

Khoo S.G. (1964)  Studies on the biology of stoneflies.  Unpublished
    thesis, University of Liverpool, England

Klee C.A. (1974)  *Some aspects of the ecology of the River Thet*, 27 pp.
    Great Ouse River Authority

Klee C.A. (1976)  *Report on Fish Ecology and the Groundwater Develop-
    ment Scheme*, 80 pp.  Great Ouse River Division, Anglian Water
    Authority

Ladle M. & Bass J.A.B. (1975)  A new species of *Metacnephia* Crosskey
    (Diptera: Simuliidae) from the south of England, with notes on its
    habitat and biology.  *Hydrobiologia* 47, 193-207

Ladle M. & Bass J.A.B. (1981)  The ecology of a small chalk stream and
    its responses to drying during drought conditions.  *Arch. Hydrobiol.*
    90, 448-66

Le Cren E.D. & Lowe-McConnell R.H. (1980)  *The Functioning of Fresh-
    water Ecosystems*, 588 pp.  Cambridge U. P., Cambridge, England

Legier P. & Talin J. (1973)  Comparaison de Ruisseaux permanents et
    temporaires de la Provence Calcaire (Comparison of permanent and
    temporary streams of calcareous Provence).  *Annls Limnol.* 9, 273-92

Mann R.H.K. (1971)  The populations, growth and production of fish in
    four small streams in southern England.  *J. Anim. Ecol.* 40, 155-90

Mann R.H.K. & Orr D.R. (1969)  A preliminary study of the feeding
    relationships of fish in a hard-water and a soft-water stream in
    southern England.  *J. Fish Biol.* 1, 31-44

Ratcliffe D.A. (Ed.) (1977)  *A Nature Conservation Review.*  Vols. 1 &
    2, 401 & 320 pp.  Cambridge U. P., Cambridge, England

Reynolds J.W. (1977)  The earthworms (Lumbricidae and Sparganophilidae)
    of Ontario.  *Roy. Ont. Mus. Life. Sci. Misc. Publ.* 1-141

Sneath P.H.A. & Sokal R.R. (1973)  *Numerical Taxonomy*, 573 pp.  W.H.
    Freeman & Company

Solomon D.J. & Templeton R.G. (1976)  Movements of brown trout *Salmo
    trutta* L. in a chalk stream.  *J. Fish Biol.* 9, 411-23

Southern Water Authority (1979)  *The Candover Pilot Scheme, Final
    Report,*  165 pp.

Water Resources Board (1973) *Water Resources in England and Wales.* Volume 1: Report, 67 pp. Volume 2: Appendices, 56 pp. H.M.S.O., London

Westlake D.F., Casey H., Dawson H., Ladle M., Mann R.H.K. & Marker, A.F.H. (1972) The chalk stream ecosystem. In: (Eds. Z. Kajak & A. Hillbricht-Ilkowska) *Proc. I.B.P. UNESCO Symposium on Productivity Problems of Freshwaters,* Kazimierz Dolny, pp. 615-35. Warszawa-Kraków

Wiggins G.B., Mackay R.J. & Smith I.M. (1980) Evolutionary and ecological strategies of animals in annual temporary pools. *Arch. Hydrobiol., Suppl.* 58, 97-206

Williams D.D. & Hynes H.B.N. (1976) The ecology of temporary streams. I. The faunas of two Canadian streams. *Int. Rev. ges. Hydrobiol.* 61, 761-87

Williams D.D. & Hynes H.B.N. (1977) The ecology of temporary streams. II. General remarks on temporary streams. *Int. Rev. ges. Hydrobiol.* 62, 53-61

Williams D.D. & Williams N.E. (1975) A contribution to the biology of *Ironoquia punctatissima* (Trichoptera: Limnephilidae). *Can. Ent.* 107, 829-32

Wright J.F., Hiley P.D., Cooling D.A., Cameron A.C., Wigham M.E. & Berrie A.D. (1984) The invertebrate fauna of a small chalk stream in Berkshire, England, and the effect of intermittent flow. *Arch. Hydrobiol.* 99, 176-99

Wright J.F. (1984) The chironomid larvae of a small chalk stream in Berkshire, England. *Ecol. Ent.* 9, 231-8

# 8 · Lot

H. DÉCAMPS, J. CAPBLANCQ & J. N. TOURENQ

8.1    Introduction                                              208

8.2    General data                                              209

8.3    Geography and hydrology                                   210

8.4    Transport of materials                                    214

    8.41   Dissolved substances                               214
    8.42   Particulate substances                             217

8.5    Phytoplankton                                             219

    8.51   Development of populations                          219
    8.52   Primary production                                  221
    8.53   Phytoplankton growth                                225

8.6    Fauna                                                     230

8.7    Hydraulic management of the Lot                           230

8.8    Conclusion                                                232

References                                                       232

## 8.1 ¯INTRODUCTION

The Lot provides a typical example of a river domesticated by man.
Diverse developments have profoundly modified flow in order to make the
river navigable and to produce electricity.  Since the 14th century,
the construction of navigation routes has transformed the middle and
lower Lot into a succession of 62 stages.  Nowadays the passage of
water through one-third of these routes is used for electricity production and there are nine large hydroelectric dams in the upper part of
the catchment.

   All these transformations have been accompanied by wastes originating
from domestic, industrial and agricultural sources.  The river has thus
changed rapidly during recent decades.  Towards the end of the 1960s,
diverse observations showed the existence of zones of doubtful water
quality with, in numerous places, a turbid or greenish appearance, due
apparently to the proliferation of algae.  It was important to understand the causes of these events.  Such has been the principle objective of the studies made on the development of materials in suspension,
particularly the phytoplankton, in waters of the Lot (Bordes *et al.*,
1973; Dauta, 1975a, b; Massio, 1976; Denat, 1977).

*Fig. 8.1A*  Entry into the
Lot of the grossly pollu-
ted Riou Mort; note the
sharp transition between
the two waters.

*Fig. 8.1B*  Weir on the
Lot downstream of the
Riou Mort; the boulders
have frequent growths of
the moss *Rhynchostcgium
riparioides*.

## 8.2   GENERAL DATA

### DIMENSIONS

| | |
|---|---|
| Area of catchment | 118240 km$^2$ |
| Length of river dealt with in chapter | 313 km below confluence of Truyère and Haut-Lot |
| Confluence of Truyère and Haut-Lot | |
| Highest point in catchment | 1858 m (Plomb du Cantal) |
| Drop in height down 313 km stretch | 230 m |

### CLIMATE

Precipitation                                         mean 986 mm yr$^{-1}$ (1971-1975)
Occurs mainly in winter, with snow at the higher altitudes

### GEOLOGY

Crystalline and intrusive rocks in the upstream part, calcareous rocks
in the middle and Tertiary rocks covered by alluvial formations in the
lower part of the basin.

### POPULATION

Mean density                                         31 people km$^2$
Principal conglomerations in the catchment
   Decazeville    26000 people
   Cahors         19000 people
   Fumel          14000 people

### LAND USE

Pasturage in the upper and middle parts of the catchment
Agriculture (fruit, cereals) in the lower catchment
Industrial zones at Decazeville and Fumel

### PHYSICAL FEATURES

Discharge at selected points (for period 1962-1975: all as m$^3$ s$^{-1}$)

| | mean annual mean | minimum mean annual | maximum mean annual |
|---|---|---|---|
| Truyère - Haut-Lot | | | |
| confluence (km 167) | 110 | 70 | 170 |
| Cahors (km 330) | 150 | 110 | 200 |

Seasonal changes in discharge (for period 1962-1975)
At km 330 mean monthly rate ranged from 14 m$^3$ s$^{-1}$ in August 1967
                                    to   515 m$^3$ s$^{-1}$ in January 1966

Influence of regulation
The reservoirs in the upper part of the catchment are filled by the end
of the winter and may be restored again during summer. According to
the year, variation can reach as much as 10$^8$ m$^3$ s$^{-1}$ depending on the
natural flow. In summer peaks of 40 - 60 m$^3$ s$^{-1}$ are superimposed in a
basal flow of 7 m$^3$ s$^{-1}$ at km 178.

### TEMPERATURE

In summer 1976, at the top of one stretch of downstream river (km 330),
temperatures reached 29 $^{\circ}$C near the surface and 23 $^{\circ}$C at the bottom (6 m).

## 8.3   GEOGRAPHY AND HYDROLOGY

The lot rises at 1295 m altitude, flowing for 491 km before its entry to the Garonne at an altitude of 22 m (Fig. 8.2).  The basin contains three main geological regions (Fig. 8.3):

Basin of the Truyère and Haut-Lot, with crystalline and intrusive
rocks;  leached brown soils, flanked by calcareous ones on
the souther border;
Basin of the middle Lot, with calcareous Jurassic rocks from
Causses of Quercy;
Lower basin with Tertiary rocks covered by alluvial formations;
soils of calcareous marl or poorly permeable calcareous clay.

The Atlantic, Mediterranean and montane influences combine to give rain, particularly in winter, this being accentuated in the mountains by snow, notably in the north-east part of the basin (Fig. 8.3).

The populations density of the Lot basin (31 inhabitants $km^{-2}$) is well below the national average.  Uncultivated zones, with few people, occur in the mountains and the Causses, limestone plateaus essentially given over to breeding animals.  Agriculture follows the alluvial axis and spreads out in the lower basin where fruit and cereal production dominate.  As for industry, which is relatively little developed, this is concerned especially with agro-food (milk in upstream areas, fruit and vegetables downstream) and, in association with the coal-mining regions of Decazeville and Fumel, the extraction of materials and metallurgy.  These zones are, together with Cahors, the source of the principal pollutants of the Lot (Fig. 8.2).

The valley of the Lot is thus very diverse.  The elongated form of the basin and the localization of important centres near the river lead to conflicting interests:  the needs for drinking water, irrigation, industries, electricity production, fishing and leisure.  The growth of these uses, aggravated by the weakness of low river flows and pollution of all  sorts, makes it apparent that improvement in water quality is an essential preliminary for all economic activity in the basin (Anonymous, 1973).

With a mean annual discharge of 150 $m^3$ $s^{-1}$ at its confluence with the Garonne, the Lot carries 46% of the precipitation falling on the catchment (986 mm $yr^{-1}$).  The natural flow regime corresponds to that of a Mediterranean type river:  high from December to April, with a maximum in February, low from July to September.  In summer the rarity of rain and intensity of evapotranspiration bring about low flows, to the extent that in one year out of ten the mean discharge for August falls below 9 $m^3$ $s^{-1}$ at the junction with the Garonne.  At Cahors (km 330), the ratio of minimum to maximum discharges, which averaged 1 : 20 for the period 1962-75, can drop to 1 : 40.  Another feature is that the regime of the Lot depends closely on the zone situated upstream of the Lot - Truyère confluence, which covers only 46% of the basin, but contributes 62% of the discharge (Fig. 8.4).

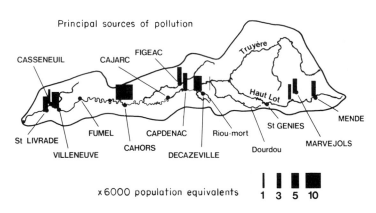

*Fig. 8.2*   The Lot basin, showing the main towns and principal sources of pollution (after Tourenq *et al.*, 1978).

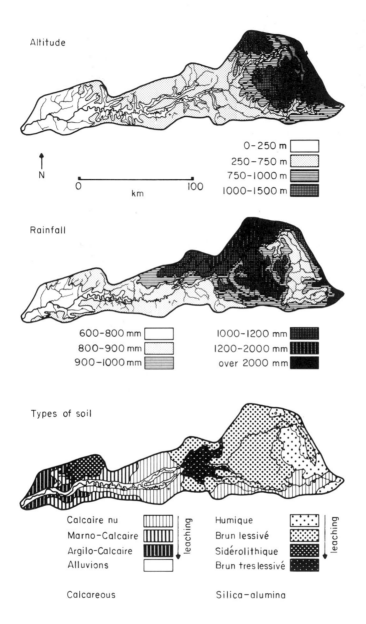

Altitude

0-250 m
250-750 m
750-1000 m
1000-1500 m

N

0        km        100

Rainfall

600-800 mm          1000-1200 mm
800-900 mm          1200-2000 mm
900-1000 mm         over 2000 mm

Types of soil

Calcaire nu                    Humique
Marno-Calcaire                 Brun lessivé
Argilo-Calcaire                Sidérolithique
Alluvions                      Brun tres lessivé

leaching                       leaching

Calcareous                     Silica-alumina

*Fig. 8.3*  The Lot basin:  altitude, rainfall and soil (after Tourenq *et al.*, 1978).

*Fig. 8.4* Precipitation and natural discharge:
(1) At the confluence of Haut-Lot and Truyère (km 178); discharges
from the Truyère are always more than that of the Haut-Lot.
(2) At Cahors (km 330), discharges added between km 178 and km 330 (b)
are low in comparison with the total discharge (a).


The natural regime is profoundly modified by the construction of
large reservoirs for hydroelectric purposes:  two on the Haut-Lot (22
x $10^6$ $m^3$ capacity) and seven on the Truyère (538 x $10^6$ $m^3$).  As a
rule these reservoirs are filled by the end of winter in order to maintain summer-flows, but their management varies and, according to the
year, there is a variation of ± $10^8$ $m^3$ in natural summer discharges.
In fact, with a summer base-flow of 6 $m^3$ $s^{-1}$ at the confluence of the
Lot and Truyère, from 40 to 60 $m^3$ $s^{-1}$ is added over a 4 - 6 h period
each day.  Further, these fluctuations are shifted in an often anarchic
fashion by small hydroelectric plants.  Consequently the times for
renewal of the water are changing constantly in consecutive stretches
of the middle and downstream course of the river; these stretches are
1 to 5 km long and 2 to 10 m deep.  To give an example, if there is a
discharge of 20 $m^3$ $s^{-1}$, the mean current does not exceed 10 cm $s^{-1}$ in
the middle and lower river and the passage of the upper water takes
one month (Fig. 8.5).

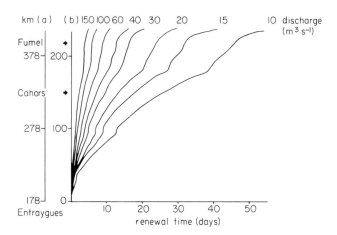

*Fig. 8.5* Change in the time for passage of water for different discharges below the Haut-Lot - Truyère confluence (after Tourenq *et al.*,
1978):
(a)   km from the source of the Lot
(b)   km from the confluence of the two rivers.


## 8.4   TRANSPORT OF MATERIALS

### 8.41   DISSOLVED SUBSTANCES

The waters of the Lot are essentially calcareous bicarbonate (Table 8.1);
$Ca^{2+}$ and $HCO_3^-$ represent on the average 62% of the cations and 82% of
the anions, calcium bicarbonate contributing from 59% (Truyère) to 79%
(Haut-Lot at km 431) of the conductivity of the water.

However, the chemical characteristics of the water differ entirely
between the Truyère and the Haut-Lot upstream of their confluence
(Table 8.1).  The Truyère drains a crystalline basin, characterized by
waters which are more acidic and three times less mineralized than those
of the Haut-Lot, whose basin is predominantly calcareous.  Besides,
Truyère water has a higher silica content, often exceeding 10 mg $l^{-1}$
$SiO_2$.  On passing downstream the water of the Truyère increases progressively in $Ca^{2+}$ and $HCO_3^-$, whilst the contents of $Na^+$, $K^+$, $Mg^{2+}$ and
$SiO_2$ remain more or less constant.  The ratio of Haut-Lot as to Truyère,
which is influenced by the hydraulic management of the large reservoirs
upstream, determines the chemical composition of water downstream of
the confluence (Angelier *et al.*, 1978; Capblancq & Tourenq, 1978).
The concentration (S) of dissolved mineral elements depends on the
ratio (X) of Truyère discharge : (Truyère + Haut-Lot) discharge at the
confluence (km 178).  The following relationships have been established:

*Table 8.1* Chemical composition of Truyère and Lot water at six stations. (Means for concentrations measured from 1971 to 1975, together with standard deviation. (After Capblancq & Toureng, 1978)

| | Truyère | Haut-Lot | middle and lower Lot | | |
| --- | --- | --- | --- | --- | --- |
| | | | km 271 | km 358 | km 431 |
| conductivity (μmhos. cm⁻¹ at 20 °C) | 56.25 ± 6.12 | 167.5 ± 13.1 | 143.6 ± 10.5 | 186.5 ± 9.8 | 207.3 ± 15.4 |
| $HCO_3^{--}$ (meq l⁻¹) | 0.46 ± 0.07 | 1.59 ± 0.08 | 1.14 ± 0.09 | 1.57 ± 0.15 | 1.87 ± 0.14 |
| $SO_4^{--}$ (meq l⁻¹) | 0.08 ± 0.02 | 0.10 ± 0.03 | 0.21 ± 0.04 | 0.32 ± 0.07 | 0.25 ± 0.05 |
| $Cl^-$ (meq l⁻¹) | 0.03 ± 0.01 | 0.06 ± 0.01 | 0.14 ± 0.02 | 0.18 ± 0.02 | 0.14 ± 0.02 |
| total anions (meq l⁻¹) | 0.58 ± 0.09 | 1.75 ± 0.12 | 1.50 ± 0.16 | 2.07 ± 0.23 | 2.26 ± 0.21 |
| $Ca^{++}$ (meq l⁻¹) | 0.27 ± 0.06 | 1.22 ± 0.23 | 0.86 ± 0.17 | 1.14 ± 0.15 | 1.64 ± 0.26 |
| $Mg^{++}$ (meq l⁻¹) | 0.14 ± 0.02 | 0.41 ± 0.10 | 0.38 ± 0.04 | 0.35 ± 0.03 | 0.40 ± 0.05 |
| $Na^+$ (meq l⁻¹) | 0.14 ± 0.02 | 0.15 ± 0.02 | 0.18 ± 0.02 | 0.17 ± 0.02 | 0.19 ± 0.02 |
| $K^+$ (meq l⁻¹) | 0.03 ± 0.01 | 0.04 ± 0.01 | 0.04 ± 0.004 | 0.04 ± 0.01 | 0.04 ± 0.01 |
| total cations (meq l⁻¹) | 0.57 ± 0.10 | 1.82 ± 0.36 | 1.45 ± 0.23 | 1.69 ± 0.21 | 2.27 ± 0.33 |
| $SiO_2$ (mg l⁻¹) | 8.42 ± 1.17 | 5.40 ± 1.0 | 6.10 ± 0.61 | 6.24 ± 0.63 | 6.40 ± 0.93 |
| $NO_3 - N$ (μg l⁻¹) | 316 ± 38 | 385 ± 55 | 380 ± 47 | 402 ± 60 | 524 ± 86 |
| $PO_4 - P$ (μg l⁻¹) | 8.96 ± 2.5 | 13.5 ± 4.3 | 10.6 ± 3.32 | 10.23 ± 2.63 | 12.94 ± 3.54 |
| total Fe (μg l⁻¹) | 310 ± 22 | 240 ± 34 | 570 ± 110 | 480 ± 55 | 390 ± 80 |

at km 220
$$(S) \text{ mg } l^{-1} = 172.5 - 112.5 \text{ X} \qquad (r = 0.852, 26 \text{ DF})$$

at km 271
$$(S) \text{ mg } l^{-1} = 223 - 88.2 \text{ X} - 24.7 \text{ Q} \quad (r = 0.776, 42 \text{ DF})$$

at km 358
$$(S) \text{ mg } l^{-1} = 254 - 80.3 \text{ X} - 21.8 \text{ Q} \quad (r = 0.623, 38 \text{ DF})$$

(where Q = logarithm of discharge at station considered).

During the year the $SiO_2$ content decreases in spring when the diatom population is growing, whilst the content of filtrable reactive phosphorous is stripped when the Chlorococcales proliferate. In the Lot the lower concentrations of nitrate (20 - 30 µg $l^{-1}$) and filtrable reactive phosphorus (0.1 µg $l^{-1}$) have been found with chlorophyll *a* concentrations of the order of 20 to 30 µg $l^{-1}$. The contents of filtrable reactive and total (filtrable + particulate) phosphorus, which are the highest in winter when drainage of the basin is intense and biological activity least, return to 5 and 30 - 90 µg $l^{-1}$, respectively, during summer. The ratio of mean nitrate to mean filtrable reactive phosphorus is of the order of 25 : 1 in winter and autumn, 60 : 1 in summer and, in certain stretches, over 800 : 1 in July and August.

Filtrable organic materials (< 0.5 µm diameter) form 27 - 86% of the total organic material transported by the Lot; their concentration is relatively constant ($\bar{x}$ = 7.73 mg $l^{-1}$, SD = 3.1). They are weakly degradable; $BOD_5$ has a mean value of 2.8 mg $l^{-1}$ and rarely exceeds 5 mg $l^{-1}$. Organic pollution is then reduced and the water of the Lot is generally well oxygenated. However, a thermal and chemical stratification, accompanied by pronounced oxygen deficits at depth, appears in certain stretches when the discharge drops below 20 $m^3$ $s^{-1}$.

At km 220, the Lot receives with the entry of the Riou Mort, which is the outlet for the Decazeville industrial basin, a heavy load of suspended materials and diverse heavy metals. In 1976-77, samples taken from the Riou Mort showed: Mn, 0.35 - 2.65 mg $l^{-1}$; Cu, 0.04 - 0.2 mg $l^{-1}$; Zn, 5.1 - 60 mg $l^{-1}$; Co, 0.11 - 1.28 mg $l^{-1}$. The influence of these additions can still be detected 40 kilometers downstream, where concentrations of 0.34 mg $l^{-1}$ Zn and 0.02 mg $l^{-1}$ Cd have been recorded (Say, 1978).

Altogether the Lot transports around 1.04 x $10^6$ tonnes $yr^{-1}$ of dissolved materials (Capblancq & Tourenq, 1978), with almost 40% coming from the lower basin which covers a fifth of the total basin and where chemical erosion is 166 tonnes $km^{-2}$ $yr^{-1}$. Nitrate and filtrable reactive phosphorus transported during periods of high flow from the upper to the lower basin are 0.85 - 3 kg N $km^{-2}$ $d^{-1}$ and 23 - 66 g P $km^{-2}$ $d^{-1}$, respectively (means from November to April).

8.42   PARTICULATE SUBSTANCES

The Lot transports mineral and organic particles in variable amounts
(Table 8.2).  During the year the relationship 'algal organic materials:
detrital organic materials:mineral materials' changes from 1 : 2 : 3 in
summer to 1 : 20 : 90 in winter.  (Algal organic materials are deduced
from counts.)  The 1 - 20 µm fraction largely predominates; it includes
the largest part of the phytoplankton and is the main cause of the
turbidity (Table 8.3).  In 1974-76, the lower part of the Lot transported about 10000 - 80000 tonnes yr$^{-1}$ of mineral and 3000 - 35000 tonnes
yr$^{-1}$ of organic materials (Décamps & Casanova-Batut, 1978).

*Table 8.2*  Concentration of suspended solids (mg l$^{-1}$ : mean, SD) at
different times of year calculated from samples collected at 15
stations on the middle and lower Lot.  (After Massio, 1975)

| | mineral particles | organic particles | |
| --- | --- | --- | --- |
| | | detritus | algae |
| 20.08.74 | 1.85 ± 0.75 | 1.18 ± 0.60 | 0.69 ± 0.46 |
| 03.12.74 | 1.40 ± 0.65 | 1.25 ± 0.62 | 0.15 ± 0.21 |
| 10.03.75 | 16.74 ± 1.59 | 3.58 ± 1.05 | 0.19 ± 0.24 |
| 22.07.75 | 4.22 ± 1.14 | 2.74 + 0.91 | 1.06 ± 0.57 |

*Table 8.3*  Size distribution of mineral and organic particles in
summer 1975, together with their influence on turbidity.  (All data
given as mean and SD)

| size range (diameter in mm) | mineral particles (mg l$^{-1}$) | organic particles | | turbidity (formazin units) |
| --- | --- | --- | --- | --- |
| | | detritus (mg l$^{-1}$) | algae (mg l$^{-1}$) | |
| total | 4.42 + 0.83 | 4.02 ± 0.65 | 1.50 ± 0.31 | 4.37 ± 0.41 |
| 32 mm | 0.77 ± 0.18 | 0.58 ± 0.10 | | 0.46 ± 0.12 |
| 32 >∅> 20 | 1.10 ± 0.17 | 0.73 ± 0.11 | | 0.38 ± 0.07 |
| 20 >∅> 1 | 2.55 ± 0.62 | 4.21 ± 0.62 | | 2.92 ± 0.33 |
| 1 > ∅ | | | | 0.62 ± 0.22 |

Mineral particles come especially from the upper part of the basin,
notably the Dourdou basin with its sandstone and red clays of the
Permian.  The Riou Mort basin is the source of metallic inputs (Table
8.4).  Further, in certain sections of the Lot, gravel extraction has

disturbed the bed of the river, undermining the banks and adding to the suspended solids. Finally, the upstream dams provoke sharp variations in discharge which lead to resuspension of sediments on the river bed (Galharague & Giot, 1974, 1975a; Décamps et al., 1976).


Table 8.4 Concentrations of various elements in sediments of the Lot ($\mu g \ g^{-1}$) above and below the entry of the Riou Mort

|  | km above entry | km | below entry | |
|---|---|---|---|---|
|  | 2 | 3 | 47 | 198 |
| Mn | 800 | 1300 | 1300 | 1400 |
| Cu | 100 | 280 | 200 | 100 |
| Zn | 350 | 6200 | 6000 | 2900 |
| Cd | 5 | 210 | 180 | 40 |
| Hg | <0.3 | 9.5 | 6 | 2.1 |
| Pb | 120 | 820 | 600 | 600 |


The organic particles include algae and detritus. Their content increases in summer, as the decreased discharge permits the development of phytoplankton (Fig. 8.6). Algae produced in the reservoirs of the Haut-Lot and Truyère feed the lower part of the river continuously. The latter offers ideal conditions for proliferation during summer.

As for turbidity (Galharague et al., 1975b, c, d), it becomes important on the one hand at times of high flow when particles, especially mineral ones, attain densities of the order of $100 \times 10^6 \ l^{-1}$ and, on the other hand, when flows are low, when particles, especially organic ones, exceed $200 \times 10^6 \ l^{-1}$ (Fig. 8.7).

Overall these suspended particles maintain a close relationship with dissolved substances and bottom sediments (Fig. 8.8). Erosion and direct input of wastes putting mineral and non-algal organic materials into suspension. In contrast, a reduction inflow leads to sedimentation of suspended materials and, in combination with increased temperature and light flux, an increase in algal biomass. Materials in suspension interact with each other in the water column. Thus, mineral particles interfere with phytoplankton growth by, for example, reducing radiant energy and adsorbing nutrients.

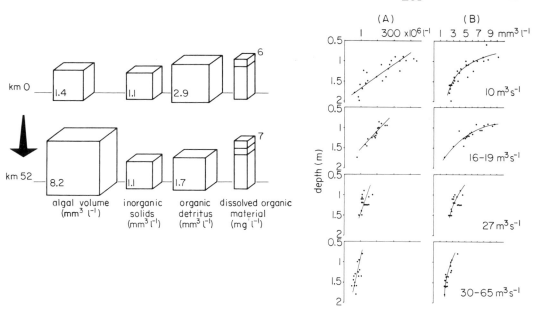

Fig. 8.6 Change in suspended materials and dissolved organic matter in the middle Lot during a passage of water over a distance of 52 km in summer 1975 (after Décamps et al., 1979)

Fig. 8.7 Variation in turbidity (Secchi disc depth) as a function of A) number, and B) volume of 2 - 20 μm diameter particles at different discharges.

## 8.5  PHYTOPLANKTON

### 8.51  DEVELOPMENT OF POPULATIONS

The upper course of the Lot is rapid and its microphytic vegetation is composed almost entirely of benthic diatoms which have been brought into suspension (Dauta, 1975a). Planktonic algae appear in the upper reservoirs; they develop further in the stretches of the middle and lower river course where their biomass varies inversely with discharge, ranging from 0.1 to 35 g m$^{-3}$ by weight and from 0.1 to 45 mg m$^{-3}$ as chlorophyll a (Dauta, 1975b; Massio, 1976; Denat, 1977; Capblancq & Dauta, 1978).

Over 90% of the plankton consists of diatoms and greens. The former, with Stephanodiscus sp., Melosira varians, M. granulata, Fragilaria crotonensis, Cyclotella sp. dominate upstream and form essentially the whole crop in spring. The latter, with nanoplanktonic Chlorococcales (Scenedesmus sp., Ankistrodesmus falcatus, Micractinium pusillum, Dictyosphaerium sp., Selenastrum sp., Coelastrum microporum, C. reticulatum, unicellular (Chlamydomonas sp.) or colonial Volvocales (Eudorina, Gonium, Pandorina morum), develop in the middle and lower

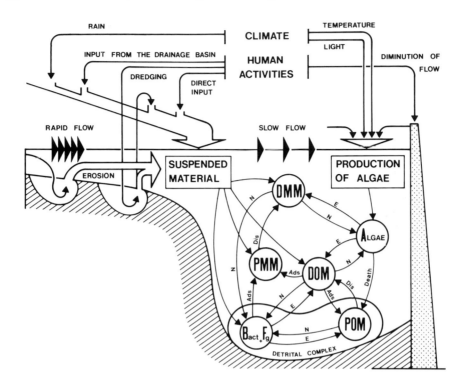

*Fig. 8.8*  Interactions in a body of water and influence of human activities (after Décamps *et al.*, 1979: by permission of Plenum Press).

stretches, forming over 50% of the biomass in summer.  Bursts of *Mougeotia* sp. and blue-green algae (*Anabaena* sp., *Aphanizomenon flos-aquae*) can appear in calmer zones in July and August.

Because of turbulence, the distribution of the phytoplankton is usually homogeneous.  However, a vertical gradient of temperature develops with discharges less than 10 $m^3$ $s^{-1}$.  The algal biomass then increases in the deepest water, as does the proportion of diatoms, the phaeophytin/chlorophyll *a* ratio and the contents of nitrate, ammonium and phosphate.  As a result of this sedimentation of dead algae and their breakdown, an under-saturation of oxygen develops near the bottom (Casanova-Batut, 1977; Capblancq & Dauta, 1978).  But this stratification is always ephemeral; an increase in flow for a few hours re-establishes  the vertical homogeneity of phytoplankton and chemistry.

The seasonal development of phytoplankton in the Lot differs little from that described for numerous meso-eutrophic lakes of the temperate zone.  However, the hydraulic management of the river increases the contrasts.  Thus, it is when renewal of the water is slow and temperatures are high, both linked with feeble flows, that there occur high concentrations of phytoplankton dominated by greens (Fig. 8.9).

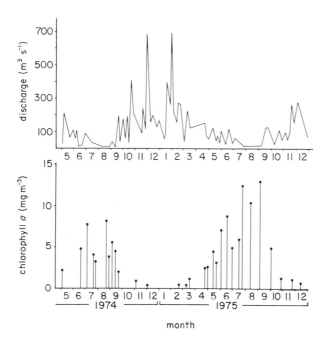

*Fig. 8.9* Changes in discharge and chlorophyll *a* content at km 278 from May 1974 to September 1975.


8.52  PRIMARY PRODUCTION

The attenuation of light by phytoplankton in the Lot is relatively weak in comparison with that due to mineral particles and organic detritus. More precisely, it has been shown that the coefficient of vertical attenuation ($\varepsilon$) is maximal in the blue ($\varepsilon$blue = 1.64 - 3.36; $\bar{x}$ = 2.65), minimal between the green ($\varepsilon$green = 0.83 - 1.6; $\bar{x}$ = 1.23) and the red ($\varepsilon$red = 0.9 - 1.5; $\bar{x}$ = 1.235). One can deduce from these values the transmission of photosynthetically active radiation (PAR) and the depth at which PAR drops to 1% of that immediately under the surface. In the middle and lower course, depth of this euphotic zone ($Z_{eu}$) ranges from 0.7 - 5.3 m; it reaches a mean of 3.1 m during the period of phytoplankton development.

Photosynthetic production by algae at various sites is shown in Fig. 8.10. Neglecting inhibition found in flasks immediately below the surface, the relation between photosynthesis and light can be expressed by the equation (Talling, 1957; Vollenweider, 1965):

$$a = a_{opt} \frac{I/I_k}{\sqrt{1 + (I/I_k)^2}} \qquad (1)$$

where $a_{opt}$ = photosynthesis at light saturation or optimal photosynthetic capacity

and $I_k$   = intensity at the threshold of light saturation

The parameter $a_{opt}$ and $I_k$ vary as a function of the structure of the
phytoplankton population. The optimum photosynthetic capacity ($a_{opt}$)
varies from 0.22 - 25 mg C mg chl $a^{-1}$ (Fig. 8.11). These extremes
correspond, respectively, to high flows and low temperatures on the one
hand and low flows and high summer temperatures on the other hand.
With the former the phytoplankton biomass is feeble, composed of diatoms
partly from the plankton of the upstream reservoirs and partly from the
benthos. With the upper extreme, the population consists of nanoplankton
(greens and cryptomonads); the photosynthetic capacity is thus quite high,
since during 64 measurements made between June and October, $a_{opt}$ reached
a mean value of 8.6 ± 2.3 mg C mg chl $a^{-1}$ $h^{-1}$.

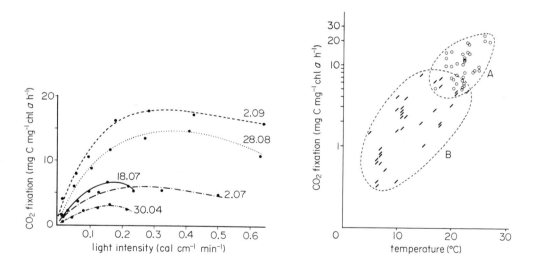

Fig. 8.10 Photosynthetic production by algae in the Lot as a function of light intensity on five different dates.

Fig. 8.11 Optimum photosynthetic capacity of Lot algae as a func-
tion of temperature:
A.  Phytoplankton predominantly of greens;
B.  Phytoplankton predominantly of diatoms.

In the Lot photosynthetic capacity varies with the nature of the
phytoplankton, apparently independently of the biomass.  Thus, the
summer nanoplankton dominated by Chlorococcales is much more active
than the spring plankton of diatoms from the upstream reservoirs (Fig.
8.11).  Further, the photosynthetic capacity of the phytoplankton varies
inversely with the mean volume of the cells (Fig. 8.12).

Fig. 8.12  Optimum photosynthetic
capacity of Lot plankton as a
function of mean cell volume

The light flux at the threshold of saturation ($I_k$) ranges from 0.03
to 0.12 cal $cm^{-2}$ $min^{-1}$, with a mean of 0.973 ± 0.02 cal $cm^{-2}$ $min^{-1}$
during the period of phytoplankton development.  For this intensity,
the ratio, $a$ / $a_{opt}$ is 0.77 ± 0.04.  The parameter $I_k$ depends not
only on $a_{opt}$, but also the initial slope K of the experimental curve
of the efficiency of conversion of light energy.  Now, contrary to
observations made in marine or lacustrine environments, K appears very
variable in the Lot (Fig. 8.9).  There are two possible explanations
for these variations:  (i) Pigment extraction by acetone which is
effective for diatoms is much less for certain Chlorococcales.  This
would explain the variations of the curves for photosynthetic assimilation as a function of light when the results are expressed relative
to chlorophyll a.  (ii) Photosynthetic activity varies inversely with
the volume of the algae (Fig. 8.12) and, therefore, in the Lot the
phytoplankton becomes more active the greater the proportion of nanoplanktonic Chlorococcales.  In support of this explanation is the fact
that algal biomass has been evaluated by two independent methods
(chlorophyll a and volume) and both give similar results.

Since the phytoplankton biomass is usually distributed homogeneously
in the water, vertical differences in photosynthesis depend only on the
influence of depth on light flux.  Under these conditions, instanta-
neous production per unit area of water surface (ΣA) obeys the equation
(Talling, 1957):

$$\Sigma A = \frac{A_{opt}}{\varepsilon} \cdot f \, (I'_0/I_k)$$

(where $A_{opt} = B.a_{opt}$).

In effect, in the Lot the relationships are straightforward
- between photosynthetic production in the water column $\Sigma A$ and the
ratio $A_{opt}/\varepsilon$   (Fig. 8.12):

$$\Sigma A = A_{opt}/\varepsilon.(2.21 \pm 0.15)$$

- between the ratio $\Sigma A.\varepsilon/A_{opt}$ and $I'_o/I_k$ (Fig. 8.13), suggesting
that daily variations in $\Sigma A$ follow the logarithm of $I'_o$ if $I_k$,
$A_{opt}$ and $\varepsilon$ remain constant.

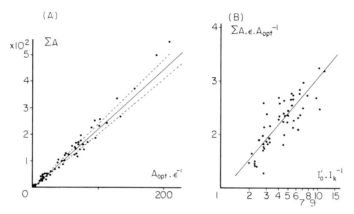

*Fig. 8.13*  Relations in the Lot between:
A.   Photosynthetic production $\Sigma A$ and the ratio $A_{opt}/\varepsilon$.
B.   The ratios $\Sigma A.\varepsilon/A_{opt}$ and $I_o/I_k$.

Therefore, the daily total of photosynthetic production can be estimated (Talling, 1957) by:

$$\Sigma\Sigma A = \Sigma A.(Ln(2\bar{I}_o/I_k)/Ln(2I_o/I_k)).0.9\Delta t$$

(where $\bar{I}_o$ is mean light flux throughout the day immediately
below the surface

$I_o$ is light flux during period of exposure

$\Delta t$ is length of day, in hours.)

The results obtained (Capblancq & Dauta, 1978) show that almost 80%
of annual algal production occurs in the three summer months.  Daily
production was:

|  | min. | max. | mean |
|---|---|---|---|
| October–May | 2 | 245 | 68 mg C m$^{-2}$ d$^{-1}$ |
| May–June | 237 | 1377 | 745 |
| July–September | 504 | 10650 | 2320 |

Overall, algal production can be estimated to be 275 g C m$^{-2}$ yr$^{-1}$, which
indicates a raised trophic level.

## 8.53   PHYTOPLANKTON GROWTH

The increase in the phytoplankton crop results from the difference
between gains measured by photosynthetic production and losses.  Zooplankton are not well developed into the stretches of the Lot, so
predators can be neglected and losses are limited to respiration and
sedimentation of dead algae.

Photosynthetic production and the development of the crop has been
followed in a single body of water during its passage downstream over a
distance of 52 km.  This was carried out from 20 to 29 August 1975 at a
time when discharge was 16 - 19 $m^3$ $s^{-1}$ (Dauta et al., 1975; Capblancq &
Décamps, 1978).  During this period phytoplankton growth consisted
almost entirely of the nanoplanktonic *Coelastrum reticulatum*, whose
density ranged from 500 to 50000 cells $ml^{-1}$.  The corresponding
increase in biomass, expressed as organic carbon calculated from the
relationships between cellular carbon and biovolume (Strathmann, 1967)
and chlorophyll and biovolume, takes the form:

$$C = C_O \cdot e^{0.135\,t} \qquad\qquad (3)$$

t being the time (in days) for passage of the water and $C_O$ the carbon
content at the beginning.  The daily rate of photosynthesis being
measured (Table 8.5) and the daily increase in biomass (P) being calculated from equation (3); the difference A - P represents losses due
to respiration and degradation.  This has been calculated as about 80%
of the carbon incorporated each day by photosynthesis.  In fact the
total loss daily reaches a mean of 1.22 ± 0.46 mg C mg chl $a^{-1}$ $h^{-1}$,
about 15 ± 3.1% of the total for optimum photosynthesis.  In order to
oxidize the carbon lost in this way, the quantity of oxygen needed
reaches only 20 - 33% of the quantity actually consumed in the Lot.  A
large part - over 50% - is thus used for degrading dissolved and organic
materials.

The scheme in Fig. 8.14 has served as the basis for the study of the
influence of different factors on the growth of the algae (Delclaux,
1980).  The assimilation of carbon as a function of temperature and
light has been calculated by integrating results from various depths
according to equation (1).  Respiration has been assumed to be constant
and estimated at 10% of the value for optimum photosynthesis.  As for
the assimilation of nitrogen $U_N$ and phosphorus $U_P$, this has been linked
with the concentration of nitrate and filtrable reactive phosphorus
through a Michaelis-Menten equation.  Algal mortality has been estimated by the increase in the phaeophytin : chlorophyll a ratio with time.
Lastly, the regeneration of nitrogen and phosphorus to  the dissolved
and particulate phases has been estimated as a function of the total
oxygen consumed, using the stoichiometric relationship C : N : P.

The model making the best use of the results obtained for the Lot
resembles those of Droop (1973), Lehman et al. (1975) and Jørgensen
(1976).  The rate of phytoplankton growth is considered there as
dependent on the internal cellular concentration of nutritional elements

*Table 8.5* Phytoplankton metabolism during passage of a water body over a 52 km distance during summer 1975. (After Capblancq & Décamps, 1978)

| time since the start (in days) | A rate of photosynthesis (mg C m$^{-3}$ d$^{-1}$) | P increase in biomass over 24 h (mg C m$^{-3}$ d$^{-1}$) | P/A | A – P respiration (converted to C equivalent) (mg C m$^{-3}$) | O nocturnal use of oxygen (converted to C equivalent) (mg C m$^{-3}$ h$^{-1}$) | $\dfrac{A - P}{O}$ × 100 min. |
|---|---|---|---|---|---|---|
| 0 | 296 | 51 | 0.17 | 245 | – | – |
| 1 | 136 | 74 | 0.54 | 62 | 13.2 | 20 |
| 2 | 644 | 89 | 0.14 | 555 | 78.1 | 30 |
| 3 | 670 | 100 | 0.15 | 570 | – | – |
| 4 | 575 | 113 | 0.20 | 462 | 81.0 | 24 |
| 5 | 390 | 108 | 0.28 | 282 | 36.0 | 33 |
| 6 | 518 | 129 | 0.25 | 389 | 83.0 | 20 |
| 7 | 676 | 159 | 0.23 | 517 | 65.5 | 33 |
| 8 | 919 | 188 | 0.20 | 731 | – | – |

(Q), which itself varies as function of cellular division and assimilation. Cellular division constitutes in effect a dilution factor and assimilation is linked to the external and internal concentrations of nutritional elements. The validity of this model has been tested on eight species in culture (Dauta, 1982a, b) and in one stretch of the Lot (Capblancq *et al.*, 1982).

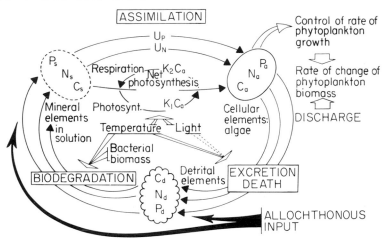

*Fig. 8.14* Nutrient cycle in the Lot. (For guide to symbols see text)

This model (Figs 8.15 - 8.17) provides an explanation of the development of the phytoplankton crop under apparently unfavourable conditions of phosphate nutrition. By storing phosphate, the algae can multiply for several days by using their reserves, thus confirming observations made in batch culture for *Scenedesmus crassus* (Delclaux & Guerri, 1980; Dauta *et al.*, 1982). It can be seen that curves simulating growth of the crop are very sensitive to the initial values of $Q_N$ and $Q_P$, the internal concentrations of nitrogen and phosphorus. In particular, variations in $Q_P$ are closely linked to those of environmental filtrable reactive phosphorus.

The phytoplankton thus increases progressively in successive stretches of the Lot when climatic conditions are favourable. During its downstream passage the algal crop increases as a function of the length of exposure to light, to the concentration of nutrient elements available and to the importance of cellular reserves. The length of exposure of the algae to light influences the net gain in carbon; for any particular site, this depends on the relationship between the depth of the euphotic zone ($Z_{eu}$) and that of the water column, more shallow sites ($Z/Z_{eu} \leqslant 1$) favouring the growth of the algal crop. However, the importance of cellular reserves must also be considered, notably those of phosphorus which is apparently a limiting element. A large proportion of the total phosphorus seems to be associated with complex organic detritus (dissolved and particulate organic materials); this is

Fig. 8.16

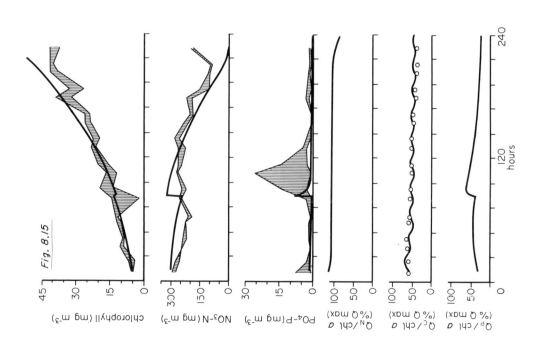

Fig. 8.15

*Fig. 8.15* Changes in phytoplankton chlorophyll a, NO₃-N and PO₄-P (shown as shaded areas) and the ratio in the algal of total C to chlorophyll a during passage of a marked waterbody down 52 km in the middle Lot; time of travel = 240 h.

The curves corresponding to the numerical simulation of changes in biomass, nitrate and phosphate and of contents of algal N ($Q_N$), C ($Q_C$) and P ($Q_P$) for the initial conditions were:

$Q_C : Q_N : Q_P = 60 : 8 : 1$ mg per mg chl $a$; $R/A_{opt} = 0.1$

*Fig. 8.16* Results of simulations showing the importance of respiration of change in biomass:

$R/A_{opt} = 0.05$ $\qquad$ (—·—·—)

$\qquad\qquad$ 0.1 $\qquad$ (——)

$\qquad\qquad$ 0.2 $\qquad$ (------)

*Fig. 8.17* Results of simulations showing the influence of the initial chemical composition of the phytoplankton:

$Q_C : Q_N : Q_P = 30 : 8 : 4$ $\qquad$ (——)

$\qquad\qquad\qquad$ 60 : 8 : 1 $\qquad$ (------)

$\qquad\qquad\qquad$ 60 : 2.5 : 0.4 $\qquad$ (—·—·—)

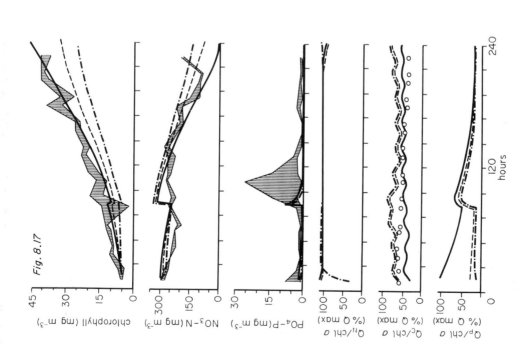

Fig. 8.17

therefore, less available for the algae.  When the concentration of available dissolved phosphorus falls below a threshold value, with up-take no longer compensating for cellular reserves, algal multiplication continues at the expense of internal reserves.  The exhaustion of these last leads progressively to cessation of growth and, with losses through mortality becoming preponderant, to a reduction in the population.  This continues until a new source of nutrient elements - from urban effluents for example - permits cellular stocks to be reconstituted.

## 8.6    FAUNA

The Lot contains various species of fish:

Carnivores:    *Salmo trutta fario, Esox lucius, Micropterus salmoides, Lucioperca lucioperca, Perca fluviatilis, Lepomis gibbosus, Anguilla anguilla, Cottus gobio.*

Omnivores:    *Ameiurus nebulosus, Alburnus alburnus, Leuciscus cephalus, Rutilus rutilus, Scardinius erythrophthalmus, Abramis brama, Blicca björkna.*

Bottom-living:  *Gobio gobio, Leuciscus leuciscus, Barbus barbus, Phoxinus phoxinus, Cobitis barbatula, Chondrostoma nasus.*

Burrowing:    *Cyprinus carpio, Tinca tinca, Chondrostoma toxostoma.*

Migratory:    *Mugil* sp.*, Petromyzon marinus, Lampetra planeri, Alosa alosa.*

The construction of large dams on the Haut-Lot and the Truyère, as well as the weirs in the middle and lower basin, have transformed the distribution of fish in the Lot (Tourenq & Dauba, 1978).  Upstream the drop in water temperature has led to a rapid diminution of numerous white fish, with *Salmo trutta fario* replacing *Leuciscus leuciscus* as the dominant.  Several species have been introduced into the reservoirs of the Haut-Lot and Truyère: *Cyprinus carpio, Tinca tinca, Rutilus rutilus, Esox lucius, Lucioperca lucioperca.*  The weirs further down-stream have stopped the passage of migrators, in particular *Salmo salar* which, although abundant until the middle of the 18th century, has disappeared today.

In spite of these important changes, however, the fish population of the Lot is still abundant and the benthos, especially of chironomids and oligochaetes, relatively rich.

## 8.7    HYDRAULIC MANAGEMENT OF THE LOT

The studies made on the Lot have served to define the objectives for

improving water quality, taking into account the diverse interests of
tourism, agriculture, industry and the valley (Décamps, 1978). The
demands on the water are varied and often antagonistic. It is essential to make sure that water resources are adequate in quantity and
quality to cover present and future needs.

Nowadays a lack of water usually appears in July and August as a
result of the increase in use, notably for irrigation (Table 8.6) and
the management of reservoirs for hydroelectric purposes. These inadequate flows lead to slow passage of the water and, as a consequence,
changes in ecological conditions. The residence time of the water
increases, temperatures rise, phytoplankton increases and domestic and
industrial pollution are concentrated. At the same time, hydroelectric
requirements disturb the regularity of flows.

*Table 8.6* Requirements for irrigation water in the valley of the
Lot. Calculation based on estimate of $1500 \text{ m}^3 \text{ ha}^{-1}$. (Anonymous,
1979)

|  | irrigated surface (ha) | water used $(\text{m}^3)$ | maximum rate of pumping $(\text{m}^3 \text{ s}^{-1})$ |
|---|---|---|---|
| 1979 | 11100 | $17 \times 10^6$ | 5.75 |
| projected increase for following 10 years | 8000 | $12 \times 10^6$ | 4.20 |
| predicted total requirement | 19100 | $29 \times 10^6$ | 9.95 |

The 1975 investigations revealed the changes likely to influence
river communities: (i) floating wastes, (ii) inadequate flows, (iii)
pulses in flow, (iv) pollution by domestic and industrial wastes,
(v) turbidity due to suspended materials, (vi) cleanliness and accessi-
bility of the banks. In fact, the most serious nuisance is the
presence of excessive heavy metal loads below the entry of the Riou Mort.

These data have led to various objectives for water for drinking,
industry and irrigation, compensation for water removed, fish life,
environmental needs and tourism (Anonymous, 1976). A set of actions
have been proposed: (i) combat industrial and domestic pollution,
(ii) maintain flows by the creation of a reservoir ensuring that dis-
charge does not drop below $12 \text{ m}^3 \text{ s}^{-1}$ at the Haut-Lot/Truyère confluence,
(iii) regulate flows by a more rational management of hydroelectric
uses, (iv) control the exploitation of gravel, (v) eliminate floating
wastes, (vi) strengthen the banks.

This series of complementary actions was drawn up to reconcile the diverse uses of a river long domesticated by man. This programme for the hydraulic management of the Lot is now being realized.


## 8.8  CONCLUSION

The hydroelectric management of the Lot has led to reduced discharges in a river already transformed by the needs of navigation. These conditions have favoured the development of phytoplankton: algae produced in spring in the reservoirs of the Haut-Lot and the Truyère are continuously inoculated into the middle and lower sections of the Lot. These algae proliferate when conditions of light and temperature are favourable. Above all, low flows, which lead to slow passage of the water, favour the increase of algal populations. These also assist the sedimentation of suspended materials. In certain stretches a thermal stratification can even appear, with a reduction in dissolved oxygen at depth.

For practical purposes it is apparent that the improvement of flows is essential to improve the quality of the water. The example of the Lot shows in effect that the mode of functioning of river and lake ecosystems can be very similar. In the middle and lower sections of the Lot, more or less pronounced periods of stagnation are frequent in July, August and September; they are accompanied by a simultaneous increase of detrital and phytoplankton organic matter. The principal problem posed is thus to determine the minimum discharge that can maintain biological equilibrium in the river. On what criterion should this be made? Developments occurring in some stretches of the Lot during summer provide one approach: the minimum discharge permissible should be that which prevents the appearance of a stratification in temperature and oxygen in the stretches most threatened by eutrophication. It still remains to determine precisely the geomorphological conditions which produce this phenomenon.


*Acknowledgements*

We thank A. Dauta for calculations made using the numerical simulation model. The figures were produced by H. Casanova and A. Païs.


## REFERENCES

Angelier E., Bordes J.M., Lucchetta J.C. & Rochard M. (1978) Analyse statistique des paramètres physico-chimiques de la rivière Lot. *Annls Limnol.* 14, 39-57

Angelier E. & Décamps H. (1978) Le fonctionnement des ecosystèmes dans les eaux aménagées. *Acad. Agricult. fr.* 741-48

Anonymous (1973) *Livre Blanc de la Vallée du Lot*, 74 pp. Association pour l'Aménagement de la Vallée du Lot.

Anonymous (1976) *Propositions pour un Schéma d'Aménagement Hydraulique de la Vallée du Lot*, 123 pp. SODETEG.

Anonymous (1979) *Propositions d'Actions en Vue de l'intégration de la Vallée du Lot au Plan Décennal de Développement du Grand Sud-Quest*, 31 pp. Mission Interdépartementale et Association pour l'Aménagement de la Vallée du Lot.

Bordes J.M., Lucchetta J.C. & Rochard M. (1973) Etude d'un écosysteme d'eau courante : le Lot. Thèse 3e cycle, Toulouse, 152 pp.

Capblancq J. & Dauta A. (1978) Phytoplancton et production primaire de la rivière Lot. *Annls Limnol.* 14, 85-112.

Capblancq J., Dauta A., Caussade B. & Décamps H. (1982) Variations journalières de la production du phytoplancton en rivière : modélisation d'un bief du Lot. *Annls Limnol.* 18, 101-32.

Capblancq J. & Décamps H. (1978) Dynamics of the phytoplankton in the river Lot. *Verh. int. Verein. theor. angew. Limnol.* 20, 1479-84

Capblancq J. & Tourenq J.N. (1978) Hydrochimie de la rivière Lot. *Annls Limnol.* 14, 25-37.

Casanova-Batut Th. (1977) Etude préliminaire en vue de la modélisation de l'écosysteme Lot. Thèse 3e cycle, Toulouse, 109 pp.

Dauta A. (1975a) Etude du phytoplancton du Lot. Thèse 3e cycle, Toulouse, 107 pp.

Dauta A. (1975b) Etude du phytoplancton du Lot. *Annls Limnol.* 11, 219-38.

Dauta A. (1982a) Conditions de développement du phytoplancton. Etude comparative du comportement de huit espèces en culture. I. Determination des parametres de Croissance en fonction de la lumière et de la température. *Annls Limnol.* 18, 217-62.

Dauta A. (1982b) Conditions de développement du phytoplancton. Etude comparative du comportement de huit espèces en cultures. II. Rôle des nutriments: assimilation et stockage intracellulaire. *Annls Limnol.* 18, 263-92.

Dauta A., Brunel L. & Guerri M.M. (1982) Détermination expérimentale des parametres liés à l'assimilation de l'azote et du phosphore par *Scenedesmus crassus*. *Annls Limnol.* 18, 33-40.

Dauta A., Décamps H., Denat D., Galharague J., Giot D., Massio J.C. & Trolet J.L. (1975) Etude de la turbidité et des matières en suspension dans les eaux du Lot et de leurs relations avec les écoulements. 3. Evolution des matières en suspension et de la turbidité dans une masse d'eau marquée. Rapport B.R.G.M., 75 S.G.N. 388 M.P.Y., 31 pp., 31 plates.

Décamps H. (1978) Qualité des eaux et développement de la vallée du Lot. *Annls Limnol.* 14, 163-79.

Décamps H. & Casanova-Batut Th. (1978) Les matières en suspension et la turbidité de l'eau dans la rivière Lot. *Annls Limnol.* 14, 59-84.

Décamps H., Capblancq J., Casanova H. & Tourenq J.N. (1979) Hydrobiology of some regulated rivers in the Southwest of France. In: Ward J.V. & Stanford J.A. (Eds) *The Ecology of Regulated Streams*, pp. 273-88. Plenum Press, N.Y.

Décamps H., Capblancq J., Casanova H., Dauta A., Laville H. & Tourenq J.N. (1981)  Ecologie des rivières et developpement: l'expérience d'aménagement de la Vallée du Lot.  In: Lefeuvre J.C., Long G. & Ricou G. (Eds)  *Ecologie et Développement*, pp. 219-34.

Décamps H., Massio J.C. & Darcos J.C. (1976)  Variations des teneurs en matières minérales et organiques transportées dans une rivière canalisée, le Lot.  *Annls Limnol.* 12, 215-37.

Delclaux F. (1980)  Production primaire en milieu thermiquement stratifié: modélisation et application au Lot.  Thèse Docteur-Ingénieur, Toulouse, 177 pp.

Delclaux L. & Guerri M.M. (1980)  Biocinetiques de la production primaire, étude expérimentale des fonctions de base de la production algale.  Thèse 3e cycle, Toulouse

Denat D. (1977)  Etude de la production primaire dans le réservoir de Cajarc (Lot).  Thèse 3e cycle, Toulouse, 94 pp.

Droop M.R. (1973)  Some thoughts on nutrient limitation on algae.  *J. Phycol.* 9, 274-82.

Gagneur J. (1976)  Etude des diptères du Lot et étude de la retenue de Cajarc.  Thèse 3e cycle, Toulouse, 189 pp.

Galharague J. (1974)  Exploitation des sables et graviers de la vallée du Lot.  Rapport B.R.G.M., 74 S.G.N. 239 M.P.Y., 30 pp.

Galharague J. & Giot D. (1974)  Etude des matières en suspension dans les eaux du Lot et de leurs relations avec les écoulements.  1e partie: résultats des études qualitatives.  Rapport B.R.G.M., 74 S.G.N. 392 M.P.Y.  33 pp., 2 annexes, 8 plates.

Galharague J. & Giot D. (1975a)  Etude des matières en suspension dans les eaux du Lot et de leurs relations avec les écoulements.  2e partie: résultats des études quantitatives.  Rapport B.R.G.M., 75 S.G.N. 154 M.P.Y.  44 pp., 6 annexes, 21 plates.

Galharague J. & Trolet J.L. (1975b)  Etude de la turbidité et des matières en suspension dans les eaux du Lot et de leurs relations avec les écoulements.  1. Suivi de la turbidité dans les eaux du Lot et liaison avec le débit.  Rapport B.R.G.M., 75 S.G.N. 386 M.P.Y.  17 pp., 33 plates.

Galharague J., Giot D. & Trolet J.L. (1975c)  Etude de la turbidité et des matières en suspension dans les eaux du Lot et de leurs relations avec les écoulements.  2. Evolution des profils en long de Saint-Parthem à Castelfranc à différents débits régularisés.  Rapport B.R.G.M., 75 S.G.N. 387 M.P.Y. 12 pp., 16 plates.

Galharague J., Giot D. & Trolet J.L. (1975d)  Etude de la turbidité et des matières en suspension dans les eaux du Lot et de leurs relations avec les écoulements.  4. Etude des éléments en traces dans les vases pendant l'été 1975.  Rapport B.R.G.M., 75 S.G.N. 389 M.P.Y.  6 pp., 7 plates.

Jørgensen S.E. (1976)  An eutrophication model for a lake.  *Ecological Modelling* 4, 253-78.

Labat R., Roqueplo C., Ricard J.M., Lim P. & Burgat M. (1977)  Actions écotoxicologiques de certains métaux (Cu, Zn, Pb, Cd) chez les poissons dulcaquicoles de la rivière Lot.  *Annls Limnol.* 13, 191-207.

Laur C. (1975)  Etude faunistique du Lot et étude du sédiment de la retenue de Cambeyrac (Truyère).  Thèse 3e cycle, Toulouse, 122 pp.

Lehman J.T., Botkin D.B. & Likens G.E. (1975)  The assumptions and
   rationales of a computer model of phytoplankton population dynamics.
   *Limnol. Oceanogr.* 20, 343-64
Massio J.C. (1976)  Facteurs d'évolution des matières en suspension
   minérales et organiques dans les eaux du Lot.  Thèse 3e cycle,
   Toulouse, 146 pp.
Say P.J. (1978)  Le Riou-Mort, affluent du Lot pollué par métaux lourds.
   I. Etude préliminaire de la chimie et des algues benthiques.  *Annls
   Limnol.* 14, 113-31
Strathmann R.N. (1967)  Estimating the organic carbon content of phyto-
   plankton from cell volume or plasma volume.  *Limnol. Oceanogr.* 12,
   411-18
Talling J.F. (1957)  The phytoplankton population as a compound photo-
   synthetic system.  *New Phytol.* 56, 133-49
Tourenq J.N., Capblancq J. & Casanova H. (1978)  Bassin versant et
   hydrologie de la rivière Lot.  *Annls Limnol.* 14, 9-23
Tourenq J.N. & Dauba F. (1978)  Transformation de la faune des poissons
   dans la rivière Lot.  *Annls Limnol.* 14, 133-38
Vollenweider R.A. (1965)  Calculation models of photosynthesis depth
   curves and some implications regarding day rates estimates in
   primary production measurement.  *Memorie Ist. ital. Idrobiol.,
   Suppl.* 18, 427-57

# 9 · Estaragne

P. LAVANDIER & H. DÉCAMPS

9.1   Introduction                                        238

9.2   General data                                        239

9.3   Geographical, physical and chemical features        240

9.4   Vegetation                                          244

9.5   Benthic invertebrates                               244

    9.51   Composition and distribution                244
    9.52   Autecology of the most abundant species     247

        9.521   *Baetis alpinus*                      247
        9.522   *Rhithrogena loyolaea*                248
        9.523   *Isoperla viridinervis*               248
        9.524   Chironomids                           249
        9.525   Taxa other than insects               250

    9.53   Population dynamics                          250

        9.531   Life cycles                           250
        9.532   Growth and period of development       251
        9.533   Loss of individuals                   251
        9.534   Movement of the benthos               252

9.6   Functioning of the ecosystem                        254

    9.61   Invertebrate feeding                        254
    9.62   Invertebrate production                     254
    9.63   Energy flow                                 255

9.7   Growth of trout in the torrent of Estibère          258

9.8   Conclusions                                         261

References                                                263

## 9.1    INTRODUCTION

The Estaragne shows the rigorous conditions of a torrential high mountain stream: a long period of snow cover, low temperatures, rapid
changes in flow.  As it is situated for the large part above the forest
line, it receives light energy directly when not covered by snow.  It
shelters a rich and diverse community of benthic invertebrates specialized for montane conditions.  The Estaragne drainage basin is characterized also by having only a very slight human influence, a feature of
particular interest at a time when numerous mountain streams have
succumbed to pollution resulting from the development of tourism.

A range of studies have been devoted to the ecology of the Estaragne
and other high mountain streams in the Pyrenees (Angelier, 1980;
Berthélemy, 1966; Dawson, 1973; Décamps, 1967a,b, 1968; Décamps &
Elliott, 1972; Décamps & Lafont, 1974; Décamps & Lavandier, 1972;
Décamps & Pujol, 1977; Dumas, 1976; Elliott, 1973; Giani & Lavandier,
1977; Lavandier 1974, 1975, 1976, 1979, 1981, 1982a,b; Lavandier & Mur,
1974; Lavandier & Pujol, 1975; Laville & Lavandier, 1977; Thomas, 1970).
This chapter is based on the results of these studies.

*Fig. 9.1* Upstream Estaragne.

## 9.2   GENERAL DATA

DIMENSIONS

| | | |
|---|---|---|
| Area of catchment | 6.5 km$^2$ |
| Length of stream | 2.8 km |
| Highest point in catchment | 3006 m |
| Drop in height | 600 m |

LOCATION

42$^\circ$ 48'N   0$^\circ$ 9'E

METEOROLOGY

Snow covers the upper part of the stream for 6 to 8 months and the lower part for 4 to 5 months

LAND USE

Population: none.  Sheep grazing in summer.  Tourist visits.  Basin is included in the Parc National des Pyrénées.

PHYSICAL FEATURES OF STREAM

Discharge at selected points in summer 1972 (in $l\ s^{-1}$):

| | minimum | maximum |
|---|---|---|
| km 0.05 | 50 | 350 |
| km 1.1 | 80 | 1050 |
| km 2.8 | 90 | 1350 |

Seasonal changes in flow - for 1969-1971 mean daily flow at km 1.1:

| | |
|---|---|
| in winter, under the snow | 20 $l\ s^{-1}$ |
| at snow-melt | 1200-1400 $l\ s^{-1}$ |

Substratum - ranges from sheets or rocks to pebbles, with local pockets of sand and under low-flow conditions.

Temperature - maximum between 1971 and 1973:

| km 0.05 | 1.1 | 2.8 |
|---|---|---|
| 4.5 $^\circ$C | 8 $^\circ$C | 13 $^\circ$C |

*Fig. 9.2*
Downstream Estaragne
at Lac Oredon

9.3  GEOGRAPHICAL, PHYSICAL AND CHEMICAL FEATURES

**Estaragne**

| | |
|---|---|
| ♤ | *Abies* |
| ♠ | *Pinus uncinata* |
| ⊕ | " " *sylvestris* |
| ♀ | *Betula* |
| ⊤⊤ | *Rhododendron* |
| ‖ ○ | grassland , peat-bog |
| ⫝̸ | firn |
| ▲····· | altitude (m) |
| ▦ | metamorphic limestone |

Fig. 9.3  The Estaragne and its catchment.

The Estaragne is part of the Neste d'Aure system and thus ultimately of the Garonne.  It lies in the high mountains of the central Pyrenees. It rises at 2450 m and ends at 1850 m at Lac Oredon.   It is only 2.8 km long, with an average slope of 20%.  The width of the stream is mostly between 1.5 and 5 m.  Of the 6.5 km$^2$ catchment area, 75% lies above the 2200 contour (Fig. 9.3).

The basin consists largely of schists and granites, but a calcareous strip forms a break on the slope shortly after the source.  The vegetation cover which is reduced by the source, increases progressively to form a turf of *Festuca eskia*, with *Rhododendron ferrugineum* and *Abies pectinata* in the lower part.  The basin lies within the Pyrenees National Park and human influence is confined to the passage of walking tourists and to the pasturing of a few sheep.

The high altitude and orientation towards the north lead to a rigorous climate.  Thus a permanent bank of consolidated snow (névé) feeds the source and snow covers the upper part of the stream for 6 - 8 months of the year and the lower part for 4 - 5 months.  This snow can reach 5 m in depth towards 2200 m.

Between 1969 and 1971, discharge during winter was as low as 20 l $s^{-1}$ under the snow and frequently exceeded 600 l $s^{-1}$ during snow-melt (Fig. 9.4).  In the absence of snow cover, rain has an immediate, often violent, effect on discharge; its influence on stream water level continues for several days.

*Fig. 9.4* Mean daily discharge of the Estaragne at 2100 m altitude over two hydrological years.

The chemistry of the Estaragne was studied regularly from 1971 to 1973.  The mean values discussed here are taken from the paper by Lavandier & Mur (1974).  Temperature (Fig. 9.5) is close to 0 °C throughout winter.  During summer, it remains close to 5 °C at the

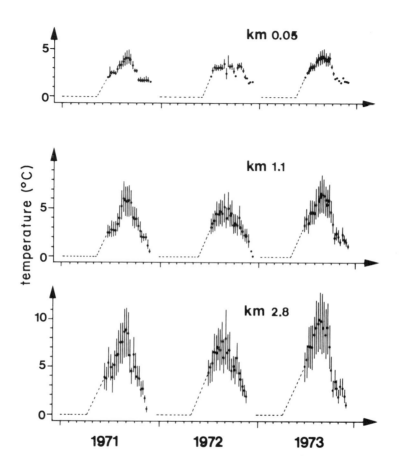

*Fig. 9.5* Mean temperature and weekly range at three stations on the Estaragne.

source and rarely exceeds 12 °C downstream although daily variations can be more than 7 °C. Between 1971 and 1973 the number of degree-days above 0 °C ranged from 450 to 540 near the source and from 940 to 1160 at the mouth.

The concentration of dissolved oxygen is always near saturation. The ionic composition of the water is dominated by $Ca^{2+}$ among the cations and by $HCO_3^-$ and $SO_4^{2-}$ among the anions. Conductivity drops steadily on passing from upstream to downstream (means from 114 µS $cm^{-1}$ to 80 µS $cm^{-1}$ in 1972). The total content of dissolved salts drops because of the input of less mineralized water as soon as the basin becomes granitic (means from 78.7 mg $l^{-1}$ to 53.7 mg $l^{-1}$ in 1972).

The content of most of the elements studied thus drops on passing down-stream (Table 9.1).

*Table 9.1*  Mean concentrations of water chemistry variables (mg $l^{-1}$) in the Estaragne during summer 1972.

| distance from source (km) | Na | K | Mg | Ca | Fe total |
|---|---|---|---|---|---|
| 0.05 | 0.48 | 0.34 | 0.37 | 22 | 0.140 |
| 2.8 | 0.49 | 0.30 | 0.31 | 15.5 | 0.015 |

| distance from source (km) | $HCO_3$ | $NO_3$-N | $PO_4$-P | $SiO_2$-Si | $SO_4$-S |
|---|---|---|---|---|---|
| 0.05 | 57.5 | 0.188 | 0.004 | 1.20 | 2.2 |
| 2.8 | 42.9 | 0.171 | 0.0022 | 1.20 | 1.3 |

The concentrations of individual ions - except $K^+$ - are significantly correlated with flow.  Concentrations of $HCO_3^-$, $SiO_2$, $Ca^{2+}$, $Na^+$, $Mg^{2+}$ at all stations, and in addition $SO_4^{2-}$ and $NO_3$-N near the source, vary inversely with flow.  In contrast, $PO_4$-P at all stations, as well as $SO_4^{2-}$, $NO_3$-N and Fe well away from the source, vary directly with flow. Concentrations of $Cl^-$ (of the order of 0.5 to 0.75 mg $l^{-1}$) appear to vary inversely with flow.

These correlations have made it possible to estimate the quantity of dissolved substances transported by the stream.  For the hydrological year 1970-71, the volume of water which flowed was 7.9 x $10^6$ $m^3$ (7.2 x $10^6$ $m^3$ in 1969-70) and the total content of dissolved substances 497 tonnes (451 tonnes in 1969-70).  About 65% of this load was trans-ported during snow-melt, 30% during the period without a snow cover and 5% in winter.  When calculated per unit area, these values correspond to 765 kg $ha^{-1}$ $yr^{-1}$ in 1970-71 and 694 kg $ha^{-1}$ $yr^{-1}$ in 1969-70. Estimates for individual components for the hydrological year 1970-71 (kg $ha^{-1}$ $yr^{-1}$) are:

|  | mean | confidence limits |
|---|---|---|
| $HCO_3^-$ | 487 | (338 - 769) |
| $Ca^{2+}$ | 177 | (123 - 246) |
| $SO_4$-S | 17 | ( 5 - 38) |
| $SiO_2$-Si | 14.5 | (10.7 - 21.5) |
| $Na^+$ | 5.4 | (3.8 - 7.7) |
| $Mg^{2+}$ | 3.8 | (2.3 - 5.4) |
| $K^+$ | 4.6 | (1.5 - 7.7) |
| $NO_3$-N | 2.3 | (1.2 - 4.6) |
| Fe total | 0.2 | (0.15 - 0.4) |
| $PO_4$-P | 0.03 | (0.01 - 0.06) |

## 9.4 VEGETATION

Right up to the melting snow the bed of the Estaragne is covered with *Hydrurus foetidus* and diatoms, among which the dominants are *Diatoma hiemale, Ceratoneis arcus, Meridion circulare* and *Cymbella ventricosa*. Mosses, such as *Brachythecium rivulare, Cratoneuron commutatum* and *Hygrohypnum molle,* are locally very abundant, especially in the tributaries. This aquatic vegetation plays a fundamental role, since it is an endogenous source of food for herbivores and detritivores.

*Hydrurus foetidus*, the most abundant alga proliferates in the main stream. During its growth, fragments of varying length are stripped by the current, especially in August when the algal population is most developed and in October when it completes its vegetative cycle. Part of the aquatic vegetation of the Estaragne and its tributaries is thus regularly moved by the current. However, after fragmentation, an important part accumulates on stones or is deposited in pools. The fine vegetable debris thus mixes with the mineral particles and is progressively incorporated into the substratum.

In total, the plant biomass transported by the current in 1972 was estimated to be:

> 80 kg fresh weight, 50 m from the source (2370 m)
> 210 kg fresh weight, 1.1 km from the source (2150 m)
> 270 kg fresh weight, 2.8 km from the source (1850 m)

At the two higher stations, in the subalpine meadow, the majority of this biomass is endogenous in origin; at the third station, in the *Rhododendron* - pine zone, it also includes exogenous leaves and woody debris.

## 9.5  BENTHIC INVERTEBRATES

### 9.51  COMPOSITION AND DISTRIBUTION

The benthic community of the Estaragne includes nearly 200 species of invertebrates (Lavandier, 1979). Ephemeroptera (5 species) and chironomids (60 species) form more than 80% of the total number of larvae (Table 9.2). *Baetis alpinus* is by far the most abundant species (Table 9.2).*. Among the more than 600000 individuals collected to provide samples representative of the various conditions in the stream, the two dominant species make up over 50% of the population and the nine most abundant species over 70%.

The temperature regime, length of snow cover and current speed all

---

*Nomenclature used is that of Illies (1978)

*Table 9.2*  Relative abundance of the various groups and main species in the benthic community of the Estaragne.

| | no. of species | percentage |
|---|---|---|
| **Groups:** | | |
| Ephemeroptera | 5 | 47 |
| Diptera Chironomidae | 60 | 35.2 |
| Plecoptera | 28 | 6.7 |
| Oligochaeta | 17 | 4.7 |
| Trichoptera | 27 | 3.1 |
| Planaria | 2 | 1.3 |
| Diptera Empididae | 7 | 0.8 |
| Diptera Simuliidae | 8 | 0.6 |
| other Diptera | 17 | 0.4 |
| other groups | 24 | 0.2 |
| total | 195 | 100.0 |

| **Main species:** | | |
|---|---|---|
| *Baetis alpinus* | Ephemeroptera | 39.6 |
| *Diamesa hamaticornis* | Chironomidae | 10 |
| *Rhithrogena loyolaea* | Ephemeroptera | 7 |
| *Eukiefferiella calvescens* | Chironomidae | 3 |
| *Micropsectra bidentata* | Chironomidae | 3 |
| *Trichodrilus macroporophorus* | Oligochaeta | 2.2 |
| *Orthocladius rivicola* | Chironomidae | 2.9 |
| *Corynoneura lobata* | Chironomidae | 1.6 |
| *Isoperla viridinervis* | Plecoptera | 1.5 |

determine the distribution of invertebrates in the Estaragne.  The coldest upstream stations harbour a stenothermal high montane fauna, while downstream the fauna is more eurythermal.  Species typical of middle or low montane regions appear only in the downstream Estaragne or in some tributaries where thermal and flow conditions are less severe.  Thus, three tendencies can be recognized when interpreting species distribution.

(i)  Certain species decrease in number progressively on passing downstream and disappear altogether at the lowest stations e.g.  the stone-fly larva *Arcynopteryx compacta*, the caddis-fly larvae *Drusus rectus* and some chironomids mainly in the subfamily Diamesinae.

(ii)  Certain species colonize the whole stream.  These include the may-fly larvae *Baetis alpinus* and *Rhithrogena loyolaea*, the oligochaete *Trichodrilus macroporophorus*, the stone-fly larva *Isoperla viridinervis*, all species which are among the most abundant of the Estaragne benthic community (Table 9.2).  Others are the planarian *Crenobia alpina*, the caddis-fly *Drusus discolor* and chironomids of the subfamily Orthocladiinae.

(iii)  Species of the mid-montane zone increase progressively downstream: *Polycelis felina, Leuctra lamellosa, Allogamus auricollis, Nais alpina, Protonemura vandeli*.

Overall the fauna increases in diversity on passing downstream (Fig. 9.6).  Its density, everywhere around or above 10000 individuals m$^{-2}$, is greatest in the subalpine zone (about 2000 m).

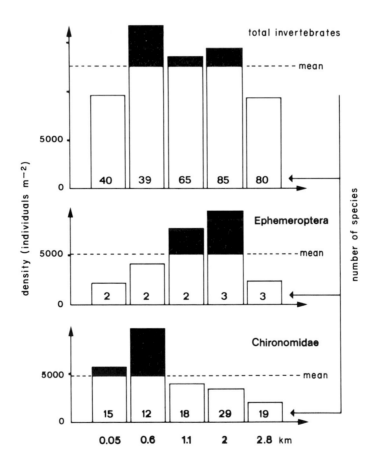

*Fig. 9.6* Mean population density and number of species at five stations on Estaragne, showing total fauna, Ephemeroptera and Chironomidae.  Parts of populations which exceed the mean for all stations are in black.

The adults play an essential role in the distribution of the benthic insects. Reproductive females provide a real mixing of the community within the modest dimensions of this hydrographical system. This distribution of egg-laying tends each year to enlarge the territory of the species: the young larvae invade the whole of the network, but then disappear at stations less suited for their development.

Sandy substrata, which form where the current is moderate during periods of low flow, but disturbed frequently when the flow increases, shelter only forms capable of burying themselves in the interstices without harm: small larvae, oligochaetes, chironomids, empidids. The mobility of the substratum prevents the development of algae and food is restricted to fine exogenous vegetable debris. At downstream sites some detritivores of larger size appear in pools where sand and vegetable debris accumulate. This is the situation with the (Plecoptera) Nemouridae.

The underside of pebbles and rocks subject to rapid currents form more populated biotopes and shelter the principal rheophilous invertebrates of the Estaragne. The stability of the substratum permits the growth of *Hydrurus foetidus* on various materials. Both intra- and inter-specific competition is severe, especially among carnivores. Thus territorial needs appear to limit the number of large individuals of *Perla grandis* to only one or two in the small cascades of one of the Estaragne tributaries. Similarly, certain sets of species, which have close ecological niches, tend to exclude each other during the course of their development e.g. the caddis larvae *Rhyacophila intermedia* and *R. evoluta* or the stone-fly larvae *Isoperla viridinervis* and *Arcynopteryx compacta*.

Altogether the population is diverse, and enriched in moderate to rapid currents. Of course the distribution - and consequently the interrelations between individuals - changes constantly in space and time as a function of the biology of the species and environmental conditions. Thus the majority of species become more rheophilous during growth. The fauna is however concentrated in the axis of the flowing water during the low winter flows under the snow, but when the cover breaks up, the animals shelter under rocks, in crevices or in the deeper layers of the stony bed. In each of these situations new relationships are established, with the formation of more or less ephemeral micro-communities.

## 9.52  AUTECOLOGY OF THE MOST ABUNDANT SPECIES

### 9.521 *Baetis alpinus*

This species colonizes the whole hydrographical network. The females lay their eggs indiscriminantly in the Estaragne and its tributaries, including zones unfavourable for the development of the species. After hatching the young larvae move in large numbers into the deeper layers

of the stony bed.   They then colonize various biotopes and tend to move
towards faster currents during their growth; they are thus frequently
transported downstream.   At any given station the biomass lost in this
way can exceed one-third of the annual production.   These losses are
compensated for by a high reproductive ability (700 - 2300 eggs per
female) and by migration of reproductive females upstream.

    The length of the life of a larva depends on local conditions of
temperature and snow cover.   At the lowest stations, the life cycle
takes one year; the full-grown larvae arc small (5 ɱ mean length) and
carry a maximum of 1800 eggs.   At the highest stations development
takes two years; the larvae reach a mean length of 7 mm at emergence
and can lay more than 2500 eggs.

    In fact, over the whole of the Estaragne the number of larvae dimi-
nishes strongly during the course of the summer.   Downstream, where
the females are small, of 1000 eggs hatched in autumn, 100 larvae
survive to the end of winter and about ten get as far as the instar
preceeding emergence after one year of larval existence.   Upstream,
where females are larger, of 1000 eggs laid by one female in autumn,
30 larvae survive the end of the first winter and two to the instar
preceding emergence after two years of larval life.   Thus upstream
population are maintained only thanks to the eggs deposited by females
flying from further downstream.   In the neighbourhood of the source,
the population of *Baetis alpinus* can even depend exclusively on this
input since the larvae appear unable to reach maturity here.   '

### 9.522 *Rhithrogena loyolaea*

This, the other may-fly abundant in the Estaragne, has a longer period
of larval development, namely two years in downstream and three years
in upstream stretches.   The mean length of mature larvae is 10 mm and
there is no significant difference in  size between upstream and downstream larvae.   Virtually all growth takes place during the period
without snow cover.   The population of *R. loyolaea* thus appears stable
from one year to the next.   Upstream, from a mean clutch of 4000 eggs
laid, about one-third hatch (Humpesch & Elliott, 1980).   400 larvae
survive after one year, 50 after two years and less than 10 reach the
end of the last instar after three years.   Drift losses are much less
than with *Baetis alpinus* and represent only 15% of the annual larva
production.   This last is greatest in the middle region of the stream,
at about 2100 m, where it ranges from 0.9 to 1.8 g $m^{-2}$ $yr^{-1}$.

### 9.523 *Isoperla viridinervis*

This stone-fly develops over three years upstream and two years downstream.   Unlike the preceding species, it is essentially carnivorous,
capturing for 9/10 of its life the larvae of chironomids and *Baetis*

*alpinus*. Its behaviour varies markedly during the course of the growth
of the animal. Young larvae move into the sediment, regularly ingesting
mineral particles and capturing diverse first instar larvae and small
oligochaetes. At an older stage the larvae of *I. viridinervis* become
very rheophilous and change their food. They capture larvae of
chironomids and *Baetis alpinus* on the *Hydrurus foetidus* or, temporarily
during the snow melt, larvae of *Rhithrogena loyolaea* which have a similar habitat.

Drift, predominantly nocturnal, affects all larval stages of *Isoperla
viridinervis*. The resulting loss of biomass can exceed one-fifth of
the annual production. In compensation the mature larvae tend to move
actively against the current, but it is mainly because of fertile
females flying upstream that the larval density is maintained at the
highest altitudes.

Each female lays about 600 eggs between June and September, depending
on altitude. Maximum numbers of larvae occur in September-October,
corresponding to when the young larvae climb to the bed of the flowing
water. The number decreases strongly during the period without snow,
especially in upstream populations.

The production of larvae is greatest ($0.84$ g m$^{-2}$ yr$^{-1}$) at 2190 m
where the population is most abundant. Downstream populations are less
productive ($0.09 - 0.16$ g m$^{-2}$ yr$^{-1}$), but, in spite of their lower
density, can produce as much or slightly more than those living near
the source ($0.11$ g m$^{-2}$ yr$^{-1}$). The mature larvae, which essentially
determine production measurements, are in effect relatively more frequent downstream because of a less rapid loss of individuals (lower
mortality, shorter life cycle).


## 9.524 *Chironomids*

Several chironomids are among the most abundant species (Table 9.2).
They are much affected by the current. During the whole period lacking
a snow cover one can observe a larval drift, this being greatest in the
middle of the day. The extent of this phenomenon is augmented by the
constant stripping of *Hydrurus* filaments to which are attached numerous
Diamesinae larvae. For these last, the annual biomass lost in this way
can exceed one-fifth of the annual larval production.

The dynamics of the Diamesinae populations reflect those of the
dominant species, *Diamesa hamaticornis*, whose eggs, laid in autum mainly
hatch at the end of winter. The Orthocladiinae and the Tanytarsini, in
contrast, hatch in abundance from autumn onwards, the numbers of larvae
decreasing much more among the former than the latter during the early
instars. In fact some Diamesinae larvae and all the Orthocladiinae
larvae hatched in autumn disappear into basal water during winter;
they reappear in large numbers after the snow melt in the *Hydrurus*
tufts. It is probable that these larvae move into the interstices of

the substratum.  The larvae of Tanytarsini, which are more rheophilous, stay in the axis of the stream in winter where they remain vulnerable to carnivores.

After the flood due to snow melt, the density of chironomid larvae diminishes in an irregular fashion.  This is accelerated, particularly for the Diamesinae and the Orthocladiinae, by the drift of old *Hydrurus* tufts at the end of summer.  At the same time the larvae increase in weight very rapidly, especially *Diamesa hamaticornis*.  Altogether the larval production of Diamesinae is 1.3 g m$^{-2}$ yr$^{-1}$ near the source, 1.5 g m$^{-2}$ yr$^{-1}$ at 2150 m and 0.5 g m$^{-2}$ yr$^{-1}$ downstream.  This population constitutes the majority of chironomid production in the Estaragne.

### 9.525 *Taxa other than insects*

Two groups of worm reach a noteworthy abundance, the oligochaetes and the planarians (Table 9.2).  The oligochaete populations appear quite stable.  The most abundant species, *Trichodrilus macroporophorus* is almost alone in colonizing the most rapid and coldest waters.  Downstream the higher temperatures and accumulation of vegetable debris favour a variety of oligochaetes.  As for the planarians, which are more numerous towards the end of summer, they have frequently been discovered to feed on adult females of *Baetis alpinus* after oviposition and also on caddis  larvae.  For both the oligochaetes and the planarians, production is higher downstream, where it reaches about 0.2 g m$^{-2}$ yr$^{-1}$.

Two groups of benthic invertebrates of small size have not been countered.  *Hydracarina* occurs particularly in mosses in these high altitude waters.  The copepod *Bryocamptus zschokkei* is very abundant in the under-flow and is certainly important as food for young insect larvae.

### 9.53  POPULATION DYNAMICS

### 9.531 *Life cycles*

The majority of Estaragne species develop in one year.  This, for example, is the case with Chironomidae, Simuliidae and Leuctridae.  It is also the case with species which come to maturity only in lower parts of the stream e.g. the caddis *Allogamus  auricollis*.  With other species development takes one year downstream and two years upstream.  One can distinguish, for example in  *Baetis alpinus*, two types of population as a function of altitude; these can co-exist at the same station with relative proportions differing from year to year as a result of climatic conditions.

Various Estaragne species are characterized by a life cycle which is always longer than one year.  Some, as the caddis *Drusus rectus* and *D.*

*discolor*, have a two-year cycle over the whole stream system; others develop over two or three years, depending on the altitude. This is the situation with the four largest species: *Isoperla viridinervis*, *Rhyacophila intermedia*, *R. evoluta* and *Rhithrogena loyolaea*. The caddis *Micrasema morosum*, which lives in mosses of the cold tributaries, has the same type of life cycle.

Three other carnivorous stone-fly larvae always develop over three years. These are *Perla grandis* restricted to the warmest tributary, *Arcynopteryx compacta* living in the highest parts of the system and *Chloroperla breviata* present along the whole stream. Finally the detritrivore stone-fly larva, *Pachyleuctra benlocchi*, has a four-year cycle in the Estaragne, one of the longest known for an aquatic insect.

It should be pointed out that oligochaetes, planarians and aquatic beetles certainly survive after their reproductive period for several years.

## 9.532 *Growth and period of development*

Growth, which is negligible or nil in winter under the snow, takes place essentially during the period without snow cover. It is then approximately exponential. It is slow for many species as the snow melts, then accelerates with the summer rise in temperature. The growth rate of an individual is related principally to the mean temperature of its surroundings and the lower temperatures near the source are most often responsible for the lower growth rate here. The length of biological cycle is thus also tied to length of snow cover and to temperature.

## 9.533 *Loss of individuals*

During the period without snow cover, there are successive population explosions due to the rapid biological cycle and high reproductive ability. The greatest changes in numbers observed between two successive cohorts are in the populations with an annual cycle, like the chironomids (5-fold increase) and those resulting from females coming from downstream, like *Baetis alpinus* in the neighbourhood of the source (13-fold increase). For species with a longer life cycle, a more or less identical number of individuals reach the end of each cycle. These are the species which constitute the stable and permanent inhabitants.

Over the year, the reduction in numbers which is relatively slow in winter, reaches its maximum in the period without snow cover when downstream drift is important. It is difficult to assess the mortality of young larvae some of whom shelter in the interstices of the substratum after hatching. Further, the rate of hatching varies according to the station as a function of mean temperature (Elliott & Humpesch, 1980). As a consequence estimates made of the numbers of eggs carried by the female are 5 to 10 times higher than the numbers of young larvae estimated by direct sampling from the stream.

After the first instar is accomplished, quite a large proportion of the fauna always seems to come to the end of its growth. The lowest mortality rates are found among the caddis with an annual cycle, the highest among the may-fly larvae, but these last benefit from the high reproductive abilities of the species. For those species, the duration of whose life cycle is variable, the elongation of their development time reduces by a factor of 3 to 8 the proportion of individuals capable of completing their cycle.

### 9.534 *Movement of the benthos*

The benthos of the Estaragne gains periodically new habitats as a result of flow conditions under flood or snow melt. The majority of the larvae particularly mature specimens, maintain themselves between rocks, pebbles or under banks. Some move into the top ten centimetres of the stream bed. These are mainly larvae of the early instars, such as *Rhithrogena loyolaea* (30%), Orthocladiinae (20%), *Isoperla viridinervis, Baetis alpinus*, Diamesinae and Tanytarsinae. These species together with *Bryocamptus zschokkei*, are abundant in the under-flow.

During the year larval density in the underflow is greatest during the snow-melt; it decreases at the end of the flood and increases again towards winter when new generations hatch. The interstices of the substratum thus constitute a refuge from which the stream can be colonized.

Downstream larval drift is particularly pronounced in a mountain torrent such as the Estaragne. The proportion of the biomass moved being greater than that in calmer waters. Thus, at 2150 m, the biomass transported in one year (2 kg dry weight = approximately $7.5 \times 10^6$ individuals) is equivalent to the whole production from an area of 300 $m^2$; at this altitude this is 100 m length of stream. According to species, the surface area needed for the production of the biomass transported ranges from 30 to over 1000 $m^2$ (Lavandier, 1979):

|  | annual drift (g) | surface needed to produce equivalent biomass ($m^2$) |
|---|---|---|
| Orthocladiinae | 6 | 30 |
| *Drusus rectus* | 80 | 130 |
| *Rhithrogena loyolaea* | 173 | 200 |
| *Isoperla viridinervis* | 240 | 320 |
| Diamesinae | 410 | 380 |
| *Drusus discolor* | 300 | 390 |
| *Baetis alpinus* | 470 | 500 |
| *Rhyacophila intermedia* | 150 | 1150 |

The last species is particularly vulnerable to drift at the end of the larval stage.

Overall the number of individuals in the drift reflects the composition of the benthic community; chironomids and may-fly larvae form more than 85% of the drift fauna, stone-fly and caddis larvae less than 3%. The likelihood of entering the drift varies according to group: Diamesinae and caddis are always relatively more frequent in the drift than in the benthos; Orthocladiinae, may-fly and stone-fly larvae are the converse, except for *Chloroperla breviata* which, at the end of its life cycle, drifts abundantly.

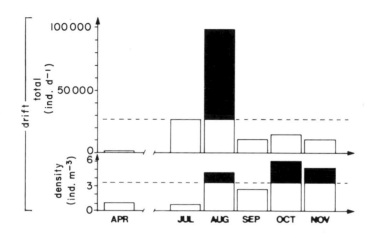

*Fig. 9.7* Seasonal changes in total drift and drift density at 2150 m. Parts which exceed the mean for all the months shown are in black.

During the course of the year, drift of the most species is greatest at the end of the snow-melt flood, at the time of recolonization of the stream beds (Fig. 9.7). Another maximum can occur in autumn when the new eggs hatch and when *Hydrurus* breaks loose. During floods, especially in spring, all benthic species can be moved. At lower water, transport by the current is related more to the behaviour of the larvae, notably their reaction to changes in light flux; it is during this period that the density of drift individuals in the water is greatest (6 individuals m$^{-3}$).

The upstream movement of the larvae concerns only mature individuals and is important for only one species, *Isoperla viridinervis*. At the time of its maximum activity, the numbers moving upstream reach one-tenth of the number and one-quarter of the biomass of those in the drift. In contrast, the flight of reproductive females upstream plays an important role in compensating drift losses for species, particularly

*Baetis alpinus* and *Rhithrogena loyolaea*.

## 9.6  FUNCTIONING OF THE ECOSYSTEM

### 9.61  INVERTEBRATE FEEDING

The majority of Estaragne invertebrates feed on vegetation more or less altered, often associated with mineral particles.  At 2000 m and above, feeding is based on *Hydrurus* and diatoms.  Detritus of endogenous or exogenous origin augments these algae at lower stations and when the stream is covered by snow.

The principal primary feeders in the Estaragne are chironomid and ephemeropter  larvae.  Some species which are primarily herbivores like *Allogamus auricollis* and *Drusus* spp. can also capture invertebrates, whilst others, primarily carnivores like *Arcynopteryx compacta*, can feed on algae and detritus to an extent which is quantitatively important.  Even the most markedly carnivorous species in the stream, *Isoperla viridinervis* and *Rhyacophila* spp., feed partly on *Hydrurus* at certain stages of their cycle.

Chironomid and ephemeropter larvae form the base for the nutrition of carnivores in the Estaragne, in varying proportions according to season.  The following percentages of prey consumed at 2150 m give an idea of the differences observed between periods without and with snow cover.

| prey | summer | winter |
|------|--------|--------|
| Chironomidae | 70% | 13% |
| *Baetis alpinus* | 15 | 68 |
| *Rhithrogena loyolaea* | 10 | 17 |
| others | 5 | 2 |

### 9.62  INVERTEBRATE PRODUCTION

The mean annual production of benthic invertebrates was studied from 1971 to 1973 (Lavandier, 1979, see Table 9.3).  It ranged from 4.9 g m$^{-2}$ yr$^{-1}$ at the source to 7.6 g m$^{-2}$ yr$^{-1}$ at the lowest station.  This production depends upstream on species whose cycle last two or more years; below 2000 m more than 80% of production depends on species which are univoltine.  This difference is not reflected in the values for the ratio $P/\bar{B}$, which always remains close to 4 (Table 9.3).  The two stations located on tributaries (slightly above their confluence with the Estaragne) present a contrast with one another.  The first, at 2150 m, has a low production; while being covered by snow too long for species living at lower altitudes, it reaches summer temperatures too high for high mountain species.  The second tributary at 1920 m, is more productive than the lowest station on the main stream: conditions ameliorated by its position make it a refuge for numerous middle montane species.

Table 9.3 Mean biomass, production and the ratio P/$\overline{B}$ for all benthic invertebrates at five stations on the Estaragne and two on tributaries for the 1971–1973 study period. P, production (g yr$^{-1}$); B, mean biomass (g).

Estaragne

| | | | | | |
|---|---|---|---|---|---|
| altitude (m) | 2370 | 2190 | 2150 | 1920 | 1850 |
| distance from source (km) | 0.05 | 0.6 | 1.1 | 2 | 2.8 |
| snow cover (month) | 6-8 | 5-8 | 5-7 | 4-6 | 4-5 |
| mean width (m) | 1.5 | 2.5 | 4 | 4 | 5 |
| max. temperature (°C) | 4.5° | 7° | 8.5° | 11° | 13° |
| mean biomass (g m$^{-2}$) | 1.18 | 1.67 | 1.63 | 1.55 | 1.92 |
| production (g m$^{-2}$ yr$^{-1}$) | 4.90 | 6.70 | 6.55 | 6.50 | 7.60 |
| P/$\overline{B}$ | 4.2 | 4.0 | 4.0 | 4.2 | 4.0 |

Tributaries

| | | |
|---|---|---|
| altitude (m) | 2150 | 1920 |
| snow cover (month) | 5-6 | 4-6 |
| mean width (m) | 0.7 | 0.9 |
| max. temperature (°C) | 15° | 17° |
| mean biomass (g m$^{-2}$) | 0.65 | 1.90 |
| production (g m$^{-2}$ yr$^{-1}$) | 3.30 | 8.10 |
| P/$\overline{B}$ | 5.1 | 4.3 |

Benthic invertebrate production in the Estaragne over 1971-73 was estimated to be 65.2 kg dry weight ha$^{-1}$ yr$^{-1}$ or 35 Kcal m$^{-2}$ yr$^{-1}$. 36% of this was due to the may-fly larvae *Baetis alpinus* and *Rhithrogena loyolaea*, 31% to the caddis *Allogamus auricollis*, *Drusus* spp. and *Rhyacophila* spp. and 16% to Diptera. Production by carnivores was 22% of the total, a high percentage, which we believe reflects the nature of energy transfer in the Estaragne community.

9.63  ENERGY FLOW

The flow of energy at three stations on the Estaragne (Fig. 9.8) reflects the differences in the environment and benthic communities on passing from upstream to downstream.

(i) The endogenous production of *Hydrurus* and diatoms, important

(A)

(B)

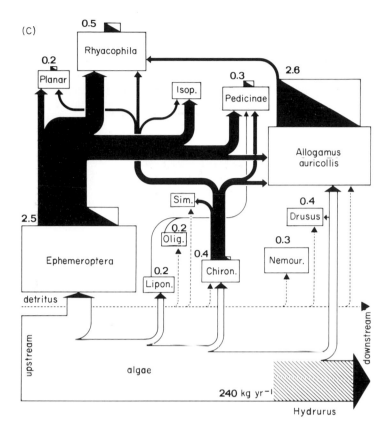

*Fig. 9.8A* Energy flux 0.05 km from the source. Taxa mentioned are those whose mean biomass is equal or more than 0.02 g m$^{-2}$; areas are in each case proportional to the mean biomass. Production is shown for values above 0.1 g m$^{-2}$ yr$^{-1}$; flux width is proportional to the amount of production.

| | |
|---|---|
| *Rhyacophila* | *Rh. contracta* + *Rh. evoluta* |
| *Isoperla* | *I. viridinervis* |
| Emp. | Empididae |
| Ephem. | *Baetis alpinus* + *Rhithrogena loyolaea* |
| Olig. | Oligochaeta |
| *Drusus* | *D. rectus* + *D. discolor* |

*Fig. 9.8B* Energy flux km 1.1 from the source.

| | |
|---|---|
| *Arcy.* | *Arcynopteryx compacta* |
| *Chlo.* | *Chloroperla breviata* |

*Fig. 9.8C* Energy flux km 2.8 from the source.

| | |
|---|---|
| Sim. | Simuliidae |

throughout the course, is supplemented downstream by exogenous material.

(ii)   The composition of the community changes at each trophic level. Among herbivores chironomids replace may-fly larvae, among omnivores *Allogamus auricollis* replaces *Drusus* spp.; among carnivores, *Arcynopteryx compacta*, which is dominant near the source, is gradually replaced by *Isoperla viridinervis*, which in turn diminishes, while *Rhyacophila* spp., planarians and Pedicinae become important.

(iii) At the same time as the number of species increases, the trophic web increases in complexity and diversification of the detritivores into micro- and macro-forms, leads to more niches being occupied, greater mean biomass and greater production (Table 9.4).

*Table 9.4* Features of the benthic invertebrate community at three Estaragne stations for the 1971-1973 study period.

| distance source (km) | 0.05 | 1.1 | 2.8 |
|---|---|---|---|
| number of species | 40 | 65 | 80 |
| diversity | 2.96 | 2.78 | 2.85 |
| evenness | 0.80 | 0.67 | 0.65 |
| mean biomass (g m$^{-2}$) | 1.18 | 1.63 | 1.92 |
| production (g m$^{-2}$ yr$^{-1}$) | 4.90 | 6.55 | 7.60 |
| P/$\bar{B}$ | 4.2 | 4.0 | 4.0 |
| mean carnivore biomass as fraction of mean total biomass | 0.4 | 0.2 | 0.2 |
| carnivore production as fraction of total production | 0.4 | 0.2 | 0.2 |
| proportion of species whose cycle is longer than one year | 0.23 | 0.11 | 0.10 |

In the course of three kilometres benthic colonization of the Estaragne changes from one dominated by its proximity to consolidated snow to one typical of a mountain stream.  However herbivores remain predominant among the primary producers and detritivores do not attain the production characteristic of streams fed by lakes or ones lower down among forest.  Overall the Estaragne has the autotrophic nature of a mountain stream coming directly from consolidated snow (névé) situated in grassland above the tree zone.

## 9.7   GROWTH OF TROUT IN THE TORRENT OF ESTIBÈRE

As trout are practically absent in the Estaragne, their ecology has been studied in a neighbouring stream, the Estibère (Elliott, 1973;

Dumas, 1976). This stream is rather similar to the Estaragne in length (3.5 km), altitude (2850 - 2450 m), mean slope (17%) and duration of snow cover (7 months). It differs especially in its passage through a series of lakes and by its S-E orientation (Décamps, 1967b). The chemistry of the Estibère also differs from that of the Estaragne, the former having a more granitic catchment. Values of 0.304 to 0.45 meq $l^{-1}$ has been recorded for the ionic content of the Estibère, as compared with 1.140 to 0.537 meq $l^{-1}$ on passing down the Estaragne. Alkalinity ($HCO_3^-$ approx. 20 mg $l^{-1}$) and Ca (approx. 6 mg $l^{-1}$) are both lower than in the Estaragne (Table 9.1).

The trout population of the Estibère has been studied at 2040 m in a zone where the slope is reduced considerably (around 2%). This population includes indigenous *Salmo trutta* and *S. gairdneri* introduced between 1958 and 1960 in large numbers.

The *S. gairdneri* population of the Estibère is characterized by a good density, but a biomass and productivity mediocre in comparison with mean values obtained for *S. trutta* at less elevated altitudes (Table 9.5). The structure of the Estibère population, with 41 and 60% of alevins in two consecutive years, makes an estimate of recruitment important. This last is essentially due to a high reproductive potential (3.2 eggs $m^{-2}$) and the absence of more mature competitors in zones where the alevins live. In effect, after reproduction, the alevins remain in shallow zones with a brisk current; they do not move into deeper zones with a slow current until the beginning of their second year. These deeper zones contain 96% of trout aged one year or more and which are for the most part sedentary. It is in these zones in contact with strongly undermined banks that the trout shelter during winter.

*Table 9.5* Comparison of features of *Salmo gairdneri* in the Estibère (data of Dumas, 1976) with means for populations of *S. trutta* in France.

| features of the populations | unit | *Salmo gairdneri* in Estibère | *Salmo trutta* in France |
|---|---|---|---|
| biomass | kg ha$^{-1}$ | 50 | 123 |
| density | (animals ha$^{-1}$) | 4300 | 3200 |
| annual survival | | 31% | 26% |
| length at 3 years | cm | 16.5 | 23.5 |
| net annual production (trout >18 cm) | kg ha$^{-1}$ yr$^{-1}$ | 21 | 46.5 |

At the end of the first year, the annual rate of survival (31%) remains high as a consequence of the absence of disease, pollution and predation. However growth remains feeble, and in spite of very rapid growth in summer (1 cm per month), the length of 3-year old trout is only 16.5 cm and the net annual production is 65 kg ha$^{-1}$ yr$^{-1}$. These low values are due to the rigorous climatic conditions of the Estibère: the water temperature is around 0.1 °C for seven months and growth is limited to summer months (Fig. 9.9).

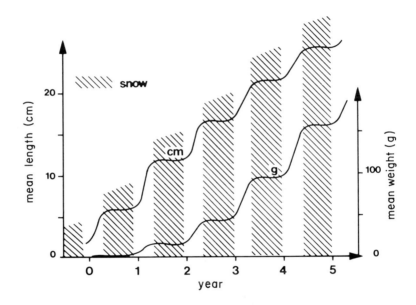

Fig. 9.9 Growth (length, weight) of *Salmo gairdneri* in the Estibère. The stream is free of snow-cover from June to October; reproduction of the trout takes place in July.

Studies have been made on the feeding by trout in the Estibère in relation to the abundance of invertebrate drift (Elliott, 1973). For three different mean temperatures (4.7 °C, 7.3 °C, 10.8 °C), the invertebrate populations in the drift and stomach contents of trout were analyzed over a 24-hour period. These showed that *S. gairdneri* and *S. trutta* were feeding principally on drift invertebrates. The two species showed a similar behaviour. At all three temperatures the principal feeding period took place in the first hours of the night, that is, when numerous benthic invertebrates were in the drift. However at 10.8 °C there was also a period of feeding during the day, at times when terrestrial invertebrates and emergent aquatic insects reach an important proportion of the drift.

In view of the abundance of invertebrates in the drift, it is probably not the availability of food but the temperature of the water which plays a major limiting role. In effect, the amount of feeding by the trout during one day was related to the mean temperature. This supplement is due to a larger stomach capacity of the trout, but especially to a more rapid gastric evacuation at a higher temperature. Using the exponential model for the rate of gastric evacuation as a function of temperature which has been established for *S. trutta* (Elliott, 1972), one can estimate the time after which no more than 15 - 10% of the stomach is filled; it is at this stage that *S. trutta* feeds once more. Applied to the observations for Estibère, these estimates show that trout with a full stomach can eat only once a day at mean temperature of 4.7 °C and 7.3 °C, but twice a day at a mean temperature of 10.8 °C. Consequently, as can be seen in Table 9.6, trout in the Estibère obtain sufficient energy for growth only when temperature conditions permit a sufficiently rapid gastric evacuation that they can feed twice a day.

*Table 9.6* Energy budget for trout at three mean temperatures in Estibère, with energy value of food consumed ($Q_c$), energy of egestion and excretion ($Q_w = Q_c/5$), energy of resting metabolism ($Q_r$), additional energy for swimming activity ($Q_a$), and energy for growth ($Q_g$). (After Elliott, 1973)

| mean temperature of water | gastric evacuation (h)* | energy budget | comments |
|---|---|---|---|
| 4.7 °C | 21 - 27 | $Q_c - Q_w = Q_r$ | one meal/day provides energy for $Q_r$ only |
| 7.3 °C | 16 - 21 | $Q_c - Q_w = Q_r + Q_a$ | one meal/day provides energy for $Q_r$ but non max $Q_a$ |
| 10.8 °C | 11 - 13 | $Q_c - Q_w = Q_r + Q_a + Q_g$ | two meals/day provide energy for $Q_r$, $Q_a$ and $Q_g$ |

*time at which 15% and 10% of the food is left in the stomach and therefore at which a new meal is possible: see text.

## 9.8  CONCLUSIONS

Situated in the high mountains, the Estaragne (and also the neighbouring Estibère) is cold and covered by snow for eight months of the year. Flows are extremely variable and the bed of the stream regularly disturbed by floods. The absence of trees along the banks reduces considerably input from exogenous materials.

Under such difficult conditions one might expect to find only a sparse benthos, limited to a few species. This is far from the situation: the benthic invertebrate community of the Estaragne survives not only by its diversity and complexity, but also by its permanence in time.

How does this come to be? Two responses can be distinguished.

(i) Certain species show an annual life cycle; they come in successive waves and exploit to the maximum the available resources. Their growth is rapid, their abundance can vary, but recruitment assures the maintenance of populations in the more extreme conditions.

(ii) Other species show a life cycle extended to two or more years when conditions become more rigorous. Several generations of the same species can thus coexist, this permits a greater occupation of diverse habitats and acts as a protection against environmental fluctuations. Thus a sudden flood will cause especial harm only to the most active age-class. Besides, the species are capable of entering into a condition of reduced metabolism at various instars of their larval development, these are adjusted permanently to the conditions in their environment and constitute the stable part of the population. The coexistence of these two strategies confers a dynamic equilibrium in the benthic community of the Estaragne.

Another important factor for survival in the Estaragne is the spatial heterogeneity of the hydrographic system. Some tributaries with temperature and flow conditions which are more clement serve as refuges for the more fragile species. The lower part of the torrent has the more diverse fauna. Adults of many species extend their habitat range by flying upstream to deposit their eggs. This results in a mixing which ensures the unity of the community and the possible occupation of all aquatic sites within the catchment.

The arrival of winter determines the rhythm of life in this community. Each year, following the snow melt flood, the benthos once again undergoes its development. Herbivores then occupy a predominant place because of the abundant algal cover on the bed and the rarity of detritus of exogenous origin.

In fact the Estaragne constitutes an ecosystem maintained in a juvenile state by its ambient conditions. It is a fragile ecosystem, with species often at the survival limit, as shown by the example of trout in the neighbouring Estibère.

The high mountain torrents thus correspond with ecosystems whose equilibrium is precarious. Protection turns out to be a delicate problem since it depends on the whole hydrographical system; any action at one point in the system could have repercussions elsewhere in it.

Thus, because of the upstream flight of fertile females, destruction of downstream habitats could affect the colonization of stations close to the source.


*Acknowledgements*

We thank J.M. Elliott for helpful suggestions. The figures were drawn by H. Casanova.

## REFERENCES

Angelier M.L. (1980)  Recherches sur l'écologie des hydracariens dans les eaux courantes, 96 pp. Thèse 3e cycle, Toulouse.

Berthelemy C. (1966)  Recherches écologiques et biogéographiques sur les plécoptères et coléoptères d'eau courante (*Hydraena* et Elminthidae) des Pyrénées. *Annls Limnol.* 2, 227-458.

Besch W.K., Backhaus D., Capblancq J. & Lavandier P. (1972)  Données écologiques sur les algues benthiques de haute montagne dans les Pyrénées. I. Diatomées. *Annls Limnol.* 8, 103-18.

Cuinat R. (1971)  Principaux caractères démographiques observés sur cinquante rivières à truites françaises.  Influence de la pente et du calcium. *Annls Hydrobiol.* 2, 187-207.

Dawson F.H. (1973)  Notes on the production of stream bryophytes in the high Pyrenees (France). *Annls Limnol.* 9, 231-40.

Décamps H. (1967a)  Introduction à l'étude écologique des Trichoptères des Pyrénées. *Annls Limnol.* 3, 101-76.

Décamps H. (1967b)  Ecologie des Trichoptères de la vallée d'Aure (Hautes-Pyrénées). *Annls Limnol.* 3, 399-577.

Décamps H. (1968)  Vicariances écologiques chez les Trichoptères des Pyrénées. *Annls Limnol.* 4, 1-50.

Décamps H. & Elliott J.M. (1972)  Influence de la mesure chimique du débit sur les invertébrés d'un ruisseau de montagne. *Annls Limnol.* 8, 217-22.

Décamps H. & Lafont M. (1974)  Cycles vitaux et production des *Micrasema* Pyrénéennes dans les mousses d'eau courante (Trichoptera, Brachycentridae). *Annls Limnol.* 10, 1-32.

Décamps H. & Lavandier P. (1972)  La faune du torrent d'Estaragne.  In: (Eds) Kajak Z. & Hillbricht-Ilkowska A. *Productivity Problems of Freshwaters, Proceedings of the IBP-UNESCO Symposium on Productivity Problems of Freshwaters,* 918 pp.

Décamps H. & Pujol J.Y. (1977)  Influences humaines sur le benthos d'un ruisseau de montagne dans les Pyrénées. *Bull. Ecol.* 8, 349-58.

Dumas J. (1976)  Dynamique et sédentarité d'une population naturalisée de truite arc-en-ciel (*Salmo gairdneri* Richardson) dans un ruisseau de montagne, l'Estibère (Hautes-Pyrénées). *Annls Hydrobiol.* 7, 115-39.

Elliott J.M. (1972)  Rates of gastric evacuation in brown trout, *Salmo trutta* L. *Freshwat. Biol.* 2, 1-18.

Elliott J.M. (1973)  The food of brown and rainbow trout (*Salmo trutta* and *S. gairdneri*) in relation to the abundance of drifting invertebrates in a mountain stream. *Oecologia* 12, 329-47.

Elliott J.M. & Humpesch U.H. (1980)  Eggs of Ephemeroptera. *Rep. Freshwat. Biol. Ass.* 48, 41-52.

Giani N. & Lavandier P. (1977)  Les Oligochètes du torrent d'Estaragne (Pyrénées-Centrales). *Bull. Soc. Hist. Nat. Toulouse* 113, 234-43.

Humpesch U.H. & Elliott J.M. (1980)  Effect of temperature on the hatching time of three *Rhithrogena* spp. (Ephemeroptera) from Austrian streams and an English stream and river. *J. Anim. Ecol.* 49, 643-61.

Illies J. (Ed.) (1978)  *Limnofauna Europaea*, 2nd Edn. xviii + 532 pp. Gustav Fischer Verlag, Stuttgart.

Lavandier P. (1974)  Ecologie d'un torrent pyrénéen de haute montagne. I. Caractéristiques physiques. *Annls Limnol.* 10, 173-210.

Lavandier P. (1975)  Cycle biologique et production de *Capnioneura brachyptera* D. (Plécoptères) dans un ruisseau d'altitude des Pyrénées Centrales. *Annls Limnol.* 11, 145-56.

Lavandier P. (1976)  Cycle vital de *Pachyleuctra benlocchi* (Navas) dans un ruisseau d'altitude des Pyrénées Centrales (Plecoptera). *Annls Limnol.* 12, 1-5.

Lavandier P. (1979)  Ecologie d'un torrent pyrénéen de haute montagne : l'Estaragne, 532 pp.  Thèse Doctorat d'Etat, Toulouse.

Lavandier P. (1981) Cycle biologique, croissance et production de *Rhithrogena loyolaea* Navas (Ephemeroptera) dens un torrent pyrénéen de haute montagne. *Annls Limnol.* 17, 163-79.

Lavandier P. (1982a)  Evidence of upstream migration by female adults of *Baetis alpinus* Pict. (Ephemeroptera) at high altitude in the Pyrenees. *Annls Limnol.* 18, 55-59.

Lavandier P. (1982b) Developpement larvaire - Régime alimentaire, production d'*Isoperla viridinervis* Piclet (Plecoptera, Perlodidae) dans un torrent froid de haute montagne. *Annls Limnol.* 18, 301-18.

Lavandier P. & Mur C. (1974)  Ecologie d'un torrent pyrénéen de haute montagne. II. Caractéristiques chimiques. *Annls Limnol.* 10, 275-309.

Lavandier P. & Pujol J.Y. (1975)  Cycle biologique de *Drusus rectus* (Trichoptera) dans les Pyrénées Centrales : Influence de la température et de l'enneigement. *Annls Limnol.* 11, 255-62.

Laville H. & Lavandier P. (1977)  Les Chironomides (Diptera) d'un torrent pyrénéen de haute montagne : l'Estaragne. *Annls Limnol.* 13, 57-81.

Thomas A.G.B. (1970)  Taxonomie et répartition des Ephémèroptères et de quelques Diptères aquatiques (Tipulidae et Psychodidae) des Pyrénées, 105 pp.  Thèse 3e cycle, Toulouse.

# 10 · Rhine

G. FRIEDRICH & D. MÜLLER

| | | |
|---|---|---|
| 10.1 | Introduction | 266 |
| 10.2 | Physiography and hydrology (including table of general data) | 268 |
| 10.3 | Water engineering measures | 275 |
| 10.4 | Ecological zonation | 278 |
| 10.41 | Introduction | 278 |
| 10.42 | Plants | 279 |
| 10.421 | Benthic algae | 279 |
| 10.422 | Macrophytes | 280 |
| 10.423 | Phytoplankton | 283 |
| 10.43 | Animals | 287 |
| 10.5 | Fish fauna and fishing | 291 |
| 10.6 | Chemistry and waste load | 294 |
| 10.7 | Self-purification and microbial activity | 301 |
| 10.8 | Concluding remarks | 308 |
| References | | 309 |

## 10.1    INTRODUCTION

The Rhine (Rhein) is a river of the superlative; there is surely no
other stream about which so much has been spoken, sung and written.
The recognition of its significance goes back to ancient Graeco-Roman
times.  As early as this, the Rhine was a navigation way of great
importance, with an extensive chain of Roman and Teutonic settlements.
Some of these have remained important cities to this day (Basel,
Strasbourg, Koblenz, Köln).  The Rhine is the only river which connects
the Alps with the North Sea.

Thanks to its beautiful scenery and vineyard slopes, it is above all
the region of the Middle Rhine whose praises poets have so often sung
and that is now, as before, a favourite recreation and holiday region.
During recent decades the Rhine basin has become one of the most
important industrial agglomerations in Central Europe, which has led to
the term "Rhine rail".  With this recent development the river has
acquired the role of a great supplier, providing drinking water for
about 20 million people, as well as industrial and cooling water for
Switzerland, France, Germany and the Netherlands.  Rhine water contributes considerably to the sprinkling irrigation of special crops, of
particular importance to the Netherlands.  The Rhine is the world's most
densely used inland navigation way with the busiest shipping traffic.
Notwithstanding the extraordinarily high water quality requirements, the
Rhine has to take up wastes to a much greater extent than most rivers.
It is due to this immense burden that the river is also called "Europe's
cloaca".

Due to its dominating role among the rivers of Central Europe, the
Rhine was the object of intensive investigations as far back as the
turn of the century, at the time of the first researches in hydrobiology
and limnology.  The works of Lauterborn (1905, 1916, 1917, 1918) are of
particular interest, representing landmarks in the history of limnology.
The river has since been investigated over and over again.  The main
emphasis has always been on determining and documenting the degree and
load of pollution.  A comprehensive limnological study on the Rhine,
such as written recently on the Danube (Liepolt, 1967), is still
awaited.  As more recent and extensive descriptions on the Rhine the
following may be mentioned:  "Umweltprobleme des Rheins" (Environmental
Problems of the Rhine : Bundesminister des Innern, 1976); this is a
study of the Council of Experts.  This work, prepared on the initiative
of the Federal Minister of the Interior, lays the main emphasis on
problems resulting from disturbances in the ecological equilibrium and
the pollution by waste waters.  The hydrology of the Rhine is presented
comprehensively and discussed in "Das Rheinstromgebiet - Hydrologische
Monographie" published by the Internationale Kommission für die Hydrologie
des Rheins (1978).  The regions of the Rhine between Basel and the German-
Dutch frontier which have preserved their natural landscale to some degree,
and as such are worth particular protection, have been listed by Solmsdorf
*et al.* (1975); this work includes many maps.

Owing to the manifold and intensive utilization of the Rhine there
is a great number of national and international organizations created
specially for its protection and improvement.  Here are some which are
especially important.

IKSR - Internationale Kommission zum Schutze des Rheins
       gegen Verunreinigungen (Member States: France, Germany,
       Luxembourg, The Netherlands, Switzerland).
IAWR - Internationale Arbeitsgemeinschaft der Wasserwerke im
       Rheineinzugsgebiet (International Working Group of the
       Waterworks in the Rhine Basin) with the participating
       working groups:

       AWBR - Arbeitsgemeinschaft der Wasserwerke Bodensee –
              Rhein (Working Group of Waterworks, Bodensee-Rhine)
       ARW  - Deutsche Arbeitsgemeinschaft der Rheinwasserwerke
              (German Working Group of Rhine Waterworks)
       RIWA - Niederlandische Rijnkommissie Waterleidingbedrijven
              (Dutch Rhine Commission of Waterworks)

83 waterworks from five countries belong the IAWR.  They supply 20
million people with drinking water and a considerable part of the
industry with service water.

Regular publications on the results of Rhine water analyses are
issued both by IAWR and IKSR.  The German Working Group of Rhine Waterworks (ARW) presents additionally regular information on the most
essential requirements from the standpoint of water supply enterprises.
In addition, there have been special international agreements concluded
in order to protect the Rhine against the introduction of dangerous and
harmful substances (Rhine-Chemistry Agreement of the Neighbouring
States in 1976) and to mitigate the chloride load (Chloride Agreement
between France, Switzerland, Germany and the Netherlands in 1972).
The guidelines worked out by the European Community on the protection of
surface waters against the release of dangerous and toxic substances to
the Rhine may also contribute to the water quality improvements in the
river.  This Guide includes a "black list" forbidding the release of
particularly dangerous matter to the rivers and lakes, as well as a
"grey list" of substances which ought to be avoided.

In the following presentation it is possible to deal with only a few
important aspects of the ecology of the river.  The account is
restricted to the region between the Bodensee in the south and the
approach area of the Rhine-Delta in the north; that is essentially the
stretch flowing through the territory of the Federal Republic of
Germany.  The Rhine downstream of the Dutch border is discussed in
Chapter 17.

## 10.2   PHYSIOGRAPHY AND HYDROLOGY

With an overall length of 1320 km, the river extends over 700 km from
south to north (Fig. 10.1). The source is in the Alps with its two
headstreams "Vorder Rhein" and "Hinter Rhein". The Vorder Rhein rises
from Lake Toma (2344 m); the Hinter Rhein has its source in the
"Paradies Glacier" (2216 m). The two headstreams which join at an
elevation of about 600 m, have already flowed 70 km (Vorder Rhein) and
60 km (Hinter Rhein), respectively. The river then goes north, under
the name "Alpine Rhine", towards the Bodensee. Before entering the
Bodensee, the Rhine slopes only slightly, which would result in considerable sedimentation if the river bed had not been narrowed by dams.
Sedimentation in fact takes place in the lake, with the mouth of the
river "growing" into the lake, so that Bregenz bay will get cut off
more and more from the rest of the Bodensee.

The origin of the whole Rhine basin dates back to the glacial epoch.
Accordingly, the Rhine in its present-day form can be regarded as
geologically young. The Bodensee, with its 542 km$^2$ surface area and
276 m depth, represents a huge balancing reservoir counteracting discharge fluctuations. In addition, the lake has a significant influence
on water quality. From the Bodensee the Rhine flows westwards and
breaks through the Jura Mountains, where it bears the name "Hochrhein".
The swift and turbulent flow in this stretch, which includes the Rhine
Falls near Schaffhausen with a drop of 24 m, indicate that erosion is
still dominant here. The Swiss river Aare flows into the "Hochrhein".
The Aare drains an even larger area than the Rhine at this point. At
Basel there is a northward change in direction; here the river enters
the Upper Rhine plain with a length of 300 km. The entry of the Neckar
and the Main subjects the Rhine to a heavy pollution load. The Neckar,
Main and the Upper Rhine have all been made navigable by weirs and
transformed into a chain of river impoundments (10.3, 11.33). In the
region of the Neckar and Main there are important settlement and
industrial agglomerations. The Nahe joins the northern end of the
Upper Rhine. This is the start of the Middle Rhine (see frontispiece),
where the river flows through the Rhenish Slate Mountains both westwards
and eastwards. At the narrowest place, with its famous Lorelei Cliff,
the whole valley is no more than 200 m wide. At the city of Koblenz,
two tributaries, the Lahn and Mosel, discharge into the Rhine. The
Mosel, with a basin of 28000 km$^2$, represents the largest tributary of
the Rhine. It is an important international waterway, transformed into
a chain of river impoundments. After passing the "Mittelgebirge", the
Rhine enters the lowlands at Bonn. The wide meanders of the Lower
Rhine have formed a "garland" of old branches here. The main tributaries of the Lower Rhine are the Wupper, Ruhr and Lippe. In addition,
the Rhine meets the Emscher, a river heavily loaded by wastes from the
industrial "Ruhr District".

The Rhine then flows with wide curves in a northerly direction to
the German-Dutch frontier. Immediately after leaving the frontier it
branches into three rivers, the Waal (Fig. 10.3), Neder-Rijn and

*Fig. 10.1*  Map of Rhine (Rhein), showing the Bodensee, the main towns and sections of the River.

*Table 10.1*  General data on Rhine catchment and river.

---

DIMENSIONS
    Area of catchment                           185300 km$^2$
    Source of river                                2344 m altitude
    Length of river (i) total (ii) in chapter   (i) 1320 km
                                                (ii) 850 km to Dutch border

CLIMATE
Precipitation
    Alpine and Prealpine Region:  summer wet, up to 3000 mm, mostly 1200 - 1800 mm,
        dry in winter.  Above 1800 m more than 50% of precipitation as snow.
    Mountains and Upper Rhine valley:  mountains wet in summer, 800 - 1200 mm.
    Upper Rhine valley, 400 - 800 mm.
    Lower Rhine typical maritime climate, 600 - 800 mm.

Temperature
    Upper Rhine valley is warm in summer and holds the German record for a maximum
        temperature (39.2 °C).

GEOLOGY
    The recent Rhine is young.  Until the Oligocene the "Urrhein" flowed in the
older tectonic valley, the Aare belonged to the Rhone-system and the "Alpenrhein"
went to the Danube-system.  The recent system of Rhine and tributaries formed during
the Pleistocene glaciation.

POPULATION
    Population and industry are concentrated along the "Rheinschiene" from Basel to
Rotterdam, with many large cities:  Basel, Strasbourg, Karlsruhe, Ludwigshafen/
Mannheim, (Frankfurt am Main), Mainz, Koblenz, Bonn/Koln, Düsseldorf, Krefeld,
Duisburg, (Ruhrgebiet), Wesel.

LAND USE
    Above Bodensee:  agriculture and forestry are dominant.
    Upper Rhine valley:  intensive agriculture, wine, fruits, agriculture, forestry.
    Middle Rhine:  agriculture, wine, forestry in the mountains.
    Lower Rhine:  intensive agriculture.
    There is intensive industrial activity from Basel to the Lower Rhine and also
        intensive gravel mining in the flood beds of the Upper and Lower Rhine regions.

PHYSICAL FEATURES OF RIVER
    On-stream lakes and weirs
        Bodensee and three lakes in Switzerland (Aare-catchment).  Several weirs
    in the Upper Rhine: Grand Canal d'Alsace, Dam Gambsheim and Iffezheim.

    Discharge at selected points (based on data for 1931-1960)

|                      | Kaub | Rees |
|----------------------|------|------|
| lowest ever recorded | 482  | 590  |
| mean annual minimum  | 700  | 990  |
| mean                 | 1560 | 2210 |
| mean annual maximum  | 4040 | 6660 |
| highest ever recorded| 6150 | 9500 |

    Influence of regulation
        Marked in connection with landscape use, especially in Upper Rhine valley
    and Lower Rhine, mainly to protect land against high floods.

    Substratum
        High Rhine    :   rocks, limestone
        Upper Rhine   :   gravel
        Middle Rhine  :   rocks, gravel
        Lower Rhine   :   gravel, mud in breakwater fields

    Temperature (1971-1980):  selected points - from Deutsches
    Gewässerkundliches Jahrbuch (1983)

|         | km    | minimum | mean | maximum |
|---------|-------|---------|------|---------|
| Walshut | 102.8 | 1.2     | 10.6 | 23.2 °C |
| Worms   | 443.4 | 2.7     | 13.7 | 26.2    |
| Köln    | 688.0 | 0.0     | 12.5 | 25.2    |
| Rees    | 837.4 | 1.8     | 12.8 | 26.2    |

---

*Fig. 10.2*  Rhine at Mannheim/Ludwigshafen, looking north.  On left, Ludwigshafen with industrial plants; on right, Mannheim with port and mouth of Neckar.  (Kommission für die Hydrologie des Rheinge-bietes; photo Landesbildstelle Rheinland-Pfalz)

*Fig. 10.3*  Bifurcation of the Rhine at Millingen and Pannderden. Left branch = Waal; right branch = Lek, Neder-Rijn (Ernst, 1972).

flow (m$^3$s$^{-1}$)

Fig. 10.4  Winter and summer discharge in Rhine for period 1936-1965 (Bundesminister des Innern, 1976).  MHQ = mean of annual maxima; MQ = mean flow; MNQ = mean of annual minima.

Fig. 10.5  Monthly discharge in Rhine and tributaries, based on long-term observations on the Basel-Kaub stretch (Deutsches Gewässerkundliches Jahrbuch, Rheingebiet, 1974).

## 10.3  WATER ENGINEERING MEASURES

Until the early years of the 19th century the Rhine was relatively free
from human interference.  Where the stream was not forced by mountains
into a narrow channel, alluvial plans were formed, with a width of
several kilometres.  These valley lowlands were either divided into
several river branches (furcations) or the river flowed through them in
large meanders.  Extensive alluvial meadows existed mainly on the Alpine
Rhine, the Upper Rhine plain and the Lower Rhine.  In these regions the
Rhine often changed its bed, devastating agriculture, forestry and
settlements near the stream.  This could even happen on some partial
reaches where the "thalweg" (the line following the lowest part of the
riverbed, which is used as the downstream shipping route) was the
national frontier.  Such changes in the channel dislocated the
nationality of important riparian places and handicapped shipping
traffic.  Mosquito plagues occurred in the Upper Rhine plain, where the
lowland forests were flooded for long periods in summer.  The relative
importance of towns and settlements was also influenced considerably in
the Lower Rhine region by natural displacements of the river course
(Hoppe, 1970).

Since the mid-19th century the Rhine and its tributaries have been
subjected to changes by engineering measures, which have partly led to
serious consequences in the ecological system.  The most significant of
these interventions was the so-called Tulla Rhine rectification (1817-
1876), whose main objective was to extend the agricultural use of the
alluvial valley plain.  The first stages in this project were to
concentrate the stream flow in a confined channel, cutting straight the
meander loops (Bensing, 1966).  In the Upper Rhine plain these rectification measures caused an increase in current velocity which led to
intense bottom erosion, amounting at Basel to about 7 m and on the
Basel-Mannheim stretch to about 1.5 m (Schäfer,  1973).  This has
resulted, especially in the southern part of the Upper Rhine plain, in
a considerable lowering of the ground-water table and in the withering
of some riparian forests.  Thus, the original natural unity of the Rhine
river and the Rhine valley has become destroyed.  Several old branches
and backwaters have ceased to exist.  Suggestions for restoring the
landscape have come from Schäfer (1973), among others, and work is now
in progress.

Subsequent river training operations have dealt with flood control,
generation of hydroelectric power and shipping traffic. In the Swiss
river basin the retention capacity has been increased by constructing
reservoirs for hydro-power utilization and by diverting the Aare river
and one of its tributaries through three lakes in the Alpine foothills.
The Aare and the Hochrhein in the stretch between Schaffhausen and
Basel have been developed for hydro-power utilization and, as a result,
considerable changes in water depth and flow velocity have taken place.

In the stretch from Rheinfelden to Basel the Rhine has been developed
for shipping traffic using up-to-date ship units.  In the section up to

*Fig. 10.6* Upper Rhine at Philippburg and Speyer (in background).
The character of landscape is determined by the old Rhine branch of
the former meandering system with its alluvial forest and by the
regulated river.

*Fig. 10.7* Backwater of the Gamsheim river dam downstream of Strasbourg. Structures from left to right: fixed transverse dike, power
stations, locks and the Ill River diverted into backwater of dam
(Internationale Kommission für die Hydrologie des Rheingebietes (1978).

Breisach there is in French territory a lateral canal with navigation
locks and hydro-power plants.  In the old Rhine channel, the "Rest
Rhine", the minimum discharge amounts to not more than 15 to 30 $m^3$ $s^{-1}$.
In the section up to Strasbourg the shipping route and the main stream
flow run alternately in the old Rhine channel and in short lateral
canals equipped with hydro-power plants and navigation locks.  Here, too,
there are extremely low flows in the sections of the "Rest Rhine".  The
final stretch up to Rastatt has also been subjected to river impoundments.  In order to maintain the ground-water level, to preserve the
remaining areas of natural lowland forests and to retain floodwater,
large-scale engineering operations have been carried out.  Currently, a
project to establish large flood retention spaces is envisaged.  Fig.
10.8 indicates the alterations in the Rhine channel over the
"Taubergiessen" area since 1827. This last region of well-preserved
lowland forest is to be placed under nature protection.

*Fig. 10.8* Changes in the channel of the Rhine since 1827 in the
Taubergiessen region (near Freiburg).

In order to prevent scour in the river bed downstream of the last
dam, erosion activities are counterbalanced by addition of gravel from
riverside gravel pits ('bed load supply').  As compared with the
construction of further impounding weirs, this method has proved to be
particularly economical.  Further, ecological effects resulting from
this technique are less than those entailed by dam construction.

On the Middle Rhine and the German Lower Rhine, the channel has been
deepened in order to facilitate navigation.  The resulting ecological
effects are relatively insignificant compared with the engineering
measures taken further upstream; they have led, nevertheless, to a
considerable impoverishment of the fields and meadows.  The Dutch
section of the Lower Rhine and some of its river mouth branches have
suffered in particular.  The intensive gravel dredging operations in
the Upper Rhine valley and on the Lower Rhine can be regarded as
drastic interventions in the relations of water and environment.  An
area of 61 $km^2$, that is about 2.7% of the alluvial plain of the German
Rhine, had been "used up" by excavations by 1973 (Solmsdorf *et al.*,
1975).

As a consequence of river impoundments on the southern reach of the
Upper Rhine, flood waves proceed more rapidly nowadays.  This means an
augmented flood danger downstream of the development section (Karlsruhe,
Mannheim, Worms).  To reduce the flood danger again, new hydraulic
engineering measures are needed.  The aim is to make the probable
maximum 200-year flood pass without damage, as was the case before
intensive river development.  This objective is  to be achieved by
deliberately inundating parts of the alluvial plains which had
earlier served as flood retention areas.  The total area to be subjected
to artificial flooding extends over 75 $m^2$.

The floods to be attenuated by retention measures usually occur
during the winter semi-annual periods.  The flood-retaining areas
(polders) will be inundated for several days in every single flooding
operation.  This is expected to favour the alluvial forest vegetation
by deposition of sediment as well as a rise of the ground-water level
beneficial to the formation of moist biotopes.  Such control measures,
however, are not expected to cause essential changes in ecological
conditions in the river itself.


## 10.4  ECOLOGICAL ZONATION

### 10.41  INTRODUCTION

In conformity with its length and the multiform character of the landscale through which it flows, the Rhine has a well-marked subdivision
into biological zones (Lauterborn, 1916-1918; Knöpp, 1957; Schwoerbel,
1959).  Whereas Lauterborn could report on a relatively natural
biological zonation, Knöpp had to conclude that all sensitive species
had gone from the Upper Rhine (in the region of Basel) down to the
Lower Rhine (German-Dutch frontier).  Only some few forms which are
definitely tolerant to pollution have survived; a "forced" biological
zonation has come into being, governed and determined mostly by the
point-source effluents and the subsequent zones of self-purification.

10.42    PLANTS

*10.421  Benthic algae*

As with most rivers, there is no comprehensive account of the benthic
algae.  Besides the usual taxonomic difficulties, problems arise due to
the use of rivers for navigation and due to the changing water level;
in particular it is difficult to use artificial substrates for periphyton studies.  The macroscopic species of the Hochrhein and Upper
Rhine are known from the studies of Lauterborn (1910), Jaag (1938) and
Krause (1976).  Krause investigated the waters of the Upper Rhine valley
with respect to the impoverishment of the landscape by river regulation
and eutrophication.  He concluded that originally there was probably a
considerable diversity of aquatic environments and their algal communities.  He recognised (p. 227) five types of community, based partly
on modern observations made by him on the relatively few undisturbed
waters.

(a)  *Rivularia haematites* in waters of 20-30 cm depth and fast currents,
but which generally lack water in winter.  The water-courses are exposed
to sunlight.

(b)  *Hildenbrandia rivularis, Lithoderma fontanum, Cladophora glomerata*
in moderately fast streaming water of 50 - 100 cm depth.  The watercourses are shaded and rarely fall dry.

(c)  *Batrachospermum ectocarpum, Draparnaldia glomerata, Bangia
atropurpurea* in very fast currents under narrow bridges.

(d)  *Chara hispida, C. aspera, C. contraria* in ponds coming from groundwater and with high light penetration, but nearly anaerobic conditions
on the bottom.

(e)  *Tolypella intricata* in pools drying out periodically.

Waters of types a - c have more or less disappeared due to anthropogenic causes.  Nevertheless the rare brown alga *Pleurocladia lacustris*
could be found by intensive collecting.  The species grows in the so-
called "Taubergiessen", in the last almost undisturbed rivulets of the
Upper Rhine valley.

Only a few typical examples of the Characeae-ponds remain and the
*Tolypella*-vegetation is much impoverished.  In dammed waters nowadays,
*Cladophora crispata, Hydrodictyon reticulatum* and *Enteromorpha* are
visible signs of eutrophication.  In the river bed itself *Hildenbrandia*
and *Lithoderma* were conspicuous till 1961.  The flat incrusting layers
of these algae have been overgrown, especially by the diatom *Melosira
varians*.  *Thorea ramosissima*, in earlier times already a rare species,
is known from recent times only from a restricted area.

According to Lange-Bertalot (in press), *Ulothrix zonata* is one of the dominant benthic algae of the Upper and Middle Rhine. *Bangia atropurpurea* is rare, but Chantransia- forms are frequent down to as far as the Lower Rhine. *Schizomeris leibleinii* occurs in the splash zone from the middle to the upper Lower Rhine (Lange-Bertalot, in press) and also in the lower part of the Lower Rhine (G.F., unpublished). After construction of some sewage plants, the species disappeared, though it is uncertain whether or not this was due to the improved water quality. Lange-Bertalot & Lorbach (1979) investigated diatom communities from the mouth of the Aare to Bonn. Benthic diatoms often formed massive growths, mainly widespread species of *Navicula*, *Gomphonema*, *Cymbella* etc. The use of glass slides exposed in the Lower Rhine has shown that the incrusting algae *Protoderma* sp. and *Sphaerobotrys fluviatilis* together with *Stigeoclonium* sp., are abundant; *Chamaesiphon incrustans* and the iron bacterium *Siderocapsa treubii* were frequent as well (Schiller, pers. comm.).


*10.422  Macrophytes*

The macrophytic vegetation of the Rhine cannot be studied properly without taking the whole alluvial valley into consideration. It was predominantly in the old branches of the Upper and Lower Rhine plain and in their surrounding bottom-land regions that large populations of helophytes and hydrophytes developed, greatly dependent upon the river. As a result of the extensive river regulation, many species have been destroyed and others are threatened by silting. Solmsdorf, Lohmeyer and Mrass (1975) investigated the whole of the Rhine between the Bodensee and the German-Dutch frontier and divided the present-day alluvial valley, which is characterized by embankments, into three main zones:

> river bed with the islands
> flood zone (dike foreshore) with intact (i.e. periodically flooded) old branches
> flood-free zone with old branches tending to silt up.

As mentioned in 10.2, floods occur on the Hochrhein in June/July and November and on the Lower Rhine mainly in February and somewhat less intensively in July. In the flood zone the high water stages may exceed the mean water stages by 7 metres or more. It was in this zone that the large riparian forests once grew.

A vegetation close to the natural state, and still dependent on the Rhine today, can be found only on the so-called "Restrhein" stretch of the Upper Rhine, on some islands in the Rheingau and on the Middle Rhine, as well as in a few places scattered along the whole river. Thanks to the recognition of nature reserves, a number of old river branches with natural vegetation are still preserved, though even these are endangered in many places, as on the Lower Rhine, by sludge accumulation or by the hazards of being trampled or bitten by cattle.

The alluvial meadow is of decisive importance as living space for
the animal world as well.  The old Rhine branches which remain are
therefore indispensable biotopes in every respect.  Two such regions are
the Taubergiessen in the Upper Rhine plain and the Xanthen old river
branch on the Lower Rhine.  Besides these wetlands of international
importance on the Rhine are the regions around Bodensee Untersee, the
Upper Rhine area between Lörrach and Kehl, the Rhine meadows at Bingen-
Erbach and the lower Lower Rhine region from Bislich to the frontier
(Solmsdorf et al., 1975).

## Hochrhein and Upper Rhine

According to the early studies of the Upper Rhine by Lauterborn
(1910, 1917), the "flowing" Rhine was almost free from phanerogams.
Only aquatic bryophytes, predominantly *Fissidens grandifrons* in the
Hochrhein and *Fontinalis antipyretica* in the Upper Rhine, were present
in the clear and cool water, forming luxuriant growths.  It was not
before the Mainz region that *Potamogeton* (*P. pectinatus* and *P.
perfoliatus*) also appeared.  Koch (1926) and Kummer (1934) referred to
the occurrence of *Ranunculus fluitans*.  Since 1970 this weed has become
a great problem, particularly upstream and downstream of the
Schaffhausen Falls and led to a total overgrowth in some reaches of the
Rhine.  As a result of excessive weed growth the river bottom once
covered by gravel has silted up very rapidly.  To make it worse, the
abundance of aquatic weeds has caused water level rises on the Upper
Rhine of up to 30 cm.  In one quite short stretch of the Swiss Rhine,
1000 tonnes of *R. fluitans* was removed in 1974 with a submersible mowing
unit within 4 weeks, in order to prevent operation troubles at a downstream water power plant (Knecht, 1982).  According to Thomas (1975),
this mass growth is due to aqueous phosphate exceeding a threhold of
about 0.07 mg $l^{-1}$ $PO_4$-P.

Of particular scenic attraction are the abandoned river courses in
the Upper Rhine region with their diverse vegetation.  Lauterborn had
already described them enthusiastically in his reports as sites of
botanical treasures.  In spite of considerable losses, due to river
training and deep erosion as a result of desiccation, and notwithstanding the filling up of old beds, there are still large areas with
remarkable aquatic and alluvial vegetation (Philippi, 1973, 1977, 1978a,
1978b).  Since the beginning of the century species which have been lost
include a large number of *Potamogeton* and Characeae species e.g.
*Tolypellopsis stelligera*.  Other species possess an extremely limited
ecological amplitude and may, therefore, easily disappear as a
consequence of even relatively slight disturbances e.g. *Naias marina*,
*N. minor*, *Trapa natans*, *Nymphoides peltata* and *Salvinia natans*.  These
still exist today, but usually in very restricted areas.

As stated by Philippi (1978, p.132), in just the Upper Rhine region
of Baden fifty species of aquatic plants may now be considered as
extinct or threatened.  He recognised the following categories:

extinct or disappeared
   e.g. *Marsilea quadrifolia, Oenanthe fluviatilis, Stratiotes aloides*

acutely endangered
   e.g. *Salvinia natans, Potamogeton coloratus, P. gramineus,
   Utricularia bremii*

endangered
   e.g. *Potamogeton densus, P. acutifolius, P. angustifolius,
   Hydrocharis morsus-ranae, Sparganium minimum, Nymphaea alba,
   Trapa natans* (acutely endangered?), *Hottonia palustris.*

In contrast to the losses resulting from pollution, development of
the Rhine and construction of gravel pits, the following aquatics have
increased:
   *Potamogeton pectinatus, P. crispus, P. perfoliatus, P. lucens* (?),
   *Elodea nuttallii, Lemna gibba, Ceratophyllum demersum, Myriophyllum
   spicatum, Utricularia neglecta* (?). (*E. nuttallii* is frequent in
   Upper Rhine and Nordrhein-Westfalia, but *E. ernstiae* grows in restricted places where water does not freeze: G. Philippi, pers. comm.)

Distributional data have also been obtained for *Phragmites*, whose retro-
gression is due primarily to water engineering measures. Under particularly favourable conditions, man-made waters such as gravel pits may
become replacement biotopes for plants in danger, as shown by Krause
(1975) for some *Chara* species, different Vaucheriaceae and *Myricaria
germanica*. Favourable conditions could be created in many places by
using suitable materials to cover the bottom and edges of such ponds.

## Middle Rhine

As a result of its confined river bed, the Middle Rhine has
apparently always been poor in aquatic vegetation; Lauterborn's
description (1918) includes only some species of aquatic bryophyte.
The literature on the river flora also includes records for *Butomus
umbellatus, Potamogeton compressus, P. densus, P. pectinatus, Ranunculus
fluitans* and *Zannichellia palustris*. *Butomus* and *Potamogeton pectinatus*
have been found in recent times (A. Krause, pers. comm.). If the
construction of the river bank were to leave a few gaps above the line
of the mean water level of the river, a rich perennial vegetation would
be likely to develop, even under present-day conditions; this might
contribute, even if only moderately, to the ecological regeneration of
the river basin (Krause, 1981).

## Lower Rhine

In the region of the Lower Rhine, the river bends which dry periodically, the oxbows and backwaters, are all habitats with a rich flora
and vegetation. In addition, dry areas between the bottom of the Rhine-
valley and the lower terrasse of the slopes are well known habitats for
so-called "Stromtalpflanzen", thermophilic species which extend north
only in the Rhine valley e.g. *Koeleria pyramidata, Poa bulbosa, Cynodon*

*dactylon, Thalictrum minus, Silene conica, Holosteum umbellatum, Isatis
tinctoria, Diplotaxis tenuifolia, Brassica nigra, Ononis spinosa,
Euphorbia sequeriana, Eryngium campestre* and *Peucedanum chabraei.
Cucubalus baccifer*, a liane relative abundant in the south, can also be
estimated as a "Stromtalpflanze" in the Lower Rhine region as well
(Burckhardt & Burgsdorf, 1964, 1965; Knörzer, 1960, 1963).

The forest stands rich in elm trees, which were once a significant
part of the vegetation, have been almost totally destroyed; the rare
persistant elms are now endangered by disease.  Endangered as well are
the thermophilic grasslands, because of the growing intensity of
agriculture and camping activities.

Backwaters, oxbows and "Kolke" of the Lower Rhine, especially in the
lowest part near Wesel, contain some well developed stands of reeds and
other macrophytic vegetation rich in very rare and endangered species.
*Nymphoides peltata* is a typical example which lives only directly by
the river.  These waters beside the Rhine are endangered especially by
cattle and fertilisers and by the smaller water bodies being filled in
deliberately.  The waters may also be filled by intensive sludge growth,
an effect of eutrophication (Burckhardt & Burgsdorf, 1964, 1965;
Hild, 1960, 1968; Hild & Rehnelt, 1965a, 1965b; Solmsdorf *et al.*, 1975).

Of special interest are the plains of sand and gravel between middle
and low water, which regularly fall dry in summer.  This is a zone
where a scattered vegetation of the Polygono-Chenopodietum-Association
grows, rich in nitrophilous and thermophilic species, the most abundant
of which belong to *Polygonum, Chenopodium* and *Bidens.*  This association
is also found in other big rivers of central Europe, which have a
changing water level, like the Elbe and Weser (Lohmeyer, 1970a, 1970b).

In harbours and their vicinity there can occur a lot of adventive
plant species imported with shipping and goods from overseas e.g.
*Solanum nititibacatum, Abutilon theophrasti* and *Ricinus communis.*
Tomatoes (*Lycopersicon esculentum*) and thorn-apple (*Datura stramonium*)
are regularly to be found (Stieglitz, 1980).  On sludge, which dries
in summer but remains wet for the rest of the year, stands of a
Nanocyperion-vegetation grow more or less regularly, especially
*Limosella aquatica* and *Potentilla supina.*  Where the sludge is very fine
and remains wet all the time, the vegetation closest to the water
develops a green cover of the macrophytic alga *Botrydium granulatum.*

## 10.423 Phytoplankton

In a river like the Rhine, flowing through several lakes and impoundment
zones, a correspondingly great variety in plankton species is to be
expected.  Yet, relatively few studies have been made of the plankton of
the river (Lauterborn, 1910; Kolkwitz, 1912; Seeler, 1938; Czernin-
Chudenitz, 1958).

In any case, the question often discussed in earlier times, whether
the Rhine plankton represents a real "potamoplankton" or not, can be
seen today to be merely a dispute about words.  Eutrophication of the
river has led to so great a phytoplankton growth that even over the
relatively short stretch from Bonn to Emmerich a considerable increase
in chlorophyll can be measured, which is not likely to be due exclusively to the plankton-rich affluents.  Table 10.3 gives information
on maximum phytoplankton densities since 1912.  The phytoplankton plays
a very important role in the ecosystem of the Rhine, as already pointed
out impressively by Knöpp (1957).  On the other hand, zooplankton has
never occurred in the stream in significant amounts.  Both the earlier
and the more recent studies indicate a zooplankton density, which for
Crustacea, is far below one individual per litre (Seeler, 1938; Seibt,
1980) and for Rotatoria less than 500 individuals per litre.  The
following description is therefore restricted to phytoplankton.

*Table 10.3*  Maximum standing crop of phytoplankton in Rhine.

| section | date | individuals $(x\ 10^3\ 1^{-1})$ | month | reference |
|---------|------|-------------------|-------|-----------|
| Hochrhein | 1912 | 22-23 | August | Kolkwitz 1912 |
| | 1955 | 1800 | March | Czernin-Chudenitz 1958 |
| Upper | 1912 | 44 | August | Kolkwitz 1912 |
| Rhine | 1936 | 500 | September | Seeler 1936 |
| | 1955 | 1800 | March | Czernin-Chudenitz 1958 |
| | 1974 | 11000-15000 | April | Backhaus & Kemball 1978 |
| Lower | 1936 | 4000 | September | Seeler 1936 |
| Rhine | 1955 | 2000 | March | Czernin-Chudenitz 1958 |
| | 1970/74 | 14000 | August | Heuss 1975 |
| | 1979 | 59000 | August | G. F. (unpublished) |
| | 1980 | 46000 | May | "            " |
| | 1981 | 56000 | April | "            " |
| | 1982 | 63200 | May | "            " |

*Hochrhein and Upper Rhine*

Until the late 1930s the phytoplankton growth and occurrence in the
Hochrhein and Upper Rhine had been governed by the plankton of the
great lakes through which the Rhine and Aare rivers flow.  This
plankton consisted almost exclusively of diatoms, chrysophytes and
cryptomonads.  Abundant species were: *Cyclotella bodanica, C. melosiroides, C. socialis, C. comta*, cryptomonads, *Ceratium hirundinella,
Asterionella formosa, Fragilaria crotonensis, Diatoma elongatum,
Synedra acus* including var. *angustissima, S. delicatissima, Dinobryon
sertularia* and *Sphaerocystis schroeteri*; *Oscillatoria rubescens,
Tabellaria fenestrata* and *Melosira islandica* var. *helvetica* had come
into the Rhine from the Zürichsee through the R. Aare (Lauterborn,
1917; Backhaus & Kemball, 1978).

The first investigations at the beginning of the 20th century (Marsson, 1907-1911 ; Kolkwitz, 1912) showed very low cell numbers with less than 100 cells ml$^{-1}$. Seeler (1936) had already found cell numbers up to 1000 ml$^{-1}$ in the Upper Rhine and distinct indications of pollution in form of *Sphaerotilus* flocs drifting with the current. At that time, however, such pollution loads had been local phenomena, counterbalanced again by the self-purification of the stream. By 1941 *Fragilaria crotonensis* and *Asterionella* were largely displaced in the Hochrhein by *Tabellaria fenestrata*. In addition, there was an increase in the following diatoms: *Diatoma vulgare, Nitzschia acicularis, Synedra ulna* var. *oxyrhynchus, Melosira varians, M. islandica* ssp. *helvetica* and *Fragilaria crotonensis*. These results were confirmed by Czernin-Chudenitz (1958).

The Upper Rhine had earlier shown a plankton greatly similar to that of the Hochrhein both in composition and density. It was only in the old branches that plankton could develop to high densities. That is why these waters were of vital importance to the fishery. At km 173 (approximately Franco-Swiss border) Czernin-Chudenitz (1958) found preponderantly *Melosira varians, Nitzschia acicularis, Cyclotella* spp., *Chroomonas* sp., *Oscillatoria rubescens, Synedra ulna* and *Tabellaria fenestrata*.

In the Upper Rhine from about Germersheim (km 384) there is already a population of *Stephanodiscus hantzschii,* which takes over the role of an "indicator form for the Rhine plankton" here (Backhaus & Kemball, 1978). Its occurrence is associated with the saline waste discharges from the Alsace. Large populations of chlorococcalean green algae and μ-algae were also found. *Rhodomonas minuta* var. *nanoplanctica, Melosira islandica* ssp. *helvetica, Gloeotilopsis planctonica, Goniochloris pulchra* and *Nitzschia fruticosa* all have originated from the lakes through which the river had passed. *Nitzschia acicularis, Asterionella formosa* and *Fragilaria crotonensis* are also characteristic species in the Upper Rhine today.

*Middle Rhine*

Lauterborn (1918) did not find obvious changes in the plankton of the Middle Rhine as compared with the Upper Rhine. Seeler (1936), however, could show that the planktonic biocoenoses were changed in this section under the influence of the waste load. Czernin-Chudenitz (1958) also found that centric diatoms were dominant, but the overall species composition had become impoverished as a result of heavy pollution. The entrance of the Mosel leads to new planktonic elements in the Rhine. Different bodies of water flow side-by-side with a different content. Such stripe formation is characteristic of the Rhine and makes detailed studies difficult.

*Lower Rhine*

At the beginning of the century, "The plankton of the Lower Rhine

shows far into the Delta region a great similarity in its essential
features to the plankton of the Middle Rhine ..." (Lauterborn, 1918).
The main constituents of the plankton were diatoms such as *Fragilaria
crotonensis, Asterionella formosa, Synedra delicatissima, Tabellaria
fenestrata* var. *asterionelloides* ; somewhat less abundant were *Synedra
actinastroides, Melosira, Stephanodiscus hantzschii, Cyclotella* as well
as, sporadically, *Coscinodiscus lacustris, Attheya zachariasi* and
*Rhizosolenia longiseta*. Among the Cyanophyceae the occurrence of
*Oscillatoria rubescens* is not very rare ...". The list continues with
a long series of Chrysophyceae and some Chlorophyceae. Cell densities
were very low at that time. During the last decade extensive studies
on the phytoplankton of the Lower Rhine have been conducted until 1974
by Heuss (1975) and subsequently by G. Friedrich. The species composition of phytoplankton has not changed markedly during these years, but
populations are much higher with maxima well over $4 \times 10^7$ cells $1^{-1}$ in
1979-1982. The plankton of the Lower Rhine is dominated by centric
diatoms e.g. *Stephanodiscus hantzschii, S. tenuis, S. subsalsus, S.
astraea* inc. var. *minutula, S. dubius, Melosira granulata* var.
*angustissima, M. islandica* and *Cyclotella meneghiniana*. The first two
are overwhelmingly predominant. Even *Asterionella formosa* may attain
densities of $1.5 \times 10^6$ cells $1^{-1}$. The seasonal changes during 1981 of
the most important groups are shown in Fig. 10.9. Filamentous
Cyanophyceae, and in particular *Oscillatoria rubescens,* which had often
been mentioned in the earlier literature, have disappeared. This is
another clear indication of intensive changes in the "Rhine water"
biotope. These changes are without doubt due to the influence of the
saline waste load and to eutrophication. The considerable increase in
plankton density has also been favoured by decreases in concentrations
of toxic substances and to the heavier preloading of plankton because of
impoundments in the Upper Rhine and from the major tributaries, the
Neckar, Main (Lange-Bertalot, 1974) and Mosel.

Although there is a considerable net increase in plankton density
between Bonn and the German-Dutch border, there is, nevertheless, a
significant depression in the Greater Duisburg area, which may be due
to toxic waste. This interpretation has been corroborated by measurements on oxygen production potential and other biochemical parameters.
As soon as this toxic component is eliminated through further developments in waste treatment, a still more intensive plankton growth may be
expected in the lower Lower Rhine (Friedrich & Viehweg, in press).

*Seasonal periodicity*

Investigations on benthic algae in different rivers have shown that
the seasonal periodicity of algal coenoses decreases with increasing
waste load; this occurs apparently because species with distinct
periodicity are affected more intensively than those without it
(Backhaus, 1968; Friedrich, 1973). In the case of the Rhine the widely
different origins of the water are an additional factor to be considered.
In the far past it can be assumed that the seasonal periodicity of the
plankton was governed and affected by that prevailing in the Alpine and

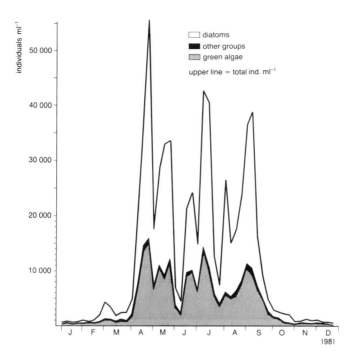

*Fig. 10.9* Changes in phytoplankton density and relative proportions of main taxa at Bimmen during 1982, based on weekly sampling. Relative proportions of groups are indicated by areas between lines (i.e. not total height in case of diatoms).

and Alpine foothill lakes. Backhaus & Kemball (1978) measured a winter minimum in March in the Upper Rhine, followed by rapid growth in April leading to an annual maximum. Subsequently the crop fell slowly, decreasing irregularly until August. The Lower Rhine shows a somewhat different picture. There are generally two maxima here (Heuss, 1975; G. F. unpublished; Fig. 10.10) a spring maximum from March to May and a summer maximum from June/July to October. It is due mainly to the intensive rainfall in sub-catchments, and the subsequent increased discharges in the Lower Rhine that the plankton fluctuation is affected. Centric diatoms are permanently predominant, Chlorophyceae may almost reach the cell density of diatoms, though a high proportion of these may be μ-algae.

## 10.43    ANIMALS

*Headwaters and the Alpine Rhine extending to the Bodensee*

As is characteristic of high-mountain streams, both the headwaters and the upper section of the Alpine Rhine are typical salmonid waters with

*Fig. 10.10* Changes in phytoplankton density and chlorophyll *a* concentration at Bimmen in 1979-1980, based on weekly sampling. (Compare with observations on individual groups in 1981 shown in Fig. 10.9)

rheophile and rheotolerant species.  In the Swiss Vorarlberg valley, the Rhine forms a lowland stream with heavy bed load transport, which has been still further increased by canalization.  This has resulted inevitably in an impairing effect on the environment for benthic organisms.  According to Bloesch (1977), because of the gravelly-sandy bottom, this Rhine section is considerably poorer in species than the R. Aare which has a similar water quality but a more silty substratum. The Aare has more insect larvae and molluscs.

The dominant fish in this section are trout and grayling.  The abandoned river channels in the alluvial area and the small watercourses flowing towards the Rhine ("Giessen") are biotopes of their own.

*Hochrhein*

After flowing through the Bodensse, the Rhine reaches a section (Hochrhein) whose aquatic bryophytes and calciphilic algae provide a favourable environment for a diverse fauna (Caspers, 1980a) consisting predominantly of rheophile insect larvae.  Caspers found 158 macro-zoobenthic species in 1978/79 (Table 10.4).  Owing to the great progress in taxonomy it is difficult to compare faunistic lists with ones made at the beginning of the century, but the totals in 1908, 1910 and 1916 for Ephemeroptera, Plecoptera and Trichoptera were 19, 13 and 54, respectively, indicating obvious losses in all three orders.  Examples of individual species which have been lost are the Trichoptera *Chimarra marginata, Glossosoma boltoni* and *Stactobiella risi* and the Ephemeroptera *Oligoneuriella rhenana,Rhithrogena* spp., *Ecdyonurus* spp.

and *Heptagenia* spp. There has been an increase in species characteristic of eutrophic still-water sections, such as the Ephemeroptera
*Cleon dipterum* and *Caenis luctuosa,* the Trichoptera *Limnephilus lunatus*
and a large number of chironomid species. Lauterborn (1916-1918)
discovered only three species of Gastropoda (*Lymnaea peregra, Ancylus
fluviatilis, Bithynia tentaculata*) in 1916. In 1978 and 1979 Caspers
found 18 such species, whose appearance he attributed to the transformation of the Hochrein into a chain of impoundments, as well as to
a growing eutrophication of the Bodensee.

*Table 10.4* Composition of benthic fauna in Hochrhein near Bad
Säckingen in 1978/1979 (Caspers, 1980a).

| | number of species | | number of species |
|---|---|---|---|
| Porifera | 2 | Ephemeroptera | 7 |
| Hydrozoa | 1 | Plecoptera | 0 |
| Tricladida | 4 | Odonata | 1 |
| Gastropoda | 10 | Heteroptera | 2 |
| Lamellibranchiata | 8 | Coleoptera | 4 |
| Oligochaeta | 5 | Trichoptera | 21 |
| Hirudinea | 3 | Diptera | 85 |
| Isopoda | 2 | (Chironomidae only | 73) |
| Amphipoda | 2 | Bryozoa | 1 |

## Upper Rhine

The increased uniformity of the river bed and the increasing waste
load have led to a considerable change in the fauna of the Upper Rhine
(Kinzelbach, 1978). The impoverishment characterized by total
disappearance of pollution-sensitive species reached its peak during
the period 1970/1971. Insofar as permitted by the impact of waste load,
epilithic species such as *Ancylus fluviatilis* and *Glossiphonia
complanata* usually inhabiting the Hochrhein and the Middle Rhine had
spread over the whole stretch. In recent years (since 1975) more
species have been observed again. In addition to the reappearance of
indigenous species, however, an increasing "alienation process" caused
by the penetration of species new to the Rhine, has become apparent.
Of the 32 new species found by Kinzelbach (1978), some come from other
rivers in the region, but some from other continents.

## Middle Rhine

The Rhine breaks through the Rhenish Slate Mountains here and flows
in a narrow valley with steep banks and cragged rocks. The dominant

epilithic species are typical mountain-species forms, because of the
rise in water temperatures during the summer period.  Barbel and nase
are the prevalent first species.  The effects of effluents have led to
a considerable decline in fauna species diversity here as well
(Schwoerbel, 1959).  Quite recently, however, the conditions for
colonization have become somewhat better again, as with the Upper
Rhine.  These improvements are due primarily to the reduction in waste
load and the resulting higher oxygen concentrations in the Rhine
(Conrath et al., 1977).  An impressive example spotlighting the biological situation in the Rhine is given by the phenomenon of the
appearance of abundant caddis flies (*Hydropsyche contubernalis*) in the
Koblenz region during 1975-1977.  Increased oxygen concentration, the
existence of flat-water zones, favourable substratum and the inflow of
Mosel water rich in phytoplankton led to the widespread propagation of
this insect.  Disturbed trophic conditions (absence of consumers and
nutritional rivals for the caddis flies) probably played an equally
great role in this process (Tittizer, 1977).  The general regeneration
in the ecological state of the Rhine has led to an extension of the
distribution range for this species (Caspers, 1980b).  As a result of
growing diversity in trophic relations, however, a decrease in the mass
occurrence of this species has taken place.  Further information on
the distribution of chironomids in the Rhine is given by Wilson &
Wilson (in press), who analyzed the composition of exuviae collected
from sites down the whole river.

Soppe (1983) has conducted several studies with the help of a
diving bell on the benthos in the zone of the Lorelei Cliff where the
Rhine has only a width of 130 m but its deepest spot with a depth of
23 m.  Of the two types of substrata distinguished, solid boulder
gravel was almost free from faunal colonization, whereas the native
bedrock served as a biotope mainly for *Hydropsyche* larvae and *Ancylus
fluviatilis*.  In current-free places, however, the boulder gravel
substratum had about 30 species of macroinvertebrate.  Similar data
have been obtained near Bonn, i.e. the transition from Middle Rhine to
the Lower Rhine (Caspers, 1980b).  The disappearance of pollution-
sensitive species, caused by the increasing pollution load and bank
development, has become obvious here as well.  The river regulation
measures have led to the disappearance of macrophytes as microbiotopes
e.g. for the dragon-flies *Calopteryx splendens* and *C. virgo*.  Bless
(1981) reports on the reappearance of the lamellibranchs *Unio pictorum*
and *Anodonta cygnea* in the lower section of the Middle Rhine and in
the upper section of the Lower Rhine.  Both have now spread again far
into the Lower Rhine (G.F., unpublished).

*Lower Rhine*

The Lower Rhine is characterized by an extreme uniformity in
environmental conditions for the fauna.  The whole of the bank is
protected, mainly with basalt stones.  The "Greater Köln area" adds a
heavy point-source pollution load to the upper section of the Lower
Rhine.  Modern studies are not yet complete enough to make a detailed

comparison with those of Lauterborn.   However, it is known that the
Plecoptera and Ephemeroptera, as well as about half of the Mollusca
species, had disappeared between the beginning of the century and 1955
(Knöpp, 1955).   On the 200 km stretch between Köln and Emmerich, there
were vast depopulated zones until the mid-1970s in which no macroinvertebrate community could live (Herbst et al., 1969, 1971; Miegel,
1963; Schiller, 1974).   Only in the abandoned river branches through
which the Rhine was flowing at times of flood could some species survive.

   With the decrease in pollution and the improvement in oxygen supply
a gradual recolonization has taken place (Landesamt für Wasser und
Abfall Nordhein - Westfalen, 1982).   The present species include for
the most part pollution-insensitive forms with a wide ecological
valency, forming a remnant biocoenosis from the inhabitants of the hard
substratum.   The most abundant macroinvertebrates in the Lower Rhine
today are (W. Schiller, pers. comm.):   *Ancylus fluviatilis, Asellus
aquaticus, Bithynia tentaculata, Dreissena polymorpha, Dugesia lugubris,
D. tigrina, Glossiphonia complanata, Fredericella sultana, Erpobdella
octoculata, Hydropsyche* sp.*, Plumatella* sp.*, Radix peregra,
Rheotanytarsus* sp.*, Sphaerium corneum, Spongilla lacustris.*   Less
frequent are: *Acroloxus lacustris, Dendrocoelum lacteum, Gammarus pulex,
Physa fontinalis, P. acuta, Helobdella stagnalis.*

   Bed load transport also plays a role in the spatial and temporal
distribution of benthic macroinvertebrates.   Petran & Kothé (1978) have
shown that the majority of organisms of the bottom fauna (Porifera,
Platyhelminthes, Gastropoda, Bivalvia, Hirundinea, Crustacea, Isopoda,
Insecta, Bryozoa) is rather sensitive to bed load transport.   In
contrast to these, mud sabellariid worms and chironomid larvae, which
are able to dig into the river bottom, are found even in stream
sections with intense bed load transport.   A biological inventory
executed by Tittizer (1980) on part of the Lower Rhine with the help
of a diving bell, indicates that near the left bank, where redeposition
of the river bottom does not occur, the greatest richness in species
and density of individuals can be found.   Towards the middle of the
river, where frequent or permanent bed load movements may be expected,
there is a rapid decrease in both species and individuals.   The right
bank zone is colonized by macrofauna coenoses poor in species and
individuals (Fig. 10.11); bed load transport takes place here only at
higher stream-flow.


## 10.5   FISH FAUNA AND FISHING

The fish fauna of a river like the Rhine cannot be understood without
knowing the history of fishing in the region.   The fish population has
always indicated the water quality better than anything else and fishing
had been the basis of existence for whole towns and villages on the
Rhine throughout the centuries.   Most of the information about the fish
fauna has been gathered by professional fishermen and sport anglers.
Changes in the fish fauna show seven phases of the history of the Rhine
(chiefly according to Trahms, 1954).

*Fig. 10.11* Cross-section of Rhine km 752. 'Relative colonization density' gives a very approximate estimate of density of individuals (T. Tittizer, pers. comm.).

### Phase 1 - before 1765

The Rhine was clean and flowed freely in its natural channel, with a main stream and a large number of lateral and old branches. Extensive flood plains provided important spawning grounds. Mature salmon migrated upstream in hundreds of thousands each year and found spawning grounds in the tributaries, which were accessible without any obstruction. Despite intensive fishing, the fish population and production remained constantly high, some place names still remind us of that time and it is reported in old documents that the servants were complaining about having to eat salmon too often.

### Phase 2 - after 1765

As a result of river engineering measures taken to make effective agricultural use of the alluvial meadow and in the interest of navigation, the tributaries of the river were cut off and through the construction of groynes the stream bed became confined. Several tributaries became inaccessible for migratory fish. As a result of dike construction important spawning and resting places got lost. First of all, spawning grounds for the salmon became restricted. In their place, due to the newly produced groyne fields, some new spawning grounds were available for other fish. Because of the intensive deep erosion, these biotopes mostly lost in the meantime

their significance for the aquatic fauna.  The introduction of steam-
ship navigation caused heavy damage to the fish fauna by disturbing the
fish and especially by destroying the fry.  The propellers of ships
caused a considerable proportion of the young fish living in the flat
groyne fields to be thrown on land.  In addition to these changes in
the living space the first serious waste loads attached to the "founder
years" contributed to an impairment of conditions.

*Phase 3 - from 1915*

   The growing urbanization and the rapidly developing large-scale
industry at the close of the 19th century had already given rise to
pollution which led to a sudden decrease in the fish production about
1915.  Besides the decrease in the salmon stock, there was also a
reduction in the "mayfish" (*Alosa alosa*) population, whose migration
upstream had been until then a special occasion for public festivals.
In 1924, after a 12-year period, the rents for the lease of fishing
sections on the Lower Rhine had to be fixed anew and reduced to a half-
or even to one-third, because of the lower fish yields.  The last
sturgeon (*Acipenser sturio*) was caught in the Rhine in 1931.  By then
the "mayfish" had also disappeared and in the most polluted part of the
Lower Rhine the remaining salmon became inedible because of their strong
phenolic taste.  The eel, however, began to grow in importance, becoming
the "bread and butter fish" of the fishermen.

*Phase 4 - from 1933*

   Industrial production increased by leaps and bounds but without
adequate waste treatment, leading to great and rapid damage to the fish
fauna.  There was a short recovery after the end of the war, due to the
postwar stagnation in production and shipping.

*Phase 5 - from about 1948*

   As the towns and industries destroyed during the war were reconstruc-
ted, pollution increased rapidly.  Long stretches began to show zones of
biological devastation.  The fish became inedible because of bad taste.
Fish mortality and large oil carpets on the surface of the water
occurred so often that long sections of the Lower Rhine were practically
free of fish.

*Phase 6 - from about 1954*

   In this phase intensive efforts were started in order to improve
water quality on the Rhine; a characteristic measure was the rule
obliging ship-owners to discharge their waste oil only at special
collecting places.  Construction of clarification plants was intensified.
Nevertheless, the waste load of the river increased to such an extent
that it practically paralysed the Rhine fishery as a branch of trade.
Only in the section above Karlsruhe could professional fishermen still
exist.  The most serious fish mortality ever, extending from the Main

to the Netherlands in the summer of 1969 (Ant, 1969), falls also in this
phase of the history of the Rhine.

*Phase 7 - since about 1975*

   The first signs of a recovery in the fish stock in the Rhine have
been apparent since 1975.  At least sport fishing has improved considerably.  Single catches have been reported of some sensitive species
that were thought to have been extinct in the Rhine (Bless, 1981; Böving,
1981).

   Originally, there were about 40 species of fish and Cyclostomata
living in the Rhine.  Out of these 8 species have disappeared (state in
1978): *Acipenser sturio, Salmo salar, S. trutta, Coregonus oxyrhynchus,
Alburnus bipunctatus, Leuciscus souffia agassizi, Leucaspius delineatus,
Alosa alosa. Petromyzon marinus* occurs only as a few specimens in the
Netherlands Rhine section.  (According to Bless, 1981, some salmon have
sporadically been caught again in the Rhine.)  In contrast to this loss,
there has been an introduction of foreign species.  This occurred first
with the North American species, *Oncorhynchus tshawytscha, Salmo
gairdneri, Ictalurus nebulosus, Micropterus salmoides* and *Lepomis* sp.
More recently, the "grass carp" (*Ctenopharyngodon idella*) from Eastern
Asia has appeared as well.  Similar changes in the fauna have been
observed in other large river basins; according to Lelek (1978) there
are now twelve naturalized foreign species in the Danube, seven in the
Elbe and eight in the Main.  In a 100-km long section on the northern
Upper Rhine and southern Middle Rhine Lelek found that of the 39 species
originally present, not more than 19 could be detected in 1976.  Still
water regions were of great importance as residence areas for the
remaining fish.  The European roach (*Rutilus rutilus*) was the most
frequent fish and is now also the dominant on the Lower Rhine with some
40 to 50% of individuals (Böving, 1981).


## 10.6   CHEMISTRY AND WASTE LOAD

Knowledge of the chemistry of the Rhine is especially important for an
understanding of its ecology, because of the presence of toxic substances from industrial and household sources and also from numerous
non-point sources originating from streets or agriculture.

*Inorganic salts*

   The water of the Rhine is moderately hard, neutral to slightly alkaline and with a high buffering capacity.  As a result the relatively
high pollution load of inorganic acids has much less effect than it
would do in a poorly buffered water.  Table 10.5 shows the changes in
concentrations of various elements in 1982 on passing down the river.

*Table 10.5* pH range and mean concentrations (mg $l^{-1}$) of various elements in the Rhine during 1982. (After DKRR, 1983)

|  | Hochrhein Oehningen km 22.9 | Upper Karlsruhe km 362.3 | Middle Koblenz km 590.3 | Lower Rhine Kleve-Bimmen km 865 |
|---|---|---|---|---|
| pH | 7.9 - 8.8 | 7.7 - 8.5 | 7.3 - 7.7 | 7.1 - 7.6 |
| Na | 5 | 63 | 61 | 68 |
| K | 2 | 5 | 5 | 6 |
| Mg | 8 | 8 | 10 | 12 |
| Ca | 41 | 48 | 64 | 66 |
| Cl | 7 | 100 | 99 | 136 |
| $SO_4$-S | 9 | 10 | 17 | 22 |

Since the Rhine is a very important source of freshwater for the Netherlands, its loading with salts discharged from potash works, mining drainage and the chemical industry has considerable significance. The first important increase in the chloride load is due to the waste discharge of the Alsatian potash industry, whereas increase on the Lower Rhine is a result of mining drainage operations and effluents from the chemical industry. A reduction in the chloride loading is economically feasible only in the case of potash industry; an international agreement was made to deal with this problem (see 10.1).

*Oxygen content and organic load*

Because of its key role in the river ecosystem, the oxygen content of the Rhine water has been a focus of public interest. After the second world war the concentration decreased until 1969 due to the high loads of poorly or untreated sewage effluent. Since that time there has been a tremendous programme of building or improving sewage works and the oxygen content has increased again (Fig. 10.12). In 1982, for instance, the lowest concentration reported (by Deutsche Kommission zur Reinhaltung des Rhein, 1983) was 4.9 mg $l^{-1}$ $O_2$. There has been a comparable decrease in the concentration of slightly degradable organic compounds, measured as BOD (Fig. 10.13). The improvement in water quality during 1970 to 1980 (Fig. 10.14) cannot, however, be attributed entirely to developments in waste disposal. It is due also to the downward economic trend in industrial production.

The waste load of the Rhine should be reduced in the coming years as a result of further developments at waste treatment plants and by internal preventive measures in industry. The aim is to arrive at a stage where biochemical oxygen demand and oxygen concentration meet the required quality standards (quality class II, BOD <5 mg $l^{-1}$, oxygen > 6 mg $l^{-1}$). Problems associated with the stretch below the Neckar (still evident in 1980) have recently been solved; particular problems that still exist are the stretch below the Main and the whole of the lower Rhine.

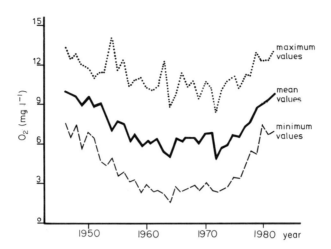

*Fig. 10.12*  Dissolved oxygen in the Lower Rhine at the German-Dutch border.  (Modified from RIWA, 1983)

........ Village-Neuf/Kembs (km 183),   ---- Seltz (km 340),   ——— Koblenz (km 590),   —·—· Bimmen/Lobith (km 865),
——— Gorinchem (km 955), for $BOD_5$ and $NH_4^+$-N from 1970 mean values from Gorinchem, Vreeswijk and Kampen

*Fig. 10.13*  Changes in water quality at five sites on Rhine during 1959-1980.

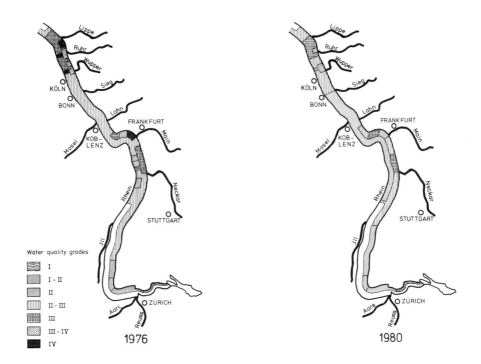

*Fig. 10.14* Water quality maps of the Rhine for 1976 and 1980 (LAWA, 1976, 1980). Grades (in terms of pollution): I, very low; I-II, low; II, moderate; II-III, critical; III, strong; III-IV, very strong; IV, overloaded.

*Nutrients*

An indication of the changes in phosphate which have taken place may be gained by considering the Bodensee. Up to about 1950 the concentrations of phosphate were near the detection limit of the method used for routine monitoring, but since that time there has been a very rapid increase. This is due to agricultural, industrial and domestic uses of phosphate. At the end of winter stratification in 1970 phosphate-P reached 40 µg l$^{-1}$ (Lehn, 1972: see also Fig. 10.15). Increases in phosphate are evident all down the Rhine, as well as increases in combined nitrogen (nitrate and ammonium) (IAWR, 1981; RIWA, 1983). Correlated changes shown by plants are the great increases in *Ranunculus fluitans* in the Hochrhein (10.422) and phytoplankton from the Upper to the Lower Rhine (10.423).

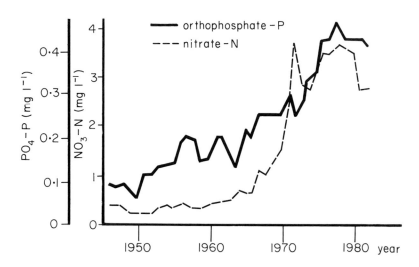

*Fig. 10.15* Orthophosphate-P and nitrate-N concentrations in the Lower Rhine at the German-Dutch border. (Modified by RIWA, 1983)

*Heavy metals*

   The natural content of heavy metals in the Rhine is low, but because of industrial and other sources concentrations in the water and especially the sediments can reach potentially harmful levels. Because of the multiform interactions between liquid and solid phases (water - seston - sediment) and because of problems at all steps in sampling and analysis, data on heavy metals need to be interpreted with particular care. Dissolved heavy metals play a less important role than those associated with suspended particles and sediments. Heavy metals in sediments are of great economic importance because it is usually impossible to dispose of dredged materials on to agricultural land. Reduction in heavy metal pollution depends largely on industry, since waste and water treatment do not give fully satisfactory results. An extensive body of data and interpretation can be found in the works of Förstner & Müller (1974), Müller (1979) and Malle & Müller (1982). In general many of the heavy metals showed peak concentrations and loads at the end of the 1960s and beginning of the 1970s. Since that time the loads for most elements have remained little changed, but those for chromium and cadmium have shown a marked decrease. Table 10.6 gives some indication for 1982 of the increases taking place on passing down the river.

*Table 10.6*  Heavy metals in the Rhine during 1982: mean concentrations ($\mu$g l$^{-1}$) and mean load (g s$^{-1}$).  (After IKSR, 1983)

|  | Hochrhein Oehningen km 22.9 | | Upper Karlsruhe km 362.3 | | Middle Koblenz km 590.3 | | Lower Rhine Kleve-Bimmen km 865 | |
|---|---|---|---|---|---|---|---|---|
|  | conc. | load | conc. | load | conc. | load | conc. | load |
| Cr | <2.0 | 0.6 | <2.0 | 2.0 | 5.3 | 11.9 | 7.9 | 22.3 |
| Ni | 5.3 | 2.0 | 9.6 | 14.1 | <5.0 | 8.1 | 9.2 | 25.4 |
| Cu | 2.8 | 1.2 | 2.9 | 4.4 | 7.4 | 16.5 | 25.7 | 69.0 |
| Zn | <30.0 | 6.3 | <30.0 | 23.8 | 45.0 | 99.1 | 108.8 | 291.5 |
| Cd | <0.3 | 0.1 | <0.3 | 0.2 | <0.3 | 0.3 | 0.6 | 1.6 |
| Hg | <0.05 | 0.02 | <0.05 | 0.06 | 0.08 | 0.2 | <0.2 | 0.59 |
| Pb | <5.0 | 1.0 | <5.0 | 4.0 | <5.0 | 8.8 | 11.6 | 32.4 |

*Micropollutants*

As the input of easily degradable substances and obvious sources of coarse pollution have been much decreased, increasing attention has been paid to substances which are poorly degradable, some of which are subject to bioaccumulation and some of which are known to be toxic, carcinogenic or mutagenic.  Because of their long-term effects, organo-halogens are of special interest.  A tremendous range of these substances which enter the river either as industrial wastes or following their use of pesticides etc.  For instance, during assessment of Rhine water quality in the Nordrhein-Westfalia region, there are 65 compounds which are detected regularly.  These include:

> chlorohydrocarbons of low molecular weight, e.g. chloroform;
> chlorohydrocarbons of high molecular weight, e.g. dichlorbenzol;
> polycyclic, aromatic hydrocarbons, e.g. 3,4 benzpyren;
> chlorophenols;
> biocides, e.g. hexachlorcyclohexane.

Because of this complexity, only some compounds can be assessed quantitatively, so 'total organic chlorine' is used as an additional summarizing parameter.  As with many other pollutants, its concentration increases in passing downstream (Table 10.7).

The concentrations of these substances are in general - looking at the Rhine as a whole - low and below toxic thresholds.  Further, there is some indication of a decrease.  The mean concentrations in 1982 were lower than in previous years, though peak loadings were still similar (KIWA, 1983).  Nevertheless there is a considerable risk that substances of biological activity are overlooked (Sontheimer, 1983).  The use of trout as test fish can still demonstrate chronic toxic effects and the AMES-Test, mutagenic effects (KIWA, 1983).

*Table 10.7*  Total organic chlorine ($\mu$g $1^{-1}$ Cl) in the Rhine during 1975 (IAWR, 1978).

| | km | 1975 | | 1976 | | 1977 | | 1978 | |
|---|---|---|---|---|---|---|---|---|---|
| | | $\bar{x}$ | max | $\bar{x}$ | max | $\bar{x}$ | max | $\bar{x}$ | max |
| Basel | 160 | 50 | 130 | 40 | 100 | 50 | 140 | 20 | 60 |
| Karlsruhe | 360 | 40 | 120 | 30 | 80 | 45 | 150 | 50 | 140 |
| Köln | 680 | 100 | 340 | 40 | 100 | 60 | 170 | 100 | 270 |
| Wittlaer | 750 | 100 | 290 | 80 | 270 | 120 | 330 | 100 | 240 |

*Waste heat*

Since the end of the sixties the waste heat load has also become an urgent problem. The situation at present is as follows (M. Wunderlich, pers. comm.). Due to the construction of more and higher-capacity power plants, the limits of a straightforward once-through system of cooling have been reached. The concentrated discharge of large heat loads can lead to restriction on other uses of the water and affects the ecological state of the surface waters.

The loading of the German Rhine section with waste heat discharges is now regulated with the aid of recommendations by the Working Community of the German Federal States (LAWA, 1977). The maximum temperature rise permitted generally in rivers is equal to $\Delta T = 5$ K; the maximum water temperature should not exceed 28 $^\circ$C in summer-warm waters and 25 $^\circ$C in summer-cold waters. These temperatures represent calculated values obtained with the assumption of a complete mixing of cooling water and river water. In the Rhine, however, a stronger margin is set for the temperature rise  ($\Delta T = 3$ K). This is justified by the assumption that, owing to the large width of the river, complete mixing is not possible over long stretches.

The five states bordering the Rhine decided in 1972 to take measures to prevent the aggravation of the thermal load in the river. A resolution was agreed that only power plants with cooling towers or similar cooling systems should be permitted. To the best of present knowledge, it is possible to avoid adverse consequences for the ecology and water resources management of the Rhine with the existing temperature limit values and by adequate associated measures (artificial aeration of cooling water).

In assessing the effects of the waste load on the ecology of the Rhine an important aspect is the insignificant transverse mixing. Waste water discharges and tributaries can be regarded in some sections of the Rhine as completely mixed only after at least 30 to 100 km. It is on the mountainous stretch between Koblenz and Bingen that faster mixing may occur. There is a similar situation on the impounded parts of the Hochrhein and the Upper Rhine, where most of the mixing takes place in the turbines of the power stations.

## 10.7  SELF-PURIFICATION AND MICROBIAL ACTIVITY

For a stream so heavily loaded with wastes as the Rhine, self-purification is of decisive ecological and economic importance.  Considering
self-purification in the light of sewage engineering, its economic
value can be quantified.  During the period of heavy pollution a waste
load of 50 x $10^6$ inhabitants (P) and population equivalents (PE) was
discharged into the Rhine before it reached the Dutch frontier.  At
Bimmen the $BOD_5$ load corresponded to 35 x $10^6$ P + PE ; this means that
15 x $10^6$ P + PE could be eliminated by self-purification.  Taking a
yearly average expenditure of 40 DM per P + PE on preventive measures,
this corresponds to a saving of about 600 million DM per year
(Bundesminister des Innern, 1976).  Without the influence of toxic
substances the effects of self-purification would have been about 30%
higher (Knöpp, 1970).

Summation parameters like TOC (total organic carbon), COD (chemical
oxygen demand) and BOD (biochemical oxygen demand) are particularly
suited to characterize the self-purification process.  Sampling for the
longitudinal profiles of the Rhine between Karlsruhe and Emmerich
shown in Fig. 10.16 was carried out by helicopter for the Upper and
Middle Rhine on 22 October 1979 and for the Lower Rhine on 25 October
1979; this assured a delay of only a few hours before analysis.  The
natural discharge of the Rhine is usually quite low in the second half
of October, which makes the effects of waste effluents particularly
serious.  Nevertheless, discharge was relatively high during the
October 1979 study, amounting to about 1250 $m^3$ $s^{-1}$ at Kaub (between
Mainz and Koblenz).  This corresponds to a value between MLQ (mean low
discharge) and MQ (mean discharge).  The pollutant concentrations
depending on the waste discharge were about one third lower than those
to be expected at MLQ, because of the relatively high degree of
dilution.  However it is the fundamental changes along the course of
the river that are emphasized here, rather than the magnitude of
individual parameters.

The longitudinal profiles of the various waste parameters in the
Rhine are not at all similar (Fig. 10.16).  The degradable organic
matter content, characterized by $C-BOD_5$ (determined by applying a
nitrification inhibitor) is relatively low along the whole stretch.
The maxima on the Lower Rhine are about twice as high as the minima on
the Middle and the Upper Rhine.  The concentrations of degradable
organic substances in the Main, Wupper and Emscher are considerably
higher than in the Rhine; those in the Mosel are less than in the Rhine.
In contrast, the effect of effluents from large-scale industry is much
higher judged by COD (oxidation with bichromate).  The Neckar, Mosel
and Ruhr show similar concentrations of COD to those in the Rhine.
On the whole, there is an increase in COD on passing down the river.

*Fig. 10.16* Changes in COD (chemical oxygen demand), TOC (total organic carbon) and C-BOD5 (5-day BOD in presence of allyl thiourea) along the Rhine. Measurements made on 22 and 25 October 1979; the break in the lines from km 640 - 670 is due to observations at the lower sites being made on the second day; all concentrations in mg $l^{-1}$.

In general, differences similar to those observed with COD are found for TOC in the Upper and Middle Rhine. It is remarkable, however, that in the Lower Rhine there is no similar increase in TOC corresponding to that which occurs with COD. Correlations between these three para-meters permit conclusions to be made concerning the "biochemical degree of purification" (C-BOD5/COD) and the "specific oxygen demand of organic substances " (COD/TOC). The biochemical degree of purification, which is characterized in the Rhine section investigated  mostly by values between 0.05 and 0.1, indicates that self-purification is well-advanced. For fully degradable substances the quotient of the biochem-ical degree of purification is usually equal to 1; fresh domestic sewage and wastes subjected to biological treatment are generally characterized by quotients equal to 0.7 and 0.25, respectively. Only the Wupper and Emscher rivers show an essentially lower degree of purification than the Rhine.

The quotient for the specific oxygen demand of organic substances in the Upper and Middle Rhine also indicates no peculiarities. It does show, however, a considerable increase in the Lower Rhine.  From this it can be inferred that, although the substances present in the Lower Rhine are largely oxidized in a biochemical respect, this is not so from a chemical point of view.

In addition to the oxidation of organic substances, nitrification takes an essential part in the oxygen consumption in the rivers.  The ammonium content comes from waste waters or gets released from protein compounds during their degradation.  Whereas C-BOD$_5$ increases relatively little on passing down the river, the oxygen consumption by nitrification increases along the whole stretch.  Upstream of Karlsruhe oxygen consumption by nitrification amounts to not more than 20% of the total BOD, whereas on the Lower Rhine it is 60 to 80%.  If the yearly average is considered, however, the proportion of nitrification in the BOD$_5$ may not be quite so high.

The increase in oxygen consumption by nitrification with the extension of the river reach is influenced by the ammonium concentration in the water of both ammonium and nitrifying bacteria.  At the beginning of the test section the former is mostly 0.3 - 0.5 mg l$^{-1}$ and on the Lower Rhine mostly 0.5 - 1.0 mg l$^{-1}$.  However, the increase in nitrifying bacteria is likely to be of great influence, this increase being characterized by "the maximum oxygen consumption due to nitrification under standard laboratory conditions" (Müller & Kirchesch, 1981a).  The biomass of nitrifying organisms increases more than ten-fold by the time the river reaches the German-Dutch border.  Tributaries with a high degree of self-purification, like the Neckar, Mosel and Ruhr, also contribute to the biomass of nitrifying bacteria.  In the Main and Wupper, which are loaded with fresh waste waters, the concentration of nitrifying bacteria is very low.  The R. Emscher differs from all the other tributaries because it combines a high concentration of nitrifying organisms with a low degree of purification.  This is due to the fact that the Emscher is the effluent of a large-scale treatment plant (river sewage works), where the sludge is regulated so as to facilitate the development of nitrifying bacteria.

The diel fluctuations in the oxygen content determined by the primary production range up to 3.0 mg l$^{-1}$, depending on the stretch, season and radiation intensity.  Oxygen contents below 2 mg l$^{-1}$ at Mainz and below 4 mg l$^{-1}$ on the Lower Rhine were not unusual during the period of the maximum waste load.  Conditions are far more favourable today and they need to be improved only on the Lower Rhine.

A river like the Rhine, receiving large amounts of waste waters from the chemical industry, may always be suspected of containing toxic substances which have inhibitory effects on the biota.  It is usually very difficult to produce causal-analytical evidence for such interference, but this becomes easier with individual point sources of pollution.  Knöpp (1970) was the first to deal with this problem.  His results for September 1966 showed that in the section of greatest importance for water management, there was a marked discrepancy between the load profile computed from emission data and the longitudinal profile of BOD determined from water samples. This was probably due to the toxic effects of effluents. This interpretation has been confirmed by data on algal photosynthetic activity (see below). Knöpp (1968) found that among the samples of water tested, six were highly toxic for microbial degradation and thirteen reduced algal photosynthesis.

In investigations on the effects of rise in water temperature in October 1972 and 1979, Müller (1975) and Wunderlich (1978) applied a quite different method to determine the inhibition ratios of self-purification on the Lower Rhine. In order to study the effects of calefaction, Rhine water was incubated both at river temperature and 10 $^{o}$C higher for seven hours. For the Upper and the Middle Rhine there was an increase in oxygen consumption at the higher temperature. In the Lower Rhine, where Knöpp's investigations had already revealed an inhibitory influence, the rate of increase in oxygen consumption was either very low or even negative. This suggests that the effects of poisons increases when the water is warmed. As the waste load of the Rhine has strongly decreased during recent years, a reduction in toxic wastes may also be assumed. This has been corroborated by studies on the rate of increase in oxygen consumption when water temperature rises due to heat discharge (Wunderlich, 1980). Nevertheless there is still a measurable inhibitory influence in parts of the Lower Rhine (Friedrich & Viehweg, in press).

Thanks to the strict control regulations, the calefaction of Rhine water by cooling water discharges has not led to serious ecological effects. Nevertheless, the results of some studies may be of interest. As a result of the rise in temperature, the rate of metabolic processes will be increased. The degree of increase plays an important part in determining the substances contained in the river water. Wunderlich (1980) concluded for the Rhine that the increased rates of oxygen consumption due to biochemical degradation and nitrification are subject to both marked spatial and temporal fluctuation. Low rates of increase often occur in the zones loaded with waste discharge from fresh communal sewage. Particularly high rates of increase are typical of zones where self-purification is well advanced.

In the degradation of organic substances, psychrophilic and meso-philic bacteria i.e. those active even at low (6 $^{o}$C) or high temperatures (35 $^{o}$C), respectively, play an essential part. Plate techniques have shown that psychrophilic bacteria are dominant during winter. A short period of warming, as often occurs when water passes once through power plants with continuous cooling, exerts no detrimental effect on psychrophilic bacteria.

Some of the hydro-electric power plants on the Rhine are operated with cooling towers having an outlet cooling system. Power plants of this type may lead to a considerable rise in oxygen consumption, even if the outflow and inflow temperatures are equal (e.g. oxygen consumption doubles within 24 hours during incubation at 20 $^{o}$C after a short

period at 40 °C (Wunderlich, 1977; Friedrich, 1979).  This presumably
results from short-term activation of bacteria in quiescent stages
(hypnospores, hunger forms).  As only a relatively slight part of the
discharge is used for cooling purposes and water is subjected to
aeration while flowing through cooling towers and outlet structures,
this increase in oxygen consumption has not led to special problems.

Outlet cooling systems and closed-circuit cooling systems show
positive features from the viewpoint of water resources management,
because they act as fixed  bed reactors in the nitrification process.
In a cooling system with outlet the ammonium contained in the river
water will be oxidised up to about 40% and in a closed-circuit cooling
system this oxidation is almost complete.  Less important but still
quantifiable is the decrease in organic carbons after the water has
passed through a cooling system (Wunderlich, 1980).

Although heterotrophic organisms play the major role in a river
with heavy waste load, autotrophic organisms must also be considered.
The remainder of this  chapter deals with phytoplankton metabolism,
based on investigations of the "oxygen production under laboratory
conditions" referred to as "OPL" (DIN 38412, Part 14).

Fig. 10.17 shows longitudinal profiles of OPL and chlorophyll a in
late October 1979.  On the Upper Rhine at Karlsruhe there is almost
no active phytoplankton at this season, although in the anabranches
(e.g. the Otterstadt old Rhine branch), OPL is still high.  Oxygen
production increases more or less constantly along this stretch of
river, with the impounded tributaries making an important contribution.
From the longitudinal profiles of OPL and chlorophyll, it is possible
to estimate to what extent primary production or phytoplankton biomass
are involved in C-$BOD_5$ or COD.  In the Upper Rhine the contribution of
algae to the BOD is more or less insignificant; it is about 20% in the
Middle Rhine in autumn, eventually rising to about 30% in the Lower
Rhine near the German-Dutch border.  The corresponding algal contribution to the COD is less.  Phytoplankton is more important in summer.

Before dealing with annual changes in phytoplankton activity, two
terms need to be explained.  "Direct algal BOD" denotes the biological
oxygen demand needed for the degradation of a momentarily present
biomass of algae.  "BOD of algal origin" denotes all products resulting
from algal primary production (breakdown stages, extracellular products).
The latter is higher than the former and the relative difference between
these two quantities increases further downstream.

The annual course of algal activity may be characterized by the OPL,
measurements of which are made for the most important sections of the
Rhine.  Of the two years shown in Fig. 10.18, 1979 was relatively "dry",
whereas 1980 was "wet".  The dynamics of algal activity on the Hochrein
at Reckingen still correspond well with those of the Bodensee.  A
characteristic feature is the spring maximum, which then breaks down,
presumably due to nutrient deficiency and grazing.

*Fig. 10.17*  Changes in algal activity measured as OPL (oxygen production in laboratory) and chlorophyll *a* along Rhine.  Measurements
made on 22 and 25 October 1979 ; the break in the lines from km 640 -
670 is due to observations at the lower sites being made on the
second day.  Assay made in laboratory using light (200 μmol photon
$m^{-2}$ $s^{-1}$)-dark bottle technique at 20 °C.

Algal activity shows a more or less constant increase, any drops
being correlated mostly with hydrological events.  For example, flow at
Koblenz amounted to about 2000 $m^3$ $s^{-1}$ in May 1980, a period with high
OPL values in the Middle and Lower Rhine.  It then rose to 4000 $m^3$ $s^{-1}$
in August 1980 and this resulted in a reduction in the OPL to 25% of its
initial value.

Studies on actinomycetes, which are of interest as producers of
odorous substances, have suggested (Müller, 1980) that, at least in the
Rhine, these belong to the "sediment bacteria".  Owing to their slow
growth and adherence of their filaments to particles, they are hardly
capable of colonizing the flowing water body actively.  Consequently,
their concentration in waters appears to depend to a large extent on
the content of resuspended sediment.  Bacteria  responsible for cresol

*Fig. 10.18* Changes in algal activity measured as OPL (oxygen production in laboratory) in 1979-1980 based on weekly measurements at four sites.  (Details given in legend to Fig. 10.17)

degradation appear to have a similar behaviour.  During the course of the year, the concentration of such sediment bacteria is correlated negatively with that of the algae.  It is clear however that a great deal more research needs to be carried out on the sediment bacteria. Knowledge about microbial activities and self-purification in the flowing water of the Rhine may now be considered quite good.  Knowledge about the contribution of the sediment bacteria and sessile organisms in general to the metabolism of the Rhine should be expanded.

## 10.8  CONCLUDING REMARKS

Although the Rhine and the land along its banks is heavily used for
settlements, industry and traffic, the river has conserved or regained
some of its original biological characteristics and has not lost its
attraction for tourists from all over the world.  Its ecosystem is on
the way of recovery from very severe damage, but the possibility of
future danger remains because of accidents connected with a river used
so intensively.

   There are some special measures which would help ensure better
conditions in and around the river.  In the Upper Rhine plain the
ecology of the alluvial valley could be reactivated by the construction
of connections between old river branches, gravel pits and tributaries
with high water quality.  These measures would augment the ground water,
support the typical biotope of the alluvial forest, improve the quality
of the side waters and make some improvement to water quality of the
main river.  Similar measures could be taken along the Lower Rhine, but
the situation there is more difficult due to the deep erosion of the
river and the dense settlements.  The gravel workings in the alluvial
plain of the Upper Rhine and of the Lower Rhine are especially
important.  Future operations should include much better reclamation
measures to improve the ecological situation in the river and the
alluvial plain.

   When the river has reached the quality objectives (German water
quality class II) which are based on biological analyses of the benthos,
water quality management has then to evaluate and describe the waste
load of the river by methods capable of differentiating various components.  Such methods should permit the concentrations of substances
hazardous for drinking and other water uses to be assessed.  They
should also monitor components which influence adversely the ecology
of the river and its self-purification capacity.  Water quality
management has to handle micropollutants with more care.  Better
measures to avoid pollution are needed; this means that industry has to
develop waste water-free techniques and to take care that the consumers
of their products do not spread so many harmful substances into the
environment and the waters.

   With the progress of mechanical-biological waste water treatment,
the input of nutrients for algae is getting more important, as these
nutrients cause eutrophication problems in the slow-flowing stretches
of the Rhine in the Netherlands.  In a river, which has been so much
improved, it will be worthwhile to evaluate its suitability for bathing
from the hygienic point of view.

   Looking back in history, one can see a correlation between good
times for man and bad times - pollution - for the Rhine.  Only when
the ecological crisis of the Rhine from 1950-1970 caused severe problems
in the use of the water by man, did sanitation programmes start to
uncouple this correlation.  Let us hope that the national and inter-

national efforts for improving the condition of the Rhine will
continue till the end of the century, so that the condition of the
river will match its importance as the artery of central Europe.

*Acknowledgements*

We have to thank our many co-workers who helped us to compile this
paper and in particular, Frau Marlies Viehweg and Frau Ingrid Busch.
Thanks also to Dr D. Backhaus (Landesanstalt für Umweltschutz, Baden-
Württemberg) for unpublished data.

REFERENCES

Anhut U. (1977)  Untersuchungen zur Molluskenfauna des rechten unteren
    Niederrheins im Raum zwischen Wesel, Rees und Emmerich. *Gewässer
    und Abwässer* 62/63, 17-62
Ant H. (1969)  Biologische Probleme der Verschmutzung und akuten
    Vergiftung von Fliessgewässern, unter besonderer Berücksichtigung
    der Rheinvergiftung im Sommer 1969. *Schriftenreihe für Landschafts-
    pflege und Naturschutz* 4, 97-126
ARGE Rhein (Arbeitsgemeinschaft Rhein) (1968)  Die Verunreinigung des
    Rheins und seiner wichtigsten Nebenflüsse in der Bundesrepublik
    Deutschland - Arbeitsgemeinschaft der Länder zur Reinhaltung des
    Rheins. *Denkschrift, Stand* 1965. Ministerium fur Ernährung,
    Landwirtschaft und Forsten, Düsseldorf
Backhaus D. (1968)  Ökologische Untersuchungen an den Aufwuchsalgen der
    obersten Donau und ihrer Quellflüsse, II.  Die räumliche und
    zeitliche Verbreitung der Algen. *Arch. Hydrobiol., Suppl.* 34 1/2,
    24-73
Backhaus D. & Kemball A. (1978)  Gewässergüteverhältnisse und Phyto-
    planktonentwicklung im Hochrhein, Oberrhein und Neckar. *Arch.
    Hydrobiol.* 82, 166-206
Bensing W. (1966)  Gewässerkundliche Probleme beim Ausbau des
    Oberrheins. *Gewässerkundliche Mitteilungen* 10, 85-102
Bless R. (1981)  Beobachtungen zur Muschelfauna des Rheines zwischen
    Köln und Koblenz. *Decheniana (Bonn)* 134, 234-43
Bloesch J. (1977)  Bodenfaunistische Untersuchungen im Aare und Rhein.
    *Schweiz. Z. Hydrol.* 39, 46-68
Böving M.-P. (1981)  Die Fischfauna des Rheinstromes und seiner direkt
    angrenzenden Altwässer im Niederrheingebiet. *Decheniana (Bonn)*
    134, 360-73
Bundesminister des Innern (1976)  Umweltprobleme des Rheins. *Sonder-
    gutachten 1976 des Rates von Sachverständigen für Umweltfragen,
    herausgegeben vom Bundesminister des Innern (BMI)*. Verlag Heger,
    Bonn-Bad Godesberg
Burckhardt H. & Burgsdorf H.L. (1964, 1965)  Flora des Altrheins bei
    Xanten und seiner Umgebung.  I Floristische Untersuchung.  II.
    Pflanzengesellschaften des Xantener Altrheins. *Gewässer und
    Abwässer* 37/38, 7-45; 43, 7-47

Caspers N. (1980a)   Die Makrozoobenthos-Gesellschaften des Hochrheins
    bei Bad Säckingen.   *Beiträge Naturk. Forsch. Sudw. Dtl.* 39, 115-42.
Caspers N. (1980b)   Die Makrozoobenthos-Gesellschaften des Rheins bei
    Bonn.   *Decheniana (Bonn)* 133, 93-106.
Conrath W., Flakenhage B. & Kinzelbach R. (1977)   Übersicht über das
    Makrozoobenthos des Rheins im Jahre 1976.   *Gewässer und Abwässer*
    62/62, 63-84.
Czernin-Chudenitz C.W. (1958)   Limnologische Untersuchungen des
    Rheinstromes.   III Quantitative Phytoplanktonuntersuchungen.   *Forsch-*
    *ungsberichte des Wirtschafts und Verkehrsministeriums Nordrhein-*
    *Westfalen* 536, 422 pp.   Köln & Opladen.
Deutsches Gewasserkundliches Jahrbuch, Rheingebiet (1974)   Landesamt
    für Gewässerkund.   Mainz.
Deutsches Institut fur Normung (1982)   Bestimmung der Sauerstoffproduk-
    tion mit der Hell-Dunkelflaschen-Methode unter Laborbedingungen, SPL.
    *DIN 38 412* Teil 14.   Deutsches Institut   fur Normung e. V., Berlin.
Ernst E. (1972)   *Der Rhein - eine europäische Stromlandschaft im Luft-*
    *bild* 146 pp.   Konkordia-Verlag, Bühl.
Förstner U. & Müller G. (1974)   *Schwermetalle in Flussen und Seen als*
    *Ausdruck der Umweltverschmutzung.*   Springer Verlag, Berlin.
Friedrich G. (1973)   Okologische Untersuchungen an einem thermisch
    anomalen Fliessgewässer (Erft/Niederrhein).   *Schriftenreihe der*
    *Landesanstalt fur Gewässerkunde und Gewässerschutz der Landes*
    *Nordrhein-Westfalen* 33, 1-125 + 16 Tables.
Friedrich G. (1979)   Biologische Untersuchungen am Rhein in Nordrhein-
    Westfalen.   Vortrag auf der Jahrestagung der deutschsprachigen
    Limnologen 1979 in Schlitz (Hessen).   (Unpublished: data available
    from author.)
Friedrich G. & Viehweg M. (in press)   Recent developments of the phyto-
    plankton and its activity in the lower Rhine.   *Verh. int. theor.*
    *angew. Limnol.* 22.
Herbst H.V., Friedrich G. & Heuss K. (1969)   *Biologische und chemische*
    *Untersuchungen zur Kennzeichnung der Beschaffenheit des Rheines am*
    *Ende einer extremen Niedrigwasserperiode,* 90 pp.   Landesanstalt für
    Gewässerkunde Nordrhein-Westfalen, Krefeld.
Herbst H.V., Friedrich G. & Heuss K. (1973)   Biologische und chemische
    Untersuchungen des Rheines bein Niedrigwasser im Oktober 1971.
    *Schriftenreihe der Landesanstalt fur Gewasserkunde und Gewasserschutz*
    *Nordrhein-Westfalen* 35, 1-44.
Hess F. (1959)   Probleme des Hochwasserschutzes am Niederrhein.
    *Gewässer und Abwässer* 23, 1-88.
Heuss K. (1975)   *Das Phytoplankton der Ströme Rhein und Donau - ein*
    *Vergleich. 18. Jahrestagung der Internationalen Arbeitsgemeinschaft*
    *Donauforschung der Societas Internationalis Limnologiae. Kurzrefe-*
    *rate* 1, pp. 217-229.   Bundesanstalt für Gewässerkunde, Koblenz.
Hild H.J. (1960)   Hydrobiologische Untersuchungen in einigen abwasser-
    belasteten niederrheinischen Gewässern.   *Hydrobiologia* 16, 203-14.

Hild H.J. & Rehnelt K. (1965a)  Öko-soziologische Untersuchungen an
    einigen niederrheinischen Kolken. *Ber. deutsch. bot. Ges.* 78,
    289-304.
Hild H.J. & Rehnelt K. (1965b)  Hydrobiologische Untersuchungen an
    niederrheinischen Gewässern. *Hydrobiologia* 25, 422-65.
Hild (H.) J. & Rehnelt K. (1967)  Öko-soziologische Untersuchungen am
    Boetzelaerer Meer (Niederrhein). *Ber. deutsch. bot. Ges.* 79,
    647-68.
Hoppe (1970)  Die grossen Flussverlagerungen des Niederrheins in den
    letzteon 2000 Jahren und ihre Auswirkung auf Lage und Entwicklung
    der Siedlungen. *Forschungen zur deutschen Landeskunde* 189.
IAWR (Internationale Arbeitsgemeinschaft der Wasserwerke im Rhein-
    einzugsgebiet) (annual)  *1 Arbeitstagungen. 2 Jahresberichte.*
    Sckretariat der IAWR, Postfach 8169, NL-1005 AD, Amsterdam, The
    Netherlands.
IKSR (Internationale Kommission zum Schutze des Rheins gegen Verun-
    reinigungen) (1977)  *Bericht über die physikalisch-chemische
    Untersuchung des Rheinwassers* VII, 1972-1975. Koblenz. Also
    annual reports: *1 Zahlentafeln. 2 Tätigkeitsberichte.*
Internationale Kommission für die Hydrologie des Rheingebietes (1978)
    *Das Rheingebiet Hydrologische Monographie*, 279 pp.  Staatsuitgeverij,
    Christoffel Plantÿnstraat 1, The Hague, Netherlands.
Jaag O. (1938)  Die Kryptogamen-Vegetation des Rheinfalls und des
    Hochrheins von Stein bis Eglisau. *Mitt. Naturf. Ges. Schaffhausen*
    14, 1-158.
Kinzelbach R. (1978)  Veränderungen der Fauna des Oberrheins.  Beitr.
    Veröff. Naturschutz. *Landschaftspflege Baden-Würtemberg* 11,
    291-301.  (Karlsruhe).
KIWA (1983)  Toxikologische Untersuchungen des Rheinwassers.  The
    Netherlands Waterworks Testing and Research Institute KIWA Ltd.,
    Rijkswijk, Netherlands.
Knecht A. (1982)  Das Wachstum von Wasserpflanzen im Hochrhein.
    *Deutsche gewässerkundliche Mitteilungen* 26(4), 109.
Knöpp H. (1957)  Die heutige biologische Gliederung des Rheinstromes.
    *Deutsche Gewässerkundl. Mitt.* 1(3), 56-63.  Koblenz.
Knöpp H. (1960)  Untersuchungen über das Sauerstoff-Produktions-
    Potential von Flussplankton. *Schweiz. Z. Hydrol.* 22, 152-166.
Knöpp H. (1968)  Stoffwechseldynamische Untersuchungsverfahren für
    die biologische Wasseranalyse. *Int. Rev. ges. Hydrobiol.* 53,
    409-41.
Knöpp H. (1970)  Neuere Studien zur Biologie und Biochemie des Rheins.
    *Gewässerschutz Wasser - Abwasser* 3, 67-82.  Institut für Siedlungs-
    wasserwirtschaft der Rhein, Westf. Techn. Hochschule, Aachen.
Knörzer H.K. (1960)  Die Salbei-Wiesen am Niederrhein. *Mitt. flor.
    soz. N.F.* 8, 209-21.
Knörzer K.H. (1963)  Pflanzenwanderungen am Niederrhein. *Der
    Niederrhein* 30, 134-37.  (Krefeld).
Koch W. (1926)  Die Vegetationseinheiten der Linthebene. *Jb. St.
    Gallener Naturf. Ges.* 61, 1-144.

Kolkwitz R. (1912)  Quantitative Studien über das Plankton des Rhein-
    stroms von seinen Quellen bis zur Mündung.  Mitteilungen aus der
    Kgl. *Prüfungsanstalt für Wasserversorgung und Abwasserbeseitigung*
    16, 167-209. (Berlin).  See also *Ber. deutsch. bot. Ges.* 30, 205-26.
Krause A. (1981)  Über die Flora und Vegetation der dem Uferschutz
    dienenden Bruchsteinmauern - Pflastern und - Schüttungen am
    nördlichen Mittelrhein.  *Natur und Landschaft* 56(7/8), 253-60
    (Stuttgart).
Krause W. (1975)  Siedlungen gefährdeter Pflanzen in Baggerseen der
    Oberrheinebene.  *Beitrage zur naturkundlichen Forschung Südwest-
    deutschlands* 34, 187-199.  Karlsruhe.
Krause W. (1976)  Veränderungen im Artenbestand makroskopischer
    Süsswasseralgen in Abhängigkeit vom Ausbau des Oberrheins.
    *Schriftenreihe f. Vegetationskunde* 10, 227-37.  (Bonn-Bad Godesberg).
Krebs F. (1978)  *Vergiftung der Selbstreinigung durch toxische Abwässer*
    182 pp.  Bundesanstalt fur Gewässerkunde, Koblenz.
Kummer G. (1934)  Die Flora des Rheinfallgebietes.  *Mitt. Naturf. Ges.
    Schaffhausen* 11.  (see pp. 68-73, Die Wasserflora des Rheines und
    seiner Ufer).
Landesamt für Wasser und Abfall Nordrhein-Westfalen (1982)  *Gewässer-
    güterbericht* 1981.  (1983) Gewässergüterbericht 1982, Düsseldorf.
Lange-Bertalot H. (1974)  Das Phytoplankton im unteren Main unter dem
    Einfluss starker Abwasserbelastung.  *Cour. Forsch. Inst. Senckenberg*
    12, 1-88.
Lange-Bertalot H. (in press)  The algal communities of the Rhine.  In:
    Kinzelbach R. *The Rhine*.  Junk, The Hague.
Lange-Bertalot H. & Lorbach K-D. (1979)  Die Diatomeenbesiedlung des
    Rheins in Abhängigkeit von der Abwasserbelastung.  *Arch. Hydrobiol.*
    87, 347-63.
Lauterborn R. (1905)  Die Ergebnisse einer biologischen Probenunter-
    suchung des Rheins.  *Arbeiten a.d. kaiserlichen Gesundheitsamt* 22,
    630-52.  (Berlin).
Lauterborn R. (1910)  Die Vegetation des Oberrheins.  *Verh. naturhisto-
    risch-medizinischen Vereins Heidelberg N.F.* 10, 450-502.
Lauterborn R. (1916-1918)  Die geographische und biologische Gliederung
    des Rheinstromes.  *Sitzungsbericht d. Heidelberger Akad. Wissensch.
    Math.-Nat.* Abt. 1: 6, 1-61; 2: 5, 1-70; 3: 7, 1-87.
LAWA (Länderarbeitsgemeinschaft Wasser) (1976)  *Die Gewässergütekarte
    der Bundesrepublik Deutschland*.  16 pp. + map.  Published by LAWA
    under chairmanship of: Minister für Landwirtschaft, Weinbau und
    Umweltschutz des Landes Rheinland-Pfalz.
LAWA (1977)  *Grundlage für die Beurteilung der Wärmebelastungen von
    Gewässern*.  Verlag Krach, Mainz.
LAWA (1980)  *Die Gewässergütekarte der Bundesrepublic Deutschland*.
    16 pp. + map.  Published by LAWA under chairmanship of: Minister
    fur Ernährung, Landwirtschaft, Umwelt und Forsten des Landes Baden-
    Württemberg.
Lehn H. (1972)  Zur Trophie des Bodensee.  *Verh. int. Verein. theor.
    angew. Limnol.* 18, 467-74.
Lelek A. (1976)  Veränderungen der Fischfauna in einigen Flüssen
    Zentraleuropas (Donau, Elbe und Rhein).  *Schriftenreihe für
    Vegetationskunde* 10, 295-308.

Lelek A. (1978)  Die Fischbesiedlung des nördlichen Oberrheins und des südlichen Mittelrheins. *Natur und Museum* 108, 1, 1-9.

Liepolt R. (1967)  *Limnologie der Donau. Eine monographische Darstellung.* Schweizerbart, Stuttgart.

Lohmeyer W. (1970a)  Zur Kenntnis einiger nitro- und thermophiler Unkrautgesellschaften im Gebiet des Mittel- und Niederrheins. *Schriftenreihe fur Vegetationskunde* 5, 29-62. (Bonn-Bad Godesberg.)

Lohmeyer W. (1970b)  Über das Polygono-Chenopodietum in Westdeutschland unter besonderer Berücksichtigung seiner Vorkommen am Rhein und im Mündungsgebiet der Ahr. *Schriftenreihe für Vegetationskunde* 5, 7-28. (Bonn.)

Malle H.G. & Müller G. (1982)  Metallgehalt und Schwebstoffgehalt im Rhein. *Wasser Abwasser* 15, 11-15.

Marsson M. (1907-1911)  Bericht über die Ergebnisse der vom 14. bis zum 21.10.1905 ausgeführten biologischen Untersuchung des Rheins auf der Strecke Mainz bis Koblenz. *Arbeiten aus dem kaiserlichen Gesundheitsamt (Berlin)* 25, 140-63 (1907); 28, 29-61 (1908); 28, 92-124 (1908); 28, 549-71 (1908); 30, 543-74 (1909); 32, 59-88 (1909); 33, 473-99 (1910); 36, 260-90 (1911).

Miegel H. (1961)  Die Molluskenfauna des Xantener Altrheins. *Gewässer und Abwässer* 29, 7-12.

Miegel H. (1963)  Untersuchungen zur Molluskenfauna linksrheinischer Gewässer im Niederrheinischen Tiefland und des Rheingebietes. *Gewässer und Abwässer* 33, 1-75.

Müller D. (1975)  Zum Einfluss der Aufwärmung auf die mikrobiologen Stoffumsetzungen in Gewässer. *Deutsche Gewässerkundliche Mitteilungen* 19, 76-82.

Müller D. (1980)  Die Bedeutung der Actinomyceten für die Geruchstoffbelastung des Rheins und des Rheinuferfiltrats, 122 pp. Dissertation, Universität Stuttgart.

Müller D. & Kirchesch V. (1981a)  Bestimmung des "maximalen Nitrifikations-Sauerstoffverbrauchs unter standardisierten Laborbedingungen, MNSL" als Analogwert für die Biomasse der Nitrifikanten. *Vom Wasser* 56, 145-51.

Müller D. & Kirchesch V. (1981b)  Nitrifikation in Fliessgewässern (Bedeutung - Messung - Berechnung). *Vom Wasser* 57, 199-214.

Müller G. (1979)  Schwermetalle in den Sedimenten des Rheins. Veränderungen seit 1971. *Umschau in Wissenschaft und Technik* 79, 778-83.

Petran M. & Kothé P. (1978)  Influence of bedload transport on the macrobenthos of running water. *Verh. int. Verein. theor. angew. Limnol.* 20, 1867-72.

Philippi G. (1973)  Zur Kenntnis einiger Röhrichtgesellschaften des Oberrheingebietes. *Beitr. naturk. Forsch. Südw-Deutschland* 32, 53-95. (Karlsruhe).

Philippi G. (1977)  Vegetationskundliche Luftbildinterpretation als Mittel zur Erfassung von Trophiestufen im Gewässerbereich am mittlere Oberrhein im Vergleich zur mittleren Saar. In: *Gewässerüberwachung durch Fernerkundung. Der mittlere Oberrhein im Vergleich zur mittleren Saar,* pp. 33-48. Bundesanstalt für Landeskunde und Raumordnung, Bonn-Bad Godesberg.

Philippi G. (1978a)  Die Vegetation des Altrheingebietes bei Russheim.
    In: *Der Russheimer Altrhein, eine nordbadische Auenlandschaft.*
    *Natur- u. Landschaftsschutzgebiete Bad.-Württ.* 10, 103-267.
    (Karlsruhe).

Philippi G. (1978b)  Veränderungen der Wasser- und Uferflora im
    badischen Oberrheingebiet. *Beih. Veröff. Naturschutz u. Landschafts-
    pflege Bad.-Württ.* 11, 99-134.  (Karlsruhe)

RIWA (Rijncommissie Waterleidingbedrijven) (1983) Jahresbericht 1982.
    Teil A. Der Rhein.  Den Haag, Netherlands.

Schäfer W. (1973)  Altrheinverbund am nördlichen Oberrhein. *Cour.*
    *Forsch.-Inst. Senckenberg* 7, 63 pp.  (Frankfurt/Main)

Schiller W. (1974)  *Der Gütezustand des Rheines in Nordrhein-Westfalen*
    *1972-1974.*  Internal publication of Landesanstalt für Wasser und
    Abwasser Nordrhein-Westfalen, Krefeld.

Schmitt A. (in press)  Das Makrozoobenthos des Rheinkanals.  Verh. d.
    11. *Jahreshauptversammlung deutschen Gesellschaft Ökologie,* März
    1981.

Schwoerbel J. (1959)  Die biologische Gliederung des Rheinstromes.
    *Das Gas - und Wasserfach* 100(44), 1-6.

Seeler T. (1938)  Uber eine quantitative Untersuchung des Planktons
    der deutschen Ströme unter besonderer Berücksichtigung der
    Einwirkung von Abwässern und der Vorgänge der biologischen
    Selbstreinigung. II Der Rhein. *Arch. Hydrobiol.* 30,

Seibt D. (1980)  *Erfassung der organischen Drift im Rhein - ein Beitrag*
    *zur faunistischen Bestandsaufnahme am Niederrhein.*  Diplomarbeit,
    Universität Köln.

Solmsdorf H., Lohmeyer W. & Mrass W. (1975)  Ermittlung und Unter-
    suchung des schutzwürdigen und naturnahen Bereiche entlang des
    Rheins (Schutzwürdigen Bereiche im Rheintal). *Schriftenreihe für*
    *Landschaftspflege und Naturschutz* 11, 186 pp.  (Bonn-Bad Godesberg).

Sontheimer H. (1983)  *Der Rhein im Jahre 1982.*  Vortrag auf der 9 Arbeit-
    stagung der IAWR in Köln, April 1983.

Soppe E. (1983)  Verteilung des Makrozoobenthos im Querprofil des Rheins
    bei der Loreley.  Verh. d. 11. *Jahreshauptversammlung der deutschen*
    *Gesellschaft für Ökologie,* März 1981.

Stieglitz W. (1980)  Bemerkungen zur Adventivflora des Neusser Hafens.
    *Niederrhein, Jahrb.* 14, 121-28.

Thomas E.A. (1975)  Kampf dem zunehmenden Wasserpflanzen bewuchs in
    unseren Gewässern. *Wasser. u. Engergiewirtschaft* 1/2, 1-8.

Tittizer T. (1977)  *Untersuchung über die Verbreitung und die Ursachen*
    *des Larvenvorkommens der Köcherfliege* Hydropsyche contubernalis
    *McLachlan im Rhein bei Koblenz,* 4 pp.  Ergebnisbericht Bundesanstalt
    für Gewässerkunde Koblenz.

Tittizer T. (1980)  Die Biologie als Hilfsmittel bei der Erfassung des
    Geschiebetransportes in Gewässern. *BfG-Informationen* 2, Koblenz.

Trahms O.K. (1954)  Der Rhein: Abwasserkanal oder Fischgewässer.
    *Mitt. Rhein. Ver. Denkmalpflege und Heimatschutz* 3/4, 1-8.

Wilson R.S. & Wilson S.E. (in press)  A reconnaisance of the  River
    Rhine using Chironomidae pupal exuviae (Diptera, Insecta). *Proc.*
    *8th International Symposium on Chironomidae,* Jacksonville, Florida.

Wunderlich M. (1977)   Bewirken die herkömmlichen Kühlverfahren von
   Kraftwerken eine Aktivierung bakterieller Ruhestadien?  In:
   (Ed. I. Daubner) *II. Int. Hydrobiol. Symp. Smolenice 1975*, pp. 287-300
   Verlag slowak. Akad. Wissenschaften, Bratislava.

Wunderlich M. (1978)  Introduction of waste heat and effects on
   microorganisms.  *Verh. int. Verein. theor. angew. Limnol.* 20,
   1855-60.

Wunderlich M. (1980)  *Kühlwasserbelastung der Gewässer - Forschungs-
   bericht UFO-PLAN-Nr. 10204033*, 94 pp.  Bundesanstalt fur Gewässerkunde, Koblenz.

# 11 · Neckar

## I. PINTÉR & D. BACKHAUS

| | | |
|---|---|---:|
| 11.1 | Introduction | 318 |
| 11.2 | General data | 319 |
| 11.3 | Physical and geographical background | 321 |
| | 11.31 Physiography and geography | 321 |
| | 11.32 Hydrology | 321 |
| | 11.33 Canalization | 322 |
| | 11.34 Land use and population | 323 |
| 11.4 | Water quality problems | 326 |
| 11.5 | Physico-chemical features | 329 |
| 11.6 | Biological composition | 332 |
| | 11.61 Biocoenotic zonation | 332 |
| | 11.62 Open water | 332 |
| | 11.63 River bed | 335 |
| 11.7 | Biological processes and the oxygen content of the river | 337 |
| | 11.71 Introduction | 337 |
| | 11.72 Physical input of oxygen | 337 |
| | 11.73 Biological input of oxygen | 337 |
| | 11.74 Microbial oxidation of organic compounds | 339 |
| | 11.75 Ammonium oxidation and nitrate reduction | 340 |
| | 11.76 Plant and animal respiration | 341 |
| | 11.77 Oxygen balance in the river | 341 |
| 11.8 | Concluding comments | 342 |
| | References | 343 |

## 11.1   INTRODUCTION

The Neckar and the Main are the largest tributaries on the right bank of the Rhine.  The position of the Neckar in central western Europe makes it a river typical of those with a moderately atlantic climate - that is, its flow regime is not subject to extreme climatic factors. The Neckar is also a typical middle mountain river; it lacks both high mountains and a long lowland region.  The downstream stretch is however canalized, a feature shared with two neighbouring rivers, the Main and the Mosel.  The Neckar has a relatively large catchment area in relation to river length, so that a large part of south-west Germany belongs to it.

The lower valley of the Neckar has been populated for a very long time.  In the neighbourhood of Heidelberg there are remains of Stone Age settlements and in the upper stretches there are numerous places whose origin starts with Roman times.  The beauty of the Neckar valley is told in many stories and songs.  It is, above all, the city of Heidelberg, which has spread the fame of the river throughout Europe.

## 11.2 GENERAL DATA

DIMENSIONS
| | |
|---|---|
| Area of catchment | 14000 km$^2$ |
| including: Enz | 2228 km$^2$ |
| Kocher and Jagst | 3840 km$^2$ |
| Length | 370 km |
| Highest point in catchment (Hohloh-Berg) | 980 m |
| Source | 706 m |
| Mouth | 86.5 m (i.e. drop = 619.5 m) |
| Slope | 1.67‰ |

CLIMATE
   Precipitation
   The catchment mostly lies in the rain shadow of the Schwarzwald (Black Forest), with annual totals in the lower regions in the range 600-800 mm and reaching 900-1200 mm locally (Schwarzwaldrand, Frankish Schild). The climate is more or less of Atlantic type, with SW winds predominating

GEOLOGY
   Predominantly in the "Südwestdeutsches Schichtstufenland".
   (See Fig. 11 and text)

LAND USE
   About three-quarters of the catchment is used for agriculture, and the remainder (excluding conurbations) for forestry or similar activities

POPULATION
| | |
|---|---|
| Whole catchment | 4.6  x 10$^6$ people |
| including central Neckar region | 2.36 x 10$^6$ people |

The main conurbations of Stuttgart, Heilbronn, Pforzheim, Heidelberg and Mannheim have a density of 1000 people km$^{-2}$.

PHYSICAL FEATURES OF THE RIVER

Discharge ($m^3 s^{-1}$) at selected points

| km from mouth | place | catchment ($km^2$) | absolute minimum | mean lowest | mean | absolute maximum |
|---|---|---|---|---|---|---|
| 318 | Oberndorf | 694.7 | 0.76 | 1.98 | 7.9 | 288 |
| 202.6 | Plochingen | 4002 | 3.70 | 9.7 | 44.1 | 1150 |
| 125.1 | Lauffen | 7910 | 14.1 | 23.4 | 84.7 | 1270 |
| 61.4 | Rockenau | 12710 | 20.0 | 35.4 | 126.0 | 2150 |

Mean monthly discharge ($m^3 s^{-1}$) at Rockenau gauging station for 1951-1979

|     |     | influence |
|---|---|---|
| Nov | 71 | autumn rain |
| Dec | 141 | |
| Jan | 163 | |
| Feb | 430 | snow-melt |
| Mar | 340 | |
| Apr | 220 | spring |
| May | 149 | |
| Jun | 91 | |
| Jul | 64 | rain — poor summer and early autumn |
| Aug | 58 | |
| Sep | 53 | |
| Oct | 54 | |

Regulation

27 weirs on canalized stretch between mouth and river - km 200 (upstream from mouth)

Substratum

In uncanalized stretch:  blocks, boulder, cobbles, pebbles

In canalized stretch:  silt and gravel

Temperature

In uncanalized stretch:  annual range from 0 to 22 $^{\circ}$C

In canalized stretch downstream of influence of cooling waters:
winter temperatures never below 4-5 $^{\circ}$C
summer temperatures reaching 28-30 $^{\circ}$C or 26-27 $^{\circ}$C after complete mixing

*Fig. 11.1* Neckar, showing various geological formations, gauging stations and power stations.

## 11.3  PHYSICAL AND GEOGRAPHICAL BACKGROUND

### 11.31  PHYSIOGRAPHY AND GEOLOGY

The Neckar arises in a small area of woodland not far from the town of
Schwenningen.  Geographically its origin belongs with the eastern edge
of the Black Forest, itself in the southern part of the 'Schwäbisch -
Fränkisches Stufenland'.  It is also a region well known geologically
due to the Triassic and Jurassic rocks whose various layers come to the
surface within a narrow region (Fig. 11.1).  From its source the river
first flows north-east with a mean slope of $0.28^{o}/oo$, but after 170 km
it bends sharply to the north-west, eventually reaching the Rhine at
Mannheim after 370 km, with a mean slope of $0.55^{o}/oo$.  The Enz,
Kocher and Jagst all have important effects on flow.

During its course the Neckar successively flows through, among other
rock types, limestone, the Liassic covered Keuper rocks, Odenwald red
sandstone, limestone again and, finally, the alluvial deposits of the
Rhine valley.  The river valley changes according to type of rock from
V-shaped to broad U-shaped or sometimes with cliffs on one bank.  Soils
and land use also depend on the geology and these in turn affect the
river.  The predominantly calcareous materials of the catchment have
an influence on river chemistry.

### 11.32  HYDROLOGY

Of the 14000 $km^2$ catchment, about 4000 $km^2$ belong to the uncanalized
stretch down to Plochingen.  The entry of the Enz doubles the catchment area, while the catchments of the Kocher and Jagst together total
about the same as the Enz.  The remainder of the catchment consists of
small tributaries, mostly from the Odenwald.  The main collecting
ground of the Neckar lies therefore partly in the north-east part of
the Black Forest and adjacent 'Schichtstufenland' as far as the
Schwäbische Alb (upper Neckar, Enz) and partly in mountain and hill
country of the so-called 'Fränkisches Schild' (Internationale Kommission für die Hydrologie des Rheingebietes, 1978).

The mean discharge per unit area as a result of precipitation within
the catchment down as far as the gauging station at Plochingen )km
202.6 from mouth, catchment area = 4002 $km^2$) is only about 0.011 $m^3$
$s^{-1}$ $km^{-2}$.  The contributions per unit area from the Enz, Kocher and
Jagst are similar.  This may be compared with values for the western
part of the Black Forest, which are in the 0.02 - 0.04 $m^3$ $s^{-1}$ $km^{-2}$.
The mean discharge of the Neckar rises from 44 $m^3$ $s^{-1}$ at Plochingen to
126 $m^3$ $s^{-1}$ at Rockenau and remains similar as far as the mouth (Fig.
11.2).

The Neckar shows very variable flows.  At Rockenau gauging station
discharge ranges from 20 to 2150 $m^3$ $s^{-1}$, with a mean of 126 $m^{-3}$ $s^{-1}$
(Landesanstalt für Umweltschutz, 1980:  see Fig. 11.3).  Low waters

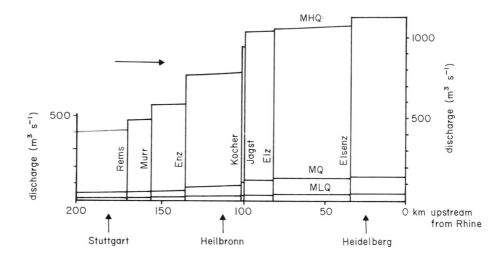

*Fig. 11.2* Discharge in the canalized part of the Neckar for 1966-1975. (Distance measured from entry in to Rhine). MHQ = mean of annual maxima; MQ = mean; MLQ = mean of annual minima)


are the most important ecologically. These occur regularly in summer and autumn. During this period the ratio between Neckar water and water from outside sources (effluents, cooling water) is in the region of 1 : 1 or even as low as 0.8 : 1 (Schaal, 1981).


11.33  CANALIZATION

In 1920 the first steps were taken towards the development of a shipping route, which was finally completed in 1968. This affects 200 km of river from Plochingen (where the R. Fils enters) to the mouth (Fig. 11.4). This stretch of the river forms a class IV waterway ("Waterway of international importance"). The 160.7 m drop in height requires 27 locks, each  of which has a double sluice and a weir; almost all also have a small hydro-electric plant. The shipping canal is at least 2.5 m deep and mostly 3 m.

The length of time the water takes to pass down the canalized stretch depends markedly on flow conditions. At the mean annual flow minimum this is 800 hours, but only 95 hours at the mean annual flow (Schmitz, 1971). This would result theoretically in a change of current speed from 7 to 50 cm s$^{-1}$, if local variations could be neglected.

Construction along the lower Neckar was carried out without fore-

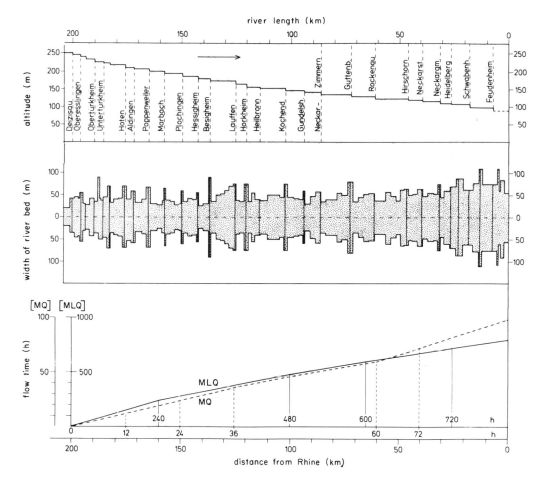

*Fig. 11.3* Hydrological profile of Neckar.    A) altitude; B) width
of river bed, based on measurements; C) mean time for passage of
water under different flow conditions in 1966-1975.  MQ = mean flow;
MLQ = mean of annual minima.  (C, after Schmitz (1971))

thought for the likely effects on the river ecosystem.  Such effects
are likely to be particularly pronounced, in view of the high input of
waste and warm water effluents.  Thus river management nowadays has to
pay a lot of money to remove or ameliorate the problems caused (see
11.4).

## 11.34  LAND USE AND POPULATION

As a rough approximation almost one-quarter of the Neckar catchment is
covered by forest.  The largest continuous areas are in the Black

*Fig. 11.4* Aerial view of Neckar at km 47 (near Hirschhorn). Note canalized banks of river, weir with sluice for ships and turbulence created by ships. (Photo by permission of Regierung von Oberbayern Nr. G 7/88 048 Hersteller "Photogrammetrie GmbH")

Forest and Odenwald - that is, in the periphery of the region. Further, the Keuper heights are forested over a large area. In contrast, the central Lias and limestone are covered with good, heavy soils. These are the favoured area for settlements. Except for a few stretches with steep rocky slopes, the river flows here through a landscape which is used intensively.

The use of the land depends partly on soil quality and climate, partly on the type and size of farm. In the Neckar region there are various agricultural zones. The upper region as far as the vicinity of Stuttgart and also the higher limestone slopes along the Jagst are characterized by cereals. In the district north-east of Stuttgart there is a zone of root crops, which reaches as far as Odenwald. There are large areas of intensive farming (vegetables, fruit, berries, vineyards) around Stuttgart and Heilbronn which contribute an additional nutrient input into the river (Fig. 11.4). In the downstream area (below Heidelberg) there are more root-crops, especially sugar-beet.

The structure of the settlements also differs in the various types of landscape.  Large villages predominate in the regions which are favoured for farming, with up to about 2000 inhabitants.  Such villages are often very old.  The Black Forest and Keuper area were colonized later and the villages are mostly smaller and often little more than a cluster of farmhouses.  The great part of the Neckar region is however transformed by industrialization.

There is an industrialized strip already in the upper Neckar, while other industrial regions follow the valleys of the Fils, Rems, Murr, Kocher and Enz.  Some of them come together in the broader conurbation of Stuttgart with its industrial concentration and dense population. In hardly any other part of Germany are towns packed so densely (Fig. 11.5).  Other industrial zones are around Heilbronn and Pforzheim, as well as in the Rhine - Neckar triangle at Heidelberg and Mannheim. 4.6 million people live in the Neckar catchment, of whom about half live in the middle Neckar region.

Fig. 11.5 Forests and agricultural land in the Neckar region. Hg, Heidelberg; Hn, Heilbronn; M, Mannheim; St, Stuttgart. (Modified from Internationale Kommission für die Hydrologie des Rheingebietes, 1978)

The characteristics of the economic structure of the Neckar region is the broadly scattered mix of agriculture and industry and the fact that the latter is specialized - heavy industries like coal and steel are lacking. Key industries are machine tools, car manufacture, electrical components, clocks, jewellery and textiles. Many of these industries have a high requirement for electrical power, which is supplied from the middle region of the Neckar itself. The need for cooling water leads to many problems in water management. Other critical problems resulting from the density of population and industry are the need for drinking and industrial water and waste disposal. The need for water is too great for the Neckar catchment alone, so drinking water is brought from the Bodensee and other districts at the rate of $8 - 10 \text{ m}^3 \text{ s}^{-1}$ at peak demand. Most of this extra water eventually reaches the Neckar as waste water.

## 11.4   WATER QUALITY PROBLEMS

Water quality in the Neckar is dominated by the need to use it for removal of wastes, its role as a coolant for power plants and the control of flow for energy supply and shipping. At times of low flow, sewage effluent contributes 30% of total flow below Stuttgart. What is more, periods of low flow coincide with warm weather. Heating is increased in the canalized stretch both because of damming and the requirement for cooling water. It is not unusual for temperatures to reach 25 °C here. The higher temperatures favour microbial, oxidative breakdown of organic materials, thus reducing the oxygen content of the water. The natural oxygen supply from the air is reduced because of lower current speeds and the water depth of 3 m. Depletion of oxygen is particularly marked in autumn when temperatures are still high and algal photosynthesis is less effective.

A further effect of damming is the marked deposition of sediments which takes place in the middle stretch of the Neckar. The often relatively low efficiency of the sewage treatment plants leads to a relatively high organic content in these sediments (something up to 20%) and this results in an additional demand. Every year $1.8 \times 10^6$ $\text{m}^3$ of sediments are deposited in the canalized stretch in addition to the sediments which are already there and are estimated to be about $2 \times 10^6 \text{ m}^3$. The extra sediments have to be removed by dredging, especially in the middle part, and their quality leads to further problems of how to get rid of the material which has been dredged. In addition to a high organic content, the sediments are enriched by heavy metals from the metal industry, cadmium being a particular problem. In recent times the manganese content has also increased at some sites, perhaps due to efforts to increase the oxygen content of the water. An increased lead content may also result due to its co-precipitation with MnII (G. Müller in symposium discussion: see Institut für Sedimentforschung der Universität Heidelberg, 1981).

The use of sediments for agriculture is impossible in view of its potentially toxic composition.  It has been suggested that they should be deposited and the metals recycled.  A favourite suggestion is to incorporate the dredged material with the materials normally used to make bricks, but up to now research has not shown how this can be done satisfactorily.

In order to deal with the problems of water quality, a cleaning-up programme was decreed by the Landesregierung of Baden-Württemberg in 1976.  Priorities are put forward to make optimal use of financial resources and improve the situation in the water.  Among possible approaches to improvement are:

  (i)   reduction in pollution;
  (ii)  addition of oxygen;
 (iii)  addition of extra water to the river.

The first was chosen for the initial goal up to 1980.  The second approach has been used in some cases, with the oxygen enrichment of water used by hydro-electric plants; this approach is however adopted only at times of low flow, because of its high energy requirement.

The daily load of organic effluents added to the river is shown in Table 11.1.  Most of the waste is directed through sewage treatment plants, there being only forty-two firms with their own treatment plant and separate effluent.  These consist of ten metal works, three aluminium surface treatment works, seven chemical works, four cement and gravel works, eleven paper manufacturers, two glue factories and five concerned with food manufacture.  The development of efficient sewage treatment (judged by $BOD_5$) in the effluent demands a high annual capital investment for building sewage treatment plants (Table 11.2). $1765 \times 10^6$ deutschmarks have been spent to get a 40% improvement in capacity, which has resulted in a 56% improvement in effluent quality.

*Table 11.1*  Daily addition of $BOD_5$ to various sections of the Neckar (Ministerium für Ernährung, Landwirtschaft Umwelt Baden-Württemberg, 1976).

| section | $BOD_5$ (tonnes $d^{-1}$) |
|---|---|
| above Fils | 96.1 |
| Fils to mouth of Enz | 181.5 |
| Enz | 62.9 |
| below Enz to mouth of Kocher | 29.8 |
| Kocher + Jagst | 55.8 |
| below Jagst to mouth | 45.7 |

*Table 11.2* Annual expenditure in the Neckar region on treatment works over the period 1975-1980.   (P, person; PE, person equiv.)

| year | capacity $(P + PE) \times 10^6$ | expenditure $(DM \times 10^6)$ | mean quality of waste $(mg\ l^{-1}\ BOD_5)$ |
|------|------|------|------|
| 1975 | 7.3  | 185 | 54 |
| 1976 | 8.1  | 290 | 46 |
| 1977 | 8.7  | 250 | 36 |
| 1978 | 9.3  | 340 | 33 |
| 1979 | 9.4  | 360 | 25 |
| 1980 | 10.2 | 340 | 24 |

The potential energy in the dammed region is used to generate electricity.  There are 24 hydro-electric plants, each with a fall in river height of 3 - 8 m.  The flow requirements of individual plants range from 46 to 100 $m^3\ s^{-1}$.  The annual output of all plants is around 55 MW (Neckar AG und Wasser-und Schiffahrsdirektion Stuttgart, 1971). Besides this direct requirement for water energy from the river, cooling water is needed for eight oil, coal or nuclear power plants with a maximum output of 3500 MW.  These plants are confined to a stretch of only 122 km.  The two nuclear plants lie at km 129 (Neckarwestheim: 858 MW) and km 77 (Obrigheim: 345 MW).  If all the heat had to be removed by water alone, it would require 130 $m^3\ s^{-1}$ at Rockenau gauging station.  Such a flow is not reached during 270 days a year, so quite apart from the ecological impact of heating, it is essential to use cooling towers.

Some of the power plants have already got cooling towers, which allow up to 70% of the additional heat energy to be lost to the atmosphere when meteorological conditions are favourable.  The nuclear power plant at Neckarwestheim, which commenced in 1976, is fitted with cooling towers which use water in three different ways to maximize the efficiency of heat loss.  At present the older power plants have been ordered to improve their cooling systems.  This is very urgent if the present plans are to be fulfilled by the end of the 1980's, because by then the maximum output of all plants will reach 6000 MW.  The maximum temperature permitted for warm water effluents is 30 - 35 °C (depending upon origin), the maximum rise in river temperature is 5 K and the maximum temperature in the river itself is 28 °C.  State government regulations control the rise in temperature permitted by each of the eight plants.  A complication of these plants is that the extent of pollution is determined by changes in oxygen (see below) as well as temperature.

For the cooling requirements of the Neckarwestheim power plant (40 $m^3\ s^{-1}$), there are both temperature and oxygen regulations.  Sufficient oxygen must be added by the power plant to compensate for the decrease that would occur due to increased microbial activity as a result of the temperature rise. When the oxygen concentration in the Neckar is very low, then an exception to the rules is permitted. The whole river at Neckarwestheim is allowed to rise by an extra 2 K, but the oxygen must be increased by the extra amount corresponding to this temperature rise. This regulation was drawn up as a result of ecological considerations.

Energy planning on the Neckar has involved voluminous studies and calculations (Ministerium für Ernährung, Landwirtschaft and Umwelt|, Baden-Württemberg, 1973). These have resulted in decisions that power plants built in the future must have circulating cooling water and that the existing power plants must reduce their heat waste at critical times. Because cooling towers cause a considerable loss of water by evaporation, energy-producing firms in Baden-Württemberg are obliged to compensate for such loss. This will probably require further reservoirs to be built.

## 11.5  PHYSICO-CHEMICAL  FEATURES

The annual amplitude in water temperature in the Neckar is marked; it follows the natural yearly changes in radiant energy and air temperature. For the non-canalized Neckar a mean annual temperature around 13 $^\circ$C (at Plochingen) is characteristic. This rises to about 16 $^\circ$C near the entrance to the Rhine. If it is considered that flow increases four-fold along this stretch and the flows from tributaries tend to reduce the temperature of the main river, then it will become clear that the additional heat load due to warm-water effluents is considerable. There are no effluents in the last 77 km of the river and therefore this part can be regarded as a cooling stretch. Another indication of the effects of the heat inputs is that the winter temperature seldom drops below 5 $^\circ$C in the canalized stretch. In summer temperatures reach 24 - 25 $^\circ$C here, or exceptionally even 27 $^\circ$C, as during the extremely hot summer of 1976.

The ground level of the main mineral components ($Ca^{2+}$, $Mg^{2+}$, $Na^+$, $Cl^-$, $SO_4{}^{2-}$, $HCO_3{}^-$) is high as a result of the geology of the catchment. This results in the high conductivity value of 750 µS cm$^{-1}$ where the Neckar enters the canalized stretch (Fig. 11.6). This is reduced slightly by the rivers from the Odenwald which have a low mineral content, but their flows are too little to have much effect. More important for the chemistry of the river  are the effluents from Stuttgart sewage treatment plant (km 174) and predominantly those from the salt industry at Heilbronn (km 111). These latter lead to marked increases in $Cl^-$, $Na^+$ and $Ca^{2+}$, with resulting mean conductivities of at least 1250 µS cm$^{-1}$. Most of these contents are conservative, with little loss on passing downstream due to chemical or biological activities.

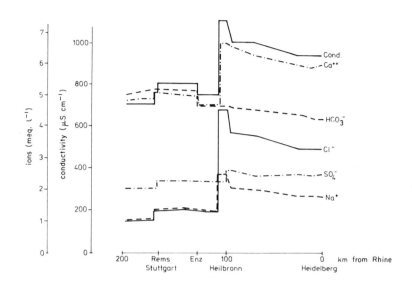

*Fig. 11.6* Changes in conductivity and concentration of various ions on passing down the canalized Neckar in 1978-1979.


The changes in pH taking place annually in the canalized stretch are strongly influenced by primary production, with a range of 7.0 - 8.6. The mean value along this stretch varies only little (7.6 - 7.8).  The diverse sewage effluents have little effect on the strong buffering capacity of Neckar water.

The Neckar is strongly influenced by the addition of organic wastes. Because of the unfavourable hydrological conditions in the canalized stretch (little turbulence, great mean depth, strong sedimentation), this leads to a marked reduction in oxygen content, as discussed in more detail below.  The load of persistent organic compounds is however low in comparison with the Rhine and lower Main, because chemical industry is sparse in the Neckar region.  In contrast, the concentrations of most heavy metals are - especially in the sediments - higher than in the Rhine and Main, as a result of the metal treatment industries.

The Neckar is in practice free of effluents for only a few kilometres.  It is polluted as soon as it enters the town of Villingen-Schwenningen where it receives effluent from the town sewage treatment plant (115000 persons + person-equivalents).  In the 170 km of the upper stretch, and indeed until the beginning of the canalized stretch, the tributaries, Eschach, Eyach and Ammer contribute a considerable organic load in addition to effluents reaching the Neckar directly from

the towns of Rottweil, Horb and especially Tübingen (165000 persons +
person equivalents). The situation was particularly critical until
sewage treatment plants were built at Villingen-Schwenningen (in 1979)
and Tübingen (in 1978). $BOD_5$ values of 15 mg $l^{-1}$ were no exception
when river flows were low; dissolved organic carbon was around 8 mg $l^{-1}$.
However self-purification is generally adequate because of the good
oxygen supply, small river depth and strong slope. The improved
effluents resulting from the new treatment plants have halved the
organic load in the river. A further drop in $BOD_5$ (to values less than
6 mg $l^{-1}$) is predicted when tertiary treatment is added to the Villingen-Schwenningen plant. In the canalized stretch the critical
points are in the region of Stuttgart and Heilbronn, but the entry of
the Fils, Rems and Murr are also quantitatively important. The sewage
treatment plant of Stuttgart was for long particularly inadequate for
the 1.25 million persons + person equivalents and the 3 - 4 $m^3$ $s^{-1}$
effluents which were always a very heavy load for the Neckar until
sewage treatment was extended in 1981.

In the lower stretch of the Neckar from the entry of the Kocher to
the mouth the density of population is less. Both because of this and
because the necessary improvements to sewage treatment plants have
mostly been made already, the additional load added to the river is
lower. Heidelberg is still the critical part on this stretch. The
results for dissolved organic carbon in 1973-1976 and 1979 (Fig. 11.7)
show this clearly, though there was a considerable decrease in 1979
due to improvements made in the intervening years. This change was due
mainly to the reduction in biologically degradable substances.

*Fig. 11.7* Comparison for dissolved organic carbon (DOC) between the
mean for 1973-1976 and that for 1979. (Data based on water taken
continuously, frozen and analyzed once every two weeks)

*Fig. 11.8* Changes in $NO_3$-N and $NH_4$-N on passing down the canalized
Neckar in 1979. (Data based on water taken continuously, frozen
and analyzed once every two weeks)

Ammonium reaches a mean concentration of 25 mg $l^{-1}$ N in some effluents and there are marked rises in the river below Stuttgart and Heilbronn (means of 2.5 and 2 mg $l^{-1}$ $NH_4$-N, respectively). There is intensive nitrification (Fig. 11.8) below these effluents and denitrification probably also plays an important role below Stuttgart. The $NH_4$-N load in the river shows characteristic annual changes, being lower in summer and higher in winter, because nitrification is less at lower temperatures.

The concentration of inorganic phosphate is strongly influenced by anthropogenic factors. The mean average concentrations in the canalized stretch lie between 1 and 1.5 mg $l^{-1}$ $PO_4$-P. Phosphate in the river shows a quasi-conservative character. Although it is removed by organisms, especially plants, the amount removed is not enough to have a marked effect on aqueous concentrations. Light is the main factor limiting plant growth in the canalized stretch of the Neckar, not nutrients.

## 11.6  BIOLOGICAL  COMPOSITION

### 11.61  BIOCOENOTIC ZONATION

Because of its geographical position in the south-west German middle mountain region, the slope and temperature regime of the Neckar mean that it belongs naturally to the group of summer-warm rivers, with a mountain stream character in the upper and middle stretches. On passing downstream about half the river would fall into the trout and grayling zones. In the upper and middle stretches, the natural biocoenotic zonation is disturbed by human and industrial effluents and, although the condition of the river has been improved in recent years, it is probably still the sewage load which is most responsible for the absence of benthic species that would otherwise be expected. In particular, many species of stenoecious Plecoptera, Ephemeroptera and Coleoptera have either disappeared completely or are present in very low populations. Fish populations were disturbed for a long time, but are now beginning to build up again.

The physical features of the present-day canalized stretch (Fig. 11.9) indicate that it would belong formally to the barbel zone, but in practice this species is not present. This stretch of the river has changed even more than those upstream, with its lentic conditions, raised temperatures and high sediment load all being important biotic factors.

### 11.62  OPEN WATER

As would be expected in a dammed river, there is a marked development of phytoplankton. Interpretation of the physical and chemical data

*Fig. 11.9* Typical canalized stretch (km 137, near Besigheim).

for this part of the river suggest that nutrients should never be a limiting factor for algal growth (Pintér & Bauer, 1980). Algal populations, especially those dominated by diatoms, are often sufficiently dense to colour the water, but surface blooms do not form. Populations are apparently limited by the poor light penetration. Secchi depths are mostly less than 50 cm and the compensation point is around 2 m on bright summer days, less under other conditions. The possibility, that the concentration of toxic materials in effluents is sometimes sufficiently high to influence phytoplankton growth, requires research.

There is usually a distinct development of planktonic algae in March, with populations reaching around 1000 individuals $ml^{-1}$ (Fig. 11.10). By April this is about one order of magnitude larger and in summer densities are still higher, with extreme records of about 60000 individuals $ml^{-1}$ (Backhaus & Kemball, 1978).

Besides the high nutrient concentrations, other factors which affect phytoplankton growth are input of light energy, water temperature and flow. Phytoplankton density shows a positive correlation with the first two factors and a negative one with the last (Backhaus & Kemball, 1978). Increases in flow above 50 $m^3$ $s^{-1}$ (at Plochingen) bring about a marked reduction in population density.

It is possible to distinguish three different phytoplankton regions

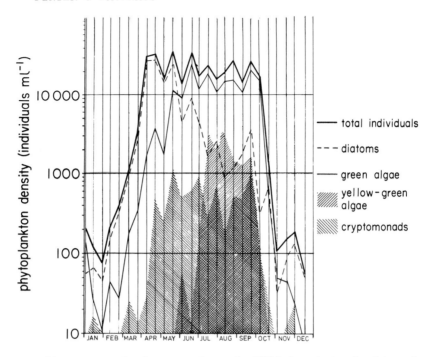

*Fig. 11.10* Temporal changes through 1974 in phytoplankton density in Neckar at km 94.2 (near Gundelsheim). (Data based on samples taken at 2-weekly intervals)

on passing down the canalized stretch. Above Stuttgart there is a relatively high biomass with 2 - 2.3 mg fresh weight $l^{-1}$. From Stuttgart to Heilbronn this drops to about 1 mg $l^{-1}$, but below Heilbronn it increases again, with values locally as high as 3.5 mg $l^{-1}$. The decrease between Stuttgart and Heilbronn is probably due to the heavy sewage effluent loading. Chloride and conductivity are two obvious variables which increase below Heilbronn, but it is an open question whether these are in any way responsible for the increase in algal biomass.

Some 300 algal taxa have been recorded for the Neckar (Backhaus, 1980), but only a few make up the bulk of the biomass. Among these are *Stephanodiscus hantzschii, S. subsalsus, Scenedesmus* spp. and various other Chlorococcales. *Stephanodiscus* forms almost 90% of the biomass in spring, but by July green algae form more than 80% of the biomass (based on fresh weight) (Fig. 11.11). Coenobial forms typically grow as single cells, a feature well known in algal culture studies.

Among the grazers, rotifers and ciliates hold first place. Only during periods of summer low flow, when the residence time of the water behind the dams is long, do large populations of cladocera and

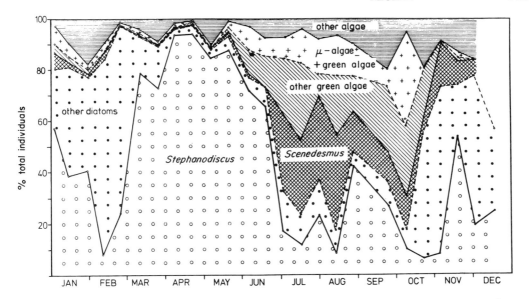

*Fig. 11.11* Relative proportions of various algal groups in Neckar phytophyton at km 94.2 (near Gundelsheim). (Data based on samples taken at 2-weekly intervals; compare with Fig. 11.10)

copepods build up. Relatively little is known about the species composition of the zooplankton, apart from a detailed study of the rotifers (Rapp, 1973). Over 20 species of rotifer have been recorded, but only *Rotaria rotaria, R. neptunia, Brachyonus calyciflorus, Chlanis dilata, Asplanchna brightwelli* and some species of *Synchaeta, Polyarthra* and *Filina* build up high populations. Although there have been no detailed studies, it seems probable that grazing plays a relatively minor role in reducing phytoplankton populations.

As mentioned above, fish populations have started to increase since 1978 when improvements in water quality were made. During the period of worst pollution, 1960-1970, fish-kills were a regular happening. The critical oxygen conditions and toxic materials permitted only 'white fish', a species-rich group predominantly of cyprinids. Today the fish populations of the upper Neckar approach those of forty years ago, before the period of economic expansion. So now at least in the upper stretch there are again carp, tench, eel, pike, zander, as well as trout and nase (Schoch, 1981). Most of them are stocked periodically by angling associations.

11.63  RIVER BED

Few studies have been made of either the macro- or micro-vegetation of the river bed, so it is impossible to give a general picture. There is sometimes a strip of *Potamogeton pectinatus* in the canalized stretch (Hoffman, 1980), mostly between 9.5 and 2 m wide. Stones in the wave zone of the river bank are often covered with *Cladophora glomerata* and

*Vaucheria* sp. or a thick layer of filamentous blue-green algae.  The bulk of this last is made of species of *Phormidium*, such as *P. autumnale*, *P. favosum* and *P. tenue*.

The periphyton consists largely of diatoms, with about a hundred species recorded.  Of these, the following become abundant at times: *Achnanthes lanceolata*, *Cocconeis placentula*, *Diatoma vulgare*, *Gomphonema angustatum*, *Navicula avenacea*, *N. mutica*, *Nitzschia capitellata*, *N. palea*, *N. paleacea*, *Surirella ovata*, *Synedra ulna*.  A re-invader is the red alga *Thorea ramosissima*; although known from old records this had disappeared until recently (D.B., unpublished).

The macrobenthos of the upper Neckar shows no obvious differences from those of other large river systems in the region, such as the headwaters of the Danube.  The species composition of the littoral region of the middle and lower stretches is that typical for lowland rivers.  The communities differ relatively little whether they are subject to strong or weak influences of sewage.  *Asellus aquaticus*, various bryozoa, sponges, Hirudinea, Gastropoda and Turbellaria are abundant everywhere.  Communities can be differentiated only by differences among the less common species.  Heterogeneity arises because of differences in substratum.  If the larval forms of nematodes, which are recognizable only with difficulty are excluded, then the numbers of species range in various situations from 12 to 24.  Especia- lly typical of the shipping stretch of the Neckar are *Asellus aquaticus*, masses of *Spongilla lacustris*, *Ephydatia fluviatilis*, *Plumatella repens*, *P. fungosa*, *Fredericella sultana*, *Bithynia tentaculata*, *Acroloxus lacustris*, *Planaria torva*, *Dendrocoeleum lacteum* and - recently - *Dugesia tigrina*.

The benthic organisms of the sediments in the middle of the river are known less well.  There is no stretch of water which is totally still and some of the above species also occur here.  The most obviously different community is that of the soft sediments just above the dams and in harbours.  Such sediments, with their high organic content, are dominated by members of the *Chironomus thummi* group.  This is especially the case where the influence of sewage effluents is somewhat stronger, such as below Stuttgart and Heilbronn.  Tubificids dominate where the organic component of the sediment is less, while *Lumbriculus variegatus* can form large masses in intermediate conditions.  In contrast to the other two groups, *L. variegatus* is influenced particularly by the chemical condition of the sediment, while mineral composition, size and structure of particles are less important.  The mussel *Dreissena polymorpha* has recently invaded the river, following the improvements in water quality.  Insect larvae of the genera *Hydropsyche* (Trichop- tera) and *Ischnura* (Odonata) appear in noticeable numbers only in lateral channels with faster flows.

## 11.7  BIOLOGICAL PROCESSES AND THE OXYGEN CONTENT
## OF THE RIVER

### 11.71  INTRODUCTION

Canalization, heavy loading with sewage effluents and the requirements
for cooling water all have a marked influence on biological processes
in the river.  Optimal conditions for biological activity occur when
three factors operate together:

   (i)    long residence time
   (ii)   high content of organic and inorganic nutrients
   (iii)  high water temperature.

The oxygen supply is much influenced by processes of $O_2$ consumption,
which is sometimes hardly compensated by physical and biological $O_2$
input.

### 11.72  PHYSICAL INPUT OF OXYGEN

In the canalized stretch the passage of oxygen from air to water is
slow, especially under low flow conditions when turbulence is also low.
Hoffmann's (1980) calculations showed an oxygen input of about 0.03 mg
$l^{-1}$ $h^{-1}$.  Accordingly the velocity constant for natural reaeration
($K_2$) is also very small at times of low flow, with values between 0.15
and 0.25 $d^{-1}$ (Pintér, 1980).  When flows are higher, reaeration is
better because of the increased turbulence and because more water
passes over the weirs.  This last takes place in flow reaches from 45
to 100 $m^3$ $s^{-1}$, depending on the size of the turbines.  The strong
influence of turbulence on the Neckar has been shown by correlations
between oxygen load on the one hand and flow and water temperature on
the other (Pintér, 1980).  Multiple correlations show high values for
this calculation ($r^2 \geqslant 0.76$).

### 11.73  BIOLOGICAL INPUT OF OXYGEN

Among processes adding oxygen to the water, phytoplankton phytosynthesis
is predominant in summer and oxygen production depends on algal biomass
and factors influencing its photosynthesis.

   The relationship between the amount of alga per unit water column
determined by the standard laboratory photosynthesis test ($Ph_{OPL}$), the
OPL test:  see Fig. 11.12, photosynthetic active radiation (PAR) and
the rate of photosynthetic input of oxygen to the Neckar ($Ph_{Neckar}$)
are given by the empirical equation (Hoffmann, 1980):

$$Ph_{Neckar} = F \cdot Ph_{OPL} \cdot a \cdot PAR^b$$

   where F is a calibration factor

        a, b are calculated constants.

(Hoffmann's symbols have been changed for use with English text).   For
fine days in summer, $Ph_{Neckar}$ was calculated to be 1.75 mg $O_2$ $l^{-1}$ $h^{-1}$
in the upper 50 cm; this value dropped to 0.125 mg $l^{-1}$ $h^{-1}$ at a depth
of 1 m.   On dull days the oxygen input drops rapidly to zero (Hoffmann,
1980).

*Fig. 11.12* Changes in chlorophyll *a* and OPL (oxygen production in
laboratory) through 1979-1980.  (Data based on samples taken at 2-
weekly intervals; oxygen production in laboratory measured at 20 $^{O}C$,
using light and dark bottles)

Measurement of the net oxygen production-potential i.e. photosynthesis under standard laboratory conditions ($Ph_{OPL}$) showed a peak of

15 mg $O_2$ $l^{-1}$ $d^{-1}$ during a two-year study (Fig. 11.12).  With very
dense algal populations, this value can reach 18 mg $l^{-1}$ $d^{-1}$.  There are
marked differences not only between months but also between years.  In
the cold and wet summer of 1980, the peak value algal biomass and
$Ph_{OPL}$ reached half that of the previous warm, dry year.

In comparison with phytoplankton the contribution of macrophytes to
the the overall oxygen load of the river can be neglected:  Hoffmann
(1980) calculated a mean value of 0.1 - 0.2 mg $O_2$ $l^{-1}$ $d^{-1}$.

## 11.74  MICROBIAL OXIDATION OF ORGANIC COMPOUNDS

The high concentrations of organic substances in the Neckar result from
the degradation both of the biomass which is produced autochthonously
and also of allochthonous materials such as waste effluents.  These
substances are in general broken down according to a first order
reaction (Pintér, 1980).  A large part of the easily oxidizable organic
material is broken down in sewage treatment plants before it reaches
the river.  The mean value for $BOD_5$ in the river in 1979/80 was about
8 mg $l^{-1}$ over much of its length (Fig. 11.13).  Peak values were much
higher, especially in 1980.  These can result from storm run-off
disturbance of the bottom sediment or the failure of a sewage treatment
plant.  The results are quite similar to those found for the Main
(Tobias, 1976).

*Fig. 11.13*  Changes in $BOD_5$ on passing down the canalized Neckar in
1979-1980.  (Data on samples taken at 2-weekly intervals).  Maxima
for 1979 and 1980 shown separately to indicate of high flow in 1980

Large areas of the river bed, especially upstream of lock-gates, are covered with fine sediment. A heavy deposition of sediments is continually taking place. As well as old sediments, which can be re-suspended by the movements of ships, sediments recently brought into the river become deposited as the river passes downstream. Microbial oxidation in these sediments adds to the overall oxygen-deficit in the water column. The intensity of oxygen consumption by these sediments has been found to be correlated positively with the oxygen concentration in the free water (Hoffmann, 1980). Oxygen consumption was found experimentally to range between 1 and 9 g $m^{-2}$ $d^{-1}$, depending on the oxygen concentration of the water. This leads to decreases in oxygen concentration in the free water of between 0.35 and 3.0 g $m^{-3}$ $d^{-1}$. Because of this relationship between oxygen consumption and oxygen in the environment, the efficiency of artificial aeration is reduced considerably.

## 11.75  AMMONIUM OXIDATION AND NITRATE REDUCTION

As can be seen from Fig. 11.8, the ammonium content of the river water is especially high between Stuttgart and Heilbronn. This is influenced to a marked extent by the rate at which bacterial nitrification takes place. This process proceeds according to the equations:

$$NH_4 + 1\tfrac{1}{2} O_2 \xrightarrow{\;Nitrosomonas\;} NO_2^- + H_2O + H_2 + 76 \text{ kcal.}$$

$$NO_2^- + \tfrac{1}{2} O_2 \xrightarrow{\;Nitrobacter\;} NO_3^- + 24 \text{ kcal.}$$

Noticeable nitrification takes place only when water temperature exceeds 15 °C and the rate increases markedly with a further rise in temperature ($Q_{10}$ = 2-3). In the middle Neckar, where the natural physical supply of $O_2$ is particularly inadequate, ammonium oxidation stresses the $O_2$ budget in the warm months, especially when flows are low. It is difficult to quantify the relative demands for $O_2$ of the oxidation of organic materials and of the oxidation of ammonium, but the extremely high ammonium concentration in some stretches suggests the latter process is sometimes even more important than the former.

Reductive reactions can also occur, whereby nitrate is reduced to ammonium or $N_2$ (denitrification):

$$NO_3 \longrightarrow NO_2 \longrightarrow NO \begin{array}{l} \nearrow N_2O \longrightarrow N_2 \\ \searrow NH_2OH \longrightarrow NH_3 \end{array}$$

The Neckar appears to resemble other rivers with a high suspended matter and detritus content in that denitrification occur while oxygen is still present in the water. This takes place on oxygen-free microzones around particles of sediment. Denitrification probably plays an important role below Stuttgart, where there is a marked drop

in the nitrate concentration of the water (Fig. 11.8); partially
anaerobic conditions exist in the lowermost part of the water column.


## 11.76  PLANT AND ANIMAL RESPIRATION

Because of the high phytoplankton density in the Neckar, account
must be taken of its potential for high respiratory activity.  Negative
balances are most likely to occur at times (as in autumn) when high
temperature is combined with low light intensity.  On the other hand
the respiratory requirement for oxygen of the littoral and benthic fauna
is so slight that it can be ignored when drawing up an oxygen budget.


## 11.77  OXYGEN BALANCE IN THE RIVER

The balance of $O_2$-producing and $O_2$-requiring processes differs along
the length of the Neckar.  The balance shifts towards the latter,
particularly under the influence of effluents in the greater Stuttgart
region (km 150 - 180), where summer oxygen-deficits are frequent and
extend well into the water column.  The occurrence of low flows during
the warmer months is particularly likely to lead to deficits (Fig.
11.14).  The high rate of oxygen production in summer during periods of
high light intensity hinders $O_2$ concentrations from dropping to zero,
but there is an obvious alteration in concentrations.  Completely
anaerobic conditions used to occur regularly in autumn; photosynthetic
activity is low at this time of year, yet water temperatures are still
high enough for effective biodegradation of dead plankton.  Such
conditions seldom occur nowadays because of the general improvements
in effluent quality.

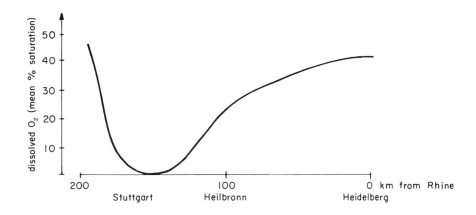

*Fig. 11.14* Changes in dissolved oxygen on passing down the canalized
Neckar in 1972-1976.

There is a notable improvement in oxygen conditions if flows at
Plochingen gauging station exceed about 50 m$^3$ s$^{-1}$.  Besides natural
aeration along this stretch and the higher flow/effluent ratio, aeration during passage over weirs is also important.  Most power stations
on the middle Neckar have been designed to work for a flow of 45 - 50
m$^3$ s$^{-1}$.  Any surplus is led downstream by a weir overfall, where considerable re-oxygenation can occur (Pintér, 1979).  As a result, at
times of high flow in winter oxygen concentrations approach saturation.

## 11.8   CONCLUDING COMMENTS

Even among European rivers, the influence of man on the Neckar has been
especially pronounced.  The long canalized stretch has a profound effect
on the ecology of the river, with its locks, shipping channel, power stations and associated thermal enrichment, and municipal and industrial
effluents.  When development was taking place along the middle and lower
Neckar, the effects of effluents and calefaction were either unknown or
underrated.  The many problems have led to steps to eliminate, or at
least reduce, harmful features along the canalized stretch, although the
question of economic needs versus the ecological health of the river must
always be kept in the foreground.  In order to maintain conditions which
are at least reasonable it is important that the ratio between volumes
of freshwater and of the various effluents does not drop below about 1 : 1.

Among the most important problems is the extremely high rate at which
sediment is deposited in the canalized part of the river (180000 m$^3$ yr$^{-1}$).
This not only reduces the cross-sectional area of the river, but also
leads to greatly increased microbial activity, which is itself enhanced
by the warm water, and hence is an important factor leading to reduced
oxygen concentrations.  It is particularly difficult to deal with these
sediments because of their relatively high content of heavy metals,
making it unsuitable for use on agricultural land.  Up to now no satisfactory solution has been found for this problem.

The drop in oxygen concentrations during the warmer months is another
serious problem.  This results from the influence of microbial activity
on organic materials and ammonia.  Although there has been a considerable improvement in recent years in the quality of organic materials
entering from effluents, levels still remain high.  Ammonia is also
important because of its potential toxicity to fish, especially if pH
values are high.  Further measures are needed to deal with this problem.

Other aspects may become increasingly important in the future.  A
reduced sediment load would lead to increased light penetration, which,
combined with high concentrations of plant nutrients, would in turn
favour very high phytoplankton crops.  The first signs are already at
hand.  A stage may even be reached when the main biological problem is
not a saprobic one, but one resulting from eutrophication.

*Acknowledgements*

We thank Dr H. Buck (Stuttgart) for help with the account of zoobenthos.

REFERENCES

Backhaus D. (1980)  Phytoplanktonuntersuchungen im Freiwasser des
    Neckars.  In:  Studien zum Gewässerschutz 4, 312 pp.  *Landesanstalt
    für Umweltschutz Baden-Württemberg - Institut für Wasser- und
    Abfallwirtschaft,* pp. 170-229, Karlsruhe·

Backhaus D. & Kemball A. (1978)  Gewässergüteverhältnisse und Phyto-
    planktonentwicklung im Hochrhein, Oberrhein und Neckar. *Arch.
    Hydrobiol.* 82, 166-206·

Gradmann R. (1931)  Süddeutschland.  2. Die einzelnen Landschaften,
    553 pp.  Verlag J. Engelhorns Nachf., Stuttgart.

Hoffmann M. (1980)  Biologischer Sauerstoffumsatz im Neckar/Photosyn-
    these, tierische Atmung, mikrobielle Sauerstoffzehrung im Sediment.
    In: Studien zum Gewässerschutz 3, 276 pp.  *Landesanstalt für
    Umweltschutz Baden-Württemberg - Institut für Wasser- und Abfall-
    wirtschaft,* pp. 2-195, Karlsruhe·

Internationale Kommission für die Hydrologie des Rheingebietes (1978)
    Das Rheingebiet. - Hydrologische Monographie, Kartenteil.  Bonn,
    Paris (no fixed office)

Landesanstalt für Umweltschutz Baden-Württemberg (1980)  Reglement für
    den Kühlbetrieb des KKW Neckarwestheim, 42 pp.  *Landesanstalt für
    Umweltschutz Baden-Württemberg,*  Karlsruhe.

Landesanstalt für Umweltschutz Baden-Württemberg - Institut für Wasser-
    und Abfallwirtschaft (1980)  Deutsches gewässerkundliches Jahrbuch,
    Sonderheft Land Baden-Württemberg, Donau- und Neckargebiet
    Abflussjahr 1979, 215 pp.  *Landesanstalt für Umweltschutz Baden-
    Württemberg - Institut für Wasser- und Abfallwirtschaft,* Karlsruhe·

Ministerium für Ernährung, Landwirtschaft und Umwelt Baden-Württemberg
    (1973)  Wärmelastplan Neckar.  Wasserwirtschaftsverwaltung Heft 3,
    18 pp.  *Ministerium für Ernährung, Landwirtschaft und Umwelt Baden-
    Württemberg,* Stuttgart.

Ministerium für Ernährung, Landwirtschaft und Umwelt Baden-Württemberg
    (1976)  Sanierungsprogramm Neckar, Plochingen bis Mannheim.
    Wasserwirtschaftsverwaltung Heft 5, 45 pp. + 13 maps. *Ministerium
    für Ernährung, Landwirtschaft und Umwelt Baden-Württemberg,*
    Stuttgart.

Neckar AG und Wasser- und Schiffahrtsdirektion Stuttgart (1971)  Der
    Neckarausbau zwischen Mannheim und Plochingen - Fünfzig Jahre Neckar-
    Aktiengesellschaft, 39 pp.  *Neckar AG und Wasser- und Schiffahrts-
    direktion Stuttgart,* Stuttgart.

Pintér I. (1979)  Naturmessungen zum Sauerstoffeintrag durch verschiedene
    Belüftungsverfahren.  In: Natur- und Modellmessungen zum künstlichen
    Sauerstoffeintrag in Flüsse, Symposium Darmstadt 1979.  Schriften-
    reihe des Deutschen Verbandes für Wasserwirtschaft und Kulturbau
    (DVWK), Heft 49, pp. 119-25.

Pintér I. (1980)  Mikrobielle Sauerstoffzehrung und physikalischer
    Sauerstoffeintrag im Freiwasser des Neckars.  In: *Landesanstalt für
    Umweltschutz Baden-Württemberg - Institut für Wasser- und
    Abfallwirtschaft,* Karlsruhe.  Studien zum Gewässerschutz Nr. 3,
    276 pp., pp. 196-254.

Pintér I. & Bauer L. (1980)  Gütezustand des Freiwassers des Neckars.
    In: Studien zum Gewässerschutz Nr. 4, 312 pp., pp. 73-169.
    *Landesanstalt fur Umweltschutz Baden-Württemberg - Institut für*
    *Wasser- und Abfallwirtschaft,* Karlsruhe.
Rapp R. (1973)  Ökologische Untersuchungen über Rotatorien in
    Stauhaltungen des Neckars von Cannstadt bis Marbach zur Beurteilung
    der Gewässerbelastung.  Zulassungsarbeit fur das Lehramt, päda-
    gogische Hochschule Esslingen, 53 pp.  (Report to Pedagogische
    Hochschule Esslingen)
Schaal H. (1981)  Der Neckar als Lebensader einer Industrieregion -
    Wasserwirtschaft im Kraftfeld vielseitiger Interessen.  In:
    Referate zum 1. Neckar-Umwelt-Symposium, 16/17 Februar 1981 in
    Heidelberg, 135 pp., pp. 1-13.  Institut für Sedimentforschung der
    Universität, Heidelberg.
Schmitz W. (1971)  Die Ausbreitung von Spaltprodukten bei Reaktorun-
    fällen in Fliessgewässern.  In: Radioaktive Stoffe und Trinkwasser-
    versorgung bei nuklearen Katastrophen.  Bericht der Arbeitsgruppe
    "Trinkwasser-Kontamination", Anhang Nr. 9, 1-15.  Bundesministerium
    des Innern, Bonn.
Schoch E. (1981)  Fischbestand und Wasserqualität des Neckars.  In:
    Referate zum 1. Neckar-Umwelt-Symposium, 16/17 Februar 1981 in
    Heidelberg, 135 pp., pp. 75-78.  Institut für Sedimentforschung der
    Universität, Heidelberg.
Tobias W. (1976)  Kriterien für die ökologische Beurteilung des
    unteren Mains.  II. Untersuchungen über den organischen Stoff-
    haushalt.  Courier Forschungsinstitut Senckenberg 18, 137 pp.


Useful sources on information not quoted in text

Dornier System (Ed.) (1972-1977)  Prognostisches Modell Neckar,
    Berichte 1-25.   Postfach 1360, D-7990 Friedrichshafen.  Dornier
    System GmbH.
Institut für Sedimentforschung der Universität Heidelberg (1981)
    Referate zum 1. Neckar-Umwelt-Symposium 16./17. Februar 1981 in
    Heidelberg, 135 pp.  Heidelberg.
Landesanstalt für Umweltschutz Baden-Württemberg (Ed.)  Studien zum
    Gewässerschutz. Heft 1: Kontrollverfahren und Massnahmen zum
    Schutze der Gewässer und des Wassers gegen Vereinigung im Rahmen
    des Umweltschutzes, 53 pp. (1971).  Heft 3: Untersuchungen über den
    Sauerstoffhaushalt des Neckars, 276 pp. (1980).  Heft 4: Unter-
    suchungen über den Gewässergütezustand des Neckars, 312 pp. (1980.
    Heft 5: Die Verfahren der biologischen Beurteilung des Gütezustandes
    der Fliessgewässer (systematisch-kritische Übersicht, 125 pp. (1980).
    Griesbachstr. D-7500 Karlsruhe.
Landesanstalt für Umweltschutz Baden-Württemberg (1979)  Umweltqualitäts-
    bericht Baden-Württemberg 1979, 217 pp.  Griesbachstr. D-7500
    Karlsruhe.

# 12 · Shannon

K. HORKAN

| | | |
|---|---|---|
| 12.1 | Introduction | 346 |
| 12.2 | Geology | 346 |
| 12.3 | Soils and land use | 348 |
| 12.4 | Climate | 348 |
| 12.5 | Towns and industry | 349 |
| 12.6 | Utilization of the river | 350 |
| | 12.61 Navigation | 350 |
| | 12.62 Power production | 350 |
| | 12.63 Amenities | 351 |
| 12.7 | Water resources and future management | 351 |
| 12.8 | Water quality | 354 |
| | 12.81 The river | 354 |
| | 12.82 The lakes | 356 |
| 12.9 | Biota | 356 |
| | 12.91 Plants | 356 |
| | 12.92 Macroinvertebrates | 359 |
| | 12.93 Fish | 360 |
| References | | 361 |

## 12.1   INTRODUCTION

The Shannon is the longest river in the British Isles, with a main channel length (including the estuary) of some 359 km.  It is a mature river (Nevill, 1972).   It flows not much above base level on a gentle gradient and cuts laterally instead of down, producing broad meanders and frequent lake-like expanses.  The source is a spring-fed pool known as the "Shannon Pot" which is located only 45 km from Donegal Bay at an elevation of 104 m on hills which rise to 600 m (Fig. 12.1).  However, because of high ground between there and the coast, the river is forced to flow south, widening into many lake-like expansions as it meanders slowly through the Central Plain.  The first of these is Lough Allen (3500 ha).  From the outflow of this lake, the main channel drops only 18 m over a river distance of 230 km, passing through further large lakes, the most important of which are Lough Ree (10457 ha) and Lough Derg (11635 ha).  Downstream of the latter the Shannon falls 30 m in only 32 km to reach tidal waters at Limerick.  The river then continues its course through 99 km of a broadening submerged valley to reach the Atlantic Ocean at Loop Head.

The catchment, including that of the estuary, drains some 14100 km$^2$ or approximately one-sixth of the area of the Republic of Ireland (McCumiskey, 1977).  The more important tributaries are: the Boyle River which flows through Lough Key (897 ha), a scenic amenity area; the River Suck; the River Brosna which drains the larger Midland lakes; the Little Brosna River and the River Inny.  The population of the whole catchment is about 500000.

## 12.2   GEOLOGY

The Shannon rises on hills which consist of rocks of the Yoredale and Pendleside Series (Upper Avonian Shales and Sandstones; Fig. 12.1). The river flows off these hills into Lough Allen which is a glacial lake, continuing from there through a narrow band of Upper Carboniferous Limestone.  Further downstream the river briefly crosses Avonian Shales and Sandstones followed by Carboniferous Limestone, then an area of Lower Avonian Shales with Carboniferous Slate and Calciferous Sandstone.  The main influence is, however, the Limestone of the Central Plain, which dominates the catchment downstream as far as the hills on the southern end of Lough Derg.  Here the rock formations are mainly Silurian with some Old Red Sandstone hills surrounded by a narrow band of Lower Avonian Shales.  West of Limerick the submerged valley that is now the estuary is floored by an expanse of Limestone.  On the western side of the Fergus Estuary the channel crosses Coal Measure uplands which are in turn replaced by Millstone Grit and Flagstones, particularly on the north bank (Co. Clare).  On the south bank (Co. Kerry) there are some Upper Avonian Shales and Sandstones.

*Fig. 12.1* Shannon Catchment, showing main tributaries, general geology and larger towns.  Numbers refer to sampling stations in Table 12.3.

Most of the important tributaries arise on either Upper or Lower
Carboniferous Limestone, apart from the Boyle River which originates on
hills formed of Lower Avonian Shales and Old Red Sandstones.  The
Shannon Lakes, apart from Lough Allen, whose origins are glacial,
originated as great solution expanses of the main channel (Praeger,
1974).

## 12.3   SOILS AND LAND USE

A variety of soils are found in the catchment.  Because some are poorly
drained and because the river is broad and slow-flowing through a
frequently waterlogged plain of low relief, flooding is common, particularly in winter.

Upstream of Lough Allen the mountainy and hilly areas are dominated
by peaty gleys with their associated climatic peats.  Further downstream these change to drumlin gleys with some associated peaty gleys,
interdrumlin peat and acid brown earths.  Around Rooskey and again south
of Athlone there is an area of basin peat with organo-mineral soils.
Lough Ree and the northern end of Lough Derg are mostly surrounded by
grey brown podzolics with associated gleys.  Between Lough Ree and
Athlone there are also some grey brown podzolics but here they are
associated with basin peat and regosols.  West of Lough Derg there are
drumlin soils, while around Killaloe there are acid brown earths and
associated brown podzolics.  Further downstream towards Limerick these
soils are replaced by gleys and associated peaty gleys.  West of
Limerick there are mudflats, especially on the Co. Clare (north) side
and these are principally regosols with some gleys.  Small areas of
grey-brown podzolics occur on the north bank while shallow brown earths
and grey brown podzolics are found on the south side.  West of Foynes
on the south shore and Ennis on the north side the principal soils are
mainly gleys and associated acid brown earths and peats.

Before farming developed, Ireland was covered in oak forests.  These
were often felled to make room for grasslands, the growing of which is
particularly suited to the mild Irish climate.  The forests may also
have declined due to climatic changes with increased rainfall which
resulted in peat formation.  In the low-lying Midland region adjoining
the Shannon large areas of peat bogs are found which are suited only
to sheep grazing and peat production.  Cattle rearing and mixed farming
are most common on the richer pastures of the catchment.

## 12.4   CLIMATE

Because of the influence of the warm Atlantic currents and mild sea
breezes the Irish climate lacks extremes.  The mean annual temperature
range is only about $10^{o}$C and varies from a mean in January of $4.4\ ^{o}$C in

the Midlands (5.5 $^{\circ}$C along the west coast) to a mean in July of 15 $^{\circ}$C.
Frost is relatively rare along the west coast but is more common in the
Midlands.  The Shannon and its lakes do not freeze, although ice may
form in sheltered bays of the larger water bodies; this seldom lasts
for more than a few days.

The prevailing winds from the Atlantic bring in rainfall which is
greatest along the western seaboard and declines eastwards.  On the
west side of the Shannon annual precipitation may be as high as 1750 -
2250 mm in hilly areas, whereas on the eastern side it averages 750 -
1000 mm, with a mean annual 30 year average over the Shannon catchment
of 1024 mm (An Foras Forbartha, 1977).  At Killaloe where the mean
annual rainfall over a 15 year period is 1029 mm, the average run-off
(inflow) was estimated at 539 mm and total losses amounted to 490 mm.
Losses due to evapotranspiration and deep seepage constituted 37.6% of
the mean annual rainfall for this period (McCumiskey, 1977).


                    12.5  TOWNS AND INDUSTRY

Due to the low-lying boggy nature of the surrounding land and the tendency towards winter flooding few towns are found along the main river
channel.  The more important ones are: Carrick-on-Shannon (population
2208), an important tourist centre containing the largest inland marina
in Ireland; Athlone (9778) which has manufacturing and electronics
industries as well as boat-building; Limerick (60665) which is a shipping port, as well as having numerous industries including manufacturing
and electronics; Shannon New Town (7000) which has a wide range of
industries, which are mostly light manufacturing and electronic plants
geared for export through the adjoining International Airport.

The Shannon Estuary is rapidly becoming an important industrial
development area.  The small port of Foynes is an oil rig supply base
as well as the export point for barytes.  Nearby, at Aughinish Island,
a £500 million aluminium production plant has recently been completed.
Electricity is generated at Tarbert (oil-fired) and a new coal-fired
power station is being constructed at Money Point (12.62).  Future
developments planned for the estuary include oil refineries and their
associated industries.

Anthracite is mined at Arigna and supplies the energy source for a
local power station.  Milled peat for fuel and horticulture is also
important in the catchment.  Silver, lead and zinc-bearing ores are
extracted from the Silvermine Hills.  The fishing industry in the area
is based mainly on commercial salmon and eels (the latter at Killaloe
and Athlone), with some shell-fishing in the estuary; a herring fishery
is active off Loop Head (see 12.93).

## 12.6  UTILIZATION OF THE RIVER

### 12.61  NAVIGATION

The first work on the development of the Shannon as a navigation channel began at Killaloe in 1755.  By 1787 the river had been made navigable as far upstream as Carrick-on-Shannon (Carty, 1977).  The more difficult Limerick to Killaloe stretch was not completed until 1814.  The latter, known as the Errina-Plessey Canal, by-passed different stretches of the river and contained a total of 11 locks.  This canal was abandoned in 1930 with the completion of the hydro-electric works (see below) containing a double lock with a lift of 30.5 m.

The Lough Allen Canal at Drumshanbo came into operation in 1814 but was little used and was closed in 1936.  Navigation was linked to the Erne when the Ballinamore and Ballyconnell Canal was opened in 1860. However, by 1880 this canal was derelict.  Now enthusiasts are trying to have this link with the Erne system re-opened.  The Shannon was connected to Dublin by 1806 with the completion of the Grand Canal.  A rival waterway, the Royal Canal, was opened in 1809.  Steam power came to the river in 1826 and regular cargo-carrying barges plied between Dublin and Limerick.  Periodically improvements were made to the navigation system, these including the construction of a number of weirs and deepening of the locks to accommodate a minimum draft of 2.0 m. However, the coming of the railways caused the demise of the water borne cargo trade.  Barges which were moving 100000 tons of cargo annually by the turn of the century found their trade declining.

The first signs of a revival in interest in the Shannon as a waterway occurred in 1954, when two passenger launches were purchased by C.I.E. (National Transport Authority) for day trips from Athlone.  The first hire cruiser for family holidays became available in 1960.  Now there are some 560 boats for hire and at least a similar number of private craft on the water.  At present over 7600 passages are made by boat through Athlone Lock during a summer season and this number is rising steadily (see 12.63).

### 12.62  POWER PRODUCTION

The Shannon became important for power production during the 1920s when the Dail (Irish Government) passed an act in 1925 enabling the first development of the Shannon hydro-electric works at Ardnacrusha (310 MW) (O'Leary, 1977).  This involved raising Lough Derg by about 60 cm (2 feet), which concentrated the entire fall of the river over a length of some 80 km at one point.  The average annual flow of the Shannon past Killaloe is 190 $m^3$ $s^{-1}$, of which 180 $m^3$ $s^{-1}$ is available for power generation.  Since that time several other power stations have been built in the catchment including Arigna (15 MW, coal fired from local supplies), Lanesboro (60 MW, peat fired from local supplies), Shannonbridge (80 MW, peat fired from local supplies), Tarbert (620 MW,

oil-fired) and Money Point (900 MW, coal fired, under construction).
The capacity of some of the smaller plants is being expanded at
present.


## 12.63  AMENITIES

Due to the fact that much of the Shannon is navigable, the river and
its lakes are becoming very popular with tourists, particularly in
terms of pleasure cruising holidays.  The navigation channels are well
marked and boating facilities, fuel, water and other provisions are
available in towns along the river and in small villages adjacent to
sheltered bays on the lakes.  By European standards the river is relatively uncrowded and unpolluted.  If the Shannon had the same density
of boats as the Norfolk Broads there would be 200000 boats (Ransom,
1971), compared to the 1100 or so craft found today.  The monetary
return from this type of tourism along the Shannon is valued at a
minimum of Ir£10 million (data of Irish Tourist Board).

Angling is also very important on the Shannon, as mentioned in 12.5.
Excellent salmonid (*Salmo salar* and *S. trutta*) and coarse fishing are
available with special attention paid to large pike, bream, rudd, rudd/
bream hybrids, perch, tench and eels.  The thermal effluent canal at
Lanesboro Power Station is a renowned haunt for specialist coarse
fishermen, as it supports large tench and bream.

Other aquatic based pastimes are enjoyed on the river including
canoeing, water-skiing and duck shooting.  There are safe beaches on
some of the lakes, picnic areas, a Forest Park on Lough Key, historic
sites and some holiday village developments are also found.


## 12.7   WATER RESOURCES AND FUTURE MANAGEMENT

The water resources of the Shannon, including tributaries, constitute
about 21% of the total available in the Republic of Ireland.  Of the
15 year mean annual rainfall of 1029 mm at Killaloe, the average run-
off is 539 mm and total losses due to evapotranspiration and deep
seepage accounts for the remaining 490 mm or 47.6% of the total
(McCumiskey, 1977).

Values for low flow at various points along the Shannon are given in
Table 12.2, together with the 99.5 percentile values at Limerick.  The
annual average flow at Killaloe is included for comparative purposes.
On the tributaries there are wide variations in the ratio of low flows
to average flows ranging from 1 to 10%, the higher the ratio the larger
the tributary.  Values for these low flows range from less than 100000
$1$ km$^{-2}$ d$^{-1}$ to as low as 3000 $1$ km$^{-2}$ d$^{-1}$.  The significance of the magnitude of such low flows is that they limit the extent to which the water
can be used for receiving effluents before quality problems develop.

*Table 12.1*  Distribution of water resources in the Republic of
Ireland (from McCumiskey, 1977).  Population estimates are based on
the 1971 Census.  The population of the country was 3.3 million in
1979 compared to 2.9 million in 1971.

| region | population (X10$^3$) | average run-off (m$^3$ s$^{-1}$) | low flow run-off (m$^3$ s$^{-1}$) | average run-off per capita (l person$^{-1}$ d$^{-1}$) | low-flow run-off per capita (l person$^{-1}$ d$^{-1}$) |
|---|---|---|---|---|---|
| 1 Eastern | 1160 | 114.0 | 5.8 | 8592 | 432 |
| 2 South-Eastern | 413 | 198.8 | 15.2 | 41600 | 3180 |
| 3 Southern | 430 | 235.5 | 11.5 | 47320 | 2320 |
| *4 Shannon, excluding tributaries joining estuary | 259 | 178.5 | 10.0 | 59550 | 3340 |
| 5 Mid-Western | 253 | 124.8 | 5.0 | 42600 | 1696 |
| 6 Western | 183 | 186.2 | 4.7 | 87920 | 2218 |
| 7 North-Western | 221 | 231.7 | 6.2 | 90560 | 2410 |

*
The entire catchment, including tributaries joining the estuary,
has a population of 470000.


Many public water supplies in the Shannon catchment use the main
channel, the tributaries and the lakes for their abstractions.  This
implies that future demands for surface water abstraction in the catchment requires careful control of discharges into the river system in
order to maintain satisfactory water quality conditions.  To date the
various developments in the freshwater reaches of the Shannon system
have not resulted in widespread pollution.  This position reflects the
low degree of industrialization and small population size rather than
good management practices.  However, some localised problem areas
occur on a few of the tributaries and a number of lakes (12.82).
Recent trends show that industrial growth, intensification of agriculture and the expansion of towns in the Shannon catchment are now in
progress.

As various water uses are inter-related they cannot be properly
managed in the catchment as a whole without adopting a comprehensive
approach.  This should take into account all possible uses, with
special emphasis on abstraction, discharge and amenity.  In achieving
such objectives the attainment and maintenance of sufficiently good

*Table 12.2*  Minimum flows over the 44-year period, 1932-1975, at various sites in the main channel of the Shannon and the larger tributaries (after McCumiskey, 1977).

| location | catchment (km$^2$) | minimum flow (m$^3$ s$^{-1}$) |
|---|---|---|
| **Shannon** | | |
| outlet of Lough Allen | 425 | 1.7 |
| Carrick-on-Shannon | 1255 | 2.3 |
| Rooskey | 1981 | 3.0 |
| Athlone | 4455 | 8.0 |
| Limerick | 11250 | 12.4 (99.5 percentile) |
| | | 26.4 (95 percentile) |
| **Tributaries** | | |
| Graney - Scariff | 296 | 0.23 |
| Killimor - Moate | 194 | 0.039 |
| Suck - Ballinasloe | 1373 | 1.37 |
| Mountain - Strokestown | 197 | 0.156 |
| Boyle - Boyle | 500 | 0.395 |
|     - Ballaghaderreen | 207 | 0.05 (95 percentile) |
| Camlin - Longford | 207 | 0.09 |
| Inny - Ballymahon | 1046 | 0.24 |
| Brosna - Tullamore | 106 | 0.1 |
|     - Clara | 324 | 0.85 |
|     - Ferbane | 1186 | 2.0 |
| Little Brosna - Birr | 479 | 0.84 |
|       - Roscrea | 26 | 0.026 |
| Nenagh - Nenagh | 142 | 0.22 |

water quality to safeguard public health is a priority. To achieve these standards requires the collection of good basic data, which must take into account present and future water uses, waste water discharges, and the hydrological, physical, chemical and biological characteristics of the catchment.

The standards laid down in the E.E.C. Directive (Council of the European Communities, 1978) which covers salmonid waters is perhaps the most appropriate for the Shannon and its main tributaries. If such standards were adopted and complied with they would ensure adequate protection for other uses also.

## 12.8  WATER QUALITY

### 12.81  THE RIVER

Much of the main channel of the Shannon is inaccessible by road.  As a result the river has not been monitored as regularly as some of the other large Irish rivers.  However, a number of surveys have been carried out and the main findings are discussed here.

In 1971 the chemistry of the river upstream of Dowra in the Upper Shannon to Thomond Bridge (Limerick) in the Lower Shannon was investigated at 22 stations.  The results showed that the greater portion of the length surveyed was in a 'satisfactory' condition (Flanagan & Toner, 1972), but the number of stations sampled was relatively few and the possibility of the survey missing localised polluted stretches was realised.  Hence, a more intensive survey of the main channel, the lakes and the main tributaries was carried out in the summer of 1972.  37 stations on the Shannon and 13 locations on the tributaries were examined (Table 12.3).  These results confirmed that over most of its length the Shannon was indeed in a 'satisfactory' condition but there were some stretches where slight impairments were noted.  Orthophosphate levels were relatively high downstream of Carrick-on-Shannon, Athlone and below the polluted Camlin tributary confluence.  Ammonia levels were slightly elevated at the former location and also below Rooskey.  These results were attributed to organic waste discharges from these towns and also to small discharges, floating solids and debris from cruising boats, the problem being most notable near marines and quays.  The localised enrichment observed below the towns mentioned above was mild in nature and the water quality in the river was generally 'fair' to 'good'.  Filamentous algal growths including *Cladophora* were common in the enriched stretches, but macroinvertebrate diversity was reduced only slightly.

The Lower Shannon was found to have good water quality conditions as far as Limerick City, in the vicinity of which occasional elevated BOD values were recorded.  No evidence was found of any contamination by waste discharges along this lower river reach (Flanagan, 1974).  One tributary in this stretch, the Kilmastulla River, has been deteriorating in quality in recent years as a result of toxic discharges from mining activities in the nearby Silvermine Mountains.  Heavy metal levels have caused denudation of the fauna in this tributary.  In October 1980, an overspill from the tailings pond at the mines resulted in a toxic waste load reaching the Shannon.  This was a 'one off' event, but points to the need for vigilance at all times in areas where potentially serious problems exist.  The mine has now closed.

One problem associated with some sections of the main river channel is that of peat-silt deposits, which build up particularly below Lanesboro, Shannonbridge and at points in the stretch from Banagher to Portumna (Toner *et al*., 1973).  This results in the virtual absence of macroinvertebrates in these areas due to the fine deposits of peat-

*Table 12.3* Water chemistry of the Shannon (main channel), together with biological quality rating. u/s = upstream; d/s = downstream; tr = trace. Quality ratings: Q5, good; Q4, fair.

| station | location | distance from st. 1 (km) | dissolved oxygen (% saturation) | BOD$_1$ (mg l$^{-1}$ O$_2$) | NH$_3$-N (mg l$^{-1}$) | NO$_3$-N (mg l$^{-1}$) | reactive phosphate-P (mg l$^{-1}$) | hardness (mg l$^{-1}$ CaCO$_3$) | quality rating (Q) |
|---|---|---|---|---|---|---|---|---|---|
| **Upper Shannon, Dowra Area** | | | | | | | | | |
| 1 | Metal Br. N. of Corratober | 0 | 98 - 119 | 0.8 - 1.2 | 0.01 - 0.03 | | | 55 - 75 | 5 |
| 2 | Bridge in Dowra | 6.0 | - | | | | | | 5 |
| 3 | 1 km d/s Dowra | 7.0 | 98 - 122 | 0.7 - 1.4 | 0.01 - 0.02 | | | 30 - 50 | 5 |
| **Carrick-on-Shannon Area** | | | | | | | | | |
| 4 | Hartley Bridge | 36.0 | 98 - 118 | 0.8 - 1.8 | 0.01 - 0.03 | | | 45 - 65 | 5 |
| 5 | 2 km u/s Carrick-on-Shannon | 38.0 | - | | | | | | 5 |
| 6 | Carrick-on-Shannon Bridge | 39.8 | 100 - 120 | 0.8 - 1.7 | 0.03 - 0.04 | 0.12 - 0.19 | 0.015 | 45 - 100 | 5 |
| 7 | 0.5 km d/s Carrick-on-Shannon | 40.5 | 95 - 122 | | 0.02 - 0.04 | 0.12 | 0.025 - 0.035 | | 4 |
| 8 | u/s Jamestown Bridge | 49.5 | 100 - 118 | 0.9 - 1.4 | 0.01 - 0.03 | | | 75 - 105 | 5 |
| **Middle Shannon, Rooskey Area** | | | | | | | | | |
| 9 | u/s Rooskey Bridge | 67.0 | 100 - 120 | 1.0 - 2.0 | 0.02 - 0.05 | | | 90 - 140 | 5 |
| 10 | 0.2 km d/s Rooskey Bridge | 67.2 | | | | | | | 5 |
| 11 | u/s Lock and d/s Canal | 67.6 | 94 - 116 | | 0.025 | 0.11 | 0.055 | | 5 |
| 12 | u/s Lough Forbes | 72.0 | | | | | | | 5 |
| 13 | u/s Camlin River, N. Branch | 76.0 | 94 - 112 | 1.1 - 1.7 | 0.02 - 0.06 | 0.10 - 0.12 | 0.025 - 0.065 | | 5 |
| 14 | u/s Camlin River, S. Branch | 77.0 | 94 - 116 | | | 0.19 - 0.33 | | | 4 |
| **Athlone Area** | | | | | | | | | |
| 15 | d/s Athlone Weir | 124.0 | 90 - 104 | 1.4 - 5.5 | 0.02 - 0.30 | 0.22 - 0.34 | 0.015 - 0.125 | 90 - 145 | 4 |
| 16 | d/s disused canal | 125.0 | 100 - 112 | | tr - 0.08 | 0.17 | 0.010 | 165 - 195 | 5 |
| 17 | u/s Wrens Island | 127.0 | 100 - 112 | | 0.07 | | 0.035 | | 5 |
| 18 | u/s Long Island | 129.5 | 104 - 107 | | | 0.44 | | | 5 |
| 19 | d/s Long Island | 132.0 | 104 - 108 | | tr | 0.42 | 0.005 | | 5 |
| 20 | Inchalee Island | 134.0 | 102 - 104 | | tr | 0.20 - 0.41 | 0.030 | | 5 |
| 21 | u/s Clonmacnoise | 137.5 | 80 - 89 | 1.1 - 1.6 | 0.02 - 0.06 | 0.22 - 2.50 | 0.015 - 0.050 | | 5 |
| **Lower Shannon, Lough Derg Area** | | | | | | | | | |
| 22 | Meelick | 174.0 | 78 - 118 | 0.5 - 3.5 | tr - 0.15 | 0.25 - 2.50 | 0.0C3 - 0.060 | 160 - 205 | - |
| 23 | Portumna Bridge | 187.0 | 76 - 114 | 0.8 - 4.4 | <0.01 - 0.19 | 0.28 - 1.90 | tr - 0.065 | | - |
| 24 | Killaloe Bridge | 222.0 | 86 - 110 | 0.5 - 3.9 | tr - 0.20 | 0.28 - 1.90 | tr - 0.033 | | - |
| 25 | O'Briens Bridge | 229.5 | 88 - 117 | 1.0 - 3.5 | 0.01 - 0.06 | 0.36 - 1.20 | 0.010 - 0.100 | | - |
| 26 | Castleconnell | 234.5 | 90 - 125 | 1.0 - 3.4 | 0.01 - 0.07 | 0.35 - 1.20 | 0.010 - 0.140 | | - |
| 27 | 2.5 km WNW Lisnagry Station | 238.5 | 96 - 100 | 1.0 - 2.7 | 0.01 - 0.07 | 0.40 - 1.35 | 0.010 - 0.110 | | - |
| 28 | 2.0 km W Mount Shannon | 239.0 | 100 - 120 | 1.0 - 2.8 | 0.01 - 0.11 | 0.45 - 1.40 | 0.010 - 0.140 | | - |
| 29 | Athlunkard Bridge | 244.5 | 92 - 114 | 1.0 - 4.1 | 0.01 - 0.08 | 0.43 - 1.30 | 0.010 - 0.310 | | - |
| 30 | Thomond Bridge | 248.0 | 84 - 109 | 1.0 ->7.0 | 0.02 - 0.37 | 0.31 - 1.20 | 0.010 - 0.150 | | - |

silt covering the substratum, although chemically the water quality is
not altered.  One tributary, the deep slow-flowing River Suck, also
receives peat-silt deposits, but its water quality is 'good'.


## 12.82  THE LAKES

The Shannon Lakes also have some influence on the main channel.  The
uppermost lake (Lough Allen) has soft water with significant coloration
(Table 12.4).  From the upstream to the downstream lakes there is a
gradual increase in mineral content and pH because of the change from
harder rocks to more soluble limestone.  An exception to this is Lough
Key which is more similar to the two lower lakes, Ree and Derg, because
the Boyle River flows through a limestone area (Fig. 12.1).  Lough
Allen is chemically the poorest lake, the other lakes having a relatively high conductivity and hardness.  Most of them are quite shallow
apart from Ree and Derg, where troughs of 30 m or more are found
(Charlesworth, 1963) and tend mostly to be low-lying and exposed,
apart from hilly areas adjoining Loughs Key and Derg.

In terms of trophic status the lakes range from oligotrophic (Lough
Allen) to eutrophic (Lough Derg).  Lough Derg is of particular interest
as data were collected by Southern & Gardiner (1926, 1932, 1938) and
Southern (1935) from November 1921 to March 1923.  These workers were
based in a freshwater laboratory near Portumna and this limnological
station was the first of its kind to be built in the British Isles.
Unfortunately, the station was closed down because of a lack of funds.
Examination of the plankton data of Southern and Gardiner show that
some changes have taken place, with increased numbers of some clado-
ceran and rotifer species which prefer slightly enriched conditions,
as well as an increase in the blue-green algal populations.  There
has been some reduction in water clarity.  These changes are slight
rather than excessive and as yet no serious decline in water quality
has occurred.


## 12.9  BIOTA

### 12.91  PLANTS

According to Praeger (1974) the fact that almost all of the larger
Irish rivers have no marked watersheds and have at least part of their
courses in the Limestone Plain has made it possible for an easy migration of species from one river system to the next resulting in a
comparative uniformity of flora and fauna.  Yet a few species are
restricted to only one or two river basins.  Examples of localised
species occurring in the Shannon system include *Inula salicina* which
is found only on the shores of Lough Derg, *Teucrium scordium* found
along the shores of Loughs Ree and Derg (Webb, 1963) and *Sium latifolium* found along the river bank of the Shannon and also in the Erne

*Fig. 12.2*  Lough Key, showing Castle Island and reed-covered shore-line in sheltered bays.

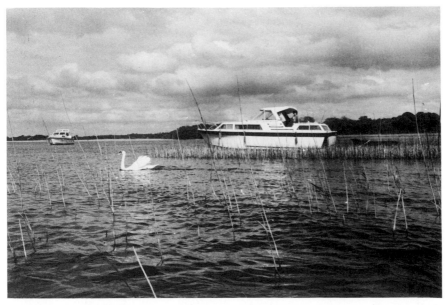

*Fig. 12.3*  The Shannon upstream of Lough Rae, showing typical low-lying banks and scattered beds of emergent plants in shallow water. (Photos:  courtesy Irish Tourist Board)

*Table 12.4* Mainstream lakes on the Shannon: selected morphological, physical, chemical and biological characteristics, showing ranges recorded during various surveys made during the period 1973-1982. tr = trace.

| lake | altitude m | size ha | conductivity µS cm⁻¹ | colour hazen units | hardness mg l⁻¹ CaCO₃ | pH | ortho-phosphate mg m⁻³ P | oxidised N mg m⁻³ N | ammonia mg m⁻³ N | chlorophyll a mg m⁻³ | Secchi disc transparency m | trophic status |
|---|---|---|---|---|---|---|---|---|---|---|---|---|
| Allen | 50.0 | 3500 | 90 – 100 | 50 | 30 – 50 | 7.1 – 7.4 | tr – 40 | 100 – 190 | tr – 40 | 1.4 – 4.2 | 1.8 – 2.5 | oligotrophic |
| Bofin | 42.0 | 408 | 190 – 315 | 40 – 60 | 80 – 140 | 7.8 – 8.1 | 10 – 40 | 140 – 590 | 10 – 30 | 1.9 – 5.5 | 1.2 – 2.3 | mesotrophic |
| Boderg | 42.0 | 430 | 155 – 310 | 50 – 60 | 70 – 145 | 7.7 – 8.0 | 10 – 50 | 140 – 650 | 10 – 20 | 1.7 – 7.5 | 1.0 – 2.0 | mesotrophic |
| Key | 44.5 | 897 | 310 – 360 | 50 – 100 | 150 – 200 | 8.0 – 8.3 | 5 – 20 | 80 – 880 | 10 – 50 | <1.9 – 9.6 | 1.5 – 4.5 | naturally eutrophic |
| Ree | 38.0 | 10457 | 300 – 395 | 30 – 60 | 165 – 195 | 7.8 – 8.4 | 10 – 140 | 90 – 900 | 10 – 170 | 0.9 – 25.0 | 1.0 – 5.5 | slightly enriched |
| Derg | 33.0 | 11635 | 410 – 510 | 20 – 70 | 205 – 210 | 7.8 – 8.6 | tr – 85 | 100 – 2000 | tr – 70 | 2.2 – 29.0* | 0.9 – 4.0 | moderately enriched |

*A special survey on Lough Derg in August 1982 reported concentrations within the 30 – 50 mg m⁻³ range.

system; *Chara tomentosa* which is unknown in Great Britain, occurs in
the Shannon basin, particularly in Lough Ree and in the Midland lakes
which drain into the Shannon.  Praeger (1974) suggested that the distribution of these rare species may signify that they are late arrivals
rather than glacial relicts and little dispersal has yet occurred.
The red alga *Bangia atropurpurea* was recorded in Lough Derg near
Portumna only recently by Scannell (1972).  It is known to have occurred
previously in the two canals which drain into the Shannon and from
whence it probably spread to the lake.  Several other aquatic plants
have spread from the Royal and Grand Canals to the Shannon basin
including *Sagittaria* spp., *Glyceria aquatica*, *Equisetum variegatum*
and several charophytes.


## 12.92  MACROINVERTEBRATES

Information on the macroinvertebrate fauna in the catchment is sparse
and is based mostly on limited sampling carried out in connection with
water quality assessments.  The Upper Shannon above Lough Allen, which
varies from torrential reaches to fast-flowing riffle areas, has a
diverse fauna typical of a clean upland stream.  Baetidae tend to dominate, with Gammaridae and Ecdyonuridae numerous.  Other taxa well
represented include *Isoperla* spp., *Leuctra* spp., *Astacus* sp., *Elmis*
spp., *Potamopyrgus*, Ceratopogonidae, *Hydropsyche* spp., and *Simulium*
spp. (An Foras Forbartha, unpublished data).

   Below Lough Allen movement of water in the main channel of the river
becomes sluggish for most of the remainder of its course to the sea.
The fauna changes to suit these conditions and is dominated in most
cases by Lamellibranchs, Oligochaeta, Chironomidae and Gastropods
(Toner *et al.*, 1973).  Other groups commonly encountered include
*Asellus* spp., Polycentropidae and other Trichoptera, Ephemeroptera
(excluding Ecdyonurids) and Hirudinea.  Less frequent are Plecoptera,
Simuliidae, other 'Diptera', Ecdyonuridae and sponges.  Some groups
are well represented but confined to particular stretches; these include
*Gammarus* spp. which occur mainly in the upper part of the river and
*Mysis* spp. mainly below Athlone.  Detailed species lists can be found
in the account of Toner *et al.* (1973).

   In general, the faunal composition reflects the deep slow-flowing
nature of the river, with burrowers and crawling types best suited to
the silted conditions.  Differences between stations can be attributed
to the nature of the substratum.  Even where slight organic enrichment
has been noted (e.g. below Athlone), both sensitive and tolerant groups
are abundant.  Larger populations of Lamellibranchs and Gastropods
featuring such genera as *Sphaerium, Anodonta, Bithynia, Potamopyrgus*
and *Theodoxus* are abundant at stations downstream of lakes.  Where the
substratum is covered in peat and silt deposits few fauna are encountered, although no adverse chemical effects have been observed at
such locations (see 12.8).

## 12.93  FISH

According to Went (1940) the Shannon, prior to the initiation of the
hydro-electric works, was an outstanding salmon fishery with a world-
wide reputation.  Following the construction of the dam near Ardnacrusha conditions were altered dramatically with the result that the
salmon fishery went into serious decline.  Net catches up to 188000 kg
were recorded before the river was harnessed for electricity supply
(O'Dowd, 1977).  Following completion of the works returns declined to
only 19000 kg (Table 12.5).  Fish passes were installed to improve the
situation and catches increased again to over 100000 kg.  Commercial
eel fishing also declined on the river for the same reason and is
recovering again.  Reduced catches of salmon in recent years are attributed to the effects of Ulcerative Dermal Necrosis (U.D.N.) disease in
the late 1960s and drift netting for salmon at sea in the 1970s.

*Table 12.5*  Salmon and eel net landings from the Shannon.  The value
for 1927 is typical of annual returns before the construction of
Ardnacrusha Power Station.  In 1939 catches were still badly
affected by the hydro-electric works and fish passes were constructed to improve the situation.  Catches then returned to 'normal'.
(Data from O'Dowd (1977) and Annual Sea and Inland Fisheries
Reports, 1963-1980)

| year | salmon landings (kg) | eel landings (kg) |
|---|---|---|
| 1927 | 188244 | 65382 |
| 1939 | 19051 | – |
| 1963 | 162822 | 33536 |
| 1964 | 157468 | 40365 |
| 1965 | 156282 | 61723 |
| 1966 | 110058 | 54184 |
| 1967 | 141771 | 87171 |
| 1968 | 127177 | 51823 |
| 1969 | 173043 | 36034 |
| 1970 | 169676 | 35054 |
| 1971 | 214889 | 31131 |
| 1972 | 132840 | 8038 |
| 1973 | 199581 | 40232 |
| 1974 | 162778 | 39200 |
| 1975 | 190193 | 34996 |
| 1976 | 118638 | 60000 |
| 1977 | 96948 | 38100 |
| 1978 | 76064 | 24000 |
| 1979 | 48752 | 42411 |
| 1980 | 43332 | 33459 |

Some excellent brown trout fishing is found on the lakes with the best catches usually in May and June when *Ephemera danica* is emerging in large numbers. The main river channel and the larger lakes provide good coarse fishing especially for pike, bream, tench, perch, rudd and rudd/bream hybrids. Whitefish are not well represented in Irish waters but one exception is the pollan (*Coregonus pollan elegans*) which is found in the Shannon lakes. This fish is of commercial importance in Lough Neagh but is not found in sufficiently large quantities to be a viable commercial proposition on the Shannon. The smelt (*Osmerus eperlanus*) has been recorded from the Shannon lakes by Kennedy (1948) and is also mentioned by Went (1948) but is usually not seen by anglers.

## REFERENCES

An Foras Forbartha (1971) *Preliminary Report on Surface Water Resources in the Shannon Water Resources Region.* An Foras Forbartha, Water Resources Division.

Carty J.J. (1977) The management of the River Shannon in the 1980s. *J. Inst. public Admin. Ireland* 25(2) Special issue, 220-30.

Charlesworth J.K. (1963) The bathymetry and origin of the larger lakes of Ireland. *Proc. R. Ir. Acad.* 63 B(3), 61-71.

Clabby K.J. (1981) *The National Survey of Irish Rivers. A Review of Biological Monitoring, 1971-1979,* 308 pp. An Foras Forbartha, Water Resources Division, WR/R12.

Council of the European Communities (1978) *Council Directive of 18 July 1978 on the Quality of Fresh Waters needing Protection or Improvement in Order to Support Fish Life.* (78/659/E.E.C.). Official Journal of the European Communities. No. L222/7.

Flanagan P.J. (1974) *The National Survey of Irish Rivers. A Second Report on Water Quality,* 98 pp. An Foras Forbartha, Water Resources Division.

Flanagan P.J. & Toner P.F. (1972) *The National Survey of Irish Rivers. A Report on Water Quality,* 213 pp. An Foras Forbartha, Water Resources Division.

Kennedy M. (1948) Smelt in the Shannon. *Ir. Nat. J.* IX, 151-52.

Lennox L.J. & Toner P.F. (1980) *The National Survey of Irish Rivers. A Third Report on Water Quality,* 104 pp. An Foras Forbartha, Water Resources Division.

McCumiskey W. (1977) The management of the River Shannon in the 1980s. *J. Inst. public Admin. Ireland* 25(2) Special issue, 192-210.

Nevill W.E. (1972) *Geology and Ireland,* 263 pp. Allen Figgis & Co. Ltd., Dublin.

O'Dowd P. (1977) The management of the River Shannon in the 1980s. *J. Inst. public Admin. Ireland* 25(2) Special issue, 179-87.

O'Leary D.P. (1977) The management of the River Shannon in the 1980s. *J. Inst. public Admin. Ireland* 25(2) Special issue, 160-80.

Praeger R.L. (1974) *The Botanist in Ireland,* 587 pp. EP Publishing Ltd.

Ramson P.J.G. (1971)  *Holiday Cruising in Ireland.  A Guide to Irish Inland Waterways*.  David and Charles, Newton Abbot, England.

Scannell M.J.P. (1972)  "Algal paper" of *Oedogonium* sp., its occurrence in the Burren, Co. Clare.  *Ir. Nat. J.* 17(5), 147-52.

Sea and Inland Fisheries Reports (1963-1980).  Stationery Office, Dublin.

Southern R. (1935)  The food and growth of brown trout and the River Shannon.  *Proc. R. Ir. Acad.* XXII, 87-151.

Southern R. & Gardiner A.C. (1926)  Reports from the Limnological Laboratory.  I. The seasonal distribution of the Crustacea of the plankton in Lough Derg and the River Shannon.  *Fisheries, Ireland, Sci. Invest.* 1.

Southern R. & Gardiner A.C. (1932)  Reports from the Limnological Laboratory.  II. The diurnal migrations of the Crustacea of the plankton in Lough Derg.  *Proc. R. Ir. Acad.* 40 B, 121-59.

Southern R. & Gardiner A.C. (1938)  Reports from the Limnological Laboratory.  IV. The phytoplankton of the River Shannon and Lough Derg.  *Proc. R. Ir. Acad.* 45 B, 89-129.

Toner P.F., Clabby, K.J. & Bowman J.J. (1973)  *Report on the Results of a Biological and Chemical Survey of the Main River Channel and Lakes of the Shannon System, between Battle Bridge and Killaloe undertaken in July 1972*, 89 pp.  An Foras Forbartha, Water Resources Division.

Webb D.A. (1963)  *An Irish Flora*, 261 pp.  Dundalgan Press Ltd., Dundalk.

Went A.E.J. (1940)  Salmon of the River Shannon.  *J. Dept Agriculture, Dublin* 37, No. 2.

Went A.E.J. (1948)  Irish freshwater fish.  Some notes on their distribution.  *The Salmon and Trout* 118, 9 pp.

# 13 · Caragh

H. HEUFF & K. HORKAN

13.1  Introduction                                        364

13.2  General Data                                         365

13.3  Topography                                           367

13.4  Climate and hydrology                               367

13.5  Geology, soils and vegetation                       369

13.6  Land use and population                             370

13.7  Water chemistry                                     372

   13.71  Introduction                                    372
   13.72  Origin of dissolved salts                       372
   13.73  Seasonal patterns                               372
   13.74  Differences within the catchment                374

13.8  Biota                                               374

   13.81  Algae                                           374
   13.82  Macrophytes                                     379
   13.83  Macroinvertebrates                              380
   13.84  Fish                                            382

13.9  Conclusions                                         382

References                                                382

## 13.1    INTRODUCTION

The Caragh, which drains part of south-west Ireland, is an excellent example of a river system which is relatively unpolluted from headwaters to estuary.  This is unusual in Western Europe and, for precisely this reason, the river was chosen for detailed biological investigation.  The survey, on which most of this chapter is based, took place between October 1974 and February 1977 and was financed by a grant from the National Board for Science and Technology to the Botany and Zoology Departments of University College, Dublin.  The chapter describes the general ecology of this non-calcareous, fast-flowing, clean water system.

*Fig. 13.1*  Caragh River catchment.

## 13.2  GENERAL DATA

### DIMENSIONS

| | |
|---|---|
| Area of catchment | 160 km$^2$ |
| Highest point in catchment | 975 m |
| Length of stretch described (stations 1-6) | 22.4 km |
| Drop in height down this stretch | 104 m (= 4.6 m km$^{-1}$) |

### CLIMATE (Rohan, 1975 and Irish Meteorological service, pers. comm.)

#### Precipitation

Data obtained at Valencia Observatory, the nearest major one to the catchment and Glenvickee inside the catchment.

Mean annual values:

| | | |
|---|---|---|
| Valencia | (1931-1960) | 1398 mm |
| Glenvickee | (1941-1960) | 2364 mm |

#### Temperature

Mean daily value at Valencia (1931-1960)

| | |
|---|---|
| February | 6.8 $^{\circ}$C |
| August | 15.4 $^{\circ}$C |

Mean daily sunshine at Valencia (1931-1960)   3.72 h

#### Relative humidity

mean at Valencia                               82%

### GEOLOGY

Old Red Sandstone (Walsh, 1968)

### LAND USE

Rough grazing (Agricultural statistics, 1975)

### POPULATION (Census of populations 1971, 1979)

1144 people = 7.15 people km$^{-2}$

### PHYSICAL FEATURES OF RIVER

#### Discharge at Caragh Bridge (Data from Electricity Supply Board charts for 1975 and 1976)

mean daily value 7.29 m$^3$ s$^{-1}$ = 45.6 s$^{-1}$ km$^{-2}$
range            0.24-55.1 m$^3$ s$^{-1}$

River regime: highest flows generally in October-February
              lowest flows                April-September
              flash floods                throughout the year

*Fig. 13.2* A headstream on slopes of Caher mountain at about 350 m. This rocky and treeless landscape is typical of the upper catchment.

*Fig. 13.3* Stunted trees cover the sides of streams in a few areas of the upper catchment. The steep-side valley shown lies upstream of the more typical landscape in Fig. 13.2.

## 13.3   TOPOGRAPHY

The Caragh River system occupies glacially deepened valleys on the
northern side of the mountains which form the backbone of the Iveragh
peninsula.   The valley floors are relatively broad and flat and are
flanked by steep slopes rising 200 to 700 m above the floor of the
valley.   The source of the Caragh is on the southern slopes of the
Macgillycuddy Reeks at an altitude of approximately 400 m, where
several rivulets join to form a relatively permanent stream.   This, in
a distance of just over one kilometre, then descends the rocky and
peaty slopes to the valley floor at approximately 100 m.   From here it
first flows westward for 9 km through the Bridia valley and at station
3 (Fig. 13.1) is 7.3 m wide.   The current speed in this stretch is
moderate to fast and the substratum consists mostly of small stones and
gravel.   On leaving the Bridia valley it turns northwards, and is
characterised by steep riffles over bedrock and deep sandy pools (width
at station 4 is 12.5 m).   Just before entering Lough Caragh the river
becomes deep and slow-flowing with a peaty-muddy bottom.   Below Lough
Caragh the river flows westward to enter the sea (width at station 6 is
47 m).   The substratum of this stretch consists of large boulders,
coarse pebbles and some gravel.

It should be noted that the Caragh River study was mainly concerned
with the river on the valley floor.


## 13.4   CLIMATE AND HYDROLOGY

South-west Ireland has a pronounced oceanic climate.   High humidity,
cool summers and mild winters are typical.   The distribution of precipitation is fairly even over the year, but in general winters are
somewhat wetter than summers.   Dry spells (15 or more days without rain)
are extremely rare.   At Valencia Observatory, 30 km WSW of the catchment, evaporation (Irish Meteorological Service, monthly weather
report, Part I) and potential evapotranspiration (Connaughton, 1967:
computed using Penman formula) exceed rainfall for only a short period
during each year, usually in June and July, while actual evapotranspiration is always smaller (Irish Meteorological Service, pers. comm.);
this applied even during the relatively dry years (1975, 1976) when the
Caragh investigation took place.   Strong winds from the south-west
are commonplace.   Snowfall is infrequent and short-lived.   In this type
of climate primary production can take place the whole year around.
The median dates of the beginning and end of the grass growing season
lie, respectively, between the first and fifteenth of February and the
fifteenth of December and the first of January (Atlas of Ireland, 1979).

The Caragh catchment covers 160 km$^2$.   The Caragh river reaches the
sea as a 109th order stream sensu Shreve (1966), using the Irish
Ordnance Survey map, scale 1 : 126720.   The total length of the main
channel is 27.6 km.   Since 1945 the Electricity Supply Board (E.S.B.)
has maintained a water-level gauging station at Caragh Bridge (National

*Fig. 13.4* Rocky cascades in the Caragh just above station 2 at about 60 m altitude.

*Fig. 13.5* Stretch just below boulders in Fig. 13.4, with deep pool in middle ground and pebbly riffle in foreground. All three types of environment can be found for most of the length of the river.

Grid Reference V712 926), downstream from Lough Caragh.  However most
of the data sheets have not been analysed and hence the river regime
cannot be described.  Fig. 13.6 shows daily discharge at Caragh Bridge
(data sheets of E.S.B., analysed by An Foras Forbartha) and the daily
mean rainfall for four weather stations in the catchment (data of Irish
Meteorological Service) for a 6-month period.

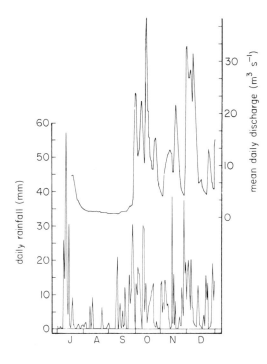

*Fig. 13.6* Daily rain-
fall in the Caragh
catchment and mean daily
discharge at Caragh
Bridge from July to
December 1976.

From this figure it is clear that river discharge shows a rapid
response to rainfall in the catchment.  This is due to the low ground
water storage capacity of the area, the steep slopes (mean slope 170 m
km$^{-1}$) and the generally impermeable nature of the soils.  Flash floods
occur at frequent intervals throughout the year.  For example, the
greatest rise in water level measured during the study period was 1.80 m
in 22 h on the 17/18 October 1974, corresponding to an increase in dis-
charge from 3.8 to 70 m$^3$ s$^{-1}$ (23.8 l s$^{-1}$ km$^{-2}$ to 438 l s$^{-1}$ km$^{-2}$).  The
mean rainfall on the 17 October for four rainfall stations in the catch-
ment was 37.8 mm.  This is 38% of the total rainfall for October of
that year.  The mean total rainfall for the previous 10 days was 20.1 mm
and was spread more or less evenly.

13.5  GEOLOGY, SOILS AND VEGETATION

The rock type in the catchment is Old Red Sandstone with conglomerates
of the Quartz-Jasper type.  Three types of Old Red Sandstone have been

recognised in the area (Walsh, 1968):

(a)    Green Sandstone, with a high content of chloride

(b)    Purple Sandstone, containing iron

(c)    Grey Sandstone, containing relatively little chloride
or iron.

The area was covered during the last Ice Age by a radiating ice cap
which had its centre to the south (Lewis, 1977). Evidence of this
glaciation can be clearly seen from the ice-scouring of the hills and
the presence of cirques and moraines. The glacial material which has
been deposited in the valleys of the catchment is derived from the Old
Red Sandstone and is, therefore, of similar chemical composition.
Although the Old Red Sandstone meets Lower Carboniferous Limestone at
the northern shore of Lough Caragh this has no effect on the ecology of
the catchment since the limestone is deeply covered by till of Old Red
Sandstone origin and nowhere does the limestone reach the surface
(Warren, pers. comm.).

The soils in the catchment are poor. They consist of peaty podzols,
lithosols and climatic peat (Soil Map of Ireland, 1974). These soils
are strongly leached and an iron pan is often present. In the case of
the peaty podzols and lithosols the parent material is Old Red Sandstone, which although a sedimentary rock, is highly resistant to
weathering. It consists mostly of silica and is, therefore, acidic and
a poor supplier of nutrients. The glacial till derived from this rock
is low in calcium; only 0.07 - 0.24% $CaCO_3$ was found in the C-horizon
(Warren, pers. comm., Schotter method). A high percentage of unde-
composed organic material is present in the soils, which are waterlogged for a considerable part of the year. The land use potential of
these soils is low. The vegetation in the area is typical of western
lowland and mountain blanket bog, flush, wet heaths and unmanured
permanent lowland wet meadow: i.e. Sphagnetalia magellanici,
Scheuchzerietea, Parvocaricetea, *Molinia* and *Erica tetralix* heaths and
Molinietalia coeruleae. There is also a small amount of woodland
present, mostly conifer plantations (Vegetation Map, Atlas of Ireland,
1979).

## 13.6   LAND USE AND POPULATION

The population density in the area is low. The majority of the people
at work are engaged in farmwork (Census of populations, 1971, 1979
and Agricultural Statistics, 1975). Settlements are dispersed and
consist of small farms with small irregular fields (pasture, hay and
crops) and largely unenclosed peat and upland areas. The main arable
crops are potatoes and oats. Fertiliser input seems to be low,
although detailed figures are not available (agricultural adviser,
pers. comm.).

*Table 13.1* Comparison for (a) land use, and (b) livestock density of Caragh catchment with that of the whole of Ireland (based on Agricultural Statistics, 1975).

| (a)  land use | % of total land area | |
|---|---|---|
| | Caragh | Ireland |
| rough grazing | 59.1 | 14.9 |
| pasture | 14.4 | 45.9 |
| meadow | 3.7 | 15.5 |
| ploughed area | 0.6 | 6.7 |
| woods, plantation | 4.3 | 4.3 |
| other land | 17.9 | 12.7 |
| (b)  livestock | density (animals per ha) | |
| | Caragh | Ireland |
| cattle | 0.26 | 1.03 |
| sheep | 0.81 | 0.53 |
| poultry | 0.18 | 1.37 |
| horses | 0.0054 | 0.013 |
| donkeys | 0.0061 | 0.0054 |
| pigs | 0.0013 | 0.11 |

Table 13.1 gives details of the agricultural land use and demonstrates the importance of grazing and sheep in the economy of the area. There are no towns or villages in the catchment. 52.8% of permanent housing units have no piped water supply of any kind (inside or outside, piped or mains), 69% have no toilet (flush, chemical or dry) (Census of Populations, 1971). From the above it is clear that nutrient input from human activities in the catchment must be on a very small scale.

In Ireland land units used in the results from the more recent agricultural statistics do not follow catchment boundaries. The above figures have been calculated from the 1975 agricultural census for an area approximately twice the size of the catchment, but this larger area can be considered to have a similar composition to that of the catchment itself. However, if a precise nutrient budget for the river were to be calculated it would be of considerable use if statistical information were made available in land area units that follow natural boundaries, as townlands often do in Ireland.

## 13.7   WATER CHEMISTRY

### 13.71   INTRODUCTION

The chemistry of the water was investigated at ten sampling stations
in the catchment at approximately monthly intervals from December 1974
to February 1977, with six stations along the main channel and four on
the tributaries (Norton *et al.*, 1978).          The water in the Caragh
river is low in electrolytes, extremely soft, slightly acidic, low in
nitrate and moderately poor in phosphate.  The major ions present are
chloride and sodium.  The water is clear and low in suspended solids,
even during floods.  Table 13.2 summarizes both data on water chemistry
and also rainwater and air from Valencia Observatory for 1975 and 1976
(Monthly Weather Report, Part I).  An unexplained imbalance exists
between the sum of cations and the sum of anions (0.51 and 0.81 meq $l^{-1}$
respectively) in the data collected on Caragh water chemistry.  Care
has been taken not to draw any conclusions from the data that require
absolute figures, but it was felt that comparisons could still be made
if it is assumed that systematic errors are involved.

### 13.72   ORIGIN OF DISSOLVED SALTS

In the Caragh catchment, the resistant rock, peaty soils, low human
population and marginal agriculture are all poor sources of mineral
ions.  Comparison of data for rain and river water (Table 13.2),
especially those for chloride, sodium and potassium, indicates that
rain probably plays an important role as a supplier of electrolytes
into the system.  The high proportion of chloride and sodium compared to
other ions indicates the marine origin of the former.  In addition to
the rainfall a further supply of marine derived ions could be carried
in by the air on the prevailing south-westerly winds, as was shown
for Connemara (Groenendael *et al.*, 1979).

### 13.73   SEASONAL PATTERNS

Seasonal patterns of chemistry are influenced strongly by the fact
that the river is subject to high flow conditions frequently and at
irregular intervals throughout the year.  Mean values of conductivity,
total hardness, calcium hardness, magnesium and calcium are significantly and inversely correlated with discharge at Caragh Bridge.
This indicates that there is a constant input (Golterman, 1975) of
magnesium and calcium, probably originating mainly from rock and till,
diluted at different times to varying degrees by the rain.  No significant correlation was found between discharge and chloride, which is
to be expected as most of the chloride is of marine origin and
supplied by the rain.

Table 13.2 Chemistry of Caragh River Water (Norton et al., 1978), precipitation and air at Valencia Observatory (Irish Meteorological Service, Weather Report, Part I). (Data on Si from An Foras Forbartha; data on $SO_4$-S, Cu, Zn, Cd and Pb from Dowling et al., 1981)

| parameters measured monthly | units | Caragh Dec. 1974–Feb. 1977 | | rainfall 1975–1976 | | air 1975–1976 | |
|---|---|---|---|---|---|---|---|
| | | $\bar{x}$ | range | $\bar{x}$ | range | $\bar{x}$ | range |
| conductivity | $\mu S\ cm^{-1}$ | 70 | 47 – 123 | 70.4 | 27 – 208 | | |
| pH | | | 5.8 – 7.6 | | 4.6 – 6.8 | | |
| alkalinity | $meq\ l^{-1}\ HCO_3$ | 0.25 | 0.05 – 0.90 | 0.14 | 0.01 – 1.04 | | |
| calcium hardness | $mg\ l^{-1}\ CaCO_3$ | 8.2 | 2 – 22 | | | | |
| total hardness | $mg\ l^{-1}\ CaCO_3$ | 16.8 | 4 – 38 | | | | |
| Cl | $mg\ l^{-1}$ | 15.2 | 6.0 – 28.0 | 13.0 | 3.1 – 51.3 | 15.8 | 3.7 – 29.4 |
| Na | " | 6.1 | 0.8 – 11.3 | 8.5 | 2.3 – 31.9 | 7.4 | 2.3 – 31.2 |
| Ca | " | 2.15 | 0.5 – 6.0 | 0.6 | 0.30 – 1.30 | 1.5 | 0.4 – 6.2 |
| Mg | " | 1.37 | 0.15 – 5.25 | 0.9 | 0.10 – 3.60 | | |
| K | " | 0.5 | trace – 3.5 | 0.5 | 0.14 – 1.30 | 1.11 | 0.43 – 4.4 |
| $PO_4$-P | " | 0.011 | 0.001 – 0.045 | 0.036 | 0.01 – 0.80 | | |
| $NO_3$-N | " | 0.050 | 0.011 – 0.200 | 0.14 | 0.01 – 0.68 | 1.83 | 0.70 – 5.4 |
| $NH_4$-N | " | 0.03 | trace – 0.10 | 1.34 | 0.61 – 2.8 | 7.1 | 4.8 – 9.6 |
| $SO_4$-S | | | | | | | |

Parameters measured occasionally in Caragh River

| | $SiO_2$-Si $mg\ l^{-1}$ | | $SO_4$-S $mg\ l^{-1}$ | | Cu $\mu g\ l^{-1}$ | Zn $\mu g\ l^{-1}$ | Cd $\mu g\ l^{-1}$ | Pb $\mu g\ l^{-1}$ |
|---|---|---|---|---|---|---|---|---|
| | $\bar{x}$ | range | $\bar{x}$ | range | $\bar{x}$ | $\bar{x}$ | $\bar{x}$ | $\bar{x}$ |
| | 0.3 | trace – 1.0 | 6 | 5 – 7 | 6 | 20.3 | 0.5 | 3 |

## 13.74  DIFFERENCES WITHIN THE CATCHMENT

In general in the Caragh, contrary to the more usual situation in which
the concentration of dissolved salts increases downstream, pH, alkalinity, total hardness, calcium hardness, magnesium and calcium actually
decrease between stations 2 and 6.  Against this general trend and in
agreement with the usual situation there is a significant increase for
calcium and calcium hardness in the Bridia Valley (between stations 1
and 2).  The decrease in concentration of these ions in the main channel
below station 2 is due to dilution by the two major tributaries, the
Owenroe and Acoose rivers, whose catchments contain a larger proportion
of upland peatland than the Bridia Valley.  Both the Owenroe and Acoose
flow through lakes, where, physico-chemical and biological processes
may perhaps favour a further decrease.  Likewise, the decrease in concentration which occurs between stations 5 and 6 is probably due to
Lough Caragh.

The pattern for chloride is different:  a general increase is seen
with proximity to the sea and exposure to westerly winds.  The Owbeg
River is significantly richer in chloride than all other sampling
sites except station 4.  This subcatchment is the nearest to the sea,
separated from it only by a relatively low saddle (365 m).  The
Meelagh River is significantly higher in chloride than the Owenroe and
Acoose and significantly lower than the Owbeg.  The Owenroe and Acoose
rivers, both relatively low in chloride, are furthest away from the
sea and the Owenroe catchment is also sheltered by a ridge of relatively high hills (600 m).

## 13.8  BIOTA

### 13.81  ALGAE

The Caragh is rich in species of algae, mainly blue-greens, greens and
diatoms:  approximately 185 taxa have been identified.  Table 13.3
lists dominant and abundant taxa as well as some rarer species.  In
general the diatoms are quantitatively the most important group, the
major genera present being *Achnanthes*, *Eunotia* and *Gomphonema*.  Blue-
green algae are common, including several heterocystous forms (and
thus presumed nitrogen-fixers).  Most algal species can be found
throughout the year although their relative abundance varies; for
example, the filamentous green are most conspicuous in summer and
autumn.  The standing crop of algae varies considerably between
different sites, being greater in shallow areas with fast currents
and  a fairly stable substratum than in pool areas with lower current
speeds.  A crude measure of algal biomass was obtained by scrubbing
a 1 $m^2$ area and measuring the volume of wet, preserved algae sedimented in a measuring cylinder.  Values ranged at station 1 from 815
ml $m^{-2}$ in April to 10 ml $m^{-2}$ in November.  Fig. 13.4 shows mean values
of biomass from 5 stations (1, 2, 3, 4, 6), all riffle areas, and
discharge (E.S.B. data, analysed by Heuff) from December 1974 to
November 1975.  Peak biomass was found during low flows, but does not
seem to be related solely to discharge, as low values of biomass

*Table 13.3* Some algal taxa in the Caragh River.

| abundant or common | other |
|---|---|
| **blue-greens** | |
| *Chaemaesiphon confervicola* | *Aphanocapsa* spp. |
| *C. incrustans* | *Aphanothece* spp. |
| *C. minutus* | *Calothrix braunii* |
| *C. polonicus* | *C. parietina* |
| *Dichothrix baueriana* | *Chroococcus* sp. |
| *D. orsiniana* | *Coelosphaerium* sp. |
| *Hydrococcus cesatii* | *Desmosiphon maculans* |
| *H. rivularis* | *Gloeocapsa sanguinea* |
| *Lyngbya aestuarii* | *Homoeothrix* sp. |
| *L. epiphytica* | *Lyngbya* spp. |
| *L. kuetzingii* | *Merismopedia* sp. |
| *Nostoc* spp. | *Oscillatoria* spp. |
| *Phormidium retzii* | *Phormidium* spp. |
| *Stigonema mamillosum* | *Plectonema* sp. |
| *Tolypothrix penicillata* | *Schizothrix* sp. |
| **reds** | |
| *Lemanea fluviatilis* | *Batrachospermum atrum* |
| *Rhodochorton violaceum* | *B. moniliforme* |
| **diatoms** | |
| *Achnanthes* spp. | *Achnanthes flexella* |
| *A. minutissima* var. *cryptocephala* | *A. minutissima* |
| *A. saxonica* | *Amphora ovalis* |
| *A. exilis* | *Anomoeoneis exilis* |
| *A. affinis* | *Cocconeis placentula* |
| *Ceratoneis arcus* | *Cyclotella* spp. |
| *Cymbella* spp. | *Diploneis ovalis* |
| *Diatoma* spp. | *Epithemia zebra* var. *saxonica* |
| *Denticula* sp. | *Melosira* sp. |
| *Didymosphenia geminata* | *Meridion circulare* |
| *Eunotia* spp. | *M. circulare* var. *constrictum* |
| *Fragilaria vaucheriae* | *Navicula gregaria* |
| *Frustulia rhomboides* | *Nitzschia* spp. |
| *F. rhomboides* var. *saxonica* f. *undulata* | *N. palea* |
| *Gomphonema acuminatum* | *Pinnularia* spp. |
| *G. acuminatum* var. *coronata* | *P. borealis* |
| *G. constrictum* | *Rhopalodia gibberula* |
| *G. olivaceoides* | *Stauroneis phoenicenteron* |
| *G. parvulum* | *S. anceps* |
| *Tabellaria flocculosa* | *Stenopterobia intermedia* |
| *T. fenestrata* | *Surirella* spp. |
| *Navicula* spp. | |

cont'd.

*Table 13.3* continued ..

| abundant or common | other |
|---|---|
| **greens** | |
| *Bulbochaete* spp. | *Aphanochaete repens* |
| *Coleochaete scutata* | *Botryococcus braunii* |
| *C. orbicularis* | *Chaetosphaeridium globosum* |
| *Gongrosira fluminensis* | Desmidiales (12 genera) |
| *Microspora amoena* | *Dictyosphaerium pulchellum* |
| *Mougeotia* spp. | *Draparnaldia* sp. |
| *Oedogonium* spp. | *Hormidium* cf. *rivulare* |
| *Rhodoplax schinzii* | *Microspora tumidula* |
| *Spirogyra* spp. | *M. wittrockii* |
| *Ulothrix zonata* | *Schizochlamys gelatinosa* |
| *Zygnema* sp. (18 - 26 μm) | |

*Fig. 13.7* Mean daily discharge at Caragh Bridge and mean algal biomass for 5 sampling stations from December 1974 to November 1975.

occurred in June and July, yet relatively high levels of biomass still occurred in November, after extensive flooding had occurred.  It is possible that extremely low flows and high temperatures during the exceptionally dry summer of 1975 may have limited growth after the spring peak.  That high biomass still occurred in November must mean that the species present are well adapted to high flow conditions.  In the upper reaches of the river (stations 1 and 2) peak biomass was due mainly to diatoms.  In the middle reaches it was due mainly to, in spring, *Ulothix zonata* (station 3) and *Microspora amoena* (station 4) and in August to co-dominance of diatoms and green filamentous forms (mainly *Oedogonium*).  Below Lough Caragh (station 6) the spring maximum was due to a mixture of diatoms, *Lemanea fluviatilis* and *Rhodochorton violaceum*, while the August maximum was mainly due to *Zygnema* sp. Several species in the Caragh form conspicuous macroscopic growths on stones and bedrock, others are common epiphytes.  The macroscopic growths can be divided into five groups according to their growth form:

(a)  Mixtures of encrusting species, closely and firmly attached to the rock.

(b)  Sheets of gliding filaments, without any obvious means of attachment (e.g. *Phormidium*).

(c)  Thick felt-like masses, attached to rocks and stones (e.g. *Didymosphenia geminata*, *Gomphonema* spp.).

(d)  Long trailing filaments, attached by the base to rocks, stones and plants (e.g. green filamentous forms and *Lemanea fluviatilis*).

(e)  Short isolated tufts (e.g. *Tolypothrix penicillata*, *Dichothrix* spp.).

Encrusting species cover rock, stones and gravel and are well adapted to fast currents, some even surviving unstable substrata.  For example, mixtures of the following encrusting species are commonly found in the Caragh: *Gongrosira fluminensis*, *Coleochaete* spp., *Stigeoclonium* bases and others, creeping Chaetophorales sheets and the lichen *Verrucaria* cf. *rivularis*.  *Rhodoplax schinzii*, a green alga containing a brilliant scarlet pigment, is well represented on water-falls.  Unlike the Upper Rhine (Jaag, 1932) and Tweed (Holmes & Whitton, 1975) the crusts are not conspicuous enough to make the rock crimson and the alga not obvious macroscopically.  This species has not been found before in Ireland, but because it is so inconspicuous may well have been overlooked previously.  Another encrusting species is *Desmosiphon maculans*, a genus new to Ireland and Britain.  It forms small, hard, dark green spots on stones in fast-flowing water and as it was found only near a bridge it may prefer shady conditions.

*Phormidium retzii* forms strong, dark green sheets on stones and plants.  It can be co-dominant with the diatoms in early spring (station 4) and is a striking feature in the river.

*Didymosphenia geminata* forms thick felt-like masses on rocks in

riffles.  The rock is covered with a layer of *Didymosphenia* attached by
their stalks, while these individuals have other *Didymosphenia*'s
growing epiphytically on them.  In this way a dense structure is built
up which also supports a community of other, smaller epiphytes, e.g.
species of *Gomphonema, Achnanthes* and *Eunotia* spp.; most of these are
also attached by stalks.  These felt-like masses usually contain a
population of invertebrates, mainly chironomid larvae.  A considerable
amount of fine sand is caught within the structure.  It seems possible
that these diatoms may obtain silica directly from the sand, rather
than from the water (see Patrick, 1977), which is relatively poor in
silica (mean 0.3 mg $l^{-1}$ Si, range trace - 1.0 mg $l^{-1}$ Si).

Long trailing filaments become obvious in the river during periods
of low flow.  These consist of green algae, e.g. *Spirogyra* spp.,
*Oedogonium* spp., *Mougeotia* spp., *Zygnema* sp. (18 - 26 μm).  However,
these algae are widespread and moderately abundant in the river the
whole year around, especially in areas with lower current speeds and
in pools.  Few of these taxa can be identified to species level since
no fruiting material has been found.  It is estimated that the 10
genera involved probably represent 24 species,  using criteria like
width of filament, chloroplast characteristics etc.

An interesting zonation is present on riffles with a lot of bedrock
and large boulders (e.g. station 5).  *Stigonema mamillosum* covers
rocks just above the average water level.  In its dry condition it has
the appearance of short, coarse, dark grey hair.  Below this zone,
usually just submerged, colonies of *Nostoc* a few millimetres in diameter cover the rock.  Below the *Nostoc* zone *Lemanea fluviatilis* grows
abundantly on the river bed, at least in spring and autumn.  Only a
few differences were found between stations based on presence or
absence of species.  For example, *Batrachospermum* spp. although not
abundant anywhere, were absent from the shallower stations (1, 2) and
from station 5, which is a cascade.  *Microspora amoena*, present at
most stations, was absent from station 1 and only occasionally present
at station 6.  Some taxa were only found downstream of Lough Caragh,
e.g. *Oedogonium undulatum, Chaetosphaeridium globosum, Coelosphaerium,*
while *Stigeoclonium* sp. was more common here than elsewhere.  *Nitella
opaca* was found only at station 3, growing in deep water  on a sandy
bottom.

The composition of the algal flora is, as might be expected, indicative of the oligotrophic nature of the Caragh system.  Species
present which are characteristic of slightly acidic water with low
salt concentrations (Patrick, 1977) are *Eunotia* spp., *Frustulia
rhomboides, Pinnularia* spp. and *Tabellaria flocculosa*.  Some are
calciphobe and can tolerate high levels of iron, e.g. *Eunotia* spp.,
*Pinnularia* spp., *Frustulia, Gomphonema acuminatum, Stauroneis
phoenicenteron* (Patrick, 1977).  Twelve genera of desmids have been
recorded in the river.  These, while never abundant, are widespread.
Many live loosely attached amongst other algae or are washed out from
surrounding bogs.  The presence of *Ulothrix zonata, Microspora amoena*

and *M. tumidula* is thought to indicate oxygen rich water (Printz, 1964).
Members of the genus *Zygnema*, with a cell width less than 30 µm are
restricted to streams with water poor in electrolytes, at least in
Scandinavia (Israelson, 1949).   In the Caragh a *Zygnema* sp. (18 - 26 µm)
is common throughout the river.   It is interesting to note that *Cladophora*, which is common elsewhere in Ireland, is absent from this river,
as the freshwater members of this genus are generally considered to be
characteristic of eutrophic conditions (Van den Hoek, 1963).

In recent times few algal studies have been carried out in oligotrophic calcium-poor rivers.   In Ireland the only case reported is a
headwater stream of the Inver system at Recess (National Grid Reference
L867 459) (McDonnell & Fahy, 1978).   When sampled in July 1976, this
stream was dominated by *Batrachospermum* spp. with the filamentous green
algae *Mougeotia*, *Spirogyra* and *Zygnema* next in abundance.   Israelson
(1949) classifies Scandinavian rivers into two major types:   "*Zygnema*
type" and "*Vaucheria* type", corresponding respectively to oligotrophic
and eutrophic waters.   The algal vegetation in the Caragh is comparable
to oligotrophic streams in Scandinavia ("*Zygnema* type").   Several of
the differential species for streams poor in minerals mentioned by
Israelson are also present in the Caragh:   *Mougeotia*, *Spirogyra*,
*Zygnema*, *Bulbochaete*, *Stigonema mamillosum*, *Batrachospermum moniliforme*,
*Schizochlamys gelatinosa*.   Israelson distinguishes two sub-types within
the *Zygnema* type:   brown water streams with much *Batrachospermum* and
clear water streams with many encrusting forms.   The Caragh would fit
in with the latter sub-type while the Inver stream would presumably fit
the first sub-type.   The Caragh also shows similarities to the Kaltisjokk, an oligotrophic stream in Sweden (Klasvik, 1974) in which
*Eunotia* spp. are  mentioned as abundant.   Klasvik found that *Achnanthes
minutissima* var. *cryptocephala* is much more abundant than *A. minutissima*,
as is also the case in the Caragh.   Round (1959) noted that the diatom
flora of the upper sediment layers of acid lakes in Ireland is similar
to  the flora of oligotrophic Scandinavian lakes.   As many of the
species mentioned by Round have also been recorded from the Caragh,
this is further evidence that the biology of the Caragh is probably
similar to Scandinavian oligotrophic waters.   Other streams comparable
to the Caragh are described by Douglas (1958) and Whitford &
Schumacher (1963).

## 13.82   MACROPHYTES

The overall cover of macrophytes in the river is extremely low and
they are entirely absent from many fast-flowing stretches.   51 species
of vascular plant and 15 species of bryophyte were recorded.   The
vascular plants can be divided into three groups based on the habitat
they occupy:

(a)   Gravel beds which are exposed during periods of low flow;

(b)   Shallow sandy areas near the river margin;

(c)   Deeper water with moderate flow.

On the gravel banks *Oenanthe crocata*, *Apium inundatum*, *Juncus bulbosus*
f. *fluitans* and *Callitriche* spp. are common and can form a dense cover.
As these plants trap sand and silt, banks may build up until swept
away by flash floods.  Shallow sandy areas near the shore, especially
in the lower stretches of the river, are sparsely vegetated by plants
such as *Eriocaulon aquaticum*, *Lobelia dortmanna*, *Littorella uniflora*
and *Elatine hexandra*.  In deeper water in the upper reaches of the
river *Scirpus fluitans* occurs, while in the middle and lower reaches
*Juncus bulbosus* f. *fluitans* and *Myriophyllum alterniflorum* are common,
accompanied very occasionally by a batrachian *Ranunculus* species.
Marsh and floating leaf zones are absent except in a few slow-flowing
stretches between station 4 and Lough Caragh, where occasional plants
of *Nuphar lutea* occur and a narrow fringe of *Phragmites australis*,
*Equisetum fluviatile* and *Carex rostrata* is developed.

Bryophytes are not an important component of the aquatic vegetation,
except in areas of bedrock, where the major species are *Fontinalis
antipyretica*, *Fissidens* sp., *Grimmia alpicola* var. *rivularis* and
*Cinclidotus fontinaloides*.  Luxurious growths of bryophytes occur on
big boulders in the river which project above the average water-level.

The species composition of the macrophytes is similar to that given
by Haslam (1978) as typical of mountain and moorland streams in S.W.
England and Scotland.  Of the trophic level indicators cited by Haslam
only those species characteristic of dystrophic and oligotrophic
conditions occur in the Caragh.

## 13.83  MACROINVERTEBRATES

Macroinvertebrates from riffle areas (stations 1, 2, 3, 6) were sampled
quantitatively on four occasions (December 1974, February, August,
December 1975).  Numerous qualitative samples were taken throughout
the survey.  Over 200 taxa were identified including 160 species
(Dowling *et al.*, 1981).  The more important observations are outlined
here.

Sponges were confined to station 6 below Caragh Lake.  Low numbers
of triclads (*Phagocata vitta*) occurred at stations 1 - 3 and 6.  Five
oligochaete families were identified, the most important species being
*Lumbriculus variegatus* and *Stylodrilus heringianus*, which were widespread.  Hirudinea were found only in low numbers.  *Gammarus duebeni*
occurred throughout the system, being most common below the lake.
*Asellus aquaticus* occurred only in silted pools near stations 4 and 5.
There were eight species of Plecoptera;  of these, *Leuctra inermis*
was most plentiful at station 1, being replaced by *Chloroperla
torrentium* and *Perla bipunctata* at station 2.  *Isoperla grammatica*
and *Leuctra hippopus* were common downstream of Caragh Lake.  Fourteen
species of Ephemeroptera were identified and of these *Baetis rhodani*

was the most important with largest numbers occurring at station 1.
Other important mayflies included *Rhithrogena semicolorata* (upper),
*Siphlonurus lacustris* (peaty pools), *Ecdyonurus torrentis* (upper),
*E. venosus* (all stations), *Caenis moesta* and *C. rivulorum* (below Caragh
Lake). Only one odonate species was found (*Coenagrion puella*, station
5). Hemiptera consisted of *Aphelocheirus montandoni* (station 6) and
small numbers of other species including *Gerris najas*.

Trichoptera are well represented, especially below the lake. The
combined observations of O'Connor (1975) and Dowling *et al.* (1978) give
a total of 43 species. The latter survey reported that the following
were most important in the headwaters: *Agapetus fuscipes*, *Diplectrona
felix*, *Philopotamus montanus*, *Tinodes machlachlani*, *Wormaldia* sp.,
*Hydropsyche siltalai*. The last was the most common net-spinning form
in the upper stretch. Trichopteran numbers were highest below the lake
where *Neureclipsis bimaculata*, *Hydroptila* spp., *H. pellucida*,
*Cheumatopsyche lepida*, *Ithytrichia lamellaris*, *Psychomyia pusilla*,
*Polycentropus flavomaculatus*, *P. kingi* and *Chimarrha marginata* were all
well represented.

Of the six species of Coleoptera recorded *Esolus parallelipipedus*
was dominant, while *Limnius volckmari*, *Oulimnius tuberculatus* and
*Elmis aenea* were less important. The Chironomidae were investigated
thoroughly and yielded 48 species. Two genera were numerically dominant, *Cricotopus* and *Eukiefferiella*, both being well represented at all
stations. Simuliidae were relatively important in the middle reaches.
Of the other Diptera only *Dicranota* and *Ephydra* occurred regularly with
highest numbers of both at station 2. Hydracarina were encountered
everywhere but did not represent a high proportion of the biomass. Of
the six species recorded, one, *Pergamasus brevicornis*, is new to the
British Isles and is not listed by Gledhill & Viets (1976). Mollusca
were poorly represented presumably because the low ionic content of
this river system. Of seven genera recorded only *Pisidium/Sphaerium*
spp. and *Ancylus fluviatilis* were ever found in numbers; they were most
common below the lake.

It is clear that the Caragh River has a wide diversity of species at
all stations, including a good representation of the most sensitive
forms. Such results are indicative of a clean unpolluted river along
the entire stretch examined and would be represented as Q5 (i.e. good
quality) by the Irish classification method (Flanagan & Toner, 1972).
The specific differences found at station 6 are caused by a lacustrine
influence carrying organic matter composed of detritus, phyto- and zooplankton from Caragh Lake and do not show any signs of a reduction in
water quality. Other differences between stations in the system are
attributed to flash flooding, which may occur in any season, resulting
in movement of the substratum and a paucity of suitable food. Evidence
in support of this is the fact that at station 1 the fauna is poorer
than elsewhere being located just downstream of a cascade. The further
downstream the organisms are located in this system the wider the river
and the more capable it is of dealing with sudden floods without
denuding the fauna.

## 13.84  FISH

The Caragh is renowned as a salmonid river, with a plentiful supply of
both salmon and sea-trout.  Brown trout are common also but they are
usually small except for a few larger specimens in the lake (E. Evans,
pers. comm.).  However, commercial fishing, including some illegal
netting may be actually reducing stocks migrating upstream to spawn.
Apparently rod and line catches had been in decline in recent years
until an improvement occurred in 1980.  The latter is mainly attributed
to a successful run of salmon during a very wet winter.

## 13.9  CONCLUSIONS

The only old biological records available for the river system concern
Lough Caragh.  Plankton was collected and described from the lake in
1906 (West & West), and a list of some macrophytes was given by Praeger
(1934).  Recent surveys of plankton and vascular plants in Lough Caragh
(Flanagan & Toner, 1975, and H. Heuff, unpublished data) have found the
lake to be practically unchanged.  In general it appears that the
Caragh system is not only relatively unpolluted from headwaters to the
estuary, but also has apparently remained hardly changed for three-
quarters of a century.  Because of this special efforts should be made
to preserve the present situation and where possible future developments of intensive farming, industry or excessive afforestation should
be avoided.

## REFERENCES

Agricultural Statistics (1975)  Central Statistics Office, Dublin,
    Ireland.
Atlas of Ireland (1979)  Royal Irish Academy, I.S.B.N. 0 901714 13 5
Census of Population (1971, 1979)  Central Statistics Office, Dublin,
    Ireland.
Connaughton M.J. (1967)  Global Solar Radiation, Potential Evapo-
    transpiration and Potential Water Deficit in Ireland, Irish Metero-
    logical Service, Technical Note No. 31.
Douglas B. (1958)  The ecology of the attached diatoms and other algae
    in a small stony stream. *J. Ecol.* <u>46</u>, 295-322.
Dowling C., O'Connor M., O'Grady M.F. & Clynes E. (1981)  A base-line
    survey of the Caragh, an unpolluted river in South-West Ireland:
    topography and water chemistry. *J. Life Sci. R. Dubl. Soc.* <u>2</u>(2),
    137-45.
Dowling C., O'Connor J.P. & O'Grady M.F. (1981)  A base-line survey of
    the Caragh, an unpolluted river in South-West Ireland:  observations
    on macroinvertebrates. *J. Life Sci. R. Dubl. Soc.* <u>2</u>(2), 147-59.
Dowling C., O'Connor J.P. & O'Grady M.F. (1978)  The Caragh River
    Survey 1974-1977, Final Report, section 3: Studies on the macro-
    invertebrates of the River Caragh.  National Science Council of
    Ireland.

Flanagan P.J. & Toner P.F. (1975) A preliminary survey of Irish lakes, An Foras Forbartha, Water Resources Division.

Gledhill T. & Vietz V.O. (1976) A synonymic and bibliographic check-list of the freshwater mites (Hydrachnellae and Limnohalacaridae, Acari) recorded from Great Britain and Ireland. *Sci. Publs Freshwater Biol. Ass. U.K.* 1.

Golterman H.L. (1975) Chemistry. In: Whitton B.A. (Ed.) *River Ecology* pp. 39-80. Blackwell, Oxford.

Groenendael J.M. van, Hochstenbach S.M.H., Mansfeld M.J.M. van & Roozen A.J.M. (1979) The influence of the sea and of parent material on wetlands and blanket bog in West Connemara, Ireland. Thesis, Catholic University Nijmegen, Netherlands.

Haslam S.M. (1978) *River Plants*. Cambridge U. P., Cambridge.

Hoek C. van den (1963) *Revision of the European species of* Cladophora. E.J. Brill, Leiden, Netherlands.

Holmes N.T.H. & Whitton B.A. (1975) Notes on some macroscopic algae new or seldom recorded for Britain. *Vasculum* 60, 47-55.

Israelson G. (1949) On some attached Zygnemales and their significance in classifying streams. *Bot. Notiser* 4, 313-58.

Jaag O. (1938) Die Kryptogamenflora des Rheinfalls und des Hochrheins von Stein bis Eglisau. *Mitt. Naturgesellschaft Schaffhausen* 14, 1-158.

Klasvik B. (1974) *Computerised Analysis of Stream Algae*. Vaxtekol. Stud. 5, Uppsala.

Lewis C.A. (1977) *South and South West Ireland, Guidebook for Excursion A15*. International Union for Quaternary Research. I.S.B.N. 0 902246 83 6.

McDonnell J.K. & Fahy E. (1978) An annotated list of the algae comprising a lotic community in Western Ireland. *Ir. Nat. J.* 19(6), 184-7.

Monthly Weather Reports (1975, 1976) Part I, general weather report. Irish Meterological Service, Dublin.

Norton M.A., Dowling C., O'Grady M.J. & Clynes E. (1978) The Caragh River Survey, 1974-1977, Final Report, Section 2: Physico-chemical Investigation Appendix 1, physico-chemical results. National Science Council of Ireland.

O'Connor P.J. (1975) Freshwater studies part II: an investigation of the trichopterous fauna and its distribution in various habitats in Ireland. Ph.D. Thesis, University College, Dublin.

Patrick R. (1977) Ecology of freshwater diatoms and diatom communities. In: Werner D. (Ed.) *The Biology of Diatoms*. Blackwell, Oxford.

Praeger R.L. (1934) *The Botanist in Ireland*. Hodges Figgis, Dublin

Printz H. (1964) *Die Chaetophoralen der Binnengewasser, Hydrobiologia* 24, 1-376.

Rohan P.K. (1975) *The Climate of Ireland*. Stationery Office, Dublin.

Round F.E. (1959) A comparative survey of the epipelic diatom flora of some Irish loughs. *Proc. R. Ir. Acad.* 60 (B,5), 193-215.

Shreve R.L. (1966) Statistical law of stream numbers. *J. Geol.* 74, 17-37.

Ordnance Survey of Ireland (1974) Soil Map of Ireland.

Walsh P.T. (1968) The Old Red Sandstone West of Killarney, Co. Kerry, Ireland. *Proc. R. Ir. Acad.* 66 (B,2), 9-26.

West W. & West G.S. (1906)  A comparative study of the plankton of some
    Irish lakes.  *Trans. R. Ir. Acad.* 23 (B,2), 77-116.
Whitford L.A. & Schumacher G.J. (1963)  Communities of algae in North
    Carolina streams and their seasonal relations.  *Hydrobiologia* 22,
    133-96.
Whittow J.B. (1974)  *Geology and Scenery in Ireland*.  Penguin Books.

# 14 · Suir

K. HORKAN

14.1   Introduction                                      386

14.2   General data                                      389

14.3   Physico-chemical features                         390

    14.31   Hydrology and water resources         390
    14.32   Water abstraction                     391
    14.33   Water quality                         391

14.4   Macrophytes and algae                             393

14.5   Macroinvertebrates                                395

14.6   Fisheries                                         396

References                                               400

## 14.1  INTRODUCTION

The River Suir has been chosen as the third representative of Irish
rivers, as it provides a useful contrast with the River Caragh
(Chapter 13).  Although it rises in the mountains and the quality of
the water in its upper stretches is excellent, the downstream part is
much more nutrient-rich, with local problems due to massive growths of
macrophytes.  The chapter attempts to give an overall picture of
current knowledge of the river.

The ecology of the Suir and its various tributaries is much influenced by the geology of the catchment.  Like other rivers in Ireland
its ancestral course was formed about 60 million years ago (Nevill,
1972) during the emergence of Europe and Asia from the Chalk Sea.
Rainfall resulted in erosion of the new land surfaces, washing silt
down the slope seawards, thus creating channels within which the first
river valleys developed.  Many rivers in Ireland, including the Suir,
originally flowed from north to south, following the gradient of the
chalk surface, but as erosion progressed deeper into the valley floor,
the waters reached harder layers such as shales.  In the case of the
Suir, this resulted in the course of the river altering to an east-
west direction wearing through synclines before turning south again
near the sea (Fig. 14.1).

The Suir rises on Borrisnoe Mountain, just north of the well-known
landmark known as the Devil's Bit, in an area of Silurian deposits.
The river falls rapidly through a narrow band of Old Red Sandstone
and from there onto Lower Carboniferous Limestone, which is the main
geological feature of the catchment.  The gradient is quite steep at
first, falling 320 m in the first 9 km, but then becomes more gradual,
descending only 60 m over the next 58 km to Clonmel.  East of Clonmel
the river skirts along the Comeragh Mountains, where the area adjacent
to the south bank consists mainly of Avonian Shales.  Near the estuary
the rocks are a mixture of Intrusive and Volcanic silica-poor rocks.

Some of the tributaries such as the Drish and Rossestown Rivers
originate in Upper Carboniferous limestone descending from there through
Lower Carboniferous Limestone.  Two other important tributaries, the
Clodiagh and Multeen Rivers, flow over Silurian rocks through narrow
bands of Old Red Sandstone and Lower Avonian Shales and Sandstones
before continuing onto the Lower Carboniferous Limestone.

Near the headwaters there are acid brown earths with peaty podzols
below these.  Much of the Suir catchment consists of flat to undulating
lowlands which are mostly grey-brown podzolics and associated gleys.

The moderate, moist climate, lacking extremes, renders the country
suitable for farming.  Apart from the hilly headwaters, the Suir catchment includes very rich grasslands with beef cattle and dairy farming
and also some cereals.

*Fig. 14.1* River Suir catchment, showing upper tributaries and general geology. The Initial Suir is also shown flowing southwards before it eventually changed to its present course (see text).

Because of the rich farming hinterland in the Suir Valley there are several important provincial towns. Agriculture related industries are to be found in these towns and include sugar refining, food processing, meat and protein production, creameries, ice cream manufacture and brewing. The historic city of Waterford lies on the estuary below the confluence of the three "sister" rivers, the Barrow, Nore and Suir. Industries to be found here include glassware, chemicals and shipping.

Many parts of the Suir Valley are popular with motoring tourists, but the main attraction is salmonid angling. Commercial salmon fishing occurs along the lower reaches of the River Suir, where best catch returns are from drift nets. However, in recent years overfishing by the latter is regarded as being a cause of decline in catches of salmon and sea-trout by anglers in the freshwater reaches. In some stretches of the Suir, usually below towns, angling is poor because of

Fig. 14.2  River Suir catchment, showing main towns, tributaries and the more important sampling stations.

the effects of domestic and industrial discharges (mostly organic, biodegradable) on water quality.  The value of salmonid catches in the River Suir during 1978 was estimated to be about IR £0.4 million (An Foras Forbartha, 1980).  Other amenities and recreational activities are best developed in the lower reaches of the main channel.  These include boating, sailing, water-skiing and other aquatic pastimes and also river walks and picnicing.

Fig. 14.3 (opposite)  R. Suir sampling station upstream of Holycross Bridge.  This is a recovery zone dominated by *Ranunculus* spp. and *Potamogeton pectinatus*.  The emergent is *Oenanthe* sp.  (Photo courtesy of Bord Failte-Irish Tourist Board)

## 14.2   GENERAL DATA

DIMENSIONS
| | |
|---|---|
| Area of catchment | 3610 km$^2$ |
| Length of main channel | 184 km |
| Altitude at source | 449 m, on Borrisnoe Mountain |

CLIMATE

Precipitation
  mostly 800-1000 mm, reaching 1800 mm in upper catchment (An Foras Forbartha, 1980).  Snow sometimes falls in the mountains, but never persists for long enough to permit a concentrated period of snow-melt in spring.

Temperature
  The mean values for Ireland (Praeger, 1974) provide a reasonable indication for the Suir catchment:

              January 4.4 $^{\circ}$C;       July 15.5 $^{\circ}$C

POPULATION
| | | | |
|---|---|---|---|
| Templemore | 2556 | Clonmel | 12418 |
| Thurles | 7439 | Carrick-on-Suir | 5511 |
| Tipperary | 4929 | Waterford | 32617 |
| Cashel | 2588 | | |

PHYSICAL FEATURES OF RIVER

Hydrometric data
Selected stations on Suir and some of its larger tributaries.  Data include long-term mean flows, highest recorded mean monthly flows per unit of time and per unit of catchment area, lowest recorded mean monthly flows per unit of time and per unit of area.  (Data from An Foras Forbartha, unpublished.)

| river | station | | long-term mean flow ($m^3 s^{-1}$) | highest recorded mean monthly flow | | | lowest recorded mean monthly flow | | |
|---|---|---|---|---|---|---|---|---|---|
| | | | | ($m^3 s^{-1}$) | ($m^3 s^{-1}$ 100 km$^2$) | | ($m^3 s^{-1}$) | ($m^3 s^{-1}$ 100 km$^{-2}$) | |
| Suir | Thurles | 6 | 4.26 | 7.75 (Jan) | 3.28 | | 0.10 (Aug) | 0.65 | |
| Suir | Breakstown Lodge | 9.5 | 9.97 | 19.87 (Jan) | 3.88 | | 0.46 (Oct) | 0.73 | |
| Suir | New Br. | 14 | 20.32 | 38.56 (Dec) | 3.44 | | 1.55 (Aug) | 0.62 | |
| Suir | Caher Park Br. | 15 | 29.41 | 55.32 (Jan) | 3.47 | | 4.19 (Oct) | 0.78 | |
| Suir | Clonmel | 17.5 | 54.20 | 92.05 (Jan) | 4.24 | | 8.83 (Aug) | 1.34 | |
| | | | | | | | | | |
| Clodiagh | Rathkennan Br. | 7A | 5.26 | 10.75 (Jan) | 4.37 | | 0.24 (Aug) | 0.67 | |
| Multeen (S)* | Aughnagross | 8A(S) | 2.17 | 3.76 (Dec) | 4.33 | | 0.21 (Aug) | 1.13 | |
| Multeen (N)* | Ballinaclough | 8A(N) | 1.55 | 3.33 (Jan) | 4.44 | | 0.26 (Aug) | 0.74 | |
| Aherlow | Killardry | 13A | 7.98 | 14.12 (Jan) | 5.17 | | 1.25 (Oct) | 1.59 | |
| Tar R. | Tar Br. | 16A | 5.92 | 10.83 (Jan) | 4.75 | | 1.43 (Oct) | 1.36 | |
| Anner | Anner | 18A | 6.18 | 11.57 (Jan) | 2.74 | | 0.91 (Aug) | 0.58 | |

*S  South Branch        N*  North Branch

## 14.3  PHYSICO-CHEMICAL FEATURES

### 14.31  HYDROLOGY AND WATER RESOURCES

Maximum flows have almost always been recorded in January and minimum flows in late summer or early autumn (see 14.2).  Hydrological studies of the time of occurrence of maximum and minimum flows in the U.K. have shown that they become later towards the south and east.  Beaumont (1975), for instance, quotes examples for the north-west of Scotland and East Anglia, where the mean maximum flows occur in December and March, respectively.  The Suir, its location in south-east Ireland, falls in between. The detailed behaviour of flows within the catchment is influenced by the geology, storage capacity of the groundwater and the proportion of precipitation reaching the rivers.

Table 14.1 summarizes two important sets of statistics important for planning.  The 95 percentile flow is perhaps the most useful guide to decision makers of the potential of receiving water to assimilate waste discharges from urban and industrial zones.  For the assessment of the water resources of a catchment when planning abstractions, it is also necessary to know the dry weather flow.  Detailed investigations are now under way to improve the basic hydrological information in the catchment with the construction of weirs, monitoring of groundwater and more accurate gauging of the system in order to plan for careful industrial and urban expansion in the future.

*Table 14.1*  Catchment area, estimated 95 percentile flow, and estimated dry weather flow at selected stations on the Suir and some larger tributaries.

| river | location | stn. no. | catchment $km^2$ | 95 percentile flow $m^{-3} s^{-1}$ | dry weather flow $m^3 s^{-1}$ | $l s^{-1} km^{-2}$ |
|---|---|---|---|---|---|---|
| Suir | Thurles Br. | 6 | 236 | 0.20 | 0.09 | 0.40 |
| Suir | Beakstown Lodge | 9.5 | 512 | 1.10 | 0.30 | 0.60 |
| Suir | New Br. | 14 | 1120 | 2.73 | 1.10 | 1.00 |
| Suir | Caher Park | 15 | 1602 | 6.10 | 3.20 | 2.00 |
| Clodiagh | Rathkennan | 7A | 246 | 0.45 | 0.17 | 0.70 |
| Multeen (S)* | Aughnagross | 8A(S) | 87 | 0.30 | 0.10 | 1.15 |
| Multeen (N)* | Ballinaclogh | 8A(N) | 75 | 0.30 | 0.10 | 1.33 |
| Aherlow | Killardry | 13A | 273 | 1.68 | 0.90 | 3.30 |
| Tar | Tar Br. | 16A | 228 | 1.68 | 1.00 | 4.39 |

*S = South branch;   N = North branch

## 14.32  WATER ABSTRACTION

The location of all surface abstractions in the catchment are shown in
Fig. 14.5.  Most of the large abstractions are from surface waters,
mainly from rivers and streams, although a few impoundments are also
involved.  Smaller abstractions tend to be provided from wells and boreholes.  In general, most of the abstractions in the Suir catchment occur
in areas remote from point waste discharges.  However, a few industrial
abstractions, notably those for the Sugar Co. below Thurles, protein and
meat plants below Caher and a chemical factory just upstream of Carrick-
on-Suir could be abstracting water of doubtful to poor quality at all
times.  Hence, regular monitoring particularly at these points is
essential to avoid providing sub-standard water for such industrial
concerns.  It would appear that in planning for the future it will be
necessary to adopt stringent water quality standards along the main
channel in order to maintain or improve present conditions to ensure
suitable water to meet future industrial need and also to allow upstream
migration of salmonids.  Such standards would be ensured by adopting
the Council of the European Communities (C.E.C.) Directive for salmonid
waters (1978) and would require a high standard of sewage treatment by
all towns presently discharging into the upper portion of the main
channel and on the tributaries of the catchment.


## 14.33  WATER QUALITY

The Suir experiences a wide range of water qualities over the length
examined (Table 14.2).  In addition there is often a considerable range
in concentrations of different chemical constituents, which is at least
in part due to differences between summer and winter flows.  The effects
upon water quality of domestic or industrial wastes show up immediately
downstream of their discharge.  However, nutrients reaching the river
from non-point sources such as runoff from fertilized agricultural land
can also have a considerable affect on water quality.

   Water quality is excellent in the upper reaches because there is
little industry, towns or extensive farming.  Initial impairment
commences below Templemore, where organic effluent is responsible for
increased BOD and also, at times, low dissolved oxygen concentrations.
Further downstream at stations 4 and 5 (Fig. 14.2), the river improves
in quality; the large weed growths sometimes lead to oxygen super-
saturation in summer.  The river then receives both industrial and
domestic organic effluents at Thurles, which cause serious impairment
in water quality over a 5 km length (Table 14.2).  Sewage-fungus is common here, especially in winter during the sugar beet campaign.

   Downstream of Holycross the river flows through a rural hinterland
where the water improves rapidly in quality, a fact reflected in the
excellent salmonid stretches that occur from Two-Ford Br. (station 11)
to just upstream of Caher.  From Caher to Clonmel there is moderate
impairment of water quality.  Downstream of Caher there is a marked

Table 14.2  R. Suir:  water chemistry (Lennox & Toner, 1980) and biological quality (Flanagan & Toner, 1972; Clabby, 1981).  Chemistry at stations 1 - 4 for 1978-79 and stations 5 - 19 for 1972-78.  Sampling usually carried out between 1000 and 1600.  Quality values based on classification system simplified as follows:  5, good; 4, fair; 3, doubtful; 2, poor; 1, bad.

| | distance from Station 1 km | dissolved oxygen (% saturation) | BOD (mg l$^{-1}$) | NH$_3$-N (mg l$^{-1}$) | NO$_3$-N (mg l$^{-1}$) | reactive phosphate-P (mg l$^{-1}$) | Quality rating (Q) 1978 | 1976 | 1971 |
|---|---|---|---|---|---|---|---|---|---|
| 1 Knockanroe Br. | 0 | 84 - 138 | 1.0 - 3.5 | 0.02 - 0.05 | 0.20 - 1.70 | 0.010 - 0.050 | 5 | 5 | 5 |
| 2 Knocknageragh Br. | 5 | 80 - 104 | 0.6 - 2.5 | 0.02 - 0.09 | 0.60 - 3.40 | 0.010 - 0.100 | 4 - 5 | 4 | 5 |
| 3 d/s Templemore, Penane Br. | 9 | 68 - 124 | 0.8 - 3.4 | 0.05 - 0.24 | 1.00 - 3.90 | 0.020 - 0.100 | 2 - 3 | 3 - 4 | 3 - 4 |
| 4 Loughmore Br. | 11 | 81 - 168 | 0.3 - 2.9 | 0.03 - 0.10 | 1.10 - 3.40 | 0.020 - 0.170 | 3 | 3 - 4 | 3 - 4 |
| 5 Rossestown Br. | 18 | 60 - 120* | 0.8 - 7.1 | tr - 0.19 | 0.46 - 2.40 | 0.013 - 0.130 | 3 - 4 | 4 | 4 - 5 |
| 6 d/s Thurles Br. | 23 | 41 - 110* | 1.3 - 7.0 | tr - 0.21 | 0.43 - 2.80 | 0.010 - 0.183 | 3 | 3 (mid channel) | 2 (mid-channel) |
| 8 Turtulla Br. | 25 | 37 - 112* | 1.4 - 8.9 | 0.02 - 1.80 | 0.58 - 2.30 | 0.013 - 0.330 | { 3 (L.H.)  2 (R.H.) | | |
| 9 Cabragh Br. | 28 | 32 - 108* | 1.2 - 20.0 | tr - 3.00 | 0.45 - 3.70 | 0.008 - 0.293 | 3 - 4 | 2 - 3 | 2 |
| 10 Holycross Br. | 32 | 44 - 146* | 0.8 - 9.3 | 0.01 - 2.65 | 0.24 - 2.20 | 0.003 - 0.200 | 4 | 4 | 4 |
| 11 Two Ford Br. | 36 | 40 - 132* | 0.9 - 9.1 | tr - 0.80 | 0.26 - 2.20 | 0.003 - 0.183 | 5 | 4 | 5 |
| 12 Ardmayle Br. | 45 | 66 - 126 | 1.0 - 6.6 | tr - 0.16 | 0.86 - 1.90 | 0.010 - 0.175 | 4 - 5 | 4 - 5 | 4 - 5 |
| 13 Golden Br. | 58 | 69 - 141 | 1.0 - 4.3 | tr - 0.14 | 0.84 - 1.80 | 0.010 - 0.070 | 4 | 4 - 5 | 4 - 5 |
| 16 Br. at Swiss Cottage | 82 | 25 - 102 | 1.6 - 8.0 | 0.04 - 0.37 | 0.84 - 1.60 | 0.010 - 0.090 | 3 - 4 | 3 - 4 | 4 |
| 17 0.5 km u/s Clonmel | 113 | 79 - 116 | 1.3 - 3.4 | 0.01 - 0.11 | 1.10 - 1.60 | 0.007 - 0.085 | 5 | 3 - 4 | 5 |
| 18 Twomile Br. | 119 | 68 - 108 | 2.1 - 3.7 | 0.06 - 3.40 | 1.10 - 1.60 | 0.010 - 0.095 | 4 | 3 | 3 - 4 |
| 19 Kilsheelan Br. | 125 | 71 - 112 | 1.3 - 2.3 | 0.03 - 0.18 | 1.10 - 1.70 | 0.020 - 0.080 | 2 - 3 | 3 - 4 | 4 - 5 |

*lower values for D.O. may have been recorded during diel surveys (see Table 14.3).

'sag' of oxygen under summer conditions and recently fish-kills have
been reported (Lennox & Toner, 1980). The latter appear to be attributable to "once off" discharges under low flow conditions rather than
overall serious impairment.


## 14.4  MACROPHYTES AND ALGAE

Some indication that a river is eutrophic can be obtained by visual
observation of relative abundances of macrophytes and macroalgae. More
detailed examination often reveals differences in the species composition of the macrophyte and algal communities of enriched and unenriched
reaches (Horkan, 1981). In recent years stretches of the River Suir
have become choked with massive weed and algal growths.

During the summer of 1978 a detailed survey was carried out in the
river near Thurles, as well as a nearby tributary, the Drish River, to
ascertain the main effects of such plant growths. This survey was
extended in 1979 upstream of Templemore and also downstream as far as
Golden; three further tributaries, the Rossestown, Clodiagh and Multeen
Rivers were also included. The more important observations are outlined
below.

Station 1 (Knockanroe Br.: 9 km downstream of the source of the Suir)
is not influenced by any large industrial or domestic effluents. Weed
and algal growth in a riffle area covered a maximum of 20% of the stony-
gravel substratum and the resulting biomass was low at all times. By
the time the river reaches Rossestown (station 5), 18 km further downstream, it has increased in volume due to the addition of some tributaries and is enriched organically by Templemore town sewage. Massive
growths of weeds and algae occur here and the dominants are *Ranunculus*
spp. and *Cladophora*.

At Thurles Br., in addition to some large growths of silt-loving
*Potamogeton* spp., there were large "streamers" of *Cladophora* and some
patches of sewage-fungus. The trend in this stretch of river of a
shift from *Ranunculus* to *Potamogeton* is presumably a result of organic
enrichment, where the effluent load produces organics which deposit
sediments. Such an effect has been noted by Haslam (1978). Two kilometres further downstream at Turtulla (station 8) the effects of town
sewage are more pronounced on the right-hand side than on the left-hand
side, the latter being diluted by a clean tributary, the Drish River.
As at Thurles Bridge, some silting occurs, most notably on the more
polluted right-hand side, where *Potamogeton* and *Cladophora* dominated.
*Cladophora* was also common on the cleaner left-hand side, together with
*Ranunculus* (Horkan, 1980, 1981).

At Cabragh Br. (station 9) the combined effects of town sewage and
industrial effluent, particularly sugar beet wastes, are felt. The
sampling point here is located in a fast-flowing stretch of water

producing a strong current which prevents extensive silt deposition and results in a reduction in *Potamogeton* growth. *Cladophora* is dominant here, but in reduced amounts, possibly because of inhibition through chemical effects and/or by light reduction due to shading and high levels of suspended solids. Westlake (1959) reported similar observations in a polluted English stream.

Further downstream the river enters a recovery zone and by the time it reaches Holycross (station 10), a river distance of 4 km, it shows improved quality. The large growths of *Cladophora* are replaced by the less problematic *Ranunculus*. At Two-Ford Br. (station 11) water quality is good; dominance by one or two genera in large quantities no longer occurs and species diversity is greater than at previous stations.

At Ardmayle Br. (station 12) the river is 40 - 50 m wide and about 0.5 m or more deep and the water takes on the characteristics of a (relatively clean) meandering lowland river. Diversity of plants remains high and mosses become more common (Horkan, 1981). At Golden (station 13) a slight reduction of water quality results from organic enrichment from Cashel entering a tributary, the Black Stream, about 4 km upstream of the sampling point. This enrichment is reduced however, because of dilution by a very clean tributary, the Multeen River, which enters the Suir about 1.5 km upstream of Golden. Nevertheless, the macrophyte biomass at this location was higher than at any other surveyed on the river during 1979, indicating very eutrophic conditions. *Ranunculus* is dominant here in the faster riffles and *Potamogeton* in the slower parts.

Of the four tributaries examined, the smallest is the Rossestown River (station 2A) which is slightly enriched from agricultural sources. Some *Cladophora* spp. growths occur here as well as mosses and *Rorippa* spp. The Drish River (station 3A) enters the River Suir upstream of Turtulla Br. (station 8) and is regarded as being relatively clean but slightly eutrophic. The flora is dominated by a mixture of mosses and *Ranunculus* spp. Long term changes in water quality may occur in this river as a peat-briquette factory (a fuel made from compressed turf) is being built nearby. This gives rise to the possibility of problems from peat silt covering the substratum and possibly interfering with invertebrates and trout spawning. The Clodiagh River (station 7A) sampling station has a loose silty gravel substratum with few stones and is regarded as being a slightly unstable stream bed. *Ranunculus* occurs in patches here with *Potamogeton* being found in the more silted areas. The fourth tributary, the Multeen River (station 8A), is clean and has a very sparse flora consisting mostly of small patches of *Ranunculus* and occasional small stands of *Potamogeton*.

The general impression gained from these observations is that balanced, stable, diverse plant communities occur in the clean stations of the Suir and some of the tributaries examined. In eutrophic areas the community is less stable, has low diversity featuring either *Potamogeton* spp. (usually *P. pectinatus*) or *Cladophora* or both, depending on the degree of siltation of the substratum.

Praeger (1974) lists a number of macrophytes of interest found along
the Suir, with the following near Clonmel:  *Rorippa sylvestris,
Ceratophyllum demersum, Lemna polyrrhiza.*   *Butomus umbellatus* provides
an interesting record near Carrick-on-Shannon.


## 14.5  MACROINVERTEBRATES

The macroinvertebrate fauna of the River Suir is described below.   The
stations sampled are those listed in Table 14.2; the substrata examined
were in riffle areas and collections made in summer.

Station 1 has a diverse fauna, reflecting the 'excellent' water quality
found there.   The main components of this fauna include Baetidae and
Helmidae, with a number of other forms being relatively common, of
which cased caddis, *Hydropsyche, Polycentropus, Ephemerella*, Ecdynurids,
*Leuctra, Gammarus,* Chironomidae and other dipteran larvae are the most
important.   At station 2 the quality of the river is 'good'.   Mites are
dominant, with lesser numbers of helmids, chironomids, Trichoptera,
*Gammarus, Potamopyrgus* and *Ancylus.*

At stations 3 and 4, there is a decline in water quality to a 'doubtful'
condition, reflecting the effects of enrichment from Templemore sewage.
The fauna reflects this condition, showing a reduction in numbers of the
more sensitive forms.   More tolerant Chironomidae (including *Chironomus*
spp.) and haliplids dominate here.   At station 5 the river, although
still eutrophic, shows some improvement in quality.   Baetidae and
helmids are dominant and a wide range of fauna are also found including
*Austropotamobius, Simulium*, chironomids including a few *Chironomus*,
cased caddis, *Leuctra* and a number of other genera including several
species of molluscs.

Downstream of station 6, the river is in 'poor' condition, with massive
weed and algal growths of *Cladophora* and *Potamogeton* (see section 14.4).
At this location chironomids dominate and Tubificidae are present in
large numbers.   Recorded here also are *Asellus*, Hirudinea (several
species), Sphaeriidae, *Lymnaea, Planorbis* and *Baetis rhodani.*   At
Turtulla and Cabragh Bridges (stations 8, 9) water quality is 'poor –
doubtful' and chironomids (with *Chironomus* spp. included) occur as
dominants.   There are also large numbers of molluscs and leeches found
here and a reduction in the numbers of tubificids.

At station 10 there are definite signs of recovery and here *B. rhodani*
dominates, the chironomids are numerous and other genera observed include
*Simulium, Asellus,* haliplids, *Lymnaea, Hydropsyche* and some cased caddis
larvae.   The fauna at station 11 is similar to that found at station 10.
Further downstream at station 12, the river deepens and widens and the
water quality is 'good'.   *Gammarus* is dominant here.   There is a diverse

fauna here with good representation of cased and uncased caddis larvae, *Leuctra*, *Ephemerella*, chironomids, *Simulium*, *Austropotamobius* and *Apheilocheirus*.   At station 13, the river shows signs of enrichment in terms of large growths of macrophytes (*Ranunculus* and *Potamogeton*), but the quality remains 'good'.   The fauna dominated by *Gammarus* also has numerous *B. rhodani*, *Simulium*, *Ryacophila*, chironomids, other Diptera, haliplids, *Gyrinus* and mites.

Downstream of Caher at station 16 the river shows 'doubtful' quality, reflecting the effects of urban and industrial sewage effluent.   Low dissolved oxygen has been recorded here at times and at least one fish-kill is recorded.   The fauna includes many tolerant forms with vast numbers of chironomids, numerous *Gammarus*, mites, Mollusca, *Asellus*, tubificids, lumbricids, *Ephemerella* and *Helobdella*.   At station 17 the water quality is improved slightly, being 'doubtful-fair'.   Current speeds are lower and fauna which are suited to such conditions such as Corixidae are common; the dominant invertebrate genus is *Valvata*.

Below Clonmel at station 18 and also at Kilsheelan the quality is 'doubtful'.   The river is very nutrient-rich, particularly at the latter station, where abundant algal growths occur.   The fauna at these locations includes large populations of *Simulium* and Tubificidae. Other invertebrates include:   Chironomidae, *Limnaea*, Sphaeriidae, *Physa*, *Planorbis*, *Valvata*, *Potamopyrgus*, *Glossiphonia*, *Baetis rhodani*, *B. scambus*, *Caenis*, *Asellus*, *Gammarus*, mites and Dytiscidae.

## 14.6  FISHERIES

In common with most large Irish rivers the commercial and angling fisheries of the River Suir are based on salmonids (*Salmo salar*, *S. trutta*, *S. trutta fario*).   Eels are numerous, but the commercial fishery is not well developed and returns amount to about 10 tonnes yr$^{-1}$ at present.   Other fish found in this catchment include pike, perch, gudgeon, stone loach, minnow and stickleback; none of these are fished commercially.   There are cockle and mussel beds in the estuary, but their full commercial potential has not been developed as yet.

The commercial salmonid fishery is confined mainly to drift net catches in the estuary.   These catches produce an annual catch currently valued at approximately £0.4 million. The general impression gained is that this is an underestimate, as some illegal fishing is carried out and many anglers do not send in catch returns.   Reduced water quality in some areas is unfavourable to game fish at present due to oxygen depletion in summer (Table 14.4, Fig. 14.4) and has resulted in complaints from angling clubs.

Although no detailed statistics are available on fish stocks in the catchment, some idea of the distribution of the more valuable stocks may be gained from Fig. 14.5, where known spawning areas of game fish

*Table 14.3* Comparison between four different stations on the Suir showing biomass, dominant macrophyte genus and approximate duration dissolved oxygen in water stayed below 50% saturation during a 24 hour survey, lowest D.O. recorded during 24h. survey and effect on caged fish (see Horkan, 1981).  L =left-hand side; R = right-hand side.

| location | situation | biomass in July g m$^{-2}$ dry weight | dominant genus | approx. time D.O. <50% (hours) | lowest D.O. (% sat.) | effect on caged fish |
|---|---|---|---|---|---|---|
| Rossestown Br. | open | 1978 / 320 | *Ranunculus* | 0 | 87% | none |
| Thurles Br. | open | 570 | *Cladophora* | 9.2 | 41 | distressed |
| Turtulla | open | 1979 L: 380 R: 320 | L: *Cladophora* R: *Potamogeton* + *Cladophora* | *15.5 | L: 48 R: 27 | L: little (some distressed) R: some died, others distressed |
| Cabragh Br. | part shaded | 240 | *Cladophora* | 15.8 | 15 | all fish died |

*D.O. in mid-channel; insufficient measurements taken on each side to determine duration of D.O. <50% on right- or left-hand sides.  However, it is known that D.O. fell to lower levels on the polluted right-hand side.

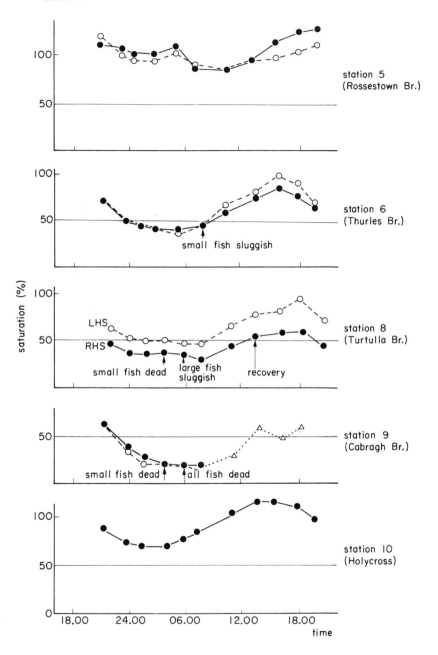

**Fig. 14.4** Diel variation in dissolved oxygen and its effect on brown trout kept in cages at different stations. RHS (right-hand side) at Turtulla is influenced by sewage, while LHS (left-hand side) by clean water. D.O. measured beside cages.

*Fig. 14.5* River Suir and main tributaries, showing locations of surface water abstractions, salmonid spawning areas and locations of private and angling club fishing waters (data from An Foras Forbartha, 1980).

are shown. The stretches owned or leased by private individuals or clubs for salmon or trout angling are also shown on this map. The general impression gained is that virtually the whole catchment could be classified as game fish waters at present because of the widespread distribution of salmonid species, although as stated above, some areas have poor water quality at times. Long term water quality management plans are now being developed to ensure that the catchment be retained as a game fishery. A new sewage works has recently been brought into use at Thurles and improvements in water quality are already evident.

# REFERENCES

An Foras Forbartha (1980) *Draft Water Quality Management Plan for the River Suir Catchment. Vol. 2.* Summary of Water Resources. 17 pp.

An Foras Forbartha (1980) *Draft Water Management Plan for the River Suir Catchment. Vol. 4. Beneficial Uses and Water Quality Criteria.* 80 pp.

Beaumont P. (1975) Hydrology. In: Whitton B.A. (Ed.) *River Ecology.* 725 pp., pp. 1-38. Blackwell, Oxford.

Bolas P.M. & Lund J.W.G. (1974) Some factors affecting the growth of *Cladophora glomerata* in the Kentish Stour. *J. Soc. Wat. Treat. Exam.* 23, 25-51.

Central Statistics Office (1980) *Census of population of Ireland, 1979. Vol. 1. Population of District Electoral Divisions, Towns and Larger Units of Area.* 95 pp. Stationery Office, Dublin.

Clabby K.J. (1981) *The National Survey of Irish Rivers. A Review of Biological Monitoring, 1971-1979.* 322 pp. Water Resources Division, An Foras Forbartha WR/R 12.

Council of the European Communities (1978) Council directive of 18 July on the quality of fresh waters needing protection or improvement in order to support fish life (78/659/EEC). Official Journal of the European Communities No. L222/7.

Edwards R.W. & Owens M. (1960) The effect of plants on river conditions. I. Summer crops and estimates of new productivity of macrophytes in a chalk stream. *J. Ecol.* 48, 151-60.

Haslam S.M. (1978) *River plants.* 396 pp. Cambridge University Press.

Horkan J.P.K. (1980) *Interim Report on Eutrophication and Related Studies of the Thurles Area of the River Suir, May-July, 1980.* 39 pp. An Foras Forbartha, Water Resources Division, WR/R10.

Horkan J.P.K. (1981) *Second Interim Report on Eutrophication and Related Studies of the Thurles Area of the River Suir, July/Sep. 1979.* An Foras Forbartha, Water Resources Division, WR/R13.

Jorga W. & Weise G. (1977) Biomasseentwicklung submerser Makrophyten in langsam fliessenden Gewässern in Beziehung zum Sauerstoffhaushalt. *Int. Rev. Ges. Hydrobiol.* 62, 209-34.

Lennox L.J. & Toner P.F. (1980) *The National Survey of Irish Rivers. A third Report on Water Quality.* 104 pp. An Foras Forbartha, Water Resources Division.

Marker A.F.H. (1976) The benthic algae of some streams in southern England. I. Biomass of the epilithon in some small streams. *J. Ecol.* 64, 359-73.

Marker A.F.H. & Gunn R.J.M. (1977) The benthic algae of some streams in Southern England. III. Seasonal variation in chlorophyll *a* in the seston. *J. Ecol.* 65, 223-34.

Nevill W.E. (1972) *Geology and Ireland.* 263 pp. Allen Figgis & Co. Ltd., Dublin.

Praeger R.L. (1974) *The Botanist in Ireland,* 587 pp. E.P. Publishing Ltd.

Westlake D.F. (1959) The effects of biological communities in polluted streams. In: *Symposium 8, The Effects of Pollution on Living Material,* pp. 25-31. Institute of Biology, London.

Whitton B.A. (1971) Filamentous algae as weeds. In: *3rd Symposium on Aquatic Weeds.* European Weed Research Council, Oxford, 1971.

# 15 · Po

## G. CHIAUDANI & R. MARCHETTI

15.1   Introduction                                         402

15.2   General data                                         403

15.3   Physical and geographical background                 405

15.4   River water uses                                     407

15.5   Estimated pollution load                             408

15.6   Physico-chemical features                            408

    15.61   Introduction                                408
    15.62   Temperature                                 408
    15.63   Suspended and dissolved solids              409
    15.64   Organic load and dissolved oxygen           410
    15.65   Toxic substances                            411
    15.66   Nitrogen and phosphorus                     413
    15.67   Other features                              413

15.7   Biology                                              413

    15.71   Introduction                                413
    15.72   Microbiology and virology                   413
    15.73   Algae                                       414
    15.74   Macrophytes and riparian vegetation         415
    15.75   Protozoa, nematodes and rotifers            416
    15.76   Macro-invertebrates                         417
    15.77   Cyclostoma and fish                         420
    15.78   Birds                                       424

15.8   General comments                                     425

References                                                  429

## 15.1   INTRODUCTION

The Po basin lies in the continental part of the Italian peninsula, of which it covers about one-quarter.  The catchment is delimited by the Alps, which fix, from west to north, the boundaries with France, Switzerland and Austria.  The southern limit is set by the northern Appennines, while towards the east the catchment faces the Adriatic. The basin runs from west to east and corresponds approximately to the 45$^\circ$ N parallel (Fig. 15.1).  The most important lakes in Italy, Maggiore, Como and Garda, lie near the transition between the two main regions of the basin, the mountain arc and the lowland plain.  The cities of Torino (Turin) and Milano (Milan) lie on the plain, with about 1.2 and 1.7 million inhabitants, respectively.  In addition there are many smaller centres of population.

*Fig. 15.1*  A) Po catchment;  B) location of main tributaries.

15.2    GENERAL DATA

DIMENSIONS

| | |
|---|---|
| Area of total catchment | 70500 km$^2$ (including 600 km$^2$ of permanent glacier) |
| Length of river | 676 km (Ministero dei Lavori Pubblici, 1968) |
| Peaks: Monte Bianco | 4810 m |
| Monte Rosa | 4633 m |
| Gran Paradiso | 4061 m |
| Source of Po on Monviso glacier | 3841 m |
| Po plain | 24450 km$^2$ |
| Average height of basin | 740 m |
| Average height of plain | $c$ 100 m |
| Gradient: Section 1 | 1785 m in first 67 km |
| Section 2 | 0.2% |
| Section 3 | 0.01% |
| Width: maximum in central part | 2.5 km (3 km at flood) |
| downstream | 0.8 km (because of embankments) |

CLIMATE

Precipitation

| | |
|---|---|
| average over whole catchment (500 rain stations) | 1100 mm |
| annual contribution from catchment | 70-80 x 10$^9$ m$^3$ |

Temperature

| | |
|---|---|
| winter average: western part of plain | $c$  0 $^\circ$C |
| near great lakes | 2 $^\circ$C |
| spring average: western part of plain | 12 $^\circ$C |
| summer average: western part of plain | 23 $^\circ$C |
| autumn average: western part of plain | 12 $^\circ$C |

Solar radiation in eastern part of plain:

| | |
|---|---|
| daily absolute maximum | 864 cal cm$^{-2}$ (July, (1966) |
| average annual | 113880 cal cm$^{-2}$ yr$^{-1}$ |
| = average daily rate | 312 cal cm$^{-2}$ d$^{-1}$ |

Foggy days: Mantova 71; Parma 65; Milano 38.

PHYSICAL FEATURES OF RIVER

Tributaries

141 second order tributaries of which 95 enter on the left and 46 on on the right

major ones (all with a catchment of at least 6500 km$^2$)

| | length |
|---|---|
| Adda | 313 km |
| Ticino | 248 km |
| Oglio | 280 km |
| Tanaro | 264 km |

Hydrological data at Pontelagoscuro gauging station
  (90 km from mouth; 8.35 m altitude; data for 1918-1980)

| | |
|---|---|
| catchment area of this section | 70091 km$^2$ (of which 0.8% glacial, 1.3% lacustrine) |
| rainfall | 1107 mm |
| runoff | 668 mm |
| loss | 439 mm |
| runoff rainfall ratio | 0.60 |
| mean monthly maximum flow | 8940 m$^3$ s$^{-1}$ = 128 l s$^{-1}$ km$^{-2}$ |
| absolute maximum flow (14.11.51)* | 10300 m$^3$ s$^{-1}$ + |
| mean flow | 1490 m$^3$ s$^{-1}$ = 21.3 l s$^{-1}$ km$^{-2}$ |
| mean monthly minimum flow | 275 m$^3$ s$^{-1}$ |

Distribution of flows throughout year:

| | Jan | Feb | Mar | Apr | May | June | July | Aug | Sep | Oct | Nov | Dec |
|---|---|---|---|---|---|---|---|---|---|---|---|---|
| rainfall (mm) | 60 | 57 | 80 | 111 | 121 | 100 | 76 | 84 | 104 | 122 | 115 | 77 |
| runoff (mm) | 45 | 43 | 57 | 58 | 73 | 69 | 48 | 36 | 45 | 63 | 73 | 58 |
| loss (mm) | 15 | 14 | 23 | 53 | 48 | 31 | 28 | 48 | 59 | 59 | 42 | 19 |
| $Q$ max (m$^3$ s$^{-1}$) | 5010 | 5810 | 5380 | 5570 | 8850 | 6990 | 4150 | 3750 | 6900 | 7290 | 8940 | 7730 |
| $Q$ med (m$^3$ s$^{-1}$) | 1180 | 1220 | 1480 | 1570 | 1920 | 1870 | 1260 | 950 | 1230 | 1650 | 1970 | 1520 |
| $Q$ min (m$^3$ s$^{-1}$) | 395 | 418 | 518 | 275 | 312 | 306 | 307 | 293 | 320 | 446 | 490 | 430 |
| $q$ (l s$^{-1}$ km$^{-2}$) | 16.8 | 17.4 | 21.1 | 22.4 | 27.4 | 26.7 | 17.0 | 13.6 | 17.5 | 23.5 | 28.1 | 21.6 |

*The embankment broke before the gauged section; estimated flow was
11580 m$^3$ s$^{-1}$. The flood covered 113000 ha (Ministero dei Lavori
Pubblici, 1979).

*Fig. 15.2*  Middle Po, showing nuclear power station.

15.3   PHYSICAL AND GEOGRAPHICAL BACKGROUND

Climate

As would be expected with a catchment which covers such a range in altitude, there are marked differences in both temperature and precipitation
in different parts.  The coldest month is January, followed by December,
while the hottest is July, followed by August.  The greatest difference
recorded on the plain between minimum and maximum temperature is 57 C
degrees (-18 $^{\circ}$C, 39 $^{\circ}$C, respectively).  In 1979 the average temperature
in the central part of the plain (Milano) was 13.5 $^{\circ}$C, with monthly
averages ranging from -4.6 $^{\circ}$C in February to 28.1 $^{\circ}$C in July.  Storms,
which are more frequent in summer, increase from west to east, with an
average yearly total of 30 storm days.  Relative humidity fluctuates
between 80% (winter average) and 50% (summer average); in the centre of
Milano the annual average is 69%.  Because of the mountain chain, strong
winds are relatively infrequent on the plain.  In winter there is
typically a complete cloud cover for a period of 10 - 15 days, followed
by 10 days or so of clear sky; in summer the number of totally overcast
days drops to 4 - 5 per month.  Fog is frequent in autumn and winter in
the valley, increasing in frequency from September to February.

Geology

While the Alps and the soils derived from them are mainly calcareous,
the Appennines are mainly sandstone.  The plain is formed by a continuous alluvial stratum, corresponding to a Quaternary basin with scarce
pre-Quaternary outcrops.  The stratum is mostly several hundred metres
thick, but reaches 2500 m in the delta; it is made of clastic materials,
which become finer the further away from the mountain slopes,
alternating with pelitic sediments.  Deep water-bearing strata consti-
tude large water reserves.

Hydrography

The initial slope of the river-bed is very steep, dropping off gradually
as the delta is approached (see 15.2).  The bed is deepest in the
terminal stretch, reaching 10 - 12 m.  Since there is also a 10 m
embankment, the river depth at peak flows is 20 - 22 m.  The course of
the river has been influenced by the deposition of alluvial materials
brought down by the tributaries.  The alpine tributaries have less
suspended solids than those on the right bank, partly because of the
presence of the lakes.

   Most tributaries can be divided into three main parts (Raffa &
Giuffrida, 1973):

   (i)   Mountain stretch, showing a narrow river-bed, with steep and
         well-defined sides;
   (ii)  Intermediate stretch, where ancient and recent alluvial
         materials penetrate into the plain; the bed increases in width

and the sides come to be determined mainly by cultivated
fields;
(iii)  A stretch through the plain which is narrow, winding and
generally embanked; the course of the river is determined
artificially.

A peculiar element of Po hydrography is the delta, which starts some
50 km from the sea and extends over 400 km$^2$. The mean altitude of this
region  is below sea level and there is a marked intrusion of saline
waters into the different branches of the delta.  The present delta
structure is due to natural events and human intervention.  There are
six branches, with the central one carrying 50 - 60% of the flow.  The
area of the delta used to increase by about 50 - 60 ha yr$^{-1}$, but this
dropped recently to about 35 ha yr$^{-1}$.

## Hydrology

The bed of the Po has been influenced by human activity to such an
extent that there is now little correlation between the nature of the
watershed and the pattern of water flow.  The first changes were
brought about by the reduction of the expansion zone due to embankment.
More recent changes which have influenced the natural hydrological
behaviour are lake regulation, construction of reservoirs, land
reclamation, extraction of water for irrigation, and the mining of
inert materials.  The key hydrological changes are the concentration of
the downflow flood and the withdrawal of water during periods of low
flow (Raffa, 1973).

Flows show an annual series of changes, of which the following are
the most important features:

(i)    Winter low flow (January, February) due to absence of thaw and
scarce rain.
(ii)   Maximum flow (May, June) caused by thaw.
(iii)  Summer low flow (August) much more marked than the winter one.
(iv)   Flood period (October, November).  A particularly ruinous
flood took place in November 1951, which covered 113000 ha
(Ministero dei Lavori Pubblici, 1979).

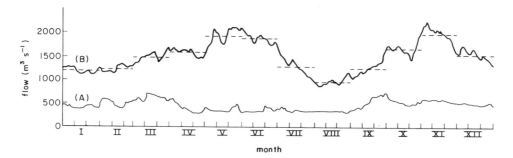

Fig. 15.3  Minimum (A) and mean (B) values for daily flow at
Pontelagoscuro closing station since 1918.

## 15.4  RIVER WATER USES

Agriculture covers 22894 km$^2$ of the basin, of which about 15000 km$^2$ are
irrigated with water direct from the Po or its tributaries.  Irrigation
canals fed by Po water were built first at the time of ancient Rome and
there was a marked increase at about the year 1000.  Since the end of
the last war, the discharge diverted for this has trebled and exceeds
240 m$^3$ s$^{-1}$ at the period of peak demand.  The direct use of Po water
for drinking purposes is limited to part of the supplies taken for the
two big urban centres of Torino and Ferrara.  Five open-cycle thermal
power stations (together producing more than 8500 MW) and two nuclear
power stations (1100 MW) take their cooling water from the river.
Reservoirs, which are sited in the upper part of the basin and were
built originally mainly for hydroelectric purposes, have an overall
storage capacity of about 1.5 x 10$^9$ m$^3$ (Ministero dei Lavori Pubblici,
1979).

Only moderate use is made of ships, mainly 1000 - 1300 tonne vessels
carrying oil, chemical products, sand and gravel.  This traffic is
expected ultimately to reach a carrying capacity of 12 x 10$^6$ tonnes
(Ugolini, 1971).  Tourist boating and angling are practiced in many
places, but swimming is banned almost everywhere because of poor
hygienic conditions.  Commercial fishing is unimportant, except for the
delta region where it is still possible to catch sturgeon.  On the
other hand, gravel and sand extraction is intense along the river.  The
total is given officially as 7 - 8 x 10$^6$ m$^3$ yr$^{-1}$, but it is probably
more than ten times as much, as demonstrated by the deepening of the
river-bed (2.5 m in 15 years, in certain places).

## 15.5  ESTIMATED POLLUTION LOAD

A study carried out by the Istituto di Ricerca sulle Acque (1977) made
an estimate of pollution by industry and the human population.  Industrial activities have been converted to inhabitant equivalents, using
current conversion equivalents (Cicioni *et al*., 1977).  The 'total
population' (TP) has been obtained by adding up the number of inhabitants (P) and the industrial equivalent (PE).  Only towns with a PE
of more than 15000, which constitute 83% of the total basin population,
have been considered.  The results were in millions:

$$TP\ (53.8)\ =\ P\ (14.9)\ +\ PE\ (38.9)$$

The river receives an average input per km length equivalent to a TP of
84000.  Most of the IE is due to the food industry, followed by the
steel and chemical industries.  The distribution of these polluting
sources is very heterogeneous.  The highest TP (35% of Po basin) is in
the Milano sub-basin, where most human and industrial wastes reach the
Po via the R. Lambro.

   The Istituto di Ricerca sulle Acque did not consider agricultural
wastes in its study.   Provini *et al*. (1979) estimated the following
totals for the catchment (all x $10^6$): cattle, 3.7; pigs, 3.5; sheep
and goats, 0.34; poultry, 80. Using conversion factors of 10.2, 3, 3.3
and 0.2, respectively (Farnesi *et al*., 1976), an estimate can be made
for the animal equivalent population of 65 x $10^6$ inhabitants.  This
raises the TP for the whole Po basin to about 119 x $10^6$ inhabitants.
Although these data are merely theoretical, they provide a useful means
of describing the distribution of the potential polluting load bordering
the Po.


## 15.6  PHYSICO-CHEMICAL FEATURES

### 15.61  INTRODUCTION

The data in this account are drawn mainly from research carried out in
1971-73 by Istituto di Ricerca sulle Acque along the river (21
stations) and by ENEL (Ente Nazionale per l'Energia Elettrica) in a
central stretch about 370 km from the source.  The ENEL research formed
part of a study to evaluate the effects of thermal discharges from a
nuclear power plant (Ioannilli, 1981).  Previous surveys include that
by Berbenni (1971) along the whole river.  Investigations prior to 1974
have been summarized by Piantino *et al*. (1977).


### 15.62  TEMPERATURE

The average annual temperature in the central stretch of the Po is about
14 $^{\circ}$C, the maximum, 29 $^{\circ}$C and the minimum, 3 $^{\circ}$C.  Although these data
are based on a relatively short period (1974-77), they can be considered

representative of this stretch (Bertonati & Ioannilli, 1981). Fluctuations in the lower stretch over the period 1961-1977 are shown in Fig.
15.4.

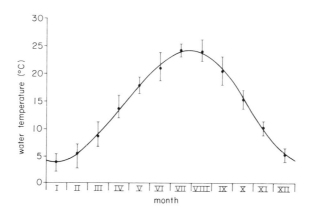

*Fig. 15.4* Mean monthly water temperature at Pontelagoscuro closing
station; values shown are the minimum, mean and maximum for the
period 1961 to 1971. (Modified from Raffa, 1973)


15.63   SUSPENDED AND DISSOLVED SOLIDS

The quantity of suspended materials transported by the Po varies
markedly and is influenced mainly by meteorological events.   The mean
values for 20 stations range from 45 to 450 mg $l^{-1}$ (105 $^{o}$C suspended
solids) with an absolute maximum of 6400 mg $l^{-1}$; the corresponding
values for volume of settling materials are 0.2, 2.1 and 40 ml $l^{-1}$,
respectively.   The concentration of dissolved substances is more
variable, with means ranging from 220 - 370 mg $l^{-1}$ and an absolute
maximum of 560 mg $l^{-1}$.   Starting from the source, the load of solid
materials increases at the rate of 1.209 kg $s^{-1}$ $km^{-1}$; 86% of this load
is of inorganic materials (Marchetti *et al.*, 1977).   In the closing
stretch (Pontelagoscuro), the load is:

| | |
|---|---|
| particulate | 12.55 x $10^6$ tonnes $yr^{-1}$ |
| dissolved | 14.14 x $10^6$ tonnes $yr^{-1}$ |
| total | 26.69 x $10^6$ tonnes $yr^{-1}$ |

The settling load, expressed in volume, is 28.4 x $10^6$ $m^3$ $yr^{-1}$.   Using
a value of 52290 $km^2$ as the area of basin subjected to erosion, this
gives a particulate solid loss of 240 tonnes $km^2$ $yr^{-1}$ (neglecting any
contribution from sources other than erosion.   This calculation based
on 1971-73 data corresponds well with an estimate based on thirty
years' observations of 258 tonnes $km^{-2}$ (Canali & Allodi, 1962).

15.64   ORGANIC LOAD AND DISSOLVED OXYGEN

$BOD_5$ values are not particularly high, with a mean for all stations of 3 mg $l^{-1}$ and COD/BOD ratio of about 5 (Provini & Marchetti, 1977). The $BOD_5$ total load in the last stretch of the basin is about 110600 tonnes $yr^{-1}$ (Marchetti *et al.*, 1977). Using data from many BOD, COD and TOC determinations in the central stretch, Bertonati (1977) and Bertonati & Ioannilli (1981) estimated the oxidation state of organic carbon with the index of 4(TOC - COD) : TOC. This index fluctuates around zero in stable hydrological conditions, but becomes negative in incipient flood phases as a result of resuspension of reduced materials in the sediments.

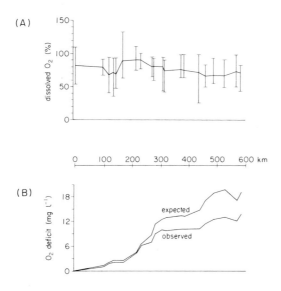

*Fig. 15.5* A) Percentage oxygen saturation (minimum, mean, maximum) along the river course for the period 1971-1973, based on monthly sampling. (Modified from Marchetti & Provini, 1977)
B) Expected and observed mean oxygen deficit along the river course. (From Marchetti & Provini, 1977)

Changes in dissolved oxygen along the river are shown in Fig. 15.5. The typical condition is that of appreciable deficit, this deficit reaching a maximum 136280 tonnes $yr^{-1}$ in the lowermost stretch. The oxygen concentrations show marked differences in summer between day and night e.g. 6 mg $l^{-1}$ in Fig. 15.6. A daily variation of about 9 mg $l^{-1}$ has been found in the central stretch of the river (Bertonati, 1977; Bertonati & Ioannilli, 1981). A vertical gradient has been noted under conditions of intense solar radiation, with 19, 16 and 14 mg $l^{-1}$ at the surface, 2.5 m and 5 m depth, respectively.

Using the method of Odum (1956), it has been possible to evaluate

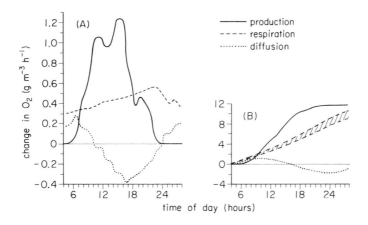

Fig. 15.6  Changes in dissolved oxygen due to photosynthesis, respiration and diffusion measured by Odum method at km 219 on 15.7.74. A) Hourly change ; B) cumulative values.  Shaded area represents plankton respiration measured by light and dark bottle method.  (From Marchetti & Provini, 1977)

gross primary productivity of the community in different parts of the river (2.28 - 15.1 g m$^{-3}$ O$_2$ in July: see 15.8).  The oxygen deficit at some stations is probably due mainly to plankton respiration and at others to benthic respiration (Marchetti & Provini, 1977).  More recently, measurements of sediment respiration made in marginal areas under low flow conditions have given values of 1.8 - 4.8 g O$_2$ m$^{-2}$ d$^{-1}$ (Bertonati & Ioannilli, 1981).


15.65  TOXIC SUBSTANCES

Studies by Vivoli & Vecchi (1977) and Prati (1977a, b) report data for arsenic, mercury, nickel, lead, copper, zinc, the pesticides α- and γ-BHC, DDT, DDE, heptachlorepoxide, dieldrin, thiodan and total phenols. Copper and lead were detectable and arsenic, nickel and zinc very frequently so (Table 15.1).  Mercury was detectable less frequently, with an absolute maximum of 17 mg m$^{-3}$ (Vivoli & Vecchi, 1977).  Among the metals listed above, zinc has the highest load in the most downstream stretch, with 2646 tonnes yr$^{-1}$.  In a more recent study (Provini et al., 1980), cadmium was detectable in 26% of samples (max. 0.8 mg m$^{-3}$) and chromium in all (2 - 107 mg m$^{-3}$).  In the central part of the Po, peak concentrations for all these elements except arsenic and zinc are usually associated with floods (Bertonati & Ioannilli, 1981).

    A study of mercury has been made for the fish Leuciscus cephalus cabeda taken from nine stations (Pennacchioni & Marchetti, 1973 ; Marchetti et al., 1975).  The concentrations of total and methylmercury expressed as mg per kg fresh weight of dorsal flesh were: 0.144 and 0.14 for the means ; 0.068 and 0.035 for the minima ; 0.380 and 0.274 for

*Table 15.1*  Concentrations of heavy metals (mg m$^{-3}$ or µg l$^{-1}$) at 21 sampling stations on Po together with annual load at the most downstream station (Pontelagoscuro).

| | Nickel | Copper | Zinc | Arsenic | Mercury | Lead |
|---|---|---|---|---|---|---|
| number of samples | 189 | 189 | 169 | 111 | 132 | 189 |
| detection limit | | | | | 0.2 | |
| positive samples (%) | 98.9 | 100 | 88.2 | 98.2 | 43.2 | 100 |
| absolute minimum | 0 | 3.5 | 0 | 0 | <0.2 | 2 |
| absolute maximum | 132 | 137 | 147 | 15 | 17 | 17.1 |
| load (tonnes yr$^{-1}$) | 89 | 1154 | 2646 | 243 | 65 | 485 |

the maxima.  Gaggino (1981) has reported a similar study for other heavy metals, but the data given as mg per kg dry weight: Cr, 0.2 - 0.4; Ni, 0.8 - 3.7; Cd, 0.005 - 0.05; Pb, 0.09 - 2.0.

In a study of 150 water samples, Prati (1977a) found that BHC was detectable in 48% and DDT and DDE in 40% of cases.  The cyclodienics (heptachlor, dieldrin, thiodan) were found in only 4.4% of cases.  The maxima were in ng l$^{-1}$:

| | | | | | |
|---|---|---|---|---|---|
| α-BHC | 180 | DDT | 372 | thiodan | 170 |
| γ-BHC | 274 | DDE | 450 | dieldrin | 32 |
| | | | | heptachlorepoxide | 59 |

In the most downstream stretch, the total load of pesticides was estimated as 6.7 tonnes yr$^{-1}$ in the period 1971-73.  More recently (1977-78), Galassi & Provini (1981) found that the concentrations of some pesticides had decreased which they ascribed to prohibitions or restrictions on use.  The concentrations of α-BHC and γ-BHC ranged from 3 - 37 and 2 - 13 ng l$^{-1}$, respectively.  Dieldrin was detectable in only one sample, while aldrin, heptachlor, heptachlorepoxide, endrin, HCB and pp'DDE were below detection limits.  In the most downstream stretch, PCB was detectable in 8 out of 18 samples, with a maximum of 42 ng l$^{-1}$ dissolved and 100 ng l$^{-1}$ adsorbed.  γ-BHC, HCB, pp'DDE and PCB were reported to be detectable regularly in flesh of *Leuciscus cephalus cabeda* and *Alburnus alburnus alborella* (Galassi et al., 1981b).

Phenols were detectable in 44% of cases (500 samples along whole river) studied by Prati (1977b), with a maximum of 0.146 mg l$^{-1}$.  The load increased on passing downstream and the annual average load in the lowermost stretch was 75 tonnes (Marchetti et al., 1977).

## 15.66   NITROGEN AND PHOSPHORUS

Except for stations close to the source, the mean ammonium concentration along the whole river is about 0.3 mg $l^{-1}$ N, although the absolute maxima are at times much higher in zones influenced by large urban discharges. Nitrate increases from the mean of 0.3 mg $l^{-1}$ N near the source to 1.7 mg $l^{-1}$ N downstream.

The mean concentration of orthophosphate in the most downstream stretch in 1971-73 was reported to be 0.151 mg $l^{-1}$ P (Istituto di Ricerca sulle Acque, 1977). A more recent study of this stretch led to estimates for the amount of nutrient flowing into the Adriatic: total N, 93091 tonnes $yr^{-1}$; total P, 11798 tonnes $yr^{-1}$ (Provini et al., 1980).

## 15.67   OTHER FEATURES

Anionic detergents and ether-extractable oils were studied in 1971-73 (Prati, 1977b). The mean concentrations were 0.05 mg $l^{-1}$ MBAS and 0.5 mg $l^{-1}$, respectively, but there were remarkable fluctuations. Total radioactivity, also measured in 1971-73, was very low and relatively constant (0.6 - 2.9 pCi $l^{-1}$) which were due exclusively to radio-nuclides occurring naturally in the water (Allegrini et al., 1977). Data on other variables ( conductivity, turbidity, pH, alkalinity) are included in 15.2.

## 15.7   BIOLOGY

### 15.71   INTRODUCTION

Knowledge about the biology of the Po is much less complete than that of physics and chemistry. Certain stretches of the river have been studied intensively, but others are scarcely known. We will try to synthetize here our knowledge, most of which derives from research either carried out by Istituto di Ricerca sulle Acque all along the river or co-ordinated by ENEL (Ioannilli, 1980) on the central part.

### 15.72   MICROBIOLOGY AND VIROLOGY

Surveys of microbial and viral components have been concerned almost entirely with hygiene. The data available concern the content of total bacteria at 20 ° and 37 °C (459 samples) of fecal and total coliforms (641), streptococci (217), sulphate-reducing clostridia (218) and Salmonella (49 samples), during surveys in 1971-75 down the whole river (Fara et al., 1977). Nardi (1977) considered the viruses Polio, Coxsackie, Echo, Reo and Adenovirus.

The overall pattern, even if showing large fluctuations, is characterised by the constant presence of all the bacterial group listed above. From the minimum, recorded at the source, values increase steadily on passing down to the delta. The number of bacteria per ml shown at 20 $^o$ ranges from $10^5$ to $10^6$ and at 37 $^o$ from $10^4$ to $10^5$. Total coliforms average around $10^3$, fecal coliforms around $10^2$, streptococci around $10^2$ and clostridia around 10 (all per ml). A survey made in 1971-1974 at 10 stations found *Salmonella* in 98% of samples, with in some cases seven different strains (species). The most frequent were *S. agona, S. panama, S. brandeburg, S. paratyphi* and *S. typhimurium*.

The survey (49 samples) was positive in 30% of cases, with a higher frequency in the central-terminal stretch. Among the 15 viral groups isolated, strains of *Coxsackie, Echo* and *Adenovirus* were the most frequent. When Po tributaries were investigated as well (85 samples), representatives of 42 virus groups were isolated with a marked preponderance of *Enterovirus* (78.5%), followed by *Echo, Coxsackie* and *Polio* (21.4%).

Yeasts have been surveyed only at four stations in the central part of the Po (1971-75, 209 water samples). There were 80 species and 15 genera. *Saccharomyces* (*S. cerevisiae* and *S. uvarum*) were the dominants, followed by *Candida, Torulopsis, Cryptococcus* and *Rhodotorula* (Viviani & Tortorano, 1977). Among these, potential pathogens (*Torulopsis glabrata, Candida albicans, C. tropicalis, C. guilliermondi* etc) occurred close to the mouth of very polluted tributaries.

## 15.73 ALGAE

Researches on Po algae have been limited to a central stretch of the river, where a quantitative study of the plankton was made from June 1974 to June 1977. Peak populations (10 - 15 x $10^3$ cells ml$^{-1}$) have been recorded in June-August. 223 species have been identified (Cironi, 1977), 16 blue-green algae, 128 diatoms, 4 dinoflagellates, 2 Euglenophyla, 67 greens, 6 chrysophytes. Only 20 species were present during the whole period at densities higher than 100 cells ml$^{-1}$ (Andreoli & Fricano, 1981). Diatoms formed 50 - 60% of the summer crop and 80% of the winter one. Abundant species were:

| blue-greens | diatoms | greens |
|---|---|---|
| *Microcystis aeruginosa* | *Melosira granulata* | *Scenedesmus quadricauda* |
| *Oscillatoria rubescens* | *Cyclotella comta* | *Actinastrum hantzschii* |
| | *C. meneghiniana* | *Coelastrum microporum* |
| | *Diatoma vulgare* | |
| | *Synedra ulna* | |
| | *Nitzschia acicularis* | |
| | *N. fruticosa* | |

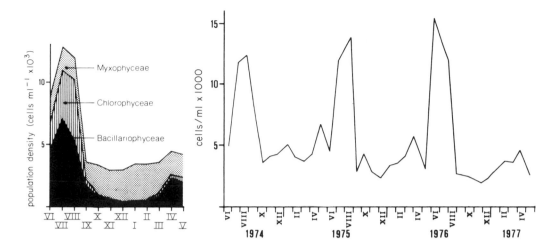

*Fig. 15.7* (left)  Mean monthly population density for three main algal groups in the central stretch of the Po for the period June 1974 to May 1977.

*Fig. 15.8* (right)  Monthly changes in density of total algal population during the same period.  (Both figures modified from Andreoli & Fricano, 1981)


## 15.74  MACROPHYTES AND RIPARIAN VEGETATION

Detailed observations about macrophytes are available only for the central stretch of the Po, but some records have been made on upper zone.  Here there are occasional stands of the Helosciadietum, Scirpo-Phragmitetum and Myriophylleto-Nupharetum associations (Corbetta & Zanotti-Censoni, 1977).  In the central stretch of the river, macrophytic vegetation is well developed, with large banks of *Phragmites australis* by itself or together with *Typha latifolia*, *T. angustifolia*, *Sparganium erectum* and *Glyceria fluitans*.  In the backwaters there are dense stands of the floating species, *Lemna minor* and *Salvinia*, and among the submerged species, *Ceratophyllum demersum*, *Potamogeton crispus*, *P. lucens*, *P. perfoliatus*, *Elodea densa*.

The Scirpo-Phragmitetum association is present in some places on the lower part of the river and, where velocity and depth are favourable, also Myriophilleto-Nupharetum and *Trapa natans*.  In the delta branches influenced by intrusion of sea water, *Phragmites australis* is overwhelmingly the most abundant macrophyte.

Corbetta & Zanotti-Censoni (1977) also studied the riparian vegetation in 1971-73; that is, the vegetation living on the wet stream margin.  In the upstream stretch, when suitable substrata are available, the Personato-Petasitetum association is present.  On passing downstream there is an increasing representation of nitrophile species,  with first *Saponaria officinalis* and then the Artemisietum velotori.  In the middle stretch the vegetation tends to be more homogeneous, with the Polygono-Chenopodietum association and the *Cyperus glomeratus* sub-association dominant.  In the lower stretch the phytosociological picture is not changed much, except for the frequency of the variant *Amaranthus* in the *Cyperus glomeratus* sub-association, which continues almost to the delta.

In the upper part of the river (from 2020 to 490 m altitude), woodland consists of *Alnus glutinosa, Fraxinus excelsior* and different species of *Salix.*  Further downstream woodlands become extremely scarce, since the areas adjacent to the river are nearly all cultivated.  *Salix alba* is quite common along the river sides, but natural stands are very reduced.  Even islands in the river are usually cultivated.  The only example of Po climax forest vegetation on one island in the lower stretch (about 550 km from the source), with specimens of *Quercus pedunculata, Populus alba, P. nigra, Ulmus campestris* and *Acer campestre* (Corbetta & Zanotti-Censoni, 1977).

## 15.75  PROTOZOA, NEMATODES AND ROTIFERS

Protozoa, especially those in the bottom sediments, were studied in 1978 at four stations from the middle zone to downstream (Madoni, 1980). Both number of species (16, 20, 22, 13 at the four stations) and population density (46, 41, 105, 36 per $cm^2$ of sediment) were low. Among Ciliophora the prevailing species were *Carchesium polypinum, Cyclidium glaucoma, Mesodinium pulex*, while among Sarcodinea the prevailing one was *Actinophrys sol.*

Information about nematodes is available for the upper (Zullini, 1969) and middle (Zullini, 1974a, b) parts of the river.  *Tobrilus longus* and *T. helveticus* are the most frequent in the former and *Paroigolaimella bernensis* in the latter.  58 species belonging to 18 families and settled on different substrata (sand, mud, vegetation) have been recognised.  In relation to this species abundance, the number of individuals is modest, probably because of the periodical washing effect of floods.  The community with the greatest number of species and individuals lies in a narrow littoral strip of land considered as an "ecotone" (soil-water-air).

Rotifers were studied in 1975 (ENEL, 1980) along the central stretch of the Po.  51 species were found, with *Brachionus calcyformis* and *Polyarthra vulgaris major* the most frequent.  Subsequent surveys have raised the total to 65, belonging to 12 sub-families.  *Brachionus, Polyarthra* and *Synchaeta* have remained the dominants, especially in August.

## 15.76  MACRO-INVERTEBRATES

### Turbellaria

Five species have been recorded from submerged and emergent macrophytes with a population density ranging from 5 to 5500 per $m^2$; *Dugesia tigrina* is the most common. Temporal changes are well characterised, with maxima in autumn and minima in summer (Cironi, 1979).

### Oligochaeta

Oligochaetes have been studied both on the river bottom (19 species: Paoletti Di Chiara, 1981) and in association with macrophytes (37 species: Samburgar, 1981). On bare surfaces, more or less subject to the influence of currents, the association is composed of *Tubifex, Limnodrilus hoffmeisteri, L. udekemianus, L. claparedeianus* and *L. profundicola*. The dominant is *L. hoffmeisteri,* which constitutes typically about 75% of the total population. The population structure is, however, broadly influenced by sediment size (Fig. 15.9). On bare surfaces, densities range from hundreds of thousands per $m^2$ to a few individuals, this being related to flood events. After floods the number of species increases to a maximum within about two months, but the number of individuals more slowly (Cironi & Ruffo, 1981). Oligochaetes associated with *Typha* and *Phragmites* are more abundant and mostly represented by Tubificidae and Naididae.

### Hirudinea

The 10 species recorded include *Thereruyzon tessulatum* (Müll.), a species new to the Italian fauna (Minelli & Mannucci Minelli, 1981). 7 species have been found on bare surfaces, among which *Erpobdella testacea* and *Helobdella stagnalis* are the most frequent. As with the oligochaetes, population densities are higher on macrophytes. The highest values have been recorded in June, with about 9000 individuals per $m^2$ of *Bratracobdella paludosa*, 3000 of *Helobdella stagnalis*, 2000 of *Erpobdella testacea*.

### Cladocera

These were studied during 1974-1977 by Rossaro (1981). The plankton of the middle part of the Po included 33 species, mostly Chidoridae and Daphnidae. Densities ranged from a maximum of about 1000 per $m^3$, usually in August, to a minimum between October and April. The most abundant species were *Moina branchiata, Bosmina coregoni, Alona rectangula*.

### Copepoda

There are 17 species of planktonic copepod (Rossaro, 1976), one Diaptomidae, one Arpacticidae and all others, Cyclopidae (Rossaro & Cotta Ramusino, 1976).

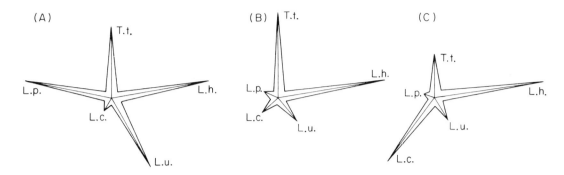

*Fig. 15.9* Relative density of various species of oligochaete associated with different sediment sizes:  A) <0.074 mm; B) 0.074 -
4.76 mm; C) >4.76 mm.  (After Paoletti di Chiara, 1981)

L.c. *Limnodrilus claparedeianus,* L.h. *L. hoffmeisteri,*
L.p. *L. profundicola,* L.u. *L. hudekemianus,*
T.t. *Tubifex tubifex*

The temporal changes in Copepoda density were much more variable than
those for the Cladocera, with summer maxima (August) over 1000 individuals per $m^3$ in 1974, reducing to less than 100 in summer 1975 and
1976, and with winter minima under 10 per $m^3$.  The most abundant species
were *Eucyclops serrulatus* and *Acanthocyclops robustus*.  *Cyclops vicinus,*
a cold stenothermal species, and *Thermacyclops crassus,* a warm stenothermal species, were both found at the same time.

Isopoda

Only one isopod, *Asellus aquaticus,* is present in the part of the
Po studied.  This species is abundant in summer months on sediments
rich in decomposing vegetation.

Amphipoda

Four species have been found in the middle Po, constituting an
Amphipoda community typical for all slow current stream of the Po
Valley.  This community, which shows a maximum density in spring and
a minimum in autumn (Cironi, 1979), is characterised by the dominance
of *Echinogammarus* (*E. veneris* and *E. stammeri*) and the absence of
*Gammarus,* the typical genus of French, German and Austrian waters.
*Innifargus elegans* and *Synurella ambulans* are also present, but in low
numbers.  These two species, characteristic of East European waters,
are at present spreading towards west in the rivers of Po Valley, but
are rare in Central Italy and absent in the south (Cironi & Ruffo,
1981).  *Echinogammarus veneris* is, on the contrary, a typical Mediterranean species, and probably of marine origin.  The two *Echinogammarus*
species, which are rarely present at the same time in other Italian

rivers, co-exist in the middle part of the Po.

## Decapoda

Research on the middle Po have shown the presence of two species, *Palaemonetes antennaris* and *Austropotamobius pallipes fulcisianus*. Both are widespread in Italian freshwaters (ENEL, 1980).

## Ephemeroptera

There are 15 species in the stretch studied. These belong to the families Siphlonuridae, Baetidae, Heptageniidae, Ephemerellidae, Caenidae, Leptophlebiidae and Ephemeridae. Querena (1980) has compared the population structure of bare sediments (with 6 species) and macrophytes (with 12 species). *Ecdyonurus aurantiacus, Heptagenia coerulans* and *Ephemera vulgata,* species associated with macrophytes, were always absent on bare substrata. The numerically prevailing species is *Cloëon dipterum* (reaching 200 m$^{-2}$), which is found only on macrophytes, followed by *Caenis* species and *Siphlonurus lacustris*.

## Odonata

22 species have been found in the central stretch of the Po, of which 9 belong to Zygoptera and 13 to Anisoptera (Galletti & Ravizza, 1977; Galletti, 1980). The most common is *Ischnura elegans*. The mean density is under 100 individuals per m$^2$, with a maximum of 550 in spring and minima in autumn (Cironi, 1979; Cironi & Ruffo, 1981).

## Hemiptera

Hemiptera are represented, in Po central reach, by 14 species, associated with macrophytes, raising 80 individuals per m$^2$ as a maximum density (Cironi & Ruffo, 1981). *Plea minutissima* is the prevailing species, present as imago all the year round with a maximum density during autumn. *Ilyocoris cimicoides, Micronecta meridionalis, Sigara dorsalis* and *S. italica* are well represented too. *Nepa cinerea* and *Ranatra linearis* are rare, and *Notonecta glauca* is very rare (Ravizza, 1976, 1981). The presence of *Mesovelia fucata* is to be pointed out because this species, considered to be typical of sub-mountainous running waters, was found associated with *Phragmites* in stagnant waters.

## Trichoptera

Information about Trichoptera from the upper zone of the river, close to the source (Moretti *et al.*, 1964) and the central stretch (G. P. Moretti, pers. comm.). There are 29 species classified but this is probably less than the real total, especially in the upper zone where only 16 species were recorded. These were: 5 *Rhyacophila* spp. (*galaerosa, intermedia, kelnerae, tristis, vulgaris*), 2 *Wormaldia* (*copiosa, occipitalis*) and others typically rheophilous and cold stenothermal (e.g. *Philopotamus ludificatus, Drusus discolor, Halesus rubricollis, Odontocerum albicorne*).

The Po central zone Trichoptera populations are clearly different. The 13 species classified are typical of lentic waters. Among these are *Hydroptila sparsa, Agraylea sexmaculata, Allotrichia pallicornis* and *Hydropsyche dissimulata*. The last one has been shown to be present in remarkable quantities both in the river and in the guts of *Leuciscus cephalus* and *Barbus barbus*. It prefers turbid waters and can be found also in polluted and oxygen depleted environments, where are often covered by ciliates.

## Coleoptera

Dytiscidae are the family most represented in the middle Po, with the most prevalent species being *Guignotus fuscillus, Noterus clavicornis, Laccophilus hyalinus* and *L. minutus*. Among the Halipida, *Halipus ruficollis* is the most frequent (Ravizza, 1980). 47 species have been identified and densities range from the abundance values 100 to 1000 individuals per $m^2$, values which may be considered to be quite low (Cironi & Ruffo, 1981). The Coleoptera population shows spring minima and autumn maxima; it colonizes *Typha* rather than *Phragmites* vegetation.

## Diptera

Diptera, together with Oligochaeta, are the group with most species in the middle Po. 109 taxa have been identified, belonging to 13 families, with many species still not classified (Ferrarese, 1980). Chironomidae clearly prevail, with 4 species of Diamesinae, 1 of Prodiamesinae, 34 of Orthocladiidae, 28 of Chironominae, 6 of Tanytarsinae, 5 of Tanypodinae. The most abundant species are: *Cricotopus (Isocladius) sylvestris, Rheosmittia spinocornis* (recorded for the first time in continental Europe), *Chironomus riparius, Chironomus* cf. *plumosus, Polypedilum cultellatum, Cryptochironomus* cf. *defectus*. The chironomid species composition is typical of a beta-mesosaprobic environment (B. Rossaro, pers. comm.).

## Mollusca

22 species have been recorded with population densities from 10 to over 1000 $m^{-2}$ and represented chiefly by *Acroluxus lacustris* (Cironi & Ruffo, 1981). The groups with most species are Sphaeriidae (5 species), Limniids (3 species) and Planorbidae (6 species).


15.77  CYCLOSTOMA AND FISH

Recent research on the Po ichthyofauna includes extensive investigations of the whole river course (Gandolfi & Le Moli, 1977a, b) and intensive investigations in the middle zone (Leone, 1981; Vitali, 1977, Vitali & Braghieri, 1979, 1981; Ioannilli, 1981; Aisa & Gattaponi, 1981) and

in the delta system (Gandolfi, 1973; Gandolfi *et al.*, 1978, 1979;
Giannini *et al.*, 1979).

44 species of fish have been identified (excluding the delta):

   27 freshwater, present already in the last century
    7 freshwater, appearing at the beginning of this century
    5 migrant anadromous
    1 migrant catadromous
    4 euryaline


The complete list and their distribution along the river is given in
Table 15.2. At the beginning of this century *Salmo gairdneri,
Salvelinus fontinalis, Micropterus salmoides, Lepomis gibbosus,
Ictalurus melas, Gambusia affinis holbrooki* (from North America) joined
the original population. *Silurus glanis*, coming from Yugoslavia,
appeared more recently (Gandolfi & Giannini, 1979).

Along with the increase in species number, there has been a change
in the population structure, resulting from the interaction of factors
such as restocking, dams and pollution. It is not possible to quantify
the changes because of the lack of reference data. However, it is
possible to document a reduction in the distribution area of some
species. In the medium-upper zone *Thymallus thymallus, Cottus gobio,*
and the Cyclostoma *Lampetra planeri zanadreai* have all disappeared;
*Cobitis taenia bilineata, Sabanejewia larvata* (Cobitida) and *Padogobius
martens* (Gobida) have decreased greatly. Sturgeon migration is
restricted to the lower half of the river and few individuals are able
to overcome the power plant dam. This dam has affected even more
drastically the reproductive migration of *Alosa fallax nilotica* (Fig.
15.10).

In the medium lower stretch, *Anguilla anguilla, Barbus barbus
plebejus, Tinca tinca* and *Esox lucius* are all on the decrease. On the
other hand, other species are increasing in abundance and areal distribution. These include the recently introduced species, *Micropterus
salmoides* and *Lepomis gibbosus*, and species subject to intensive restocking (*Salmo trutta fario* etc.) and *Alburnus alburnus alborella*
(Gandolfi & Le Moli, 1977a). *Carassius auratus* is caught occasionally,
but how this exotic species was introduced to the Po is unknown. *Salmo
gairdneri*, which was spreading until a few years ago, is now decreasing
in density and area.

In the central stretch (Vitali & Braghieri, 1981), the most important
species in the annual catch are *Alburnus alburnus alborella* (60 - 70%),
*Rutilus rubilio* (11 - 16%), *Scardinius erythroptalmus* (4 - 11%),
*Barbus barbus plebejus* (2 - 4%) and *Leuciscus cephalus cabeda* (2 - 3%).
Records in terms of weight are shown in Fig. 15.11.

*Table 15.2*  A. Fish species recently recorded in four stretches of the Po. I source - km 103; II km 103 - km 282; III km 282 - km 492; IV km 492.
Abundance:  1 rare;  2 occasional;  3 common;  4 very common (modified from Gandolfi & Le Moli, 1977a).
B. 19th century records for the Po taken from Festa (1892) and Pavesi (1896) by Gandolfi & Le Moli (1977a).  + = present.
(Most of the binomials used by these authors differ from the ones in present use and listed here)

| | | A. 1970s stretch | | | | B. | |
|---|---|---|---|---|---|---|---|
| | | I | II | III | IV | 1892 | 1896 |
| Petromyzonidae | *Petromyzon marinus* | | | 1 | 1 | + | + |
| | *Lampetra planeri zanandreai* | 2 | 2 | 2 | 1 | + | + |
| Acipenseridae | *Acipenser sturio* | | 1 | 2 | 2 | + | + |
| | *A. naccarii* | | | 2 | 2 | + | + |
| | *Huso huso* | | | | ? | ? | |
| Clupeidae | *Alosa fallax nilotica* | | ? | 2 | 3 | + | + |
| Salmonidae | *Salmo trutta m. fario* | 3 | 2 | 1 | 1 | + | + |
| | *S. trutta marmoratus* | 2 | 1 | | | ? | + |
| | *S. gairdneri* | 4 | 3 | 2 | 1 | | |
| | *Salvelinus fontinalis* | ? | | | | | |
| Thymallidae | *Thymallus thymallus* | 2 | ? | 1 | | | |
| Esocidae | *Esox lucius* | 3 | 3 | 3 | 3 | | + |
| Cyprinidae | *Rutilus rubilio* | 3 | 4 | 4 | 4 | + | + |
| | *R. pigus* | 3 | 3 | 2 | 1 | | + |
| | *Leuciscus cephalus cabeda* | 3 | 4 | 4 | 4 | | |
| | *L. souffia muticellus* | 4 | 3 | 2 | 1 | | |
| | *Phoxinus phoxinus* | 3 | 2 | 1 | 1 | + | + |
| | *Tinca tinca* | 3 | 3 | 3 | 3 | + | |
| | *Scardinius erythrophthalmus* | 3 | 4 | 4 | 4 | + | + |
| | *Alburnus alburnus alborella* | 3 | 4 | 4 | 4 | + | + |
| | *Chondrostoma soetta* | 3 | 4 | 4 | 4 | + | + |
| | *C. toxostoma* | 3 | 4 | 4 | 3 | + | + |
| | *Gobio gobio* | 3 | 3 | 3 | 2 | + | + |
| | *Barbus barbus plebejus* | 4 | 4 | 4 | 4 | + | + |
| | *B. meridionalis* | 2 | 1 | 1 | | + | + |
| | *Carassius carassius* | | 1 | 3 | 3 | | |
| | *C. auratus* | | 1 | 1 | 1 | | |
| | *Cyprinus carpio* | 3 | 4 | 4 | 4 | + | + |
| Cobitidae | *Cobitis taenia bilineata* | 3 | 3 | 2 | 1 | + | + |
| | *Sabanejewia larvata* | 1 | 1 | 1 | | | |
| Ictaluridae | *Ictalurus melas* | 2 | 3 | 4 | 4 | + | + |
| Siluridae | *Silurus glanis* | | | 1 | 1 | | |
| Anguillidae | *Anguilla anguilla* | 3 | 4 | 4 | 4 | + | + |
| Gadidae | *Lota lota* | | ? | 1 | ? | | + |
| Gasterosteidae | *Gasterosteus aculeatus* | | | 1 | 1 | | |
| Poeciliidae | *Cambusia affinis holbrooki* | | | 1 | 3 | | |
| Mugilidae | *Liza ramada* | | | 2 | 3 | | |
| | *L. saliens* | | | | ? | | |
| Percidae | *Perca fluviatilis* | 1 | 2 | 3 | 3 | + | + |
| Centrarchidae | *Lepomis gibbosus* | 3 | 4 | 4 | 4 | | |
| | *Micropterus salmoides* | 2 | 3 | 4 | 4 | | |
| Gobiidae | *Padogobius martensi* | 2 | 3 | 3 | 1 | + | + |
| Cottidae | *Cottus gobio* | 2 | 1 | ? | | + | + |
| Pleuronectidae | *Platichthys flesus luscus* | | | | 2 | | + |

*Fig. 15.10* Longitudinal distribution of the diadromous species *Alosa fallax nilotica* down the Po (modified from Gandolfi & Le Moli, 1977a).

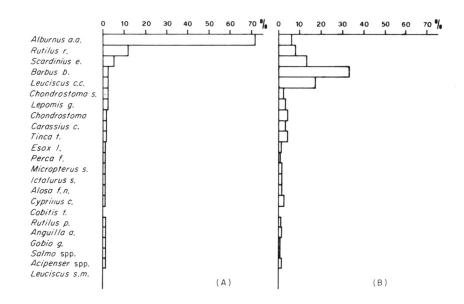

*Fig. 15.11* Species composition of fish catches in the central stretch of the Po during 1976-77 as A) % number, and B) % yield (modified from Vitali, 1977; Vitali & Braghieri, 1981).

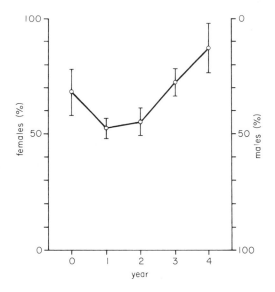

*Fig. 15.12* Age and sex distribution of *Alburnus albidus alborella* population in the central reach of the Po, showing mean values and 95% confidence limits (from Vitali & Braghieri, 1981).

The sex ratio showed a dominance of females (60 - 75%) for all the common species.  For *Alburnus alburnus* the distribution of sex ratio for age classes showed an increase in the differences between sexes with age (Fig. 15.12).

Research between 1974 and 1977 (Vitali, 1977; Aisa & Gattaponi, 1981) showed that 41 - 63% of fish suffered from diseases of internal and external organs, chiefly of an inflammatory nature.  Moreover, 65% of 878 individuals examined, were infected with parasites.  These include Protozoa (*Mixobolus, Trichodina*), Trematoda (*Asymphylodora tincae, Cariophyllaeus* sp.), Nematoda (Ascaridea), Acanthocephala, Hirudinea and Crustacea (*Lamproglena pulchella* and others).

Studies at 11 sampling stations in the delta region showed 47 fish species, some of freshwater, some euryaline and others typical of sea coastal waters (Gandolfi, 1973).  More recent research (Gandolfi *et al.,* 1978; Giannini *et al.,* 1979) has investigated annual fluctuations in species.  A gradual impoverishment in fish populations is taking place here, related to the reduction in flow and greater sedimentation.

15.78  BIRDS

The birds of stream margins of the middle Po have been characterized qualitatively in two ways, species phenology and the choice of

habitats.  Of the 70 species observed, 24 are nesting (8 resident and
16 present in summer), 36 appear more or less regularly during migrations or in the winter season and 10 are present only occasionally.

Among the nesting birds, three heron species are particularly
important, both in terms of biomass and for their close trophic
relations with the aquatic food web; these are the night heron
*Nycticorax nycticorax*, the little egret *Egretta garzetta*, and the
purple heron *Ardea purpurea*.  Their colonial nests can be found in the
residual riparian woods and their trophic territory corresponds to the
river and its tributaries.  Two gulls (the common tern, *Sterna hirundo*
and the little tern, *Sterna albifrons*) nest during their reproductive
on the ground of sandy and almost bare islands.  They catch fish
chiefly in the main stretch of the river.  The habitat of the common
moorhen (*Gallinula chloropus*) and coot (*Fulica atra*) is represented
usually by scrub at the sides and shallow waters with floating and
emergent vegetation.  In winter the resident coots are joined by other
birds and aggregate in large groups in zones of deep water.  A
community of flycatchers (Muscicapidae) and thrushes (Turdinae)
populates the reed and scrub belts in spring and summer.

The middle Po habitats are potentially, at least, among the most
important resting stations of Northern Italy for the migratory aquatic
avifauna.  However, recent changes in the physiognomy of the river and
the various anthropogenic influences have strongly reduced their
suitability.  Ducks (Anatidae), shore birds (Caradriidae), loons
(Gaviidae), grebes (Podicipedidae) and cormorants (Phalacrocoracidae)
now all occur in lower numbers and less regularly.

Species diversity is richest during the period of autumn migration
and least in winter.  The number of individuals and size of biomass
are, on the average, however, relatively high in winter (Toso & Tosi,
1981).

## 15.8  GENERAL COMMENTS

It will be clear from the above account that most attention has been
given to the central stretch of the Po and that most of the biological
data are either descriptive or quantitative accounts of populations.
To frame the Po under a "metabolic" point of view, it is possible only
to refer to data obtained by the Odum (1956) oxygen method.  This study
(Marchetti & Provini, 1977) dealt with six stations sampled over the
same 24-hour period in July 1974.  These data show a pattern of passing
down the river of first autotrophy (P/R ratio higher than 1), then
heterotrophy in the central stretch (P/R declining to 0.29) and finally
rising again in the last station (Fig. 15.13).

Although such a study does not permit a natural heterotrophic
condition to be distinguished from one due to organic pollution (Mann,

1975), it seems almost certain that waste load brought in by the R. Lambro is chiefly responsible. This contains the wastes from 31% of the inhabitants and 36.4% of the industrial equivalents of the whole basin. There was a considerable decrease in gross production below the entry of the Lambro, being five times less than in the upper part of the Po. It seems possible that inhibitory substances present in Lambro water may have influenced this result. Algal growth inhibition has been verified on several occasions by *Selenastrum capricornutum* bioassays, with growth being less than expected on the basis of nutrient concentration (Chiaudani & Vighi, 1977).

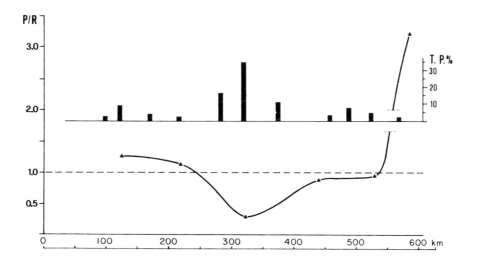

*Fig. 15.13* Longitudinal distribution down the Po of pollution loads expressed as % total.

In addition to the subsequent increase in P/R on passing further downstream, an increase in gross production was also observed, at levels typical of recovery zones (Knöpp, 1960; Hammer, 1965; Woods, 1965). This is related to reduced organic pollution and probably also improved hydrological conditions.

Although the pattern described above is based on one survey, it underlines the existence in the Po as well as other lotic environments, of a "spectrum of metabolic states" (Wetzel, 1975) involving auto- and heterotrophy. The spectrum depends on the role played by, on the one hand, autochthonous primary production, and, on the other by alloch-thonous materials supporting heterotrophic metabolism (Hynes, 1969, 1970; Mann, 1969). For a complete knowledge of the Po, it would be interesting to superimpose the structural imagine of the biological communities on this "spectrum of metabolic states". Such comparison is possible for the central Po and the results suggest that the expected relationship does not exist. The temporal trends (during

1975-77) of the diversity index values calculated by Cironi (1979)
for the macro-invertebrates on *Typha* and *Phragmites* in the central
Po lie between 3 and 4 (Fig. 15.14), values typical of moderately
polluted waters.  The redundancy values, providing an index of species
dominance, remained nearly constant during the period of observation.
The results of the 1978 studies co-ordinated by Ghetti (1980) are also
partly in contrast with the conclusions drawn from the trend of P/R
values.  The use of biotic indices (Woodiwiss, 1964; Verneau & Tuffery,
1968; Woodiwiss, 1978) all indicate good conditions in the central Po
and deterioration in the final stretch (Ghetti, 1980:  see Table 15.3).
This trend is confirmed by decreasing diversity index values on passing
downstream (Table 15.4) and also by other quality indices considered by
Ghetti's co-ordinated working group (Verneaux & Faessel, 1976;
Gardenier & Tolkamp, 1976).

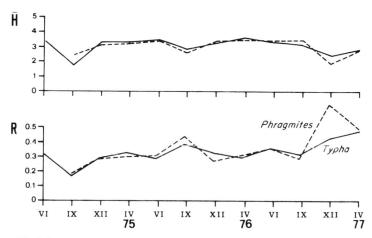

*Fig. 15.14* Changes in diversity index (H) and redundance (R) for
macroinvertebrates on *Phragmites* and *Typha* in the central stretch of
the Po during the period June 1974 to April 1977 (modified from
Cironi, 1977; Cironi & Ruffo, 1981).

The lack of correspondence between the two approaches can be
ascribed mostly to the shortage of metabolic data and the unsuitability
of structural analyses to describe the situation in a river.  If the
complete faunistic list for the central Po is considered, it suggests
a community consisting chiefly of common species typical of polluted
environments, Oligochaetes with *Limnodrilus* dominant and chironomids
of the *Chironomus tummi* group; Plecoptera are virtually absent.  The
studies co-ordinated by Ghetti (1980) and carried out by Schmitz *et
al.* (1980) and Tittizer & Wiemers (1980) on four zones of the Po
showed the dominance of β-mesosaprobic organisms, the rarity of α-
mesosaprobic and oligosaprobic ones and the absence of polysaprobic
ones.  These data are in better accord with the pattern of metabolic
data.

*Table 15.3.* Biotic index values calculated by various authors for four stations on the Po (Ghetti, 1980).

| Index | Authors | km from source | | | |
|-------|---------|------|------|------|------|
| | | 387 | 445 | 535 | 585 |
| Biotic Index | Bonazzi & Ghetti (1980) | 9 | 8 | 6 | 5 |
| (Woodiwiss, 1964) | De Pauw *et al.* (1980) | 9 | 9 | 9 | 5 |
| | Vanhooren & Dehavay (1980) | 8 | 6 | 6 | 5 |
| | Woodiwiss & Rees (1980) | 8 | 6 | 6 | 4 |
| Biotic Index | Bonazzi & Ghetti (1980) | 9 | 6 | 5 | 5 |
| (Verneau & | Debout & Tachet (1980) | 8 | 6 | 5 | 5 |
| Tuffery, 1968) | De Pauw *et al.* (1980) | 6 | 6 | 5 | 4 |
| | Mouthon & Faessel (1980) | 10 | 9 | 5 | 6 |
| | Woodiwiss & Rees (1980) | 7 | 6 | 5 | 4 |
| Extended Biotic | Woodiwiss & Rees (1980) | 8 | 6 | 6 | 4 |
| Index | De Pauw *et al.* (1980) | 9 | 10 | 9 | 5 |
| (Woodiwiss, 1978) | | | | | |

*Table 15.4.* Diversity index values calculated by various authors for four stations on the Po (Ghetti, 1980).

| Diversity Index | Authors | km from source | | | |
|-----------------|---------|------|------|------|------|
| | | 387 | 445 | 535 | 585 |
| Margalef (1951) | Vanhooren & Dehavay (1980) | 2.1 | 2.6 | 1.9 | 1.5 |
| | Woodiwiss & Rees (1980) | 2.4 | 1.4 | 1.1 | 0.5 |
| Shannon & Weaver | Madoni (1980) | 2.8 | 4.2 | 3.7 | 2.8 |
| (1963) | Woodiwiss & Rees (1980) | 1.8 | 1.2 | 0.4 | 0.8 |
| Wilhm & Dorris | Woodiwiss & Rees (1980) | 2.5 | 1.8 | 0.6 | 1.2 |
| (1968) | | | | | |

Many other elements concur to prove the importance of pollution not only for the central stretch, but for nearly the whole river. Among these are the changes in fish populations, the presence of toxic compounds in fish flesh, the inhibitory effects of waters on algal growth and the various chemical and microbiological indices. The Po, the most important Italian river is, except for the uppermost stretch, unfit

for bathing and as a source of drinking water and some restrictions
have to be placed on its use for irrigation.  It is also responsible
for the extreme eutrophication phenomena which have occurred in recent
years in the coastal waters which receive the Po loading (Chiaudani *et
al.*, 1980).

The Po is a river whose ecosystem needs to be studied much more fully
in order to characterize its structural and functional aspects.  Yet
more than enough is known to demonstrate the extent to which the river
has been compromised and to press for immediate intervention.


## REFERENCES

Aisa E. & Gattaponi P. (1981)  Indagine idrobiologica sugli effetti
    degli scarichi termici di centrali termoelettriche nel medio Po.
    Patologia e parassitologia della fauna ittica de medio Po.  *Riv.
    Idrobiol.* 20, 305-36.  (The ecosystem of the middle reaches of the
    River Po.  Pathological processes and conditions and parasitism in
    the ichthyological fauna of the mid-stretch of the river.*)
Allegrini M., Lanzola E. & Marchetti R. (1977)  Indagine sulla qualità
    delle acque del fiume Po.  Cap. 8: Radioattività delle acque.
    *Quaderni Istituto di Ricerca sulle Acque* 32, 495-501.
    (Survey on the water quality of the Po river.  Water radioactivity
    measurements.)
Andreoli C. & Fricano G. (1981)  Variazioni stagionali (giugno 1974-
    maggio 1977) del fitoplancton nel tratto medio del Po interessato
    dalla costruenda Centrale termonucleare di Caorso (Piacenza).  *Riv.
    Idrobiol.* 20(1), 139-52.  (The ecosystem of the middle reaches of
    the River Po.  Seasonal variations (from June 1974 to May 1977) of
    the phytoplankton in the section of the River Po on which the thermo-
    nuclear plant of Caorso (Piacenza) will be built.*)
Berbenni P. (1971)  Igiene e inquinamento del Po.  Atti 2[o] Congresso
    Nazionale del Po.  Mantova 9-10 ottobre 71. Vol. 1, 189-215.
    (Sanitary conditions and pollution of the Po river.)
Bertonati M. (1977)  Indagine idrobiologica per la valutazione degli
    effetti degli scarichi termici di centrali termoelettriche sull'
    ecosistema del medio Po.  Cap. 2: Qualità dell'acqua, and Cap. 3:
    Qualità dei sedimenti.  ENEL-DCO, III Rapporto annuale.  (Hydro-
    biological survey on the effects of thermàl discharges on the Eco-
    system of the middle reaches of the Po river.  Ch. 2: Water quality,
    Ch. 3: Sediment quality.)
Bertonati M. & Ioannili E. (1981)  Indagine idrobiologica sugli effetti
    degli scarichi termici di centrali termoelettriche del medio Po.
    Considerazioni sulle caratteristiche dell'acqua e dei sedimenti nel
    medio Po nel triennio 1974-1977.  *Riv. Idrobiol.* 20(1), 123-38.
    (Hydrobiological survey of the effects of thermal discharges on the
    middle Po.  Remarks about water and sediment characteristics of the
    middle Po river during the triennium 1974-1977.)

*Literal translation of title of volume:  "Hydrobiological survey of the
  effects of thermal discharges on the middle Po."

Bonazzi G. & Ghetti P.F. (1980)  see: Ghetti (1980)  pp. 241-69.

Canali L. & Allodi G. (1962)  Contribution a l'étude sur le transport solide in suspension dans le cours d'eau et sur la dégradation du sol dand le basin du Po.  Symposium AIHS, Bari.

Chiaudani G., Marchetti R. & Vighi M. (1980)  Eutrophication in Emilia-Romagna coastal waters (North Adriatic Sea, Italy): a case history. *Prog. Wat. Tech.* 12, 185-92.

Chiaudani G. & Vighi M. (1977)  Indagine sulla qualità delle acque del fiume Po.  Cap. 9: Caratteristiche trofiche e fenomeni fitotossici. *Quaderni Istituto di Ricerca sulle Acque* 32, 503-21.  (Survey on the water quality of the Po river.  Ch. 9: Trophic and toxic condition.)

Cicioni G., Marchetti R. & Spaziani F.M. (1977)  Indagine sulla qualità delle acque del fiume Po.  Cap. 3: Fonti potenziali di inquinamento. *Quaderni Istituto di Ricerca sulle Acque* 32, 69-114.  (Survey on the water quality of the Po river.  Ch. 3: Potential sources of pollution.)

Cironi R. (1977)  Indagine idrobiologica per la valutazione degli scarichi termici delle centrali termoelettriche sull'ecosistema del medio Po.  Cap. 5: Organismi inferiori del complesso trofico. ENEL-DCO, III Rapporto annuale.  (Hydrobiological survey on the effects of thermic discharges on the ecosystem of the middle reaches of the River Po.  Ch. 5: Phytoplankton, zooplankton and benthos.)

Cironi R. (1979)  Struttura delle comunità macrobentoniche in zone a macrofite nel medio Po.  XLVII Conv. Unione Zool. Italiana, Bergamo, settembre 1979.  (The macroinvertebrate community associated with macrophytes in the middle Po.)

Cironi R. & Ruffo S. (1981)  Indagine idrobiologica sugli effetti degli scarichi termici di centrali termoelettriche nel medio Po.  Il popolamento a macroinvertebrati bentonici del medio Po, in relazione alla qualità dell'acqua. *Riv. Idrobiol.* 20, 47-82.  (The ecosystem of the middle reaches of the River Po.  Some studies on the macro-invertebrate community of the Po near Caorso.*)

Corbetta F. & Zanotti-Censoni A.L. (1977)  Indagine sulla qualità delle acque del fiume Po.  Cap. 17: Cenosi macrofitiche. *Quaderni Istituto di Ricerca sulle Acque* 32, 679-722.  (Survey on the water quality of the River Po.  Ch. 17: Macrophyte distribution.)

Debout O. & Tachet H. (1980)  see: Ghetti (1980)  pp. 162-71.

De Pauw N., Verreth J. & Talloen M. (1980)  see: Ghetti (1980) pp. 34-93.

Fara G.M., Pagno A., Nardi G., Bracchi U. & Legnani L. (1977)  Indagine sulla qualità delle acque del fiume Po.  Cap. 14: Aspetti microbiologici. *Quaderni Istituto di Ricerca sulle Acque* 32, 619-37.  (Survey of the water quality of the River Po.  Ch. 14: Microbiological characteristics.)

Farnesi B., Puddu A. & Spaziani F.M. (1976)  Coefficienti di popolazione equivalente delle attività economiche. *Quaderni Istituto di Ricerca sulle Acque* 33, 106 pp.  (Industry and population equivalents as a means to estimate pollution loads.)

*See p.429.

Ferrarese U. (1981)  Indagine idrobiologica sugli effetti degli scarichi termici di centrali termoelettriche nel medio Po.  Nota sui Ditteri del medio corso del fiume Po a Caorso (Piacenza).  *Riv. Idrobiol.* 20(1), 245-54.  (The ecosystem of the middle reaches of the River Po. Diptera of the middle stream of the river at Caorso (Piacenza.*)

Festa E. (1892)  I pesci del Piemonte.  *Boll. Musei Zool. Anat. comp. Torino* 7, 1-125.  (The fish of Piemonte Region.)

Gaggino G.F. (1982)  Contenuti di alcuni metalli pesanti nella fauna ittica del fiume Po.  *Inquinamento* 24(6), 25.  (Heavy metal content of some fish in the River Po.)

Galassi S. & Provini A. (1981)  Chlorinated pesticides and PCB contents of the two main tributaries into the Adriatic sea.  *Sci. tot. Environ.* 17, 51-57.

Galassi S., Gandolfi G. & Pacchetti G. (1981)  Chlorinated hydrocarbons in fish from the River Po (Italy).  *Sci. tot. Environ.* 20, 231-40.

Galletti P. (1981)  Indagine idrobiologica sugli effetti degli scarichi termici di centrali termoelettriche nel medio Po.  Indaginin idrobiologiche sul medio Po a Caorso: *Odonata*.  *Riv. Idrobiol.* 20, 205-16.  (The ecosystem of the middle reaches of the River Po. Hydrobiological survey of the middle Po at Caorso: Odonata.*)

Galletti P. & Ravizza C. (1977)  Note sull'entomofauna acquatica del corso medio-inferiore del Po: *Odonata*.  *Rend. Ist. Lomb. Sci. Lett.* 111, 89-100.  (Notes on *Odonata* of middle part of the Po river.)

Gandolfi G. (1973)  Primi dati sul popolamento ittico nelle acque interne del delta padano.  *L'Ateneo Parmense* 9, 409-17.  (The fish population of the Delta of the River Po: preliminary observations.)

Gandolfi G. & Le Moli F. (1977a)  Indagine sulla qualità delle acque del fiume Po.  Cap. 18: Distribuzione della fauna ittica nel Po.  *Quaderni Istituto di Ricerca sulle Acque* 32, 723-45.  (Survey on the water quality of the River Po.  Ch. 18: Fish distribution in the River Po.)

Gandolfi G. & Le Moli F. (1977b)  A preliminary report on fish distribution in the Po river.  *Boll. Zool.* 44, 149-54.

Gandolfi G., Giannini M. & Torricelli P. (1978)  Osservazioni sulla biocenosi ittica della Sacca del Canarin (Delta del fiume Po).  *Boll. Zool.* 45 (Suppl.), 24.  (Observations on the fish population in the Delta of the River Po (Sacca del Canarin).)

Gandolfi G. & Giannini M. (1979)  La presenza di *Silurus glanis* nel fiume Po.  *Natura*, Milano 70, 3-6.  (On the presence of *Silurus glanis* in the River Po.)

Gardeniers J.J.P., Tolkamps H.H. & Biejer J.A.J. (1980)  see: Ghetti (1980)  pp. 312-33.

Ghetti P.F. (1980)  Biological Water Assessment Methods.  3rd Technical Seminar on Torrente Parma, Torrente Stirone, Fiume Po.  Vol. 1. Publ. Commission of the European Communities.

Ghetti P.F. & Bonazzi G. (1980)  Biological Water Assessment Methods. 3rd Technical Seminar on Torrente Parma, Torrente Stirone, Fiume Po. Vol. 2. Publ. Commission of the European Communities.

Giandotti M. (1941)  Le caratteristiche fisiche del Po nei rapporti con la sua navigabilità.  *Annali del Ministero dei Lavori Pubblici* 6. (Physical characteristics of the River Po in relation to its navigability.)

Giannini M., Vitali R. & Gandolfi G. (1979) Studio quantitativo sul popolamento ittico di un ambiente salmastro nel delta del fiume Po (Sacca del Canarin). *Atti Soc. Tosc. Sci. Nat. Mem.*, Ser. B. 86 (Suppl.), 100-103. (Quantitative studies on fish populations in brackish water of the Delta of the River Po (Sacca del Canarin).)

Hammer L. (1965) Photosynthese und Primärproduktion im Rio Negro. *Int. Rev. ges. Hydrobiol.* 50, 335-39.

Hynes H.B.N. (1969) The enrichment of streams. In: *Eutrophication, Consequences, Correctives.* Proc. Symp. nat. Acad. Sci., Washington. pp. 188-96.

Ioannilli E. (1981) Indagine idrobiologica sugli effetti degli scarichi termici di centrali termoelettriche nel medio Po. Progetto sperimentale per la valutazione degli effetti degli scarichi termici e trattamento statistico dei dati. *Riv. Idrobiol.* 20(1), 83-122. (The ecosystem of the middle reaches of the River Po. Experimental design to assess the impact of thermal discharge and statistical treatment of data.)

IRSA (Istituto di Ricerca sulle Acque, Consiglio Nazionale delle Ricerca) (1977) Indagine sulla qualità delle acque del fiume Po. *Quaderno* 32, 821 pp. (Survey of the water quality of the River Po.)

Knöpp H. (1960) Untersuchungen über das Sauerstoff-Produktions-Potential von Flussplankton. *Schweiz. Z. Hydrol.* 22, 152-66.

Leone V. (1981) Indagine idrobiologica sugli effetti degli scarichi termici di centrali termoelettriche nel medio P. Quadro faunistico del medio Po: i Vertebrati. *Riv. Idrobiol.* 20(1), 83-92. (The ecosystem of the middle reaches of the River Po. Faunistic picture of the middle river: Vertebrata.)

Madoni P. (1980) see: Ghetti (1980) pp. 295-312.

Mann K.H. (1969) The dynamic of aquatic ecosystems. In: Cragg J.B. (Ed.) *Advances in Ecological Research* 6, 1-81.

Mann K.H. (1975) Patterns of energy flow. In: Whitton B.A. (Ed.) *River Ecology.* 725 pp., pp. 248-63. Blackwell, Oxford.

Marchetti R., Garibaldi L. & Zambon G. (1977) Indagine sulla qualità delle acque del fiume Po. Cap. 7: La qualità delle acque del Po, quadro di insieme. *Quaderni Istituto di Ricerca sulle Acque* 32, 345-493. (Survey of the water quality of the River Po. Ch. 7: The overall picture.)

Marchetti R., Pennacchioni A., Ottolenghi L. & Gaggino G.F. (1973) Indagine sul mercurio totale e metile in acque interne italiane e in specie ittiche dulcicole. *Acqua Aria* 32, 41-52. (Inorganic and organic mercury concentration in water and fish of different Italian freshwater environments.)

Marchetti R. & Provini A. (1977) Indagine sulla qualità delle acque del fiume Po. Cap. 10: Ossigeno e carico organico. *Quaderni Istituto di Ricerca sulle Acque* 32, 523-49. (Survey of the water quality of the River Po. Ch. 10: Oxygen balance and its relationship to organic loads.)

Margalef R. (1951) Diversidas de species en las communidades naturales. *P. Ins. Biol. Appl.* 9, 5-27. (Species diversity in natural communities.)

Minelli A. & Mannucci Minelli M.P. (1981)  Indagine idrobiologica
    sugli effetti degli scarichi termici di centrali termoelettriche nel
    medio Po.  Irudinei del medio corso del Po a Caorso (PC).  *Riv.*
    *Idrobiol.* 20(1), 187-94.  (The ecosystem of the middle reaches of the
    River Po.  Leeches of the middle river at Caorso.)
Ministero dei Lavori Pubblici (1979)  *Per il Po e i suoi affluenti.*
    Pubbl. Magistrato per il Po, Parma.  148 pp.  (River Po and its
    tributaries.)
Moretti G.P., Vigano' A. & Taticchi Vigano' M.I. (1964)  Some informa-
    tion on the orobiontic fauna of Trichoptera of the Italian Western
    Alps above 2000 m.  *Proc. Internat. Symp. Trichoptera*, Lunz, pp.
    87-92.  (Some information on the orobiontic fauna of Trichoptera of
    the Italian Western Alps above 2000 m.)
Mouthon J. & Faessel B. (1980)  see: Ghetti (1980)  pp. 174-205.
Nardi G. (1977)  Indagine sulla qualità delle acque del fiume Po.  Cap.
    15: Aspetti virologici.  *Quaderni Istituto di Ricerca sulle Acque*
    32, 639-59.  (Survey on the water quality of the River Po.  Ch. 15:
    Viral component.)
Odum H.T. (1956)  Primary production of flowing waters.  *Limnol.*
    *Oceanogr.* 1, 102-17.
Paoletti di Chiara A. (1981)  Indagine idrobiologica sugli effetti
    degli scarichi termici di centrali termoelettriche nel medio Po.
    Gli Oligocheti del benthos del medio Po a Caorso (PC).  *Riv.*
    *Idrobiol.* 20, 173-78.  (The ecosystem of the middle reaches of the
    River Po.  Benthic Oligochaeta of the middle river near Caorso.)
Pavesi P. (1896)  La distribuzione dei pesci in Lombardia.  Fusi, Pavia.
    (Fish distribution in Lombardy.)
Pennacchioni A. & Marchetti R. (1973)  Mercury contamination in some
    Italian waters, organisms and sediments.  In: Proc. Int. Symposium,
    *Problems of the Contamination of Man and Environment by Mercury and*
    *Cadmium*, Luxembourg, July 1973.
Piantino P., Vanini G.C. & Verde L. (1977)  Indagine sulla qualità delle
    acque del fiume Po.  Cap. 2: Precedenti bibliografie sulla qualità
    delle acque.  *Quaderni Istituto di Ricerca sulle Acque* 32, 21-67.
    (Survey on the water quality of the River Po.  Ch. 2: Annotated
    bibliography on the water quality of the River Po.)
Prati L. (1977a)  Indagine sulla qualità delle acque del fiume Po.
    Cap. 12: Pesticidi.  *Quaderni Istituto di Ricerca sulle Acque* 32,
    581-603.  (Survey of the quality of the River Po.  Ch. 12: Pesti-
    cides.)
Prati L. (1977b)  Indagine sulla qualità delle acque del fiume Po.
    Cap. 13: Detergenti anionici e fenoli.  *Quaderni Istituto di Ricerca*
    *sulle Acque* 32, 605-18.  (Survey of the quality of the River Po.
    Ch. 13: Anionic detergents and phenols.)
Provini A., Mosello R., Pettine M., Puddu A., Rolle E. & Spaziani F.M.
    (1979)  Metodi e problemi per la valutazione dei carichi di
    nutrienti.  In: *Atti del "Convegno sulla Eutrofizzazione in Italia".*
    C.N.R., Roma 3-4 October 1978.  (Methods and problems for nutrients
    load calculation.)

434     CHIAUDANI & MARCHETTI

Provini A., Gaggino G.F. & Galassi S. (1980)  Po e Adige: valutazione
    statistica della frequenza di campionamento in un programma di
    monitoraggio.  *Ing. Ambientale* 9, 379-90.  (Po and Adige: statistical
    evaluation of sampling frequency in a monitoring program.)
Querena E. (1981)  Indagine idrobiologica sugli effetti degli scarichi
    termici di centrali termoelettriche nel medio Po.  Gli Efemerotteri
    nel medio Po a Caorso (PC).  *Riv. Idrobiol.* 20(1), 195-204.  (The
    ecosystem of the middle reaches of the River Po.  Ephemeroptera of
    the middle river at Caorso.*)
Raffa U. (1972)  Particolarità del regime dei corsi d'acqua padani in
    relazione ai fenomeni di inquinamento.  *Quaderni Assoc. Idrotecnica
    Italiana* (sez. Padana) 1, 15-36.  (Hydrological particularities of
    the rivers of the Po valley in relation to pollution.)
Raffa U. (1973)  L'evoluzione recente del regime idrico e termico del
    Po.  2° Convegno sullo stato di avanzamento della radioecologia in
    Italia.  Parma 24-25 May 1973.  (Recent hydrological and thermal
    changes in the River Po.)
Raffa U. & Giuffrida G. (1973)  Alimentazione delle falde attraverso
    d'alveo dei corsi d'acqua.  Atti 2° Colloquio Internazionale sulle
    acque sotterranee.  Palermo, 28-30 April.  (Artificial recharge of
    groundwater by river water infiltration.)
Ravizza C. (1976)  Note sull'entomofauna acquatica del corso medio-
    inferiore del Po: *Hemiptera heteroptera*. *Rend. Ist. Lomb. Sci.
    Lett.* 110, 55-66.  (Note on the aquatic entomological fauna of the
    mid-lower stretch of the Po: *Hemiptera heteroptera*.)
Ravizza C. (1981)  Indagine idrobiologica sugli effetti degli scarichi
    termici di centrali termoelettriche nel medio Po.  Note sugli
    Emitteri ed i Coleotteri popolanti le acque del medio Po, fra
    Caorso e Isola Serafini.  *Riv. Idrobiol.* 20, 217-230.  (The eco-
    system of the middle reaches of the River Po.  Notes on the aquatic
    Hemiptera and Coleoptera inhabiting the river between Caorso and
    Isola Serafini.*)
Rossaro B. (1976)  Lo zooplancton del Po a Caorso. *Rend. Ist. Lomb.
    Sci. Lett.* 110, 35-54.  (Zooplankton of the Po at Caorso.)
Rossaro B. (1978)  Composizione tassonomica e fenologica delle *Ortho-
    cladiinae* (Dipt. Chironomidae) nel Po a Caorso (PC) determinata
    mediante analisi delle exuvie. *Riv. Idrobiol.* 17, 277-300.  (The
    taxonomic composition and phenology of the Orthocladiinae, Dipt.
    Chir. in River Po near Caorso, as determined by analysis of pupal
    exuviae.)
Rossaro B. (1981)  Indagine idrobiologica sugli effetti degli scarichi
    termici di centrali termoelettriche nel medio Po.  Il carico biolo-
    gico del medio Po presso Caorso (PC): la sua composizione specifica
    e le relazioni con le proprieta dell'acqua esaminate mediante
    l'analisi delle componenti principali. *Riv. Idrobiol.* 20(1),
    152-166.  (The ecosystem of the middle reaches of the River Po.
    The biological load in the middle Po near Caorso: specific composi-
    tion and its relation with water characteristics, examined by
    principal components analysis.)

Rossaro B. & Cotta-Ramusino M. (1976) I Copepodi del Po a Isola
Serafini. *Riv. Idrobiol.* 15, 187-203. (Copepoda of the River Po
near Isola Serafini.)

Rossetti M. & Raffa U. (1973) Evoluzione idrologica e idrografica
della regione del Delta del Po. *Ann. Univ. Ferrara*, Sez. 1
(Ecologia) 1, 5-28. (On the evolution of Po hydrography and hydrology in the Delta.)

Sambugar B. (1981) Indagine idrobiologica sugli effetti degli scarichi
termici di centrali termoelettriche nel medio Po. Gli Oligocheti
raccolti tra i banchi di macrofite nel medio Po a Caorso (PC).
*Riv. Idrobiol.* 20, 179-186. (The ecosystem of the middle reaches of
the River Po. The Oligochaeta sampled from the bank macrophytes of
the middle Po at Caorso.*)

Shannon C.E. & Weaver W. (1963) *The Mathematical Theory of Communi-
cation*. Univ. Illinois Press, Urbana.

Schmitz W., Buck H. & Vobis H. (1980) see Ghetti (1980) pp. 383-415.

Tittizer T. & Wiemers W. (1980) see: Ghetti (1980) pp. 335-81.

Tosi G. & Toso S. (1981) Indagine idrobiologica sugli effetti degli
scarichi termici di centrali termoelettriche nel medio Po. Caratteristiche strutturali e dinamiche della popolazione ornitica del
medio Po. *Riv. Idrobiol.* 20(1), 337-82. (The ecosystem of the
middle reaches of the River Po. The bird population along the Po
near Caorso.*)

Ugolini P. (1971) La navigazione interna nella Valle Padana. Atti 2°
Congresso Nazionale del Po. Mantova, 9-10 ottobre 71, Vol. 1,
107-23. (Navigation through the Po river system.)

Vanhooren G. & Dehavay P. (1980) see: Ghetti (1980) pp. 96-139

Verneaux J. & Tuffery G. (1968) Méthode zoologique pratique de détermination de la qualité biologique des eaux courantes. Indices Bio-
tiques. *Annls scient. Univ. Besançon Zool.* 3, 79-90.

Verneaux J. & Faessel B. (1976) Quoted in Mouthon & Faessel (1978).

Vitali R. (1977) Indagine idrobiologica per la valutazione degli
effetti degli scarichi termici di centrali termoelettriche sull'
ecosistema del medio Po. Cap. 6: Popolazione ittica. ENEL-DCO,
III Rapporto annuale. (Hydrobiological survey of the effects of
thermal discharge on the ecosystem of the middle reaches of the Po.
Ch. 6: Fish stock assessment.)

Vitali R. & Braghieri L. (1979) Parametri dinamici di popolazioni di
barbo (*Barbus barbus plebejus*) e cavedano (*Leuciscus cephalus cabeda*)
nel medio Po. XLVII Conv. Unione Zool. Italiana. Bergamo,
September 1979. (Dynamic characteristics of fish populations
(*Barbus barbus plebejus*) and (*Leucisus cephalus cabeda*) in the middle
stream of Po river.)

Vitali R. & Braghieri L. (1981) Indagine idrobiologica sugli effetti
degli scarichi termici di centrali termoelettriche nel medio Po.
Caratteristiche strutturali e dinamiche del popolamento ittico del
medio Po nella zona di Caorso. *Riv. Idrobiol.* 20(1), 269-304.
(The ecosystem of the middle reaches of the River Po. Structural
and dynamic characteristics of the fish community of the middle
stream of Po river at Caorso.*)

Viviani M.A. & Tortorano A.M. (1977)  Indagine sulla qualità delle acque del fiume Po.  Cap. 16: Aspetti micologici. *Quaderni Istituto di Ricerca sulle Acque* 32, 661-78.  (Survey of the water quality of the River Po.  Ch. 16: Survey of yeasts.)

Vivoli G. & Vecchi G. (1977)  Indagine sulla qualità delle acque del Fiume Po.  Cap. 11: Metalli tossici. *Quaderni Istituto di Ricerca sulle Acque* 32, 551-78.  (Survey of the water quality of the River Po.  Ch. 11: Heavy metals.)

Wetzel R.G. (1975)  Primary Production.  In: Whitton B.A. (Ed.) *River Ecology*.  725 pp., pp. 230-47.  Blackwell, Oxford.

Wilhm J.L. & Dorris T.C. (1968)  Biological parameters for water quality criteria. *Bioscience* 18, 477-81.

Woodiwiss F.S. (1964)  The biological system of stream classification used by the Trent River Boards. *Chem. Ind.*, 443-47.

Woodiwiss F.S. (1978)  Comparability study of biological-ecological assessment methods.  2nd Technical Seminar on the River Trent and tributaries.  Commission of European Communities, Brussels.

Woodiwiss F.S. & Rees S. (1980)  see: Ghetti (1980)  pp. 417-40.

Woods W. (1965)  Primary production measurements in the Upper Ohio River. *Spec. Publ., Pymatuning Hab. Ecol. Univ. Pittsburg* 3, 66-78.

Zullini A. (1969)  Osservazioni sui Nematodi di un biotopo fluviale (Po a Trino Vercellese). *Rend. Accad. naz. Lincei, sez.* VIII, 47, 109-15.  (Notes on aquatic Nematoda in the River Po (Trino Vercellese).)

Zullini A. (1974)  The nematological population of the Po river. *Boll. Zool.* 41, 183-210.

Zullini A. (1975a)  Nematodi dello Psammon del Po. *Rend. Ist. Lomb. Sci. Lett.* 109, 79-83.  (Nematoda as a component of psammon of the River Po.)

Zullini A. (1975b)  Nematodi ripicoli del Po. *Rend. Ist. Lomb. Sci. Lett.* 109, 130-42.  (Riparian nematodes of the Po.)

# 16 · Lower Rhine-Meuse

G. VAN URK

16.1    Introduction                                           438
16.2    General data                                           439
16.3    The Meuse in the Netherlands                           442
        16.31  Hydrography                                     442
        16.32  Pollution                                       442
               16.321   Sewage discharges                      442
               16.322   Industrial waste water discharges 442
               16.323   Water quality                          442
        16.33  Plankton                                        444
        16.34  Macroinvertebrates                              444
               16.341   Before regulation                      444
               16.342   After regulation                       445
        16.35  Fish                                            445
16.4    The freshwater tidal area of the Biesbosch             445
        16.41  History                                         445
        16.42  Hydrography and hydrochemistry                  447
        16.43  Plankton                                        449
        16.44  Benthic plants                                  450
               16.441   Phanerogams                            450
               16.442   Cryptogams                             453
        16.45  Macroinvertebrates                              454
               16.451   Tidal channels                         454
               16.452   Intertidal zone                        454
        16.46  Fish                                            455
        16.47  Concluding remarks                              456
16.5    Present state of the lower Rhine                       457
        16.51  Pollution monitoring in the lower Rhine         457
        16.52  Macroinvertebrate fauna of the IJssel           458
               16.521   Description of the IJssel              458
               16.522   Macroinvertebrates on stones           459
               16.523   Macroinvertebrates in sediments        462
               16.524   Prospects                              463
References                                                     464

## 16.1   INTRODUCTION

The geomorphology of the central part of the Netherlands is largely
determined by the rivers Rhine and Meuse.

Originally, the Rhine meandered freely in its lower reaches and continually formed new courses.  Its present partition into three branches
- Waal, Nederrijn and IJssel - was effected by means of two canals, the
Pannerdens Kanaal, dug in 1707, and the Bijlands Kanaal, which dates
back to 1780; both are described in detail by Van de Ven (1976).  Further downstream the river ran through a large peat swamp, where it also
flowed through different courses.  For embankment and land reclamation
purposes the river needed to be restricted in its freedom.  The present
configuration was not reached until 1870 when the Nieuwe Merwede was
completed, which leads most of the Waal water into the Hollands Diep.

The Hollands Diep also receives the water from the Meuse by way of
the Bergsche Maas, which was completed in 1904 (Figs. 16.1, 16.3).  Until
very recently, no port or industry of any significance has grown up
along the Hollands Diep and Haringvliet.  The port of Rotterdam is
situated further north, at a former mouth of the Meuse, and was connected to the sea by the Nieuwe Waterweg in 1872.  Therefore, large parts
of the Rhine-Meuse estuary long remained fairly unspoiled.  Its freshwater tidal area was unique in Europe, in spite of the pollution of the
rivers Rhine and Meuse.

In 1970 a dam with sluice was completed across the mouth of the
Haringvliet, as part of the so-called "Delta Works", which are to protect the south-western part of the Netherlands against flooding from
the sea.  The closure has reduced considerably tidal movements in the
Haringvliet and Hollands Diep (Deltadienst of Rijkswaterstaat, 1973)
and robbed the area of most of its unique natural features.  Before the
closure a number of studies had been carried out in the freshwater
tidal area, which enables me to include a general description of this
particular habitat.  In order to understand the hydrography and hydrochemistry of the main part of this area, the Netherlands stretch of the
Meuse will be described briefly, as will the present state of one of
the branches of the Rhine.

## 16.2  GENERAL DATA
(Meuse in the Netherlands)

DIMENSIONS
Area of catchments
whole Meuse (including other countries) 33000 $km^2$
below Eysden (Belgian-Dutch border)     12000 $km^2$ (part of catchment
in Germany)

major tributaries: Roer (Rur)            2350 $km^2$ (majority in Germany)
                   Niers                  1350 $km^2$ (majority in Germany)
                   Dieze                  2800 $km^2$

Length of river
Eysden to most downstream weir (Lith)      195 km
Lith to confluence with branch of Rhine
(Nieuwe Merwede)                            60 km
final stretch to the sea (estuary of
Haringvliet)                                50 km
Drop in height of river in Netherlands      45 m

CLIMATE

Precipitation
In Dutch part of the catchment about 700 mm $yr^{-1}$; 200 mm $yr^{-1}$ carried
off by the tributaries of the Meuse.

LAND USE
agriculture                              4000 $km^2$
woodlands                                 900 $km^2$
urban                                     700 $km^2$

POPULATION
Meuse catchment in the Netherlands: $3 \times 10^6$ people = 250 people $km^{-2}$

PHYSICAL FEATURES OF RIVER
Discharge (means for the period 1951-1960: $m^3 s^{-1}$

|           | Borgharen | Lith |           | Borgharen | Lith |
|-----------|-----------|------|-----------|-----------|------|
| January   | 504       | 609  | August    | 108       | 167  |
| February  | 456       | 538  | September | 126       | 185  |
| March     | 356       | 468  | October   | 159       | 232  |
| April     | 227       | 306  | November  | 257       | 332  |
| May       | 149       | 208  | December  | 377       | 465  |
| June      | 116       | 169  |           |           |      |
| July      | 83        | 127  | year      | 243       | 317  |

These mean values do not reflect the large fluctuations in discharges
that occur because the Meuse is a rain-fed river. During dry periods
(e.g. summer 1976) the discharge at Borgharen may be less than 10 $m^3 s^{-1}$.
A treaty between Belgium and the Netherlands should guarantee a minimum
discharge of 25 $m^3 s^{-1}$ by building reservoirs upstream.

## Tributaries

|  | discharge $(m^3 s^{-1})$ | point at km entry to Meuse |
|---|---|---|
| Roer (= Rur) | 20 | 79 |
| Niers | 8 | 157 |
| Dieze (confluence of Aa and Dommel) | 18 | 220 |

## Major Weirs

|  | km sign | water level above weir (+NAP) |
|---|---|---|
| Borgharen | 15 | 43.90 |
| Linne | 71 | 20.50 |
| Roermond | 81 | 16.75 |
| Belfeld | 100 | 14.00 |
| Sambeek | 146 | 10.75 |
| Grave | 175 | 7.50 |
| Lith | 201 | 4.60 |

Regulation changed the character of the river completely except for the section between Borgharen and Linne.

## Abstraction from Meuse

Important abstraction points for Rotterdam and the Hague are near km sign 250.  Both cities use Meuse water for their drinking water supply.

## Substratum

Gravel, sand, silt, clay.

## Temperature (monthly mean values at Borgharen)

|  | 1976 | 1977 |  | 1976 | 1977 |
|---|---|---|---|---|---|
| January | 6.6 | 5.1 | August | 19.7 | 20.9 |
| February | 4.5 | 6.6 | September | 16.3 | 19.3 |
| March | 7.3 | 10.5 | October | 14.3 | 16.0 |
| April | 11.2 | 10.2 | November | 9.3 | 11.3 |
| May | 15.4 | 15.8 | December | 5.3 | 7.5 |
| June | 19.4 | 20.0 |  |  |  |
| July | 22.0 | 21.8 | year | 12.6 | 13.8 |

These temperatures are of course not natural, but influenced by regulation and cooling water discharges.

*Fig. 16.1* The Rhine-Meuse system in the Netherlands.  The Kromme
Rijn, Oude Rijn and Vecht do not form a part of the main system nowadays.  Canals are not shown.  The figures indicate km signs.

16.3   THE MEUSE IN THE NETHERLANDS

## 16.31   HYDROGRAPHY

The main data on the hydrography of the Netherlands stretch of the Meuse
are summarized in 16.2.   The regulation of the river was carried out in
the years 1919-1926, except for the river section between km 17 and km
63, which forms the boundary between Belgium and the Netherlands.   Here
a lateral canal was constructed instead.   As a result, this section of
the river has retained some of its original character (16.35).   More
data on the hydrography of the Meuse basin are found in Van der Made
(1972) and Chapter 1.   (The km values in the present chapter refer to
distance from where the Meuse first meets the Netherlands border.)


## 16.32   POLLUTION

### 16.321 *Sewage discharges*

A survey of the discharges of biodegradable organic wastes in 1970 was
presented in a report by a working group on the sanitation of the
Meuse (Werkgroep Sanering Maas, 1971).   Altogether some $3.4 \times 10^6$ population equivalents were discharged into the river, directly or indirectly, the greater part between km 12 and km 108.   The greatest
source, the river Roer, discharged some $10^6$ population equivalents into
the Meuse at km 79.   As a result of a sanitation programme, sewage discharges have been cut down considerably (Dijkzeul, 1981).


### 16.322 *Industrial waste water discharges*

In 1970, industry discharged large amounts of ammonia, cyanides and
heavy metals into the Meuse.   The river's high cadmium and zinc contents - about 4 µg $l^{-1}$ and 250 µg $l^{-1}$, respectively - are partly due
to former mining activities, but metallurgical works have also made
substantial contributions.   Mercury used to be discharged by a chlor-
alkali plant near km 74 and chromium by a plant near km 109.   In recent
years these discharges have been reduced significantly.


### 16.323 *Water quality*

Due to long residence times in the river sections between weirs, as
well as the aerating effect of the weirs, the average oxygen content of
the Meuse water is nearly always above 6 mg $l^{-1}$ (Fig. 16.2; Koolen,
1973).   The $NH_4^+$-content has a maximum of more than 5 mg $l^{-1}$ in the
unregulated section of the river between km 17 and km 52, but decreases
rapidly downstream, at least in summer, due to nitrification.   The
phosphate-P content is 0.3 - 0.7 mg $l^{-1}$, the "natural" content being
about 0.07 mg $l^{-1}$ (Zuurdeeg, 1980).   Heavy metals, except nickel, are
for the greater part adsorbed on the suspended matter and therefore

*Fig. 16.2* Mean oxygen content of Meuse in the Netherlands during summer 1970.  The km distances are those indicated on signs by the river.  (Data from Rijksinstituut voor Zuivering van Afvalwater)

accumulate in the sediments.  At high river flow the sediments may be resuspended in the water; consequently, exceptionally high metal contents may be measured, but normally the metal contents of the downstream reaches are not very high.

*Table 16.1* Trade metal concentrations in sediments from the Meuse and the Rhine, expressed as $\mu g\ g^{-1}$ at 50% <16 µm.  (Data derived from:  Salomons & Mook, 1977; Rijksinstituut voor Zuivering van Afvalwater, 1980; unpublished reports)

|                                         | Cr  | Ni | Cu  | Zn   | Cd  | Pb  |
|-----------------------------------------|-----|----|-----|------|-----|-----|
| river sediments deposited 15th-16th century | 77  | 33 | 21  | 93   | 0.5 | 31  |
| Meuse sediment about 1960               | 216 | 44 | 160 | 1516 | 28  | 382 |
| Rhine sediment about 1900               | 140 | 35 | 42  | 380  | 1.4 | 82  |
| Rhine sediment about 1960               | 640 | 54 | 295 | 2420 | 14  | 530 |
| Rhine sediment 1977                     | 830 | 75 | 290 | 1660 | 37  | 390 |

So, despite considerable waste water discharges in its middle course, the water quality of the Meuse, where it enters its tidal section (km 201), has always been relatively good, or at least far better than that of the Rhine in the freshwater tidal area. The same applies to the content of heavy metals - except cadmium and zinc- in Meuse sediments as compared with Rhine sediments (Table 16.1).

## 16.33  PLANKTON

The composition of the plankton of the Meuse, analysed since 1966 (Rijksinstituut voor Zuivering van Afvalwater, 1970), has not changed in recent years. In the river sections between the weirs algal blooms may develop. These blooms do not occur in regular seasonal patterns; their formation depends on river flow (Table 16.2).

*Table 16.2*  Algal blooms in the Meuse at km 201 in five successive years.

| year | month | biomass ($\mu$g l$^{-1}$ chlorophyll *a*) | dominant species |
|------|-------|-------------------|------------------|
| 1974 | April | 78 | *Synedra acus* |
|      | May-October | 65-107 | *Melosira granulata* |
| 1975 | June | 78 | *Diatoma elongatum* |
| 1976 | April | 53 | *Asterionella formosa* |
|      | May | 80 | *Diatoma elongatum* |
| 1977 | May-October | 56-103 | *Melosira granulata* |
| 1978 | July | 49 | *Stephanodiscus hantzschii* |
|      | August | 80 | *Melosira granulata* |

## 16.34  MACROINVERTEBRATES

### 16.341 *Before regulation*

Shortly before work on the regulation of the Meuse was started, a working group was formed to study the ecological impact of regulation. In the summer of 1918 the group went on a rowing expedition down the river in order to collect the material, and published a preliminary report the following year (Romijn *et al.*, 1919). Typical rheophilous species were still found at that time, such as the gyrinid beetle *Orectochilus villosus*, which was abundant on the stones of the jetties (see 16.521) and the water-bug *Aphelocheirus aestivalis*. On the whole, the identification of Ephemeroptera, Trichoptera and Chironomidae was not up to species level, so that no good picture of these groups can be derived from this report. Imagines of the damsel fly

*Calopteryx splendens* were reported to be abundant in the vegetation at the riverside, but it is not explicitly stated if the larvae occurred in the river itself. The closely related *C. virgo* was found only along the brooks that flow into the Meuse. Larvae of *Ophiogomphus serpentinus* were recorded in the river itself. Imagines of the water-beetle *Helophorus arvernicus* were also abundant, but again nothing was said about the occurrence of the larvae. Among the molluscs, *Unio crassus* var. *batavus* was characteristic of the Meuse.

Although the report announces subsequent studies, the difficulties encountered in the organization of more expeditions may have been too great; as it is, no reports on the Meuse prior to its regulation have appeared since then.

### 16.342 *After regulation*

Recent data on the macroinvertebrates of the Meuse are scarce; they point to a rather diverse macroinvertebrate fauna which resembles that of stagnant eutrophic waters in the Netherlands, so from the viewpoint of river ecology there is little reason to go into this any further. The crayfish *Orconectes limosus* seems to have invaded the Netherlands via the Meuse basin and is likely to replace the indigenous species *Astacus astacus* in most habitats (Geelen, 1975).

### 16.35   FISH

Although the barbel *Barbus barbus* is uncommon in the Netherlands, it still occurs in the Meuse. It has survived the regulation, especially in the section between km 15 and km 71, as mass fish kills due to excess discharges of ammonia or cyanides by a local industry have demonstrated. Another common fish in the Meuse is the beaked carp *Chondrostoma nasus*. Locally numerous in the Meuse, but also occurring elsewhere, are the dace *Leuciscus leuciscus* and the chub *Leuciscus cephalus*. For the rest, the fish stock of the Meuse resembles that of other - stagnant - waters in the Netherlands (De Groot & Muyres, 1980).

## 16.4   THE FRESHWATER TIDAL AREA OF THE BIESBOSCH

### 16.41   HISTORY

The western part of the Netherlands between the river branches (Vecht, Oude Rijn, Lek, Merwede and former branches of the Meuse situated south of the Merwede; Figs. 16.1, 16.3) originally consisted of peat swamps lying above mean river level. In the Middle Ages the swamps were embanked and cultivated as pastures. Drainage caused the peat soil to set. As a consequence, the rivers rose to levels higher than pre-

*Fig. 16.3* Map showing man-made channels in the main system of the Rhine-Meuse, together with former river channels.

viously and overflowed if a dike was breached.  Flooded areas were
usually reclaimed by repairing the dikes.

In 1421 the so-called "Sint Elizabethsvloed" inundated the Groote
Waard, a very large polder area between the Merwede and a former course
of the Meuse, which was never fully re-embanked again due to administrative disorganization.  The rivers Waal and Meuse flowed into this
newly formed sheet of water.  The sand transported by these rivers was
deposited on the original peat soils, forming sandflats which developed
a vegetation of their own, which in turn facilitated the deposition of
silt by a reduction of the currents.  Many such sand- and mud-flats
were embanked and transformed into small polders suitable for farming.
Eventually, the sand- and mud-flats and polders hampered the rivers in
their flow, increasing the risk of inundation upstream at high river
discharges.  Therefore, at the end of the 19th century two new channels
were dug, the Nieuwe Merwede and the Bergsche Maas, which were to carry
off the water of the Waal and the Meuse, respectively.  With the completion of these channels, the present configuration of the area was
achieved.  The area was called the Biesbosch, which stands for "wood
of rushes".  The development of the Biesbosch is illustrated in Fig.
16.4.  Unless stated otherwise, the following paragraphs deal with the
situation which existed before 1970 (16.1).


16.42  HYDROGRAPHY AND HYDROCHEMISTRY

The isohaline of 300 mg $Cl^-$ $l^{-1}$, which may be considered as the boundary between fresh and brackish water (Den Hartog, 1968), normally
lies in the Hollands Diep, a few kilometres west of the Biesbosch
(Peelen, 1967).  Mean high and low water levels on the Nieuwe Merwede/
Waal and on the Amer/Bergsche Maas/Meuse are given in Fig. 16.5; data
on the ebb and flood currents in the rivers are summarized in Table
16.3 (data from Rijkswaterstaat).  The currents in the tidal channels
of the Biesbosch were about the same as in the Amer.  The same applies
to mean high and low water levels.  At ebb-tide many of the smaller
creeks fell dry.

The creeks north of the Nieuwe Merwede were fed by water from the
Rhine.  The channels and creeks south of the Nieuwe Merwede were not
fed directly by the Nieuwe Merwede thanks to a dike along the south
bank of the river, but by the Amer, which carries off water from the
Meuse.  From the $Cl^-$ contents it was concluded that only the western
part received some water from the Rhine which entered at flood tide
from the Hollands Diep, as the flow of the Nieuwe Merwede is much
greater than that of the Meuse (Parma, 1968).  This resulted in short
residence times of the water in the northern part of the Biesbosch, so
that the water quality here was nearly as bad as in the Rhine itself.
In the Meuse sector, on the contrary, the oxygen content was near
saturation, except in the small creeks where oxygen consumption by
sediments was high.  The $NH_4^+$ content was lower and the $NO_2^-$ and $NO_3^-$
contents were higher than in the Meuse (Parma,  1968).

*Fig. 16.4*  Development of the Biesbosch area from 1700 to 1960.
(Based on historical maps)

*Fig. 16.5* Mean high water levels (MHW) and mean low water levels (MLW) on the Nieuwe Merwede-Waal (Rhine) and the Amer-Bergse Maas-Maas (Meuse) in 1969. (Data from Rijkswaterstaat)

*Table 16.3* Velocity (m s$^{-1}$) and direction of the maximum ebb and flood currents on the rivers in the tidal area.

| | | Rhine flow (m$^3$ s$^{-1}$) | | | | |
|---|---|---|---|---|---|---|
| | | 980 | | 2300 | | 9400 |
| | | ebb | flood | ebb | flood | ebb | flood |
| Nieuwe Merwede | 979 | 0.65 ← | 0.55 → | 0.70 ← | 0.40 → | 1.15 ← | 0.35 → |
| Boven Merwede | 960 | 0.75 ← | 0.40 → | 0.90 ← | 0.10 ← | 1.85 ← | 1.55 ← |
| Amer | 261 | 0.75 ← | 0.85 → | 0.80 ← | 0.85 → | 0.85 ← | 0.70 → |

## 16.43 PLANKTON

A typical plankton community developed only in the Meuse sector. It is however difficult to obtain a full picture of the plankton composition: in the absence of complete mixing in the channel system, different populations may have developed at different places (Ringelberg, 1968) and due to changing current velocity the plankton composition changed within a tidal cycle. Maximal concentrations of phytoplankton were found, when the flood current was maximal. At high water, part

of the phytoplankton sank to the bottom and was only partly resuspended
in the ebb current.  Different species reacted in different ways to
changes in the current velocity (Dresscher *et al.*, 1968).  At low water,
many planktonic diatoms could be found on the sand- and mud-flats
(Zonneveld, 1960, Appendix 15).

The characteristic species of the Meuse sector of the Biesbosch were
the diatom *Actinocyclus normannii* and the copepod *Eurytemora affinis*
(Dresscher, 1968; Leentvaar, 1968).  These species also occur in great
numbers in the freshwater tidal areas of other rivers, such as the
Elbe (Nöthlich, 1972) and the Weser (Behre, 1961).  In both the Elbe
and the Biesbosch, the greatest numbers of *E. affinis* were found in
spring, while *Actinocyclus normannii* reached its maximum in the summer
months.  It seems clear that there is no trophic relationship between
these species, but that each species is adapted to the current conditions in the freshwater tidal area.  Otherwise, there was a great
resemblance to the phytoplankton of the Meuse, in which *Melosira
granulata* is the most abundant species (Table 16.2).  Since the closure
of the  Haringvliet in 1970, *Actinocyclus normannii* has disappeared
(Peelen, 1975).

In the zooplankton of the small creeks, many primarily benthic
organisms (e.g. *Chydorus sphaericus*, *Alona* sp.) were found (Leentvaar,
1968).

16.44  BENTHIC PLANTS

16.441 *Phanerogams*

The early stages of development of the vegetation on an intertidal
sandflat are shown in Fig. 16.6.  All species but *Scirpus triqueter*
are also common in non-tidal habitats in the Netherlands.  However,
the morphology of the plants in the tidal area often differs from those
in the non-tidal habitats e.g. in the Biesbosch, *Caltha palustris*
forms typical "spins" of new roots and is considered as a separate
variety, *Caltha palustris* var. *araneosa* (Van Steenis, 1971).  Elsewhere, *Scirpus triqueter* is also found in non-tidal habitats e.g. in
the Upper Rhine area, but is never as common as in the freshwater
tidal areas.  Some of the species mentioned in Fig. 16.6 also grow
in brackish water e.g. *S. maritimus*.

Where the soil level was above 0.80 m + NAP (Table 16.4), willows
were usually planted.  In the natural situation no trees other than
willows would have grown; once the tidal influence had diminished to
a level where it no longer dominated, willow coppice would probably
have developed, but before this stage was reached in the Biesbosch,
the land was embanked and cultivated as farmland.  In the intertidal
willow coppices the composition of the herb layers was strongly
influenced by the way the willow coppices were exploited (Zonneveld,
1960).

Fig. 16.6 Development of vegetation on an intertidal sandflat in the freshwater tidal area. In stage IV on the left a cultured reed marsh with ditches is shown and on the right the more natural vegetation developing without any artificial drainage. (Modified from Zonneveld, 1960)

*Table 16.4* Summary of data on the Biesbosch area. (For references, see text)

| | subtidal zone | intertidal zone | | |
| --- | --- | --- | --- | --- |
| | open water | sand- and mud-flats | rush and reed marshes | willow coppices |
| area | 12 km$^2$ | together 32 km$^2$ | | 20 km$^2$ |
| level with regard to NAP | below − 0.70 m | − 0.70 m to NAP | − 0.50 m to + 0.80 m | above + 0.80 m |
| water movement | current to c 1.0 m s$^{-1}$ | decreasing exposure to water movements ⟶ | | |
| frequency of flooding | 100% | 100% | 100% | may be less than 100% |
| duration of flooding in a tidal cycle | | c 12 − 7 h | c 11 − 4 h | less than 4 h |
| main phanerogam | locally *Potamogeton* sp. | | *Scirpus* sp. *Phragmites australis* | *Salix* sp. |
| characteristic algae | planktonic *Actinocyclus normannii* | locally *Vaucheria compacta* | at banks of creeks *Vaucheria compacta* | *Vaucheria* sp. |
| macro-invertebrates more or less restricted to the habitat | Unionidae | *Trocheta bykowskii*? | at banks of creeks *Pseudamnicola confusa* | *Orchestia cavimana* |
| macro-invertebrates with wide distribution in the area | Chironomidae sp., Tubificidae sp., *Pisidium* sp., *Sphaerium corneum, S. solidum* | | | |

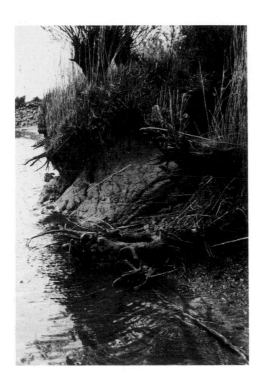

*Fig. 16.7* Tidal creek with natural levee. (Photo by O.M. van Hoorn)

The study of Zonneveld was confined to the Meuse sector of the Biesbosch, but the vegetation of the Rhine sector was almost the same (Verhey, 1961a)', which shows that it is the frequency and duration of flooding that are the key factors in the development of the vegetation, not the composition of the water.

## 16.442 *Cryptogams*

Blue-green algae and diatoms occurred on intertidal sand- and mud-flats near MLW (mean low water). Among diatoms, the planktonic species *Melosira granulata* and *Actinocyclus normannii* were common, but benthic species like *Rhoicosphenia curvata* and *Cocconeis placentula* were also found in large numbers (Zonneveld, 1960, Appendix 15). At a somewhat higher level in the intertidal zone *Vaucheria compacta* formed extensive growths (Simons, 1974, 1975). This species was also found on the banks of creeks in the rush and reed marshes, together with *Callitriche stagnalis*. Where sedimentation was not too intensive, *Rhizoclonium riparium* was abundant on all kinds of substratum, together with *Ulothrix tenerrima*, *Microcoleus vaginatus* or *Vaucheria* sp. (Nienhuis,

1975, 1980). *Cladophora okamurai* was found on hard substrata. This species seems to be confined to freshwater tidal areas, at least in the Netherlands (Den Hartog, 1967).

On the soil of the willow coppices locally rich *Vaucheria* growths were encountered, the main species being *V. frigida, V. bursata* and *V. canalicularis* (Simons, 1975). On the trunks of the willows, the bryophytes up to MHW (mean high water), *Marchantia polymorpha, Fissidens bryoides, F. taxifolius, Mnium* sp. and *Chiloscyphus pallescens,* occurred in a vertical zonation (Barkman, 1953).


## 16.45   MACROINVERTEBRATES

### 16.451 *Tidal channels*

A tidal channel is like a major river, as far as its width, depth and current velocity is concerned. Only the current is not unidirectional, but alternating. Tubificidae, Chironomidae and, in particular, bivalve molluscs occurred in the main tidal channels below MLW. In addition to ubiquitous species like *Sphaerium corneum*, species with a fluviatile distribution were found, including *S. rivicola, S. solidum, Unio crassus* var. *batavus, U. tumidus, Anodonta anatina* and a number of species of *Pisidium* (Wolff, 1968, 1970; Kuiper & Wolff, 1970). These fluviatile species were absent in channels which were directly fed by water from the Rhine.

Gammarids were represented in the Biesbosch by *Gammarus zaddachi*; between 1958 and 1960 this species was very common in the Meuse sector of the Biesbosch, whereas it occurred only sparsely in the Rhine sector. After a sudden break-down of the populations of *G. zaddachi* in 1960, some specimens of *G. pulex* were occasionally found in the tidal creeks of the Biesbosch (Den Hartog, 1964). Among Tubificidae, *Potamothrix moldaviensis* was the most numerous species, followed by *Limnodrilus claparedeianus* and *L. hoffmeisteri* (Verdonschot, 1980).


### 16.452 *Intertidal zone*

Aquatic species can only colonize the intertidal sand- and mud-flats if the substratum retains enough water during ebb-tide, as has been demonstrated by Kuiper & Wolff (1970) for *Pisidium* sp. Unionidae are however not found above MLW (Wolff, 1968). Hardly anything is known about the distribution of Tubificidae and Chironomidae on the mud-flats. As a rule, in the intertidal zone the same species of Tubificidae occur as in the sediments below MLW, as for example in the Elbe (Pfannkuche, 1977). *Sphaerium corneum* was found in small numbers on the mud-flats (Wolff, 1970). Sand does not retain much water and consequently few aquatic species can maintain themselves on intertidal sand-flats. Only a few individuals of *Gammarus zaddachi* and *Sphaerium solidum* were found in this habitat and in many places the sand-flats

looked almost barren (Wolff, 1973).  Den Hartog & Van Rossum (1957)
consider that the lower part of the intertidal zone was the favourite
habitat of the leech *Trocheta bykowskii*, which fed mainly on Tubificidae.
However, it was also found higher up in the intertidal zone (Heyligers,
1961).  None of the species, occurring on intertidal sand- and mud-
flats are completely confined to this habitat and many niches seem to
be unoccupied there (Wolff, 1973).

Higher up the intertidal zone the small snail *Pseudamnicola confusa*
was found.  This is an intertidal species,  which occurred most commonly along the banks of small tidal creeks in the willow coppices and
reed marshes, characterized by a vegetation of *Vaucheria compacta*,
*Callitriche stagnalis* and *Polygonum hydropiper*.  It lived only in
places where it was sheltered against the currents and also occurred in
brackish water (Butot, 1960; Den Hartog, 1960).  Aquatic organisms were
still dominant  in the soil of the lower willow coppices; at 0.40 m
below MHW the soil fauna consisted mainly of Tubificidae, *Pisidium* sp.
and *Potamopyrgus jenkinsi*.  Near MHW the species present were mainly
terrestrial (Heyligers, 1961).  At this level there occurred also the
semi-terrestrial amphipod *Orchestia cavimana* (Den Hartog, 1963).
Although this species is not confined to the freshwater tidal area,
it seems more numerous there than in any other habitat.  It also occurs
regularly along the major Dutch lakes (Den Hartog & Tulp, 1960) and
along the Meuse.  The condition for its occurrence seems to be that the
water level either remains the same, as in the lakes and the regulated
Meuse, or that it fluctuates in a predictable way, as in a tidal area
*O. cavimana* has not been found along the branches of the Rhine, where
changes in water level occur irregularly and where there may be a
great distance between the high and low water line.  In the soil of
the willow coppices there were also larvae of Chironomidae, Ceratopogonidae and Dryopidae, but these could not be identified to species
(Heyligers, 1961).

Among the smaller animals, species generally regarded as rare occurred, e.g. the ostracod *Candona sarsi* (Heyligers, 1961).  Boaden (1976)
presents some qualitative data on the meiofauna of the sand-flats.


16.46  FISH

Characteristic species of the lower courses of large rivers are shad
*Alosa alosa* and twait *Alosa fallax*.  Before 1970 twait occurred in
rather large numbers in the Biesbosch.  Shad had already virtually
disappeared, mainly because of overfishing (Verhey, 1961b).  The
houting *Coregonus oxyrhynchus*, previously also numerous, had already
disappeared by about 1910.  A history of fishing in the lower river
reaches is presented by Verhey (1961b).

*Fig. 16.8* A) End of creek in tidal willow coppice.  B) Intertidal sand-flat at ebb-tide.  (Photos by O.M. van Hoorn)

## 16.47  CONCLUDING REMARKS

Unfortunately, there were no studies of production and energy flow in the Biesbosch.  The ecosystem was supplied with enormous amounts of nutrients from the rivers and probably was a highly productive one. Man exploited the area, harvesting large amounts of rushes, reed and osiers.  Among the species living in the Biesbosch, the most conspicuous were a number of euryhaline ones, particularly among the benthic algae (Den Hartog, 1968).  Here again, it appears that the transition from the river community into the brackish water community was a gradual one, rather than distinctly marked.  It should be remembered that at storm floods brackish water penetrated into the Biesbosch (Peelen, 1967).  The transition from the freshwater tidal area to the brackish part of the Rhine-Meuse estuary is discussed at length by Wolff (1973). On the other hand, no clear picture can be given of the transition from the non-tidal river to the freshwater tidal area.  Nearly all freshwater species occurring in the Biesbosch are also found either in the Meuse or the Rhine, but the reverse need not be true.  In a natural river ecosystem, aquatic insects such as Ephemeroptera and Trichoptera constitute a major part of the macroinvertebrate fauna.  Surveys of the macroinvertebrate fauna of a branch of the Rhine in 1973-1977 showed that these groups were absent from the river at that time (Van Urk, 1978).  The regulation of the Meuse has certainly affected the distribution of rheophilous species of Ephemeroptera and Trichoptera in this river.  There were very few records of these groups from the Biesbosch, whereas the area suffered far less from human activities than the rivers.  This scarcity might be a natural feature of the freshwater tidal area,  as the habitat offers special problems to successful colonization. But other factors may still play a role.  With the closure of the Haringvliet, the habitat of the freshwater tidal area was not entirely lost.  Along the Oude Maas there is a river stretch that has a great resemblance with the Biesbosch.  Water flows through the Oude Maas either from the Waal or from a mixture of the Waal and the Meuse, depending on the flow of the Rhine.  The area is much smaller than the Biesbosch and is therefore vulnerable, as it is situated near the port and industry of Rotterdam.  Nevertheless, it would be of great value in order to study yet unsolved ecological problems of the freshwater tidal habitat, once the water quality of the Rhine has improved sufficiently.

## 16.5  PRESENT STATE OF THE LOWER RHINE

### 16.51  POLLUTION MONITORING IN THE LOWER RHINE

The Rhine is the major source of freshwater in the Netherlands.  In the country's water management system the Rhine water affects water quality both in the Dutch polders and in coastal waters, including the Wadden Sea (Wolff, 1978).  The Rhine used to be the source of drinking water for Amsterdam, Rotterdam and The Hague, but the latter two cities are now supplied by the Meuse.  Monitoring the water quality of the Rhine is therefore of great importance and a regular chemical monitoring programme was established in 1953.  Particular attention was given at first to the oxygen balance, but interest is now shifting to heavy

metals and organic micropollutants.  Attention is being paid especially
to the geochemical behaviour of heavy metals in estuarine areas (e.g.
Salomons & Mook, 1977; Salomons et al., 1981).  The behaviour of
organic micropollutants is followed only incidentally (Duinker &
Hillebrand, 1979).  Yet effects of organic micropollutants have been
demonstrated more clearly than those of heavy metals:  the insecticides
telodrin and dieldrin, originating from an industry near Rotterdam,
have caused a reduction of the numbers of sandwich terns in the Wadden
Sea, and polychlorinated biphenyls from the Rhine are held responsible
for the decline of harbour seals.  From their experiments with rainbow
trout, Poels et al. (1980) conclude that the Rhine water is chronically
toxic to salmonids.

   Hydrobiological research in the Rhine itself constitutes only a minor
part of the pollution monitoring programme.  It started with the analysis of the phytoplankton composition (Wibaut-Isebree Moens, 1956), at a
later stage supplemented with analyses of periphyton on artificial
substrata, but these groups have not shown much change in recent years
(Rijksinstituut voor Zuivering van Afvalwater, unpublished data).
Research into the macroinvertebrate fauna in one of the branches of
the Rhine in the Netherlands was started in 1973.


16.52  MACROINVERTEBRATE FAUNA OF THE IJSSEL

16.521 *Description of the IJssel*

The flow of the IJssel can be kept high by means of weirs in the
Nederrijn (Deltadienst of Rijkswaterstaat, 1973).  In the period of
1975-1980 the mean flow was 343 $m^3$ $s^{-1}$; the lowest flow was 139 $m^3$ $s^{-1}$
in July 1976 and the highest flow 1165 $m^3$ $s^{-1}$ in February 1980.  The
current velocity at km 885 is 0.5 - 1.1 m $s^{-1}$ and decreases to 0.3 -
0.7 m $s^{-1}$ at km 996.  The river bed consists of gravel, sand and clay,
median grain size decreasing downstream like in the Nederrijn/Lek
(Terwindt et al., 1963).

   Jetties were built on the river in the 19th century, in order to
protect its banks from erosion (Fig. 16.9).  Current velocity increased,
as a result of the constriction of the river channel by these jetties
(Jansen et al., 1979).  In some stretches the banks have recently been
consolidated with metal slag.  In the lower course, a vegetation of
*Nuphar luteum, Scirpus* sp., *Typha* sp. and *Phragmites australis* has developed at the banks.  In the upper reaches only isolated plants of
*Potamogeton nodosus* can be found.  Bryophytes and filamentous algae,
growing on the upper side of the stones at jetties and on the metal
slag, form a natural "trap" to the suspended matter in the river water.
Much silt is deposited and forms a substratum for many midge larvae
and worms.  The bottom of the stones, however, is almost clean and here
species which require a hard substratum, such as *Ancylus fluviatilis*,
can settle.

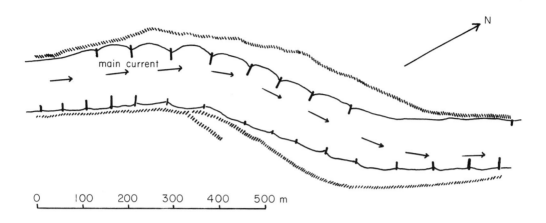

*Fig. 16.9* Section of the IJssel at km 916, showing the position of the jetties (from a river map of Rijkswaterstaat).

### 16.522 *Macroinvertebrates on stones*

The composition of the macroinvertebrate fauna on the stones from the jetties in the period 1975-1981 is presented in Table 16.5. Bryozoa and sponges were not included in this survey. The most conspicuous feature of the fauna was the appearance of a population of *Hydropsyche contubernalis* and the mass development of Chironomidae since 1978. Among the latter, *Dicrotendipes* gr. *nervosus* and *Cricotopus* sp. are the most numerous (Van Urk, 1981). At the most downstream sites *Glyptotendipes* dominates; it is also very common in the reed-beds. Less abundant on the stones are *Parachironomus* gr. *arcuatus, Xenochironomus xenolabis, Endochironomus albipennis* and *Nanocladius* sp. Tanytarsini are found in very low numbers, as are Tanypodinae (names according to Moller Pillot, 1979). These changes in numbers can hardly be considered merely as natural fluctuations. It has to be assumed that water quality factors play a role, in particular the oxygen content and perhaps also the contents of cholinesterase inhibitors such as organophosphorus insecticides (Table 16.6). Besch *et al.* (1977) reported that concentrations as low as 1 µg $l^{-1}$ of the carbamate insecticide Fenethcarb, which is also a cholinesterase inhibitor, interfered with netspinning of *Hydropsyche angustipennis*. Chironomidae are equally sensitive to cholinesterase inhibitors (Karnak & Collins, 1974). Other water quality factors may of course also play a role. It is interesting to compare this mass occurrence of *H. contubernalis* with that reported for the middle Rhine in 1975-1977 (see p.290).

Among the species found, *H. contubernalis* is the only one that is

*Table 16.5*  Composition of the macroinvertebrate fauna on jetties of the River IJssel in autumn.

| | 1975 | 1976 | 1977 | 1978 | 1979 | 1980 | 1981 |
|---|---|---|---|---|---|---|---|
| *Hydra sp.* | 3 | – | – | 136 | 111 | 383 | 58 |
| *Dugesia polychroa* | – | 29 | 22 | 86 | 136 | 89 | 260 |
| *D. tigrina* | – | – | – | – | 7 | 84 | 639 |
| *Nais bretscheri* | – | – | – | 5 | – | 30 | 5 |
| *N. elinguis* | – | – | – | 21 | – | 75 | 45 |
| *N. simplex* | – | – | – | 74 | 45 | 67 | 2 |
| *Stylaria lacustris* | 48 | 15 | 219 | 609 | 631 | 371 | 799 |
| *Glossiphonia complanata* | 16 | 14 | 12 | 19 | 20 | 23 | 30 |
| *G. heteroclita* | 9 | 17 | 35 | 14 | 34 | 9 | 38 |
| *Erpobdella octoculata* | 35 | 41 | 244 | 281 | 126 | 181 | 104 |
| *Acroloxus lacustris* | 31 | 1 | 26 | 53 | 59 | 73 | 68 |
| *Ancylus fluviatilis* | 30 | – | 104 | 376 | 74 | 349 | 122 |
| *Bithynia tentaculata* | 50 | 3 | 43 | 51 | 709 | 645 | 1626 |
| *Lymnaea peregra* | 43 | 13 | 71 | 28 | 47 | 62 | 67 |
| *Physa fontinalis* | 7 | 5 | 39 | 10 | 40 | 50 | 8 |
| *Dreissena polymorpha* | 114 | 4 | 3 | 18 | 52 | 91 | 204 |
| *Asellus aquaticus* | 142 | 345 | 4370 | 3168 | 2326 | 2373 | 1621 |
| *Hydropsyche contubernalis* | – | – | – | 10 | 253 | 2104 | 598 |
| *Cyrnus trimaculatus* | – | – | – | – | – | – | 107 |
| *Ecnomus tenellus* | – | – | – | – | – | – | 35 |
| *Chironomidae* | 19 | 56 | 254 | 4680 | 7282 | 10198 | 15287 |
| others | 33 | 2 | 19 | 39 | 95 | 313 | 234 |
| total number of animals | 580 | 545 | 5461 | 9678 | 12047 | 17570 | 21957 |

*Table 16.6* Water quality variables of the Rhine at km 862 (German-Dutch border). Mean values for April-September.

| | | 1973 | 1974 | 1975 | 1976 | 1977 | 1978 | 1979 | 1980 | 1981 |
|---|---|---|---|---|---|---|---|---|---|---|
| discharge | $m^3 \ s^{-1}$ | 1818 | 1592 | 2207 | 1128 | 2035 | 2663 | 2125 | 2674 | 2288 |
| suspended solids | $mg \ l^{-1}$ | 44 | 43 | 39 | 63 | 39 | 56 | 38 | 27 | 34 |
| dissolved oxygen | $mg \ l^{-1}$ | 4.1 | 3.9 | 6.1 | 5.4 | 6.3 | 6.8 | 6.9 | 7.1 | 7.1 |
| total organic C | $mg \ l^{-1}$ | | 14.3 | 11.4 | 15.1 | 10.0 | 10.1 | 10.2 | 8.5 | 9.8 |
| $NH_4^+$-N | $mg \ l^{-1}$ | 1.9 | 1.9 | 0.6 | 1.1 | 0.6 | 0.5 | 0.5 | 0.4 | 0.4 |
| chlorophyll *a* | $\mu g \ l^{-1}$ | | | | | 69 | 51 | 47 | 56 | 85 |
| cholinesterase inhibitors (as para-oxon) | $\mu g \ l^{-1}$ | 2.0 | 1.6 | 6.5 | 11.6 | 4.6 | 1.3 | 1.2 | 0.6 | 0.9 |

characteristic of the potamon.  At some sites it is particularly abundant, densities approaching one individual per cm$^2$ of stone surface.
It was found along the whole river from km 885 to km 1000.  The factors
determining the zonation of Hydropsychidae in a stream have been discussed by Hildrew & Edington (1979), who consider temperature as the
main factor.  In the IJssel, temperatures show a wide range, with
summer maxima exceeding 20 $^o$C.


16.523 *Macroinvertebrates in sediments*

Constriction of the river channel by jetties and the increased current
velocities have influenced the sedimentation pattern.  The river bed
of the IJssel consists of hard packed sand and gravel and here hardly
any macroinvertebrates are found.  A regular sedimentation of the suspended matter occurs only near the mouth of the river, especially near
the banks where the current diminishes.  A mainly tubificid macroinvertebrate population here reaches densities of some thousands of
individuals per m$^2$.  There is no distinct seasonal trend in their
numbers.  The dominant species are *Limnodrilus hoffmeisteri* and *L.
claparedeianus*; their proportions vary from place to place and from
year to year (Table 16.7).


*Table 16.7*  Percentage composition of Tubificidae in samples of
sediment at a depth of about 2 m from the IJssel at km 1000; a) near
the left bank; and b) near the right bank.

|  | 1976 | | 1978 | | 1979 | | 1980 | |
|---|---|---|---|---|---|---|---|---|
|  | a | b | a | b | a | b | a | b |
| *Limnodrilus claparedeianus* | 51 | 41 | 25 | 16 | 15 | 51 | 6 | 42 |
| *L. hoffmeisteri* | 22 | 22 | 68 | 73 | 72 | 30 | 89 | 56 |
| *Tubifex tubifex* | 12 | 18 | 3 | 3 | 3 | <1 | 5 | 2 |
| *Potamothrix moldaviensis* | 7 | - | <1 | 2 | 1 | 7 | - | - |
| other species | 8 | 19 | 4 | 6 | 10 | 11 | - | <1 |

   Dominance of *L. claparedeianus* has seldom been reported elsewhere,
the dominant species in this kind of habitat nearly always being *L.
hoffmeisteri*.  It is not clear what biotic or abiotic factors favour
*L. claparedeianus* here.  Relations between different species of Tubificidae may be very complex (Pfannkuche, 1977) and there is no simple
explanation for the changes in relative proportions of the various
species.  A large proportion of *L. hoffmeisteri* has often been considered to be indicative of pollution by organic wastes, but it seems
doubtful whether this also holds in this habitat.

Fig. 16.10  Frequency distribution of numbers of Chironomidae in grab samples (surface area 0.03 m$^2$) from the sediments of the IJssel at km 1000.


In the IJssel a quite different association of Tubificidae is found among vegetation near the banks.  The main species are *Psammoryctides albicola, P. barbatus* and *Limnodrilus udekemianus* (Van Urk, 1978).  *Tubifex tubifex, Potamothrix hammoniensis* and *P. moldaviensis* are found in all habitats in low numbers.  Chironomidae occur at much lower densities than Tubificidae, but appear to increase slightly (Fig. 16.10).  The species (or larval groups) that are found regularly, are: *Procladius* s.l, *Dicrotendipes* gr. *nervosus, Cryptochironomus* gr. *defectus, Parachironomus* gr. *arcuatus* and *P.* gr. *longiforceps, Microchironomus tener, Chironomus* f.l. *fluviatilis* and *reductus, Glyptotendipes* sp., *Cricotopus* sp.  Their densities are too low to detect any periodicity or trends for individual species.  Comparison of species composition with that of other rivers is difficult for the same reason.  Besides Chironomidae, occasional specimens of Unionidae, in particular of *Anodonta anatina* and *Unio pictorum*, were found in grab samples in 1978-1980.  As mentioned above (16.451), Wolff (1968) was unable to find any Unionidae in the Rhine branches.  This absence was attributed to the pollution of the river water and it seems reasonable to conclude that their present occurrence has been permitted by recent improvements in water quality, such as the increase in oxygen content.  Among the Pisidiidae only *Pisidium henslowanum* and *Sphaerium corneum* are found, both in relatively low numbers.  The fluviatile species *S. rivicola* and *S. solidum* have so far not been encountered in the IJssel.


16.524 *Prospects*

Monitoring the macroinvertebrate fauna of the IJssel has revealed some signs of restoration of the river's ecosystem. However, it is far from being unpolluted, and even further from being a natural river. Many species, such as the burrowing Ephemeroptera that were reported once to have been abundant, may have disappeared forever, since there may be no suitable refuge left to permit re-introduction. Nevertheless, biological surveillance of the Rhine is not only valuable from a viewpoint of pollution monitoring, but may also contribute to knowledge of the autecology of species whose occurrence is restricted to the lower reaches of large rivers, which form a very vulnerable habitat in a densely populated area like Western Europe.

*Acknowledgements*

I thank Miss S.M. Wiersma and A. Espeldoorn for carrying out part of the work on the IJssel reported here, L.W.G. Higler and I. Wolters for identification of the Trichoptera, and F. Repko for checking the identification of Chironomidae.

## REFERENCES

Barkman J.J. (1953)  Over de Mosvegetaties van onze getijgrienden. *Buxbaumia* 7, 42-49.  (On the moss vegetation of our tidal willow coppices.)

Behre K. (1961)  Die Algenbesiedlung der Unterweser unter Berücksichtigung ihrer Zuflüsse. *Veröff. Inst. Meeresforsch Bremerhaven* 7, 71-263.  (The algae of the Lower Weser and its tributaries.)

Besch W.K., Schreiber I. & Herbst D. (1977)  Der Hydropsyche- Toxizitätstest, erprobt an Fenethcarb. *Schweiz. Z. Hydrol.* 39, 69-85. (The *Hydropsyche* toxicity test, tried out with Fenethcarb.)

Boaden P.J.S. (1976)  Soft meiofauna of sand from the delta of the Rhine, Meuse and Scheldt. *Neth. J. Sea Res.* 10, 461-71.

Butot L.J.M. (1960  *Pseudamnicola confusa* (Frauenfeld 1863) algemeen in de Biesbosch (Gastropoda, Prosobranchia). *Basteria* 24, 60-65. (*Pseudamnicola confusa*, a common mollusc in the Biesbosch.)

De Groot A.T. & Muyres W.J.M. (1980)  Visserijkundige waarnemingen vispassages 1975 tot en met 1979. *Visserij* 33, 446-61.  (Observations on fish migration 1975-1979.)

Deltadienst of Rijkswaterstaat (1973)  The Realization and Function of the Northern basin of the Delta project. *Rijkswaterstaat Communications* 14, 73 pp.  Government Publishing Offices, Gravenhage, Netherlands.

Den Hartog C. (1960)  Verspreiding van het slakje *Pseudamnicola confusa* in het Deltagebied van Rijn en Maas. *Basteria* 24, 66-74. (Distribution of *Pseudamnicola confusa* in the Deltaic region of Rhine and Meuse.)

Den Hartog C. (1963)  The amphipods of the deltaic region of the rivers Rhine, Meuse and Scheldt in relation to the hydrography of the area.  Part II:  The Talitridae. *Neth. J. Sea Res.* 2, 40-67.

Den Hartog C. (1964)  The amphipods of the deltaic region of the rivers
    Rhine, Meuse and Scheldt in relation to the hydrography of the area.
    Part III: The Gammaridae. *Neth. J. Sea Res.* 2, 407-57.
Den Hartog C. (1967)  Brackish water as an environment for algae.
    *Blumea* 15, 31-43.
Den Hartog C. (1968)  Een karakteristiek van het zoetwatergetijdenge-
    bied en de plaats van de Biebosch in dit geheel. *Meded. Hydrobiol.
    Ver.* 2, 173-82.  (A delineation of the freshwater tidal area and the
    position of the Biesbosch in that area.)
Den Hartog C. & Tulp A.S. (1960)  Hydrobiologische waarnemingen in
    Friesland. *De Levende Natuur* 63, 133-40.  (Hydrobiological obser-
    vations in Friesland.)
Den Hartog C. & Van Rossum E. (1957)  De bloedzuiger *Trocheta bykowskii*
    in de benedenrivieren. *De Levende Natuur* 60, 228-33.  (The leech
    *Trocheta bykowskii* in the lower river reaches.)
Dijkzeul A.H.J. (1981)  De waterkwaliteit van de Maas in Nederland in
    de periode 1953-1980. *H₂O* 14, 628-34.  (The water quality of the
    Meuse in the Netherlands in the period 1953-1980.)
Dresscher T.G.N. (1968)  De soortencombinaties van de in het water
    zwevende plantaardige micro-organismen van enkele plaatsen in de
    Brabantse Biesbosch en de omgeving. *Meded. Hydrobiol. Ver.* 2,
    154-66.  (The phytoplankton of the Brabantse Biesbosch and adjacent
    rivers.)
Dresscher T.G.N., Bakker C. & Geelen J.F.M. (1968)  Planktonconcentra-
    ties en getijbeweging in het Gat van de Turfzak. *Meded. Hydrobiol.
    Ver.* 2, 192-206.  (Plankton concentrations and tidal movements in a
    small tidal creek.)
Duinker J.C. & Hillebrand M.T.J. (1979)  Behaviour of PCB, pentachloro-
    benzene, hexachlorobenzene, α-HCH, γ-HCH, β-HCH, dieldrin, endrin
    and p,p'-DDD in the Rhine-Meuse estuary and the adjacent coastal
    area. *Neth. J. Sea Res.* 13, 256-81.
Geelen J.F.M. (1975)  *Orconectes limosus* (Raf.) and *Astacus astacus* L.
    (Crustacea, Decapoda) in the Netherlands. *Hydrobiol. Bull.* 9,
    109-13.
Heyligers P.C. (1961)  De bodemfauna van de grienden.  In: Verhey C.J.
    (Ed.) *De Biesbosch, land van het levende water*, 255 pp., pp. 85-117.
    Thieme, Zutphen, Netherlands.  (The soil fauna of the willow
    coppices.)
Hildrew A.G. & Edington J.M. (1979)  Factors facilitating the co-exis-
    tence of Hydropsychid caddis larvae (Trichoptera) in the same river
    system. *J. Anim. Ecol.* 48, 557-76.
Jansen P.P., Van Bendegon L., Van den Berg J., De Vries M. & Zanen A.
    (1979) *Principles of River Engineering. The Non-tidal Alluvial
    River*, 509 pp.  Pitman, London.
Karnak R.E. & Collins W.J. (1974)  The susceptibility to selected
    insecticides and acetylcholinesterase activity in a laboratory
    colony of midge larvae, *Chironomus tentans* (Diptera:Chironomidae).
    *Bull. environ Contam. Toxicol.* 12, 62-69.
Koolen J.L. (1973)  De kwaliteit van het Maaswater in Nederland.
    *H₂O* 6, 3-14.

Kuiper J.G.J. & Wolff W.J. (1970)  The Mollusca of the estuarine region
     of the rivers Rhine, Meuse and Scheldt in relation to the hydro-
     graphy of the area.  III. The genus *Pisidium*.  *Basteria* 34, 1-40.
Leentvaar P. (1968)  Het zooplankton van de Brabantse Biesbosch en
     enige eigenschappen van de planktongemeenschap.  *Meded. Hydrobiol.*
     *Ver.* 2, 167-72.  (The zooplankton of the Brabantse Biesbosch with
     notes on the plankton association of the area.)
Moller Pillot H.K.M. (1979)  De larven der Nederlandse Chironomidae
     (Diptera).  *Nederl. Faun. Meded.* 1, 1-276.  (The larvae of the Dutch
     Chironomidae (Diptera).)
Nienhuis P.H. (1975)  Biosystematics and ecology of *Rhizoclonium*
     *riparium*  (Roth) Harv.  (Chlorophyceae, Cladophorales) in the estua-
     rine area of the rivers Rhine, Meuse and Scheldt, 240 pp.  Ph.D.
     Thesis, University of Groningen.
Nienhuis P.H. (1980)  The epilithic algal vegetation of the SW Nether-
     lands.  *Nova Hedwigia* 33, 1-94.
Nöthlich I. (1972)  Trophische Struktur und Bioaktivität der Plankton-
     gesellschaft im unteren limnischen Bereich des Elbe-Aestuars.
     *Arch. Hydrobiol., Suppl.* 43, 33-117.  Trophic structure and bio-
     activity of the plankton communities in the lower reach of the Elbe
     estuary.)
Parma S. (1968)  Hydrografie van de Biesbosch.  *Meded. Hydrobiol. Ver.*
     2, 95-145.  (Hydrography of the Biesbosch.)
Peelen R. (1967)  Isohalines in the delta area of the rivers Rhine,
     Meuse and Scheldt.  *Neth. J. Sea Res.* 3, 575-97.
Peelen R. (1975)  Changes in the plankton of the estuarine area of the
     Haringvliet-Hollands Diep-Biesbosch in the S.W. Netherlands caused
     by the dams through Volkerak and Haringvliet.  *Hydrobiol. Bull.* 8,
     190-200.
Pfannkuche O. (1977)  Ökologische und systematische Untersuchungen an
     naidomorphen Oligochaeten brackiger und limnischer Biotope, 138 pp.
     Ph.D. Thesis, University of Hamburg.  (Ecological and taxonomic
     investigations of naidomorph Oligochaeta of brackish and freshwater
     habitats.)
Poels C.L.M., Van der Gaag M.A. & Van de Kerkhoff J.F.J. (1980)  An
     investigation into the long-term effects of Rhine water on rainbow
     trout.  *Water Res.* 14, 1029-35.
Rijksinstituut voor Zuivering van Afvalwater (1971)  *Hydrobiologisch*
     *onderzoek van de Maas*, 63 pp.  Mededeling No. 9, Staatsuitgeverij's
     Gravenhage, Netherlands.  (Hydrobiological research of the Meuse.)
Rijksinstituut voor Zuivering van Afvalwater (1980)  *De waterkwaliteit*
     *van de Rijn in Nederland in de periode 1972-1979*, 71 pp.  Nota No.
     80-032, Lelystad, Netherlands.  (Water quality of the Rhine in the
     Netherlands 1972-1979.)
Ringelberg J. (1968)  De relatieve abundantie van de planktonische
     raderdieren in de Biesbosch en omgeving.  *Meded. Hydrobiol. Ver.* 2,
     183-91.  (The relative abundance of planktonic Rotifers in the
     Biesbosch and adjacent rivers.)

Romijn G., Uyttenbogaart D.L. & Heimans J. (1919)  Verslag van het
    biologisch onderzoek van de Maas en hare oevers.  In: *Natuur-
    historisch Genootschap in Limburg Jaarboek 1918*, 177 pp., pp. 93-
    145.  Maastricht, Netherlands.  (Report on the biological explora-
    tion of the Meuse and its banks.)
Salomons W. & Mook W.G. (1977)  Trace metal concentrations in estuarine
    sediments:  mobilization, mixing or precipitation.  *Neth. J. Sea
    Res.* 11, 119-29.
Salomons W., Mook W.G. & Eysink W. (1981)  *Biogeochemical and Hydro-
    Dynamic Processes affecting Heavy Metals in Rivers, Lakes and
    Estuaries*, 68 pp.  Delft Hydraulics Laboratory Publication No. 253
Simons J. (1974)  *Vaucheria compacta*: a euryhaline estuarine algal
    species.  *Acta bot. neerl.* 23, 613-26.
Simons J. (1975)  *Vaucheria* species from estuarine areas in the
    Netherlands.  *Neth. J. Sea Res.* 9, 1-23.
Terwindt J.H.J., De Jong J.D. & Van der Wilk E. (1963)  Sediment move-
    ment and sediment properties in the tidal area of the Lower Rhine
    (Rotterdam Waterway).  *Verh. Kon. Ned. Geol. Mijnbouw Genootsch* 21,
    243-58.
Van der Made J.W. (1972)  Hydrografie van het Maas-Bekken.  $H_2O$ 5,
    356-62.  (Hydrography of the Meuse basin.)
Van de Ven G.P. (1976)  *Aan de wieg van Rijkswaterstaat Wordings-
    geschiedenis van het Pannerdens Kanaal*.  Gelderse Historische Reeks
    VIII, 439 pp.  Walburg Pers, Zutphen.  (The roots of Rijkswaterstaat
    History of the "Pannerdens Kanaal".)
Van Steenis C.G.G.J. (1971)  De zoetwatergetijde-dotter van de
    Biesbosch en de Oude Maas, *Caltha palustris* L. var. *araneosa* var.
    nov.  *Gorteria* 5, 213-19.  (The freshwater tidal marsh marigold of
    the Biesbosch and the Oude Maas.)
Van Urk G. (1978)  The macrobenthos of the River IJssel.  *Hydrobiol.
    Bull.* 12, 21-29.
Van Urk G. (1981)  Veranderingen in de macro-invertebraten-fauna van
    de IJssel.  $H_2O$ 14, 494-99.  (Changes in the macroinvertebrate fauna
    of the IJssel.)
Verdonschot P.F.M. (1980)  Aquatische Oligochaeta  III Het Deltagebied.
    *Delta Instituut voor Hydrobiologisch Onderzoek Rapp Versl 1980-9*,
    115 pp.  (Aquatic Oligochaeta  III The Delta area.)
Verhey C.J. (1961a)  Vegetatie en fauna van de Sliedrechtse Biesbosch.
    *De Levende Natuur* 62, 138-44.  (Vegetation and fauna of the
    Sliedrechtse Biesbosch.)
Verhey C.J. (1961b)  De vissen en de visvangst.  In: Verhey C.J. (Ed.)
    *De Biesbosch, land van het levende water*, 255 pp., pp. 139-64.
    Thieme, Zutphen, Netherlands.  (The fish and fisheries.)
Werkgroep Sanering Maas (1971)  *De verontreiniging van de Maas,
    aanbevelingen tot sanering*.  Rapport, 68 pp.  's Gravenhage,
    Netherlands.  (The pollution of the Meuse, recommendations with
    respect to sanitation.)
Wibaut- Isebree Moens N.L. (1956)  *Rivierenonderzoek 1954-1955*.
    Report in 6 pts., 37 pp. and annexes, mimeographed.  Amsterdam.
    (Investigations in the rivers 1954-1955.)

Wolff W.J. (1968)  The Mollusca of the estuarine region of the rivers
   Rhine, Meuse and Scheldt in relation to the hydrography of the area.
   I The Unionidae. *Basteria* 32, 13-47.
Wolff W.J. (1970)  The Mollusca of the estuarine region of the rivers
   Rhine, Meuse and Scheldt in relation to the hydrography of the area.
   IV The genus *Sphaerium*. *Basteria* 34, 75-90.
Wolff W.J. (1973)  The estuary as a habitat.  An analysis of data on
   the soft-bottom macrofauna of the estuarine area of the rivers Rhine,
   Meuse and Scheldt. *Zool. Verh.* 26, 1-242.  Brill, Leiden.
Wolff W.J. (1978)  The degradation of ecosystems in the Rhine.  In:
   Holdgate M.W. & Woodman M.J. (Eds.) *The Breakdown and Restoration
   of Ecosystems*, 496 pp., pp. 169-88.  Plenum Press, New York.
Zonneveld I.S. (1960)  De Brabantse Biesbosch A study of soil and
   vegetation of a freshwater tidal delta.  Ph.D. Thesis, Agricultural
   University of Wageningen. *Versl. Landbouwk Onderz* 65.20, 3 vols.,
   Pudoc Wageningen.
Zuurdeeg B.W. (1980)  De natuurlijke chemische samenstelling van Maas-
   water. *$H_2O$* 13, 2-7.  (The natural chemical composition of River
   Meuse water.)

# 17 · Glåma

## O. M. SKULBERG & A. LILLEHAMMER

17.1   Introduction                                        470

17.2   General data                                        471

17.3   Physiographic conditions                            474

    17.31   The river system                          474
    17.32   Geological characteristics                475
    17.33   Topography                                478
    17.34   Climate and hydrology                     478
    17.35   Area and settlement                       482

17.4   Chemical and biological quality of water            482

    17.41   Water chemistry                           482
    17.42   Water fertility                           483

17.5   Plant and animal life                               485

    17.51   Classification of the natural geogra-     485
            phical region
    17.52   Benthic vegetation                        485

        17.521   Attached algae                   485
        17.522   Bryophytes                       487
        17.523   Higher plants                    488

    17.53   Seston - especially phytoplankton         488
    17.54   Benthic invertebrates                     491
    17.55   Fish                                      492

17.6   The lake-river system                               494

References                                                 495

## 17.1  INTRODUCTION

The Glåma is chosen as a representative of the rivers of Fennoscandia, a region of Archaean crystalline rocks in North Europe.  Fennoscandia, which received its name from the Finnish geologist Ramsay (1898), has a geological structure and natural environment quite different from that of the neighbouring countries.  This chapter is based partly on research carried out in recent years by the authors, but to a larger extent on publications and survey reports.

*Fig. 17.1*  River Glåma at Orvos - a streaming silver-band in the landscape.

## 17.2  GENERAL DATA

DIMENSIONS

   Area of catchment                          41767 km$^2$

LAND USE

   agriculture                          5%

   forest                               34%

   marshes and peatland                 5%

   freshwater                           4%

POPULATION

   catchment above junction of Glåma and R. Vorma          93000 people

   catchment of R. Gubrandsdalslågen and R. Vorma          205000

   catchment below meeting of the two major tributaries
   and Lake Øyeren                                          100000

   catchment below Lake Øyeren                              120000
                                                          ─────────
                                                          518000 = approx
                                                          12 people km$^{-2}$

PHYSICAL (AND OTHER) FEATURES OF RIVER

   Discharge etc

| location | catchment (km$^2$) | mean annual discharge (m$^3$ s$^{-1}$) | forest area | agricultural area | population in catchment |
|---|---|---|---|---|---|
| Glåma at Funnefoss | 20670 | 320 | 7822 km$^2$ | 580 km$^2$ | 93000 |
| Glåma at Fredrikstad | 41767 | 706 | 13600 km$^2$ | 1934 km$^2$ | 518000 |

*Fig. 17.2* The Glåma basin. Numbers
indicate size of sub-catchments (in km²)

Discharge ($m^3 s^{-1}$) at Solbergfoss, downstream of the Lake Øyeren.
(Data from Glommens og Laagens Brukseierforening)

| month | year | | | | mean for |
|---|---|---|---|---|---|
| | 1958 | 1969 | 1970 | 1980 | 1955-1964 |
| January | 355 | 406 | 367 | 414 | 395 |
| February | 332 | 379 | 335 | 386 | 380 |
| March | 300 | 309 | 295 | 267 | 354 |
| April | 308 | 475 | 326 | 427 | 419 |
| May | 673 | 1289 | 1042 | 948 | 1094 |
| June | 1178 | 1103 | 909 | 1212 | 1182 |
| July | 1132 | 405 | 789 | 993 | 1060 |
| August | 789 | 401 | 572 | 607 | 817 |
| September | 504 | 379 | 582 | 653 | 784 |
| October | 880 | 463 | 674 | 891 | 747 |
| November | 579 | 388 | 546 | 492 | 581 |
| December | 382 | 383 | 474 | 448 | 408 |
| | | | | | |
| Mean for year | 618 | 647 | 576 | 645 | 685 |

Selected waterfalls along the Glåma

| name | distance from Lake Rien | approximate vertical fall | mean discharge for 1946-1975 ($m^3 s^{-1}$) |
|---|---|---|---|
| | (km) | (m) | |
| Kuråsfoss | 34 | 48.2 | 20 |
| Røstefoss | 51 | 11.0 | - |
| Brufoss | 72 | 6.0 | - |
| Erlifoss | 77 | 8.0 | 50 |
| Barkaldfoss | 149 | 4.0 | 102 |
| Braskereidfoss | 323 | 7.6 | 247 |
| Funnefoss | 434 | 11.0 | 305 |
| Rånåsfoss | 457 | 14.5 | 628 |
| Bingsfoss | 463 | 3.0 | 650 |
| Solbergfoss | 506 | 22.6 | 656 |
| Kykkelsrud | 516 | 25.5 | 658 |
| Vamma | 520 | 27.5 | 660 |
| Sarpsfoss | 577 | 20.1 | 669 |

## 17.3  PHYSIOGRAPHIC CONDITIONS

### 17.31  THE RIVER SYSTEM

The name Glåma includes both the main river in the valley of Østerdalen, and the river below the confluence with the R. Vorma, the main tributary.  It extends down to the estuary in the archipelago of Hvaler on the eastern side of the outer Oslofjord (Fig. 17.2).  The name is interpreted linguistically to identify the brightness and clarity of its water.  River Glåma means the shining river - reflecting much light. And the Glåma is still a streaming silver band in the landscape of East Norway, close to the border between Norway and Sweden (Fig. 17.1).  It is the largest water course in Scandinavia with respect to discharge and its catchment comprises 13% of Norway's total land area (see 17.2).

A multitude of small lakes in the mountain area of Røros form the headwaters of the Glåma, but the river is not usually called by this name until downstream of Lake Rien (762 m altitude).  From L. Rien the Glåma flows southwards to L. Aursunden.  This lake (catchment area 830 km$^2$, surface area 44 km$^2$, depth 60 m) is regulated between the 684 and 690 contour lines.  The mean discharge of the Glåma at the outlet of L. Aursunden is 20 m$^3$ s$^{-1}$ (Skulberg, 1967a).

The Glåma proper is divisible into four sections (Fig. 17.3) based on the distribution of waterfalls (Norwegian *foss*); the first 178 km down to Barkaldfoss; the 185 km from Barkaldfoss to Braskereidfoss;

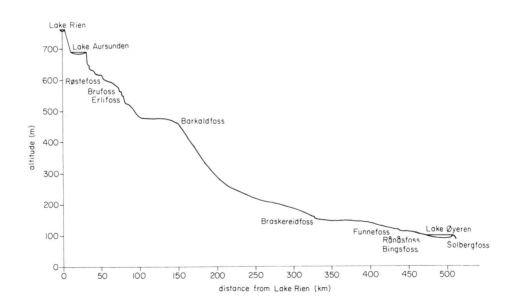

*Fig. 17.3* Longitudinal profile of the Glåma (Lake Rien to Solbergfoss), showing major waterfalls.

the 170 km from Braskereidfoss to the outlet of Lake Øyeren; the last
65 km from L. Øyeren to the sea.  Information on selected waterfalls is
included in 17.2.  In the valley of Østerdalen, the Glåma receives a
number of tributaries, including the Folla, the Atna and the Rena, the
largest of the three.  The Rena flows through L. Storsjøen (catchment area 2270 km$^2$, surface area 48.8 km$^2$, depth 309 m).  Further south
the Glåma broadens its valley.  The Flisa and the Opstad are among the
larger tributaries here.  The latter drains L. Storsjøen (catchment
area 774 km$^2$, surface area 42.5 km$^2$, depth 17 m) in the Odalen district.

The Glåma's catchment area at Funnefoss, before the confluence with
the Vorma, is 20670 km$^2$, while the catchment of the Vorma-Lågen water-
course is 17294 km$^2$.  After joining with this major tributary - and
almost doubling its size, the Glåma becomes a mighty river.  The region
here has a landscape of low hills and small plains.  The river is some
250 to 400 m broad and relatively shallow, with a slope of round about
0.39 m km$^{-1}$.  Where the river enters L. Øyeren at Fetsund, it has a
mean discharge of about 650 m$^3$ s$^{-1}$.

L. Øyern (catchment area 39964 km$^2$, surface area 85.2 km$^2$, depth
70 m) is a typical fjord lake (Strøm, 1935).  Due to the large volumes
of water entering the lake, the water mass has a short renewal time.
Based on the assumption that the throughflow occurs only in the epilim-
nion during stagnation periods, the theoretical retention time of L.
Øyeren will vary between 8-17 days, when discharge in the Glåma lies
in the 500-1000 m$^3$ s$^{-1}$ range.

From L. Øyeren (101 m altitude) down to the estuary there is a series
of four step-like waterfalls.  Tributaries are all relatively small and
enter mainly on the east bank; they include the Lekumelva and the Rak-
kestadelva.  The Glåma is slow-flowing and almost lake-like in the area
upstream of the waterfall Sarpsfossen.  Here the river is divided in
two branches, with much higher current speeds in the eastern, larger
one, which includes the waterfall Sarpsfossen.  The western branch has
lake-like extensions (L. Mingevatn, L. Vestvatn, L. Visterflo).  The
two branches meet again in the estuary at Greåker.

The estuary of the Glåma, with a salinity gradient 10-20 o/oo, is
of the halocline type.  Depending on river discharge and the force of
the tide, the water can reach up to Sarpsfossen and L. Visterflo.
The tidal range is very low, approximately 0.13 m.


17.32   GEOLOGICAL CHARACTERISTICS

The bedrock of the drainage basin dates from the Archaean era (Holtedahl,
1953a;  1960).  Granites, gneisses and other hard crystalline rocks are
dominant (Fig. 17.4).  Towards the west - in the catchment of the
Gudbrandsdalslågen - the geology is partly characterised by deposits
from Lower Palaeozoic times in the Caledonian mountain range.  The
central northern part consists mainly of Eocambrian sediments (thick

*Fig. 17.4* East Norway—solid rocks

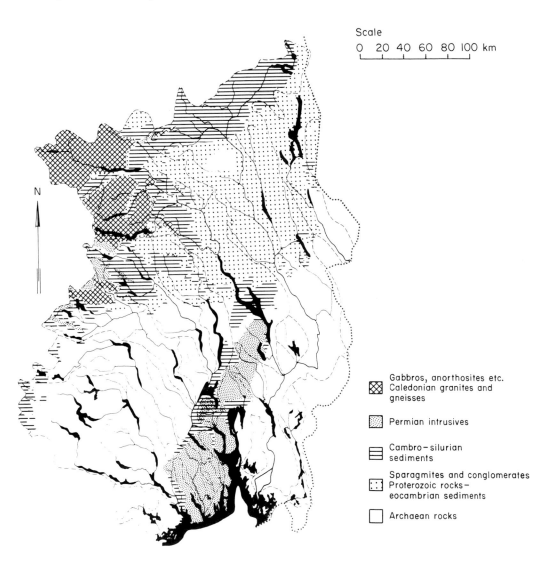

Scale
0  20  40  60  80  100 km

⊠ Gabbros, anorthosites etc.
   Caledonian granites and
   gneisses

▦ Permian intrusives

☰ Cambro-silurian
   sediments

Sparagmites and conglomerates
Proterozoic rocks-
eocambrian sediments

☐ Archaean rocks

sandstones, conglomerates etc.) known as sparagmite.  Towards the
south, rocks of gneiss and granite predominate, both types showing wide
variation with respect to mineral  composition and granular size.  This
extremely hard bedrock - belonging to the Fennoscandian Shield - is
about 1800 million years old (dating by radioactive methods).

During the Quaternary period the area was affected by glaciation

*Fig. 17.5* Superficial deposits

N

Marine sediments
– silt and clay
Moraine
Glacifluvial deposits
– sand and gravel

Scale

0   20  40  60  80  100 km

and glacial deposits are extensively present (Holtedahl, 1953b; Fig.
17.5).  Deglaciation and postglacial uplift have been reasonably well
elucidated by Scandinavian geologists (John, 1979).  The glacial
streams from the melting ice masses were heavily loaded with erosion
materials and they left important deposits in the landscape.

Above the upper marine limit (in the valley of Østerdalen 190 - 200 m

altitude) terraces, deltas and overflow gorges are numerous. They are
remnants of the ice-dammed lakes between the mountains and the glaciers
(Holmsen, 1915; Holtedahl, 1953b); moraines are particularly common.
Glacial gravel and river deposits of sand and gravel are present in the
valleys.

In areas, which were situated below sea level during the period of
deglaciation, conditions are different. Deposits - partly morainic and
partly glacifluvial - form series of conspicuous transverse ridges.
Glacifluvial deltas are frequent. The fine material of glacial streams
(clay and silt) was carried away to be deposited at a distance, partly
as varved clay in fresh or brackish water. The delta in the northern
part of Lake Øyeren is still the most considerable area of sedimentation in Norway. Extensive parts of the catchment area below the marine
limit consist mainly of clay (Bjørlykke, 1940). The late Weichselian
deglaciation in the southern part of the Glåma drainage basin has
recently been the subject of careful study (Sørensen, 1979).

## 17.33  TOPOGRAPHY

The Glåma system drains the central  and eastern side of the Scandinavian mountains - the Scandes. The highest places in the catchment area
exceed 2000 m altitude (e.g. Høgronden, Rondeslottet). More than half
lies above 500 m and about a quarter above 1000 m. The inland region
is dominated by valleys with moderate hills and more or less rounded
mountains (locally named *voler*). In the southern part of the drainage
basin a dissected lowland landscape of varying height rises to about
200 m above sea level (Fig. 17.6). Rounded hills (locally named *åser)*
of Archaean rocks and relatively low altitude (Precambrian peneplain)
are characteristic land forms (Fig. 17.7).

## 17.34  CLIMATE AND HYDROLOGY

The openness of Norway to oceanic influences and the topography of the
country combine to produce outstanding differences in physical features
and climate. This is demonstrated in the drainage basin of the Glåma,
which is relatively narrow, but extends for more than five hundred
kilometres from south to north and an altitude range of about 2000 m.
The region belongs to the Boreal Forest climatic zone, with the mean
temperature in the coldest month below -3 °C and a snow cover every
year (Birkeland, 1935). The length of the vegetative season, defined
as the number of days between the date in spring when the daily mean
temperature reaches  +3 °C and the date in autumn when the daily mean
again falls to this figure, is a good indication of seasonal variation
of temperature. This period corresponds roughly to the growing season
for grass (Lauscher *et al.*, 1955). The vegetative season at Røros
(673 m altitude) is 7 May - 29 September and at Lillestrøm (104 m
altitude) 11 April - 25 October. (These dates are averages for the
period 1901-1930.)

*Fig. 17.6*  The Glåma at Braskereidfoss - low relief is predominant.
(Photo by Widerøes Flyveselskap A/S)

*Fig. 17.7*  The Glåma at Bingsfoss during storm flood.  Lowland landscape with hills of Archaean rocks.  (Photo by Widerøes Flyveselskap
A/S)

The monthly means of temperature and precipitation for selected stations are shown in Fig. 17.8.  The Scandes mountain range forms an effective barrier against the influence of the ocean, and marked continentality is therefore found in the central  and northern part of the catchment area. Maritimity is experienced to some degree towards the south, with more wind activity and greater cloud cover.

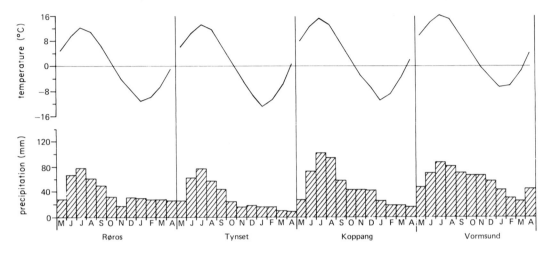

*Fig. 17.8* Mean temperature and precipitation for individual months during the period 1931-1960.

Ice conditions in the lakes give further important information about the climate, since the key factor in the formation of ice is temperature.  The average dates for freezing and break-up of the ice are given in Table 17.1 for selected lakes.  The lakes of the area are typically cold dimictic-temperate lakes, with circulation in spring and autumn. Thermal stratification is inverse in winter and direct in summer.

*Table 17.1* Duration of the ice cover.  Average dates for freezing and break-up of ice.*

| lake | period of observations | date of freezing | date of breaking |
|------|------------------------|------------------|------------------|
| Aursunden | 1923 - 1950 | November 20 | June 5 |
| Savalen | 1945 - 1950 | November 17 | May 23 |
| Atnasjøen | 1917 - 1950 | November 18 | May 5 |
| Storsjøen (Rendalen) | 1902 - 1950 | February 7 | May 2 |
| Mjøsa | 1863 - 1980 | December 10 | April 25 |
| Øyeren** | 1930 - 1960 | December 10 | April 20 |

*Data from Norges vassdrags- og elektrisitetsvesen.
**Local information.

The hydrology of the Glåma can be characterized by the annual variation in discharge (Otnes & Ræstad, 1978). Owing to the precipitation
regime and the storage of snow during winter, rivers of east Norway
have a maximum discharge in spring and a minimum discharge in late
winter and during the summer (Tollan, 1972). Two discharge maxima are
usually experienced during the spring and early summer. "Home flood"
is due to the melting of snow in the lowlands, and "mountain flood"
when melting from snow and glaciers in the highlands starts later on.

The Glåma, however, shows modifications to this general drainage
pattern (see Fig. 17.9). This is due partly to peculiarities of the
catchment and partly to extensive management of water flow for hydro-
electric power production. The spring flood is an unfailing phenomenon,
usually with two maxima. The first corresponds with the flood from the
valley of Østerdalen, the second one with the flood through the valley
of Gudbrandsdalen. This is caused by several circumstances, among
which is the delay to the water of the R. Gudbrandsdalslågen due to its
passage through L. Mjøsa.

*Fig. 17.9* Changes in discharge of the Glåma at Sarpsborg hydrometric station throughout 1970.

The hydrological data included in Section 17.2 are based on
measurements at official hydrometric stations. At Solbergfoss (downstream of L. Øyeren) the spring flood on average peaks on June 1. The
earliest and latest dates are 6th May (1959) and 7th July (1927),
respectively. The average discharge in the period 1911-1950 was 686 $m^3$
$s^{-1}$. This corresponds with a discharge per unit area of 17.1 l $s^{-1}$
$km^{-2}$ and a mean precipitation of 540 mm for the whole Glåma catchment.
Coincidence of floods in the two major valleys can give rise to catastrophic high waters (Tollan, 1967). This has taken place at intervals,
with the years 1740, 1743, 1749, 1754, 1760, 1771, 1773, 1789, 1795,
1850, 1860, 1910, 1916, 1927, 1934, 1966 and 1967 being outstanding.

The largest discharge was experienced during 1789 and 1860, when the peak was 4700 $m^3$ $s^{-1}$ and 4200 $m^3$ $s^{-1}$, respectively (at Solbergfoss, downstream of L. Øyeren). The level of L. Øyeren rose by 11 m and the surface of the lake increased from 35 to 54 km in length (Blakstad, 1939).

## 17.35  AREA AND SETTLEMENT

The natural environment accounts for the character of the human settlement (Arnesen, 1970; Holtan, 1973). The population is concentrated mainly towards the coast. The inland pattern of distribution is linear, following the valleys that dissect the mountains. Of a total of 518000 inhabitants, approximately 50% live in towns, 10% in town-like agglomerations and 40% are dispersed. Their distribution is evident from data in 17.2. The economy of the Glåma basin is centred on agriculture, forestry and manufacturing industries (Sund, 1960). Although now somewhat less important than urban industries, agriculture has been the dominant factor in forming the pattern of settlement and economic activities in the area.

The Glåma is important as an energy resource. Hydroelectric plants both with a low head and a large flow of water, and plants with a high head and a small flow are used. The first category, the lowland type, is confined to waterfalls in the lower part of the basin. The production of hydroelectricity from the Glåma amounts to 9000 GWh (9 x $10^{12}$ Wh), approximately 12% of the water-power used at present in Norway.

## 17.4  CHEMICAL AND BIOLOGICAL QUALITY OF WATER

## 17.41  WATER CHEMISTRY

The quality of run-off water depends mainly on the chemistry of the precipitation, the bedrock, the plant cover and human activity in the catchment area. The distance from the sea coast has an important influence on precipitation. The prevailing south-western air currents bring moisture laden air of oceanic origin towards Norway and there is a gradient in chloride concentrations from west to east across the Scandes - from approximately 30 to 0.3 mg Cl $l^{-1}$. The Glåma basin belongs to the part of the gradient with low concentrations.

Precipitation over southern Scandinavia contains large amounts of $H^+$, $SO_4^{2-}$ and $NO_3^-$ that originate as air pollutants in the industrialized areas of Europe. The inland waters in regions with Archaean rocks and thin cover of unconsolidated glacial till and soil have extremely low buffer capacities. Increasing acidity results in changes in chemical and biological conditions. Interference with fish reproduction and spawning, leading to decline in fish populations, are particularly well-documented (Wright et al., 1976; Drabløs & Tollan, 1980).

The bedrock has a marked influence on the composition of inland waters (Kjensmo, 1966). The drainage water from the hard-schistose and siliceous rocks of the Scandes is very poor in electrolytes and its conductivity is sometimes below 10 $\mu$S cm$^{-1}$. The softer, frequently somewhat calcareous rocks are richer in electrolytes and conductivity lies in the range 100 - 200 $\mu$S cm$^{-1}$. Areas of this type are, however, sparse and restricted in the Glåma basin (Holtan, 1975).

Examples of hydrochemical data are given in Tables 17.2 and 17.3. The highest conductivity values occur during periods of low flow. The percentage of ground water in the Glåma is normally high, and consequently the water will contain a relatively high content of dissolved inorganic substances under such conditions. Daily measurements of hydrographical parameters reveal a complex relationship between meteorological conditions and the composition of river water (Skulberg, 1968). The range for conductivity and turbidity during a 24-hour period can approach that of the yearly range. Important factors influencing the quality of Glåma water are discharge, surface water run-off from cultivated fields, pollution from sewage and transport of seston (Arnesen, 1969).

## 17.42  WATER FERTILITY

For water management and pollution control it is important to understand changes in algal vegetation of waters receiving nutrients in excess of the natural supply. The use of the algal growth potential test (AGP) for establishing qualitative and quantitative responses to the major plant nutrients has rendered possible effective regional comparisons of water fertility (Skulberg, 1967b; Källqvist, 1973; Kotai et al., 1978). The inland waters of the Glåma basin belong to several categories with regard to water fertility, ranging from a distinct oligotrophic-dystrophic type to an extreme eutrophic type. Based on land use, the predominant type of watershed may be classified into three categories; bioassay determinations of the fertility of various water show marked differences with respect to their potential for supporting algal growth (Table 17.4).

The size of algal crops in the predominantly oligotrophic waters of the Glåma has in most cases been shown to depend on the amount of available phosphorus. In some instances, however, enrichment experiments with essential nutrients have indicated that availability of nitrate as well as orthophosphate was important for the total yield. In such balanced conditions both nitrogen and phosphorus may act as limiting nutrients, either simultaneously or alternately.

*Table 17.2*  Aqueous chemistry at selected stations in the Glåma. Data of April 1967.

| variable | unit | Aursunden | Elverum | Funne-foss | Solberg-foss |
|---|---|---|---|---|---|
| temperature | $^{\circ}C$ | 1.8 | 0.2 | 3.3 | 2.5 |
| colour | mg $l^{-1}$ Pt | 8 | 40 | 112 | 55 |
| turbidity | F.T.U. | 0.3 | 1.1 | 5.4 | 8.0 |
| conductivity $20^{\circ}C$ | $\mu S$ $cm^{-1}$ | 35 | 42 | 29 | 45 |
| acidity | pH | 6.9 | 6.9 | 6.5 | 6.8 |
| magnesium | mg $l^{-1}$ | 0.8 | 1.0 | 0.8 | 1.4 |
| calcium | mg $l^{-1}$ | 4.4 | 5.2 | 2.9 | 4.3 |
| manganese | mg $l^{-1}$ | 0.005 | 0.016 | 0.031 | 0.11 |
| iron | mg $l^{-1}$ | 0.02 | 0.13 | 0.31 | 3.3 |
| nitrate | $\mu g$ $l^{-1}$ N | 65 | 50 | 265 | 330 |
| nitrogen (total) | $\mu g$ $l^{-1}$ N | 295 | 220 | 525 | 650 |
| reactive phos-phate | $\mu g$ $l^{-1}$ P | 2 | 8 | 3 | 5 |
| phosphorus (total) | $\mu g$ $l^{-1}$ P | 5 | 10 | 15 | 83 |
| chloride | mg $l^{-1}$ | 6.5 | 1.0 | 1.3 | 2.0 |
| sulphate | mg $l^{-1}$ S | 0.7 | 2.0 | 1.5 | 2.6 |
| oxygen demand ($KMnO_4$) | mg $l^{-1}$ $O_2$ | 1.4 | 6.5 | 7.9 | 5.0 |

*Table 17.3*  Water chemistry at monthly intervals during 1977 in the Glåma at Solbergfoss.

| variable / date | colour mg $l^{-1}$ Pt | turbi-dity F.T.U. | conduc-tivity $\mu S$ $cm^{-1}$ $20^{\circ}C$ | pH | nitrogen (total) $\mu g$ $l^{-1}$ N | phospho-rus (total) $\mu g$ $l^{-1}$ P |
|---|---|---|---|---|---|---|
| January 9 | 26.0 | 0.96 | 39.5 | 6.9 | 500 | 12 |
| February 19 | 21.5 | 0.60 | 41.1 | 7.1 | 420 | 8 |
| March 13 | 26.0 | 0.80 | 40.9 | 6.9 | 470 | 11 |
| April 17 | 197.0 | 6.8 | 47.0 | 7.0 | 640 | 17 |
| May 22 | 104.5 | 3.8 | 33.2 | 7.0 | 490 | 13 |
| June 12 | 95.5 | 2.9 | 34.2 | 6.9 | 400 | 12 |
| July 18 | 85.5 | 1.4 | 31.3 | 6.8 | 270 | 10 |
| August 5 | 37.5 | 0.76 | 35.0 | 7.3 | 230 | 12 |
| August 25 | 40.5 | 0.77 | 33.8 | 6.7 | 270 | 8 |
| September 25 | 46.0 | 1.1 | 38.4 | 7.2 | 370 | 7 |
| October 23 | 32.3 | 1.5 | 40.9 | 7.2 | 390 | 6 |
| November 18 | 197 | 5.8 | 40.3 | 6.8 | 380 | 15 |
| December 18 | 92 | 3.0 | 37.6 | 7.2 | 400 | 11.5 |

*Table 17.4* Examples of algal growth potential (AGP) and associated water chemistry.

| predominant watershed | locality | AGP cells $l^{-1}$ x $10^6$ | phospho-rus (total) $\mu g\ l^{-1}$ P | nitrogen (total) $\mu g\ l^{-1}$ N | conduc-tivity $\mu S\ cm^{-1}$ 20°C |
|---|---|---|---|---|---|
| forest and | Savalen | 2 | 3 | 50 | 49 |
| mountain | Ossjøen | 3 | 8 | 205 | 16 |
| areas | Storsjøen (Rendalen) | 4 | 6 | 155 | 26 |
| | Storsjøen (Odalen) | 4.5 | 8 | 210 | 25 |
| rural and | Otta | 7 | 4 | 130 | 16 |
| agricultural | Mjøsa | 18 | 10 | 400 | 39 |
| areas | Øyeren | 20 | 9 | 265 | 35 |
| | Nitelva (Glåma) | 145 | 28 | 600 | 48 |
| urban and | Svartelva | 156 | 55 | 1200 | 118 |
| industrial | Rømua | 965 | 170 | 570 | 115 |
| areas | Nitelva (Kjeller) | 1600 | 174 | 1590 | 93 |

## 17.5  PLANT AND ANIMAL LIFE

### 17.51  CLASSIFICATION OF THE NATURAL GEOGRAPHICAL REGION

By use of descriptions of the geology, climate, plant and animal life in the Glåma basin, it is possible to classify the area into a system of natural, geographical regions (Udvardy, 1975).  A subdivision of Norden into vegetation zones has been made, based mainly on floristical, phytosociological and ecological principles (Nordic Council of Ministers, 1977).  According to this the Glåma basin belongs largely to the boreo-nemoral and boreal zones of the palaearctic realm.  The alpine zone is represented in areas above the tree line (600 - 800 m altitude) in the mountain range of the Scandes.

### 17.52  BENTHIC VEGETATION

#### 17.521 Attached algae

A vegetation rich in species and sometimes abundant occurs in the Østerdalen part of Glåma (Skulberg & Kotai, 1977).  More than two hundred algal species have been reported from attached vegetation, about eighty of which may be considered common.  These consist primarily of several genera of green algae and diatoms, but species of blue-green and red algae may also be abundant (Table 17.5).  Species which frequently

*Table 17.5*  Common species of benthic algae in the upper Glåma.
++++ abundant

BLUE GREEN ALGAE

*Lyngbya* sp.
*Nostoc* cf. *verrucosum*
*Oscillatoria* cf. *bornetii*
*Oscillatoria* sp.
*Phormidium autumnale*
*Scytonema* sp.
*Stigonema mamillosum*

GREEN ALGAE

*Bulbochaete* sp.
*Closterium* spp.
*Cosmarium undulatum*
*Cosmarium* spp.
*Draparnaldia glomerata*
*Microspora amoena* ++++
*Mougeotia* spp.
*Oedogonium* sp.
*Scenedesmus obliquus*
*Scenedesmus* spp.
*Spirogyra* spp.
*Staurastrum alternans*
*S. apiculatum*
*Staurastrum* spp.
*Stigeoclonium tenue*
*Ulothrix zonata* ++++
*Ulothrix* spp.
*Vaucheria* sp.
*Zygnema* spp. ++++

CHRYSOPHYTES

*Hydrurus foetidus* ++++

DIATOMS

*Achnanthes* spp.
*Amphora ovalis*
*Ceratoneis arcus* ++++
*Cymbella* cf. *affinis*
*Cymbella* spp.
*Diatoma elongatum*
*Didymosphenia geminata* ++++
*Eunotia arcus*
*Fragilaria construens*
*Fragilaria* sp.
*Gomphonema olivaceum*
*Gomphonema* spp.
*Meridion circulare*
*Navicula* spp.
*Nitzschia palea*
*Nitzschia* spp.
*Pinnularia* cf. *brevicostata*
*P. viridis*
*Pinnularia* sp.
*Synedra ulna*
*Tabellaria fenestrata*
*T. flocculosa* ++++

RED ALGAE

*Batrachospermum* sp.
*"Chantransia chalybea* var.
   *leibleinii"*
*Lemanea fluviatilis* ++++

form mass developments are *Zygnema* spp., *Ulothrix zonata*, *Microspora amoena*, *Tabellaria flocculosa*, *Ceratoneis arcus*, *Didymosphenia geminata*, *Hydrurus foetidus* and *Lemanea fluviatilis* (Skulberg & Kotai, 1978). Measurements of conspicuous benthic communities have shown values for mass in the range 50 - 500 g organic dry weight per $m^2$ of river bottom.

The seasonal variation in the development of attached algae is pronounced, with light climate, temperature and rhythm of discharge as the key factors.  Four periods may be distinguished:

November to February: dark part of winter, low flow;
March to April: light part of winter, low flow;
May to mid-July: summer, high flow;
Mid-July to November: summer and autumn, low flow.

During the light part of winter a varied flora of diatoms, *Hydrurus foetidus* and *Ulothrix zonata* are well developed.  This is torn away from the substratum by the powerful turbulence of storm flows.  During the period of high summer flows, there is a luxuriant vegetation of *Zygnema* spp., *Microspora amoena* and *Lemanea fluviatilis*.  The dark part of winter shows a decline in algal crops, but cold-water species may persist for longer or shorter periods.

Among the diatoms of quantitative importance *Didymosphenia geminata* (Fig. 17.10) deserves mention.  This species may flourish in late summer and, where growth continues undisturbed, the grey and brown colonies reach several centimeters in diameter.  In some places this diatom forms a dense cover on boulder surfaces directly facing the current.  In rapids which are not ice-covered it can grow throughout the year.

In the lower stretches of the Glåma (downstream of its confluence with the Vorma) current speeds are lower and substrata of sand, clay and mud predominate.  The vegetation of attached algae both reflects these conditions and shows a distinct eutrophic trend.  Species of *Vaucheria* (*e.g. walzi*) develop typically as cushion-like growths close to the soft substratum.  Abundant epipelic algae include *Oscillatoria brevis, O. limosa, Phormidium uncinatum, Fragilaria capucina, Nitzschia acicularis* and *Diatoma elongatum*.    Various Chlorococcales and green flagellates occur sporadically.

The attached algal vegetation of the Glåma may be classified according to the chief floral component (Israelson, 1949), with an upstream *Zygnema*-type (relatively oligotrophic) and a downstream *Vaucheria*-type (relatively eutrophic).  Vegetation of the *Vaucheria*-type has however increased during recent years in some parts of the Østerdalen stretch.  This is the result of changes due to pollution and regulation for hydroelectric purposes.

## 17.522 *Bryophytes*

Bryophytes have not been the subject of special study.  Species identified in connection with algal studies include: *Brachythecium rivulare, Ceratodon purpureus, Dichelyma falcatum, Drepanocladus exannulatus,*

*Fontinalis antipyretica, F. dalecarlica, Hygrohypnum ochraceum, Philonotis fontana, Pohlia albicans, Schistidium alpicola.*

## 17.523 Higher Plants

Although the river may be characterized generally as oligotrophic, higher plant vegetation is abundant in some areas, especially on finer sediments (Rørslett, 1972; Arnesen, 1970). Some indication of the relative abundance of the various species is given in Table 17.7.

The most common long-shoot plants are species of *Potamogeton* and *Sparganium* (Table 17.6). *Myriophyllum alterniflorum* is typical of the more oligotrophic parts of the water course. Isöetids are usually predominant in shallow parts of the lower reaches. The zone between high and low water is colonized by e.g. *Eleocharis acicularis, Juncus filiformis, Subularia aquatica* and *Ranunculus reptans.* A helophyte vegetation can grow abundantly on fine-grained river beds: prominent species are *Equisetum fluviatile, Carex aquatilis* and *Carex rostrata.*

Vegetation dominated by *Phragmites australis* and/or *Schoenoplectus lacustris* occurs only occasionally, except in areas of slow-moving water upstream of the estuary (broads and lake-like extensions of the river). Here they are the predominant species, with the associated plants being ones indicative of highly eutrophic conditions (e.g. *Ceratophyllum demersum, Myriophyllum verticillatum*). The brackish-water area of the estuary has a vegetation of freshwater and marine species (Rørslett, 1974). Among those which are abundant include *Potamogeton perfoliatus* and two species of *Ruppia* (Rørslett, 1975).

## 17.53  SESTON - ESPECIALLY PHYTOPLANKTON

The term "seston", as used by Kolkwitz (1912), means the particulate mass of various living and non-living substances in water. The composition of the suspended matter in the Glåma is very complex, depending on the geology, climate and biology of the catchment area, and on the biology of the river itself. Man's activities greatly influence the amount of suspended material. Observations on the seston, and especially the phytoplankton, in the lake-river system Lake Mjøsa - Lake Øyeren - R. Glåma have been carried out extensively by the Norwegian Institute for Water Research (Lindstrøm et al., 1973). Several methods have been used for the measurement and identification of the seston (Baalsrud & Henriksen, 1964; Skulberg, 1978).

The phytoplankton of the Glåma is dominated mainly by diatoms and chrysophytes. A comparison of fluctuations in seston and diatoms in the water column upstream of Sarpsfossen during one year (1970) with fluctuations in discharge are shown in Fig. 17.11. Seston concentrations were highest just as the spring flood started to develop. Erosion in the lowland parts of the catchment gives rise to increased

*Table 17.6*  Examples of vascular plants in the upper, lower and estuarine Glåma.    ++++, abundant; +++, frequent; ++, occasional; +, rare.

| species | upper | lower | estuary |
|---|---|---|---|
| *Equisetum fluviatile* | +++ | ++++ | |
| *Isoetes echinospora* | ++ | +++ | |
| *Sparganium angustifolium* | ++ | + | |
| *S. ramosum* | + | ++ | ++ |
| *S. simplex* | + | ++ | |
| *Potamogeton alpinus* | ++ | | |
| *P. gramineus* | ++ | + | |
| *P. perfoliatus* | ++ | +++ | ++ |
| *Alisma plantago-aquatica* | ++ | +++ | |
| *Sagittaria sagittifolia* | + | +++ | |
| *Phragmites australis* | ++ | +++ | ++++ |
| *Phalaris arundinacea* | ++ | +++ | ++ |
| *Scirpus lacustris* | ++ | +++ | |
| *S. tabernaemontanii* | | | ++++ |
| *Eleocharis acicularis* | +++ | ++++ | |
| *E. palustris* | +++ | +++ | |
| *E. uniglumis* | + | + | +++ |
| *Carex aquatilis* | ++ | +++ | |
| *C. gracilis* | +++ | ++++ | |
| *C. rostrata* | ++++ | +++ | |
| *Acorus calamus* | | ++ | |
| *Juncus filiformis* | +++ | ++++ | |
| *Ceratophyllum demersum* | | +++ | ++ |
| *Ranunculus peltatus* | ++ | +++ | |
| *R. reptans* | +++ | ++++ | |
| *R. trichophyllus* | ++ | | |
| *Subularia aquatica* | +++ | ++++ | |
| *Crassula aquatica* | ++ | +++ | |
| *Callitriche* spp. | ++ | ++++ | |
| *Elatine hydropiper* | ++ | +++ | |
| *E. triandra* | + | +++ | |
| *Peplis portula* | + | + | |
| *Myriophyllum alterniflorum* | +++ | ++ | |
| *M. verticillatum* | | ++ | ++ |
| *Hippuris vulgaris* | ++ | + | |
| *Cicuta virosa* | ++ | ++ | |
| *Limosella aquatica* | +++ | +++ | |
| *Bidens tripartita* | ++ | +++ | |

*Fig. 17.11* Changes in relative amounts of seston and population density of diatoms at Sarpsborg throughout 1970. (Compare with Fig. 17.9 which shows discharge at the same site and through the same year)

*Fig. 17.12* Comparison of climatic conditions (radiation and precipitation), water temperature, phytoplankton population density and volume of selected species in the Glåma at Solbergfoss throughout 1976.

amounts of clay and silt in the river water during this period. Two species of diatoms dominated the plankton community. *Asterionella formosa* was largely responsible for the early summer burst (June-July) and during the late summer peak (August-September) *Fragilaria crotonensis* exceeded *Asterionella* in abundance.

In the lowland part of the river, the species composition (Table 17.7) reflects the increased fertility of the water resulting from agricultural drainage and sewage effluents. Nutrient enrichment leads both to an increase in the blue-green algal crop and in its species composition. During August and September 1976 the first bloom of *Oscillatoria bornetii* f. *tenuis* took place in the water system L. Mjøsa-R. Glåma. The population density of this species increased during August, although diatoms were still the dominant. By the end of September the density of blue-green alga was considerable (L. Mjøsa, approximately 2 g m$^{-3}$: Holtan, 1979) and it dominated the planktonic vegetation during late autumn (October-November). The development of this population resulted in an obnoxious and abnormal taste and odour of the water in the lake-river system.  Investigations made it clear that the alga was involved, directly or indirectly, in the establishment of conditions leading to taste and odour problems in the waterworks using L. Mjøsa-R. Glåma as raw water supply.  Among the substances associated with the *Oscillatoria* bloom was geosmin (trans-1,10 dimethyl-trans-9-decalol) giving the water a very distinct and offending flavour.

*Table 17.7*  Common planktonic algae in the lower Glåma. ++++ abundant

| BLUE-GREEN ALGAE | DIATOMS |
|---|---|
| *Anabaena flos-aquae* ++++ | *Asterionella formosa* ++++ |
| *Gomphosphaeria lacustris* | *Diatoma elongatum* ++++ |
| *Merismopedia tenuissima* | *D. vulgare* |
| *Oscillatoria bornetii* f. tenuis ++++ | *Fragilaria crotonensis* ++++ |
|  | *Cyclotella stelligera* |
| GREEN ALGAE | *C. comta* |
|  | *Melosira granulata* var. angustissima |
| *Ankistrodesmus falcatus* | *M. islandica* ssp. *helvetica* |
| *Chlamydomonas* spp. | *M. italica* |
| *Closterium pronum* | *Rhizosolenia eriensis* ++++ |
| *Crucigenia tetrapedia* | *R. longiseta* ++++ |
| *Dictyosphaerium pulchellum* ++++ | *Stephanodiscus hantzschii* |
| *Gemellicystis neglecta* | *Synedra acus* |
| *Scenedesmus* spp. | *S. ulna* |
|  | *Tabellaria fenestrata* var. asterionelloides ++++ |
| CHRYSOPHYTES | *T. flocculosa* |
|  | *T. fenestrata* |
| *Chrysococcus* spp. |  |
| *Dinobryon borgei* |  |
| *D. divergens* |  |
| *D. sociale* |  |
| *Kephyrion spirale* |  |
| *Mallomonas elongata* ++++ |  |
| *M. reginae* ++++ |  |
| *Synura* spp. ++++ |  |

## 17.54   BENTHIC INVERTEBRATES

Studies of the benthos of stony substrata at eleven stations from the alpine zone down to the Lake Øyeren area have shown that insect nymphs and larva completely dominate the fauna.  Chironomids are most numerous, while Trichoptera have the largest biomass.  There are, however, marked longitudinal differences in the faunistic composition.  Plecoptera are best represented in the upper stretches, whereas Ephemeroptera and Trichoptera are most abundant further downstream, where pollution from sewage and the influence of flow regulation increase the benthic algal crops. At the station above L. Øyeren, where pollution is prominent, Chironomids dominate entirely (Fig. 17.13). Reservoirs influence the faunistic composition by increasing the abundance of filter feeding Trichoptera, *e.g.* below the Barkaldfoss and Rånåsfoss reservoirs (Lillehammer & Saltveit, 1980).

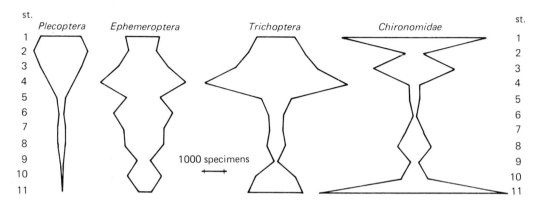

*Fig. 17.13* Longitudinal changes in main insect groups on passing down the Glåma (from stations 1 to 11).

Nymphs of the stoneflies *Capnia pygmaea* and *C. atra* occur in especially large numbers in the Østerdalen part of the Glåma, forming the basis for local ice fishing for grayling and whitefish each April. The nymphs crawl in profusion onto the river ice ready for emergence. They are then collected by fishermen, who use them as bait (Jensen, 1968; Lillehammer, 1971).

Blood-sucking blackflies are a considerable nuisance for humans and livestock in the area of Sarpsborg in the lower Glåma. A notorious example is the large numbers of *Simulium truncatum* in a tributary connected to a dam spillway with lacustrine conditions above. Development during the spring and summer floods benefits from the rich supply of organic seston in the river water. Raastad (1974) estimated that more than a billion flies emerge from the rapids in this tributary during swarming.

17.55  FISH

The freshwater fish recorded from the Glåma water-course comprise 28 species (Huitfeldt-Kaas, 1918; Halvorsen, 1968; Hansen, 1978; Pethon, 1974, 1978, 1980). The fauna is especially rich in the lower river reaches - *e.g.* the L. Øyeren area with 24 species and the river below L. Øyeren with 27 species (Table 17.8).

*Table 17.8*  Freshwater fish in different parts of the Glåma.
Records:  x, frequent;  (x), occasional.

| species | upper | middle sub-section one | middle sub-section two | L. Øyeren | lower | estuary |
|---|---|---|---|---|---|---|
| sea lamprey | | | | | (x) | x |
| lamprey | | x | x | x | x | |
| char | x | | | | | |
| trout | x | x | x | x | x | x |
| whitefish | x | x | x | x | x | x |
| cisco | | | | x | (x) | (x) |
| grayling | x | x | x | x | x | |
| salmon | | | | | (x) | x |
| smelt | | | x | x | x | x |
| pike | x | x | x | x | x | (x) |
| roach | | x | x | x | x | x |
| dace | | | x | x | x | x |
| chub | | | x | x | x | (x) |
| ide | | | x | x | x | x |
| rudd | | | | | x | (x) |
| minnow | x | x | x | x | x | |
| asp | | | x | x | x | (x) |
| bleak | | | x | x | x | (x) |
| bream | | | x | x | x | x |
| white bream | | | | x | x | |
| crucian carp | | | | x | x | |
| ten-spined stickleback | | | | x | x | x |
| three-spined stickleback | | | | | x | x |
| burbot | x | x | x | x | x | (x) |
| perch | x | x | x | x | x | x |
| pikeperch | | | | x | x | (x) |
| ruffe | | | x | x | x | x |
| bullhead | x | x | x | x | x | |

The water-course is favourable for grayling and trout over an extensive stretch.  From the alpine zone down to the Østerdalen stretch of the Glåma, grayling, bullhead and minnow are the most common species in riffles, with smaller populations of trout and burbot (Borgstrøm *et al.*, 1976).  Further downstream, pike, perch, whitefish and some cyprinids are common, and productive fishing grounds are found in L. Øyeren and elsewhere.  The Atlantic salmon (*Salmon salar*) occurs in the Glåma below Sarpsfossen and in the tributary, R. Ågårdselv.  The population is small

in recent years (e.g. Halvorsen, 1968), but at the beginning of this
century catches could be of some economic importance (an annual average
of 3500 kg during 1878-1882: Kiær, 1915).  Twelve freshwater species are
present in the estuary, with ide, roach, perch, dace, bream and ruffle
being those most commonly observed.  In this area it is possible to
fish freshwater fish in the outstreaming, surface water, and saltwater
fish in the inward-streaming bottom (marine) water.


## 17.6  THE LAKE-RIVER SYSTEM

The Glåma can be divided on faunal criteria into three main parts:

Upper  above the tree line, i.e. the alpine zone.
Middle from the tree line down to the L. Øyeren area.
Lower  downstream of L. Øyeren area, via estuary to the sea.

Upper.  Studies of Norwegian streams have shown that the benthic fauna
decreases in species above the tree line, and most of the species
present are less abundant than downstream (Lillehammer, 1974; Lillehammer et al., 1980).  The reduction is marked in most of the benthic
groups and filter-feeding Trichoptera larvae are rarely recorded above
the tree-line.  Only nine species of fish have been recorded in the
upper Glåma (Table 17.8).

Middle.  This part can be divided into two subsections, above and downstream of the central area of the Østerdalen valley.  In the former,
Plecoptera are abundant and grayling and bullhead are the most common
species of fish; only two species of cyprinid have been recorded,
minnow and roach.  The benthic fauna of the lower subsection is dominated by Ephemeroptera and Chironomidae.  Plecoptera are sparse and
the number of cyprinid species increases gradually on passing downstream.

Lower.  Two species of fish not recorded in the other parts of the
Glåma are present below L. Øyeren.  These are the Atlantic salmon and
the rudd.  In the estuary, besides the twelve species of freshwater
fish (see above), there are eighteen marine species (Pethon, 1980).
Insects are sparsely represented in the benthos, except for chironomids.

   The changes in plankton composition are demonstrated downstream from
L. Mjøsa through the Vorma and Glåma, L. Øyeren and the estuary and out
to the sea of the Hvaler archipelago.  Diatoms are in general predominant, but blue-green algal blooms can occur (17.53).  In L. Øyeren
flagellates are relatively more frequent (Skulberg, 1972).  There is a
distinct similarity between the plankton of L. Mjøsa and that of the
Vorma and Glåma down to the inlet of L. Øyeren, with Asterionella
formosa, Fragilaria crotonensis, Diatoma elongatum, Tabellaria fenestrata var. asterionelloides and Rhizosolenia spp. the dominants.  These
species also characterize to a large extent the plankton down to the
estuary (Lindstrøm et al., 1973).  This shows that the populations

developing in an upstream lake can influence the composition of plankton in the whole water-course downstream.

REFERENCES

Arnesen R. (1969) *En Undersøkelse av Glåma i Østfold. Kjemiske og Bakteriologiske Forhold.* 0-217. Norsk Institutt for Vannforskning, Oslo. (A survey of the Glåma in Østfold. Chemical and bacteriological situation.)

Arnesen R. (1970) *En Undersøkelse av Glåma i Østfold. Sammenfattende del.* 0-217. Norsk Institutt for Vannforskning, Oslo. (A survey of the Glåma in Østfold. Summary.)

Birkeland B.J. (1935) *Mittel und Extreme der Lufttemperatur.* Geofysiske Publikasjoner, Vol. XIV, No. 1, 155 pp. Det Norske Videnskaps-Akademi, Oslo.

Bjørlykke K.O. (1940) *Utsyn over Norges jord og jordsmonn.* Nor. Geol. Unders. 156, 235 pp. Aschehoug, Oslo. (Overview of Norwegian soil conditions.)

Blakstad W. (1939) *Tømmerfløtingen i Nedre Glomma gjennom de siste hundre år.* 326 pp. Thorvald Nielsens Boktrykkeri, Porsgrunn. (Timber floating in the lower Glåma over the last hundred years.)

Baalsrud K. & Henriksen A. (1964) Measurement of suspended matter in stream water. *J. Am. Wat. Wks Assoc.* 56, 1194-200.

Borgstrøm R., Brittain J.E. & Lillehammer A. (1976) Østerdalsskjønnet. Glåma mellom Auma og Høyegga. Virkning på fisket. *Rapp. Lab. Ferskv. Økol. Innlandsfiske* 25, 1-6. (Effect of stream regulation on fisheries.)

Drabløs D. & Tollan A. (Eds) (1980) *Ecological impact of acid precipitation.* Proceedings of an international conference, Sandefjord, Norway, March 11-14 1980. 383 pp. SNSF-project, ÅS-NLH.

Halvorsen O. (1968) Resultater av månedlige garnfangster i Glomma ved Sarpsborg. *Fauna* 21, 166-74. (Result of monthly net catches in the Glåma at Sarpsborg.)

Hansen L.-P. (1978) Forekomst og fordeling av noen fiskearter i Nordre Øyeren. *Fauna* 31, 175-83. (Occurrence and distribution of some fishes in northern Øyeren.)

Holmsen G. (1915) Brædæmte sjøer i Nordre Østerdalen. *Nor. Geol. Unders.* 72. (Glacier - dammed lakes in northern Østerdalen.)

Holtan H. (1973) *Glåma i Hedmark. Undersøkelser i Tidsrommet 1966-1972.* 0-138/70. Norsk Institutt for Vannforskning, Oslo. (Glåma in Hedmark. Surveys in the period 1966-1972.)

Holtan H. (1975) *Gudbrandsdalsvassdraget, Mjøsa, Vorma. Resipient-Undersøkelser i Forbindelse med Planlagte Vassdragsreguleringer 1974-1975.* 0-151/73. Norsk Institutt for Vannforskning, Oslo. (Gudbrandsdal river-system, Mjøsa, Vorma. Recipient surveys in connection with planned stream regulations 1974-1975.)

Holtan H. (1979) The Lake Mjøsa story. *Arch. Hydrobiol. Beih. Ergebn. Limnol.* 13, 242-58.

Holtedahl O. (1953a,b) *Norges Geologi.* Bind I, pp. 1-583. Bind II, pp. 585-1118. Aschehoug, Oslo.

Holtedahl O. (1960)  *Geology of Norway*.  Nor. Geol. Unders. 208.  580 pp.  Aschehoug, Oslo.

Huitfeldt-Kaas H. (1918)  *Ferskvandsfiskenes Utbredelse og Indvandring i Norge*.  106 pp.  Kristiania (Centraltrykkeriet).  (Freshwater fish in Norway and their immigration.)

Israelson G. (1949)  On some attached Zygnemales and their significance in classifying streams.  *Bot. Not.* 4, 313-58.

Jensen K.W. (1968)  Abbor.  In: Jensen K.W. (Ed.) *Sportsfiskerens Leksikon I*, pp. 1-11.  Gyldendal, Oslo.  (Perch.)

John B. (Ed.) (1979)  *The winters of the world, earth under the Ice Ages*.  256 pp.  David & Charles, London.

Källqvist T. (1973)  Algal assay procedure (bottle test) at the Norwegian Institute for Water Research.  *Nordic Symposium on Algal Assays in Water Pollution Research, Oslo 1972*, pp. 5-17.  Nordforsk Publication 2, Helsinki.

Kiær A.Th. (1915)  *Smaalenenes Amt 1814-1914*.  543 pp.  E. Sem, Fredrikshald.

Kjensmo J. (1966)  Electrolytes in Norwegian lakes.  *Schweiz. Z. Hydrol.* 28, 29-42.

Kolkwitz R. (1912)  Plankton und Seston.  *Ber. deutsch. Bot. Gesell.* XXX, 334-46.

Kotai J., Krogh T. & Skulberg O. (1978)  The fertility of some Norwegian inland waters assayed by algal cultures.  *Mitt. int. Verein. theor. angew. Limnol.* 21, 413-36.

Lauscher A., Lauscher F. & Printz H. (1955)  *Die Phänologie Norwegens. Teil I. Allgemeine Übersicht*.  Skriftet utgitt av Det Norske Videnskaps-Akademi i Oslo.  I. Mat.-Naturv. Klasse, 1955.  No. 1, 99 pp.  Jacob Dybwad, Oslo.

Lillehammer A. (1971)  Døgnfluer. Øyestikkere. Steinfluer.  *Norges Dyr* 4, 217-230. Cappelen, Oslo.  (Norwegian fauna - ephemeras, dragonflies and stoneflies.)

Lillehammer A. (1974)  Norwegian stoneflies, II. Distribution and relationship to the environment.  *Norsk ent. Tidsskr.* 21, 195-250

Lillehammer A., Borgstrøm R. & Skulberg O. (1980)  Miljøpå-virkninger i norske vassdrag.  1. Økosystembeskrivelse.  *Fauna* 33, 109-116.  (Environmental impacts on Norwegian water-courses.  Ecosystem description.)

Lillehammer A., Brittain J. & Borgstrøm R. (in prep.)  Bunnfauna og fiskeregistreringer på strykstrekninger i Glåma.  (Bottom fauna and fishery registrations in the fast-flowing stretches of Glåma.)

Lillehammer A. & Saltveit S.J. (1980)  Stream regulation in Norway.  In: *The Ecology of Regulated Streams*, pp. 201-213.  Plenum Publishing Corporation.  ...s in the fast-flowing stretches of Glama.)

Lindstrøm E.A., Skulberg R. & Skulberg O.M. (1973)  Observations on planktonic diatoms in the lake-river system Lake Mjøsa - Lake Øyeren - River Glåma, Norway.  *Norwegian J. Bot.* 20(2/3), 183-95.

Nordic Council of Ministers (1977)  *Naturgeografisk Regionindelning av Norden*, 130 pp.  NU B 1977:34.  (Nature geographical classification of regions in the Nordic countries.)

Otnes J. & Ræstad E. (1978)  *Hydrologi i Praksis*, 314 pp.  Ingeniørforlaget, Oslo.  (Hydrology in Practice.)

Pethon P. (1974)  Fiskeundersøkelsene 1973.  In: Pethon P. (Ed.)
  Øraundersøkelsene  årsrapport 1973, pp. 139-52.  Zoologisk museum,
  Oslo.  (Fisheries survey for 1973.)

Pethon P. (1978)  Age, growth and maturation of natural hybrids between
  roach (Rutilus rutilus L.) and bream (Abramis brama L.) in Lake
  Øyeren, SE Norway.  Acta Hydrobiol. 20, 281-95

Pethon P. (1980)  Variations in the fish community of the Øra Estuary,
  SE Norway, with emphasis on the freshwater fishes.  Fauna norv. Ser.
  A I, 5-14

Ramsay W. (1898)  Über die geologische Entwicklung der Halbinsel Kola
  in der Quartärzeit.  Fennia 16(1), 4.  Helsingfors

Rørslett B. (1972)  Resipientforholdene i Romeriksvassdragene Nitelva,
  Leira og Rømua.  II. Botaniske undersøkelser.  0-55/68.  Norsk
  Institutt for Vannforskning.  (Recipient conditions in the Romerike
  rivers, Nitelva, Leira and Rømua.)

Rørslett B. (1974)  Hydrobotaniske forhold i Øra-området ved Fredrikstad.  0-50/73.  Norsk Institutt for Vannforskning.  (Hydrobotanical
  conditions in Øra region at Fredrikstad.)

Rørslett B. (1975)  Potamogeton perfoliatus i Øra, et brakkvannsområde
  ved Fredrikstad.  Blyttia 33, 69-82.  (Potamogeton perfoliatus in
  the Øra, a brackish water area at Fredrikstad.)

Raastad J.E. (1974)  Outbreaks of blood-sucking black flies (Simullidae)
  in Norway.  In: Proc. Third International Congress for Parasitology
  2, 918-919.  München

Skulberg O.M. (1967a)  Beskrivelse og Undersøkelser av Vannforekomster.
  Glåma.  0-110/65.  Norsk Institutt for Vannforskning.  (Descriptions
  and surveys of water in the Glåma.)

Skulberg O.M. (1967b)  Algal cultures as a means to assess the fertilizing influence of pollution.  J. Wat. Pollut. Control Fed., 113-27.
  Washington D.C.

Skulberg O.M. (1968)  Some experimental investigations of self-purification processes.  Grundförbättring 21, 25-37

Skulberg O.M. (1972)  Resipientforholdene i Romeriksvassdragene Nitelva,
  Leira og Rømua.  Hovedrapport I, 0-55/68.  Norsk Institutt for
  Vannforskning, Oslo.  (Recipient conditions of the Romerike water
  courses, Nitelva, Leira and Rømua.)

Skulberg O.M. (1978)  Sestonobservasjoner ved vassdragsundersøkelser.
  Fauna 31, 48-54.  (Seston observations in connection with the investigation of water.)

Skulberg O.M. & Kotai J. (1977)  Utredning om Begroingsforhold og
  Vannkvalitet for Østerdalsskjønnet.  Vassdragsstrekningen fra Stai
  til Samløp med Rena.  Undersøkelsene i Vegetasjonsperioden 1977.
  Norsk Institutt for Vannforskning, Oslo.  (Survey of benthic vegeta-
  tion  and water quality in Østerdal.  Stretch from Stai to Rena.
  Investigations during the growth period of 1977.)

Skulberg O.M. & Kotai J. (1978)  Miljøfaktorer og algeutvikling i
  strømmende vann.  Noen observasjoner av innvirkningene av vassdragsreguleringer på begroingsforhold i Glåma i Østerdalen.  In: Norsk
  Institutt for Vannforskning, Årbok 1977, pp. 63-73.  Norsk Institutt
  for Vannforskning, Oslo.  (Environmental factors influencing algal
  development in flowing waters.  Some observations on the implication
  of water regulation on the state of growth in the Glåma in Østerdal.

Strøm K. (1935)   Sammensatte botnsjøer og litt om geomorfologiske
    sjøtyper i Norge.   *N. Geogr. T.* 5, 334-341.   (Glacier-eroded lakes
    and geomorphological lake types in Norway.)

Sund T. (1960)   Norway.   In: Sømme A. (Ed.) *A Geography of Norden*,
    363 pp.   J.W. Cappelens Forlag, Oslo.

Sørensen R. (1979)   Lake Weichselian deglaciation in the Oslofjord
    area, south Norway.   *Boreas* 8, 241-246.

Tollan A. (1967)   Når vannet stiger.   Om flom og flomberegninger.
    *Elektroteknisk Tidsskrift* 32, 1-6.   (When water rises: about floods
    and flood estimation.)

Tollan A. (1972)   Variability of seasonal and annual runoff in Norway.
    *Nordic Hydrology* 3, 72-79.

Udvardy M. (1975)   *A Classification of the Biogeographical Provinces of
    the World*.   IUCN Occasional Paper No. 18, 48 pp.   International
    Union for Conservation of Nature and Natural Resources, Morges,
    Switzerland.

Wright R.F., Dale T., Gjessing E.T., Hendrey G.R., Henriksen A.,
    Johannessen M. & Muniz I.P. (1976)   Impact of acid precipitation on
    freshwater ecosystems in Norway.   *Wat. Air Soil Pollut.* 6, 483-99.

# 18 · Dunajec

B. KAWECKA & B. SZCĘSNY

| | | |
|---|---|---|
| 18.1 | Introduction | 500 |
| 18.2 | General data | 501 |
| 18.3 | The catchment | 502 |
| | 18.31 Geography | 502 |
| | 18.32 Physiography | 502 |
| | 18.33 Meteorology | 504 |
| | 18.34 Geology | 504 |
| | 18.35 Land use: agriculture and forestry | 505 |
| | 18.36 Man's activity | 505 |
| | 18.37 Management | 506 |
| 18.4 | Environment | 506 |
| | 18.41 Substratum | 506 |
| | 18.42 Flow | 507 |
| | 18.43 Temperature and light climate | 507 |
| | 18.44 Water chemistry | 507 |
| | 18.45 Pollution | 508 |
| 18.5 | Plants and animals of the upper and middle (unregulated) river | 510 |
| | 18.51 Plants | 510 |
| | 18.52 Macroinvertebrates | 511 |
| | 18.53 Fish | 512 |
| | 18.54 Zonation | 512 |
| | 18.55 Seasonal changes | 516 |
| | 18.56 Particular habitats | 521 |
| 18.6 | Plants and animals of the lower (regulated) river | 521 |
| 18.7 | Concluding comments | 522 |
| References | | 523 |

## 18.1  INTRODUCTION

The Dunajec* has the most beautiful landscape of any mountain river in
Poland and is the most popular with tourists.  This landscape includes
the extraordinarily picturesque mountains around the springs of the
tributaries in the Tatra Mountains and the famous gorge in the Pieniny
Mountains, where it is traditional to float on rafts down the river.
These rides, which have gone on for more than a century, offer the parti-
cipants unforegettable and exciting moments.  It is not surprising,
therefore, that the river has been the subject of interest for naturalists
over the past century and a half.  Detailed hydrobiological studies on
the Dunajec were however initiated on a wide-scale by Karol Starmach
only in the 1960s.  Although some aspects are now understood quite well,
information about others is of only a preliminary and inventory nature.
This is because of the numerous methodological difficulties which have
been encountered and the lack of specialists for many taxonomic groups.

Although the river is only 251 km long and the direct distance from
source to its entry to the Vistula (Wisła) is 120 km, it flows through
a wide range of geographical and climatic conditions and geological
formations.  As a consequence there are marked changes in character on
passing downstream.  In the upper part of the catchment there are high
mountain streams with gradients exceeding $400^o/oo$ separated by young
post-glacial lakes.  In contrast the downstream part of the river has
almost a lowland character.  Between these there are all the intermediate forms associated with the natural development of a river.  The
variety of biotopes and the history of the region result in communities
of aquatic organisms with diverse ecologies and which frequently
include endemic post-glacial relic species.  The upper section of the
Dunajec and most parts of the tributaries are still of an entirely
natural character.

There are three national parks within the 70% of the catchment which
lies within Poland.  These are: the Tatra National Park, the Pieniny
National Park and the Gorce National Park.  Because of its striking
landscape and scientific interest, the Dunajec was included in "Project
AQUA" (Luther & Rzóska, 1971).  The river is also of great economic
importance for Poland.  In its mid-section there are two on-stream
reservoirs with dams, which are also used for recreational purposes
and production of hydro-electric energy.  Other reservoirs are planned
or under construction, including ones on tributaries.

Plans for construction of reservoirs totalling $2.32 \times 10^8$ m$^3$  capacity close to the boundary of the Pieniny National Park encountered
strong opposition not only from scientists but much of society, because
they endangered the unique nature of the Pieniny Mountains.

*English transliteration = Dunaiyetz

## 18.2   GENERAL DATA

DIMENSIONS
  Area of catchment     6798 km$^2$
  Length of river        251 km
  Highest points        2663 m Gerlach Mt in Tatras; 1310 m Turbacz Mt in Beskids
  Altitude at source    1540 m Chochłowski Stream springs in Wolowiec Mt.


CLIMATE
  Climate associated with vertical zones in Polish Northern Carpathians.  Data calculated
  from 10 year period (from Hess, 1965)

| zones | Upper limit (m) | | Mean temperature (°C) | | | Precipi-tation mean ann. (mm) | Snow cover | | Cloudiness mean ann. days | |
|---|---|---|---|---|---|---|---|---|---|---|
| | | | | | | | occurrence mean ann. days | depth mean max. (cm) | sunny | cloudy |
| | Tatras | Beskids | annual | Jan. | July | | | | | |
| Cold | 2665 | | - 4.0 | - 12.0 | 4.0 | 1625 | 290 | > 200 | 45 | 140 |
| Temperate cold (climatic snow line) | 2200 | | - 2.0 | - 10.0 | 6.0 | 1750 | 250 | > 200 | 35 | 160 |
| Very cool | 1850 | | 0.0 | - 8.5 | 8.2 | 1800 | 215 | > 200 | 40 | 165 |
| Cool (upper forest limit) | 1550 | | 2.0 | - 6.0 | 10.5 | 1600 | 180 | > 120 | 45 | 145 |
| Temperate cool | 1150 | 1100 | 4.0 | - 5.5 | 13.0 | 1400 | 140 | > 60 | 40 | 160 |
| Temperate warm | | 750 | 6.0 | - 4.0 | 15.5 | 1000 | 105 | > 25 | 55 | 175 |
| | | 280 | 6.2 | - 8.5 | 16.5 | 500 | 105 | | 20 | 110 |


POPULATION
  1.4 x 10$^6$ people


PHYSICAL FEATURES OF RIVER
  Flow in m$^3$ s$^{-1}$   (Data from Informacje G.U.S., 1979)

| km | absolute minimum | mean of minima | mean | mean of maxima | absolute maximum |
|---|---|---|---|---|---|
| 18 | 9.2 | 15.6 | 84.0 | 1210 | 3500 |
| 150 | 3.0 | 6.8 | 30.1 | 448 | 1240 |


VEGETATION ZONES IN POLISH NORTH CARPATHIANS

| zone | main associations | mean upper limit (m) | | |
|---|---|---|---|---|
| | | Tatras | | Beskids |
| | | granite | limestone | |
| Supra-alpine (subnival) | Distichetum subnivale | 2663 | | |
| Alpine zone meadows | Trifido-Distichetum Seslerion Tatrae | 2300 | 2250 | |
| Mountain pine zone, Pinus mughus | Pinetum mughi carpathicum | 1850 | 1800 | |
| Upper mountain forest zone | Piceetum tatricum | 1650 | 1590 | 1310 |
| Lower mountain forest zone | Fagetum carpathicum | | 1250 | 1150 |
| Foothills | Tilio-Carpinetum, Pino-Quercetum | | | 550 |

## 18.3   THE CATCHMENT

### 18.31   GEOGRAPHY

The Dunajec is the right-bank mountain tributary of the River Vistula.
Almost all its catchment lies within the Northern Carpathians.  The
catchment includes the highest massif of the Carpathian range, the
Tatra Mountains, and some minor ranges of the West Beskid Mountains
(18.2).  The source of the river is regarded as the head-spring of the
Chochołowski stream.  This  stream joins with the Kościeliski stream
at the base of the Tatra Mountains to form the Czarny Dunajec.  This in
turn joins with the Biały Dunajec to form the Dunajec proper (Fig. 18.2).
The Poprad, the largest tributary and over half of whose length lies in
Czechoslovakia, drains the highest granite parts of the Tatra Mountains,
the High Tatras.

   The lowest part of the drainage basin of the Dunajec lies outside
the Carpathians and constitutes part of the Sandomierska Basin.  The
Dunajec discharges to the Vistula at an altitude of only 170 m.

*Fig. 18.1*  The Dunajec near Krościenko in the Beskid Mountains.

### 18.32   PHYSIOGRAPHY

The highest part of the catchment is of an alpine character (18.2).

*Fig. 18.2* The Dunajec and its tributaries.

It is an area of bare, sheer granite rocks with glacial action evident in the form of eroded valleys and numerous cirque lakes - the High
Tatras. The unitary gradient of the streams is often considerable and
may reach 600°/oo. Falls are a common phenomenon. The West Tatras
lacks post-glacial lakes, in spite of similar post-glacial activity in
the past. Instead, there are karst phenomena associated with an accumulation of limestones.

The prevailing character of the catchment as a whole is of mountains
of medium height, with softly inclined slopes and afforested tops. The
lowest part of the catchment has a typical lowland and agricultural
character.

18.33  METEOROLOGY

The catchment of the Dunajec is characterized by a great variety of climatic type, with six zones distinguished (see 18.2). The difference
between the mean annual air temperature in the lowest and highest parts
of the catchment area is 12 °C (+8 ° to -4 °C), a decrease of 0.5 °C per
100 m of altitude. The snow-line lies at 2300 m in the Tatra Mountains.

Thermal inversion is a relatively frequent phenomenon in the valleys.
The highest precipitation and the greatest number of cloudy days occur
between 1550 and 1850 m. There is a relatively high precipitation and
a large number of cloudy days in July (Hess, 1965). The climate of the
Pieniny Mts (Romer, 1949) is slightly warmer and drier than that of the
adjacent mountain ranges, as may be expected from their lower elevation.

18.34  GEOLOGY

The bed-rock of the majority of the catchment is formed by the Carpa-
thian Flysch or the thick series of sediments of the Magurian sandstones.
The highest part of the catchment, lying within the limits of the Tatra
Mts, is built of crystalline rocks or granite. Around this granite
trunk there are various sediment formations from Permian to cretaceous.
Calcareous rocks form the east part of the High Tatra (the Bielskie
Tatra), some parts of the West Tatra and the Pieniny Mts.

From the foot of the Tatra Mts down to the Pieniny Mts, the Dunajec
flows over glacial formations (loams, sands) derived mostly from the
Tatra range. The river then falls into the gorges of the Pieniny and
Beskid Mts. Below Nowy Sącz the river flows over alluvial sediments
of its own origin combined with diluvial ones from the glaciation of
the Cracovian Stage (= Mindel). Information on soils of the catchment
are given by Pasternak (1968).

## 18.35  LAND USE:  AGRICULTURE AND FORESTRY

Over half the catchment area is under agricultural utilization, the
greatest part of which is arable land, though there are also meadows
and pastures.  Forests cover 35% of the area, but most lie within the
mountain ranges - the Tatra, Gorce and Pieniny Mts, where the national
parks are situated, and the Beskid Sądecki Mts.

   In the Carpathian part of the drainage basin the natural plant
associations form characteristic zones within the vertical range.  Six
such zones can be distinguished whose vertical ranges correspond with
the climatic zones in the particular mountain chains (see 18.2).  Three
zones, alpine meadow, supra-alpine and dwarf mountain pine, occur only
in the Tatra Mts.

   The relatively large area lacking forest cover in the upper catchment, the steep slopes and the frequent occurrence of driving rains in
summer, all combine to favour soil erosion and the linear erosion of
streams.  Soil wash-off has been estimated to be about 0.1 to 0.5 mm
$yr^{-1}$, but can sometimes reach 1.0 mm $yr^{-1}$.  Soil landslips are frequent
(Figula, 1956).  Soil wash-off results in a considerable increase in
suspended materials transported by the Dunajec during periods of flood.

   There is little peat within the catchment.  A peat bog above Nowy
Targ is 11.47 $km^2$, less than 1% of the Czarny Dunajec catchment.


## 18.36  MAN'S ACTIVITY

Among the population of 1.4 million people only 25% live inside towns.
The populations of the four largest towns are: Zakopane, 30000; Nowy
Targ, 30000; Nowy Sącz, over 61000; Tarnów, 100000.  The towns are
centres of industry.  These include food and shoe industries at Nowy
Targ, and chemical industry at Nowy Sacz and Tarnów.  Many mountain
villages are used for health or recreational purposes.  Lodging
facilities are well developed and there is a high tourist capacity.
Numerous villages and industrial plants discharge their sewages to the
Dunajec and to its tributaries.

   The middle section of the Dunajec is separated by dams associated
with two reservoirs, the Rożnów and the Czchów.  The former, which was
built in 1942, had originally a capacity of 2.29 x $10^8$ $m^3$ (Wróbel,
1965), but due to silting this is now less than 1.7 x $10^8$ $m^3$.  The
Czchów acts as a retarding reservoir for the Rożnów and has a capacity
of only 0.12 x $10^8$ $m^3$.  Other reservoirs planned for the Dunajec and
its tributaries will have a total capacity of about 7.6 x $10^8$ $m^3$ by
the end of this century.

   The dam of the Rożnów reservoir is 32 m high.  As the outlet is
relatively low, it lies below the thermocline in summer.

18.37  MANAGEMENT

The main organizations are  as follows:

| Polish | English |
|--------|---------|
| Urzędy Wojewódzkie | Voivodeship Authorities |
| Ośrodek Badań i Kontroli Środowiska | Centre for Monitoring the Environment |
| Okręgowa Dyrekcja Gospodarki Wodnej | District Office of Water Management |
| Zakłady Energetyczne Okręgu Południowego w Katowicach | Energy Plants of the Southern District of Katowice |
| Polski Związek Wędkarski | Polish Angling Union |

The Dunajec belongs territorially to Voivodeship Authorities located in Nowy Sącz and Tarnów.  They both determine the regulations for use of water and release of pollution and give any permission needed.  They also request the chemical and biological investigations to be made by the Centre for Monitoring the Environment in Kraków.  They own the results, which are official and published in the Atlas of Pollution of Polish Rivers (Atlas zanieczyszczenia rzek Polskich).

The District Office of Water Management in Kraków is engaged in river management, preservation of plant facilities and construction of buildings associated with water treatment.  The reservoirs are subject to the Energy Plants of the Southern District of Katowice.

The fish in both the river and reservoirs belong to the Polish Angling Union, who administer all aspects of stocking and angling and make claims in the case of fish mortalities due to pollution.


18.4  ENVIRONMENT

18.41  SUBSTRATUM

For most of its length, the river bed consists mostly of loosely arranged boulders and smaller rocks.  Sheets of rock occur only in short sections of the Tatrian tributaries and in gorges.  The bed of the most downstream section of the river is covered by pebbles and sand, which are found also in marginal pools along the whole river, often together with much silt.

Upstream of the reservoirs, the banks of the river are quite low and flat, formed by natural stony deposits often overgrown with willow brakes.  Downstream of the reservoirs, the banks are managed and covered with large angular stones.

## 18.42   FLOW

The Dunajec shows marked variation in flow.  In the middle section the
ratio of maximum to minimum flow exceeds 400, in the Tatrian affluents
it reaches 1230 and in Biała Tarnowska it can even reach 3300
(Ziemońska, 1965).  The highest flows occur in summer, especially in
July, and are connected with driving rains.  Spates develop rapidly,
but have a short duration time as a result of the small retention
capacity of the drainage basin (Figuła, 1965).  Relatively high flows
also occur in spring due to snow melt.  During periods of flood the
water level of the unregulated part of the river can reach 5.7 m above
the mean annual height.  Minimum flows occur in winter, when some
sections of Tatrian streams freeze up entirely.  A few streams also
have sections which are dry from late autumn through the winter,
although they will of course have a snow cover for much of the time.


## 18.43   TEMPERATURE AND LIGHT CLIMATE

The thermal characteristics of the Dunajec from Nowy Targ to the mouth
were reported in detail by Gołek (1961).  The highest monthly mean
occurred in July, August and June (Fig. 18.3).  The highest temperatures recorded anywhere were in the Pieniny section, although the
monthly means here were still less than further downstream.  This is
due to the influence of a warm local climate and perhaps also the high
turbulence of the water as it passes among the rocky rubble there.
The reservoirs Roznów and Czchów disturb the natural thermal profile
of the river, generally increasing the temperature of water in the
downstream section.

   The temperature regimes of the spring tributaries have not yet been
studied in detail.  A number of measurements have however been made in
the streams Chochołowski, Kościeliski and Czarny Dunajec during the
course of chemical (Oleksynowa & Komornicki, 1965) or hydrobiological
studies.  The water temperature of the lower Chochołowski stream
(890 m altitude) ranged from 1.0 to 14.0 $^{\circ}$C, according to season; in
the lower Kościeliski stream (930 m altitude) the range was 2.0 to
11.0 $^{\circ}$C.  The highest temperatures in the upper Czarny Dunajec did not
exceed 13.0 $^{\circ}$C.

   Many of the streams and narrower stretches of river are shaded by
evergreen forest.  A fringe of deciduous trees is common in the section
above the reservoirs, with alder further upstream tending to give way
to willows further downstream.  The section downstream of the reservoir
is relatively open, with shading likely to have little impact on the
ecology of the river.


## 18.44   WATER CHEMISTRY

Details of the chemistry of waters of the Dunajec and tributaries are

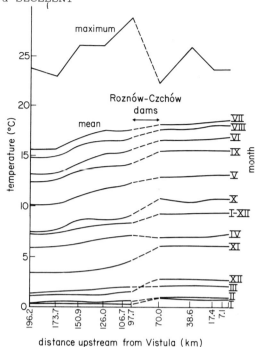

Fig. 18.3 Thermal profile of the Dunajec from 196.2 km upstream of the mouth to the mouth, showing mean monthly temperatures for 1948-1958 (except for 38.6 km) and the maximum recorded during the same period (after Gołek, 1961). Data for 38.6 km upstream based on data for 1901-1958.

given in Table 18.1. The water is mostly of calcium-carbonate type with high values for calcium and total alkalinity along the whole length of the river. The inflow of waste waters from the nitrogen industry in Tarnów via the Biała Tarnowska river considerably enriches the waters of the Dunajec in combined nitrogen (over 4 mg $l^{-1}$ N).

The physical and chemical conditions of the water change markedly during floods caused by heavy rain. The water becomes muddy and appears yellow-brown to an observer on the banks. The river transports a great amount of suspended material and rolls cobbles and boulders over the bottom. The total amount of the materials in suspension or dragged by the Dunajec in its middle section is estimated to be 486000 t $yr^{-1}$; this amount is discharged to the Roznów reservoir, where 90% of it is deposited, filling the reservoir at a rate of 0.84% of its capacity per year (Brański, 1975).

## 18.45   POLLUTION

The Dunajec is considered to be a relatively clean river in spite of

*Table 18.1*  Chemical composition ($mg\ l^{-1}$) of River Dunajec.

| river | Kościeliski Potok | Chochołowski Potok | Czarny Dunajec | Biały Dunajec | Dunajec | Dunajec |
|---|---|---|---|---|---|---|
| distance from mouth | km 2.5 | km 236.0 | km 205.0 | km 6.3 | km 197 | km 185.0 |
| conductivity ($\mu S\ cm^{-1}$) | | 177.9 -202.8 | 176.9 -292.8 | 174.1 -317.2 | 233.3 -361.9 | 138.9 -351.9 |
| pH | 7.1 - 8.1 | 7.0 - 7.8 | 8.0 | 8.0 | 8.0 - 8.1 | 8.0 - 8.1 |
| total alkalinity ($meq.\ l^{-1}$) | 1.24- 1.72 | 1.1 - 1.44 | 3.0 - 3.2 | 2.8 - 3.1 | 2.7 - 3.4 | 2.6 - 3.3 |
| Na | 0.6 - 3.0 | 3.7 | 6.2 | 11.8 | 12.2 | 10.8 |
| K | | 2.0 | 2.3 | 3.2 | 3.8 | 3.8 |
| Mg | 4.6 - 9.9 | 3.6 - 13.0 | 8.7 - 12.3 | 10.4 - 10.7 | 11.4 - 14.6 | 9.7 - 10.4 |
| Ca | 21.8 -33.3 | 19.6 - 39.4 | 39.2 - 51.3 | 39.2 - 51.3 | 42.4 - 52.9 | 43.9 - 54.5 |
| Fe | 0 - 0.03 | 0 - 0.05 | 0.15 - 0.4 | 0.2 - 0.3 | 0.2 - 0.45 | 0.2 - 0.3 |
| Cl | 0 - 0.35 | 0 - 0.35 | 6.2 - 5.9 | 8.9 - 14.4 | 11.9 - 16.5 | 9.9 - 14.4 |
| $SO_4$-S | 5.6 -13.5 | 4.6 -15.6 | 5.3 - 6.0 | 3.4 - 10.2 | 3.2 - 9.4 | 7.6 - 7.7 |
| $NH_3$-N | | 0.09 - 0.24 | 0.10 - 0.28 | 0.10 - 0.55 | 0.45 - 0.61 | 0.20 - 0.34 |
| $NO_3$-N | | 0.70 - 1.06 | 1.10 - 1.50 | 1.38 - 1.50 | 0.96 - 1.62 | 1.00 - 1.18 |
| $PO_4$-P | 0 - 0.01 | 0 - 0.005 | 0.006- 0.02 | 0.16 - 0.19 | 0.13 - 0.16 | 0.04 - 0.06 |
| $O_2$ | 10.5 -11.4 | 9.8 -11.4 | 12.7 -13.0 | 11.6 -12.5 | 11.0 -12.7 | 12.0 -13.5 |
| $BOD_5$-$O_2$ | | 1.7 - 2.0 | 2.6 - 3.7 | 4.3 - 5.1 | 4.2 - 6.3 | 3.0 - 3.3 |
| oxidability -$O_2$ | | 1.7 - 2.4 | 0.6 - 2.9 | 1.9 - 5.2 | 2.4 - 6.7 | 1.4 - 5.9 |

| river | Dunajec | | | | | |
|---|---|---|---|---|---|---|
| distance from mouth | km 120.0 | km 106.0 | km 34.0 | km 28.0 | km 22.0 | km 0.5 |
| conductivity ($\mu S\ cm^{-1}$) | 230.8 -351.9 | 288.5 -377.5 | 230.8 -363.9 | 346.2 -477.3 | 346.3 -548.8 | 243.0 -429.5 |
| pH | 8.0 - 8.1 | 7.8 - 8.0 | 7.9 - 8.0 | 7.8 - 7.9 | 7.6 - 7.7 | 7.8 - 7.9 |
| total alkalinity ($meq.\ l^{-1}$) | 2.8 - 3.1 | 3.5 - 3.8 | 3.1 - 3.6 | 3.2 - 3.8 | 3.2 - 5.2 | 3.1 - 4.1 |
| Na | 7.1 | 15.2 | 14.2 | 24.7 | 20.2 | 17.5 |
| K | 2.2 | 4.4 | 4.4 | 4.9 | 4.8 | 4.6 |
| Mg | 10.7 - 10.8 | 11.7 - 11.8 | 10.8 - 14.6 | 11.7 - 12.7 | 12.7 - 18.5 | 11.7 |
| Ca | 43.7 - 59.7 | 55.0 - 67.7 | 50.1 - 60.9 | 50.1 - 67.3 | 59.9 - 92.9 | 53.4 - 74.2 |
| Fe | 0.2 - 0.32 | 0.32 - 0.5 | 0.17 - 0.35 | 0.3 - 0.45 | 0.4 - 0.5 | 0.3 - 0.4 |
| Cl | 9.9 - 13.4 | 14.9 - 18.5 | 12.9 - 16.5 | 21.8 - 27.8 | 19.9 - 23.7 | 17.9 - 20.6 |
| $SO_4$-S | 8.6 -11.4 | 9.3 - 12.8 | 6.7 - 16.0 | 11.8 - 17.2 | 13.9 - 21.9 | 11.4 - 15.9 |
| $NH_3$-N | 0.240- 0.250 | 0.40 - 0.55 | 0.125- 0.26 | 2.400- 3.48 | 1.850- 1.90 | 0.98 - 2.00 |
| $NO_3$-N | 1.300- 1.480 | 1.46 - 2.24 | 1.44 - 1.68 | 2.040- 3.28 | 2.100- 2.48 | 2.50 - 3.28 |
| $PO_4$-P | 0.006- 0.05 | 0.16 - 0.19 | 0.01 - 0.05 | 0.006- 0.06 | 0.016- 0.04 | 0.016- 0.06 |
| $O_2$ | 13.4 -13.7 | 11.4 -12.7 | 9.8 - 11.4 | 9.6 - 10.4 | 9.6 - 10.2 | 10.0 - 11.9 |
| $BOD_5$-$O_2$ | 2.7 | 7.4 - 7.5 | 2.4 - 4.0 | 3.0 - 6.2 | 2.0 - 6.0 | 3.0 - 4.0 |
| oxidability -$O_2$ | 1.8 - 2.3 | 2.6 - 4.7 | 2.7 - 8.2 | 2.5 - 5.1 | 1.8 - 4.1 | 2.0 - 5.2 |

the fact that, together with its tributaries, it receives annually 55
x $10^6$ $m^3$ of sewage purified only moderately and 50 x $10^6$ $m^3$ of sewage
with no treatment (Informacje G.U.S., 1979).  Both minor and some more
important tributaries are heavily polluted, in some cases for considerable stretches.  This pollution is accompanied by such phenomena as
a marked oxygen deficit, complete elimination of sensitive invertebrates
(Ephemeroptera, Plecoptera, Trichoptera) and the mass development of
the sewage-bacterium *Sphaerotilus natans* and fungi.  However no degradation of environmental conditions has been observed in the Dunajec
river above the mouth of the Biała Tarnowska river i.e. neither Nowy
Targ nor Nowy Sącz lead to a lack of oxygen or elimination of sensitive
invertebrates (Dratnal *et al.*, 1979).

## 18.5   PLANTS AND ANIMALS OF THE UPPER AND MIDDLE
(UNREGULATED) RIVER

### 18.51   PLANTS

Algal communities are rich in species.  Wasylik (1971) recorded 459
taxa from the upper section of the river, while Chudyba (1965) recorded
214 taxa in the middle and lower section.  Overall, diatoms constitute
the bulk of the species (Table 18.2), though population densities are
relatively low in the highest Tatra streams.  As altitude decreases
their densities increase (Kawecka, 1980).  Other groups of algae can
form species-rich aggregations, such as locally the blue-green algae in
both calcareous and non-calcareous upper sections of Tatra streams and
the green algae in the mid-section of the river.

*Table 18.2*  Composition of algal flora in upper, middle and lower
Dunajec, expressed as percentage of total flora.  (Based on data in
Chudyba (1965) and Wasylik (1971).)

| phylum | % flora in sections of river | | |
|---|---|---|---|
| | upper | middle | lower |
| Cyanophyta | 6.9 | 11.8 | 8.0 |
| Rhodophyta | 0.4 | 0.7 | |
| Bacillariophyta | 76.9 | 81.2 | 83.1 |
| Chrysophyta | 0.7 | 0.2 | |
| Chlorophyta | 15.1 | 6.1 | 8.9 |

The golden-brown alga *Hydrurus foetidus* occurs in masses in streams of the Tatra Mts, especially in spring. The bryophyte flora of the upper part of the river and tributaries is rich. Mosses are also obvious in the middle section, near the Pieniny Mts, but much less abundant than further upstream. They presumably also occur downstream of the dams, but no surveys have been made. Rooted submerged macrophytes are apparently absent from the Dunajec, but again there have been no detailed surveys to establish this with certainty.

## 18.52  MACROINVERTEBRATES

The list of invertebrates in the Dunajec and its spring tributaries is still far from complete. Key publications are those of Dratnal and Szczęsny (1965), Dratnal *et al.* (1979), Kamler (1967), Kownacka (1971), Kownacka & Kownacki (1965), Kownacki (1971, 1977), Sowa (1975a, b), Wojtas (1964). Diptera, and especially Chironomidae, are the most numerous group. They have the greatest number of species (over 100) and, in most samples, the greatest number of specimens. However they are out-numbered in polluted sections by the Oligochaeta, on the stony bed by Naidae (mainly *Nais*) and on the fine-grained bed by Tubificidae.

Among the other groups, Ephemeroptera, Plecoptera and Trichoptera are numerous. In spite of the fact that Chironomidae and Naidae out-number other groups of invertebrates, they seldom constitute the bulk of the biomass. It seems probable that Trichoptera and Ephemeroptera are the dominant groups in this respect.

A classification has been made (Dratnal *et al.*, 1979) of macro-invertebrates in the Pieniny section of the Dunajec into functional feeding groups, based on the categories of Cummins (1974): predators, collectors, filter-feeders, grazers, scrapers, shredders. A summary of the quantitative data is given in Table 18.3. The participation of shredders is extremely low, because their source of food, pieces of vascular plant tissue, is low in that section of the river. The proportionally high participation of grazers and collectors indicates that epiphytic microphytes are the main food source.

*Table 18.3* Composition of functional feeding groups of macro-invertebrates in the Pieniny section of the Dunajec, expressed as percentage of the total. Data calculated from mean annual number of specimens (= 32395 $m^{-2}$).

| | |
|---|---|
| predators | 1.15% |
| collectors | 87.50% |
| grazers and scrapers | 9.70% |
| shredders | 0.04% |
| filter-feeders | 1.45% |

## 18.53  FISH

The fish fauna of the Dunajec drainage basin has been a subject of
interest to fish biologists from Nowicki (1880, 1882, 1883) to Kołder
(1967), Solewski (1963, 1964, 1965) and Bieniarz & Epler (1972); since
1976 studies have been carried out by the Department of Water Biology
in Kraków.  Different fish communities occur in different sections,
though angling has a marked influence on the relative composition of
the species.  In the mountain sections of the Czarny Dunajec and Biały
Dunajec there are only two species, *Cottus poecilopus* and *Salmo trutta*
f. *fario*.  Downstream, in the vicinity of Nowy Targ, where the river
loses the character of a mountain stream, the number of species
increases markedly.  In addition to the above, there occur also
*Thymallus thymallus, Phoxinus phoxinus, Barbus meridionalis, Leuciscus
cephalus, Cottus gobio* and *Chondrostoma nasus*.  Downstream of Nowy Targ,
near Knurów, further species appear: *Barbus barbus, Rutilus rutilus,
Abramis brama* and *Hucho hucho*.  This last was introduced to the Dunajec
from the catchment area of the Danube (Dunaj).  With the exception of
*Cottus poecilopus* all these species occur down to Nowy Sącz.  The
relative proportion of the various species changes gradually in favour
of fish typical of submontane rivers - *Barbus barbus, Chondrostoma
nasus* and *Leuciscus cephalus*.

Below Nowy Sącz the Dunajec loses its submontane character, probably
due to the heavy pollution.  Salmonids, *Phoxinus phoxinus* and *Gobio
gobio* are all lacking.  Instead, the following species appear:
*Leuciscus leuciscus, Alburnus alburnus, Perca fluviatilis, Acerina
cernua, Alburnoides bipunctatus, Lucioperca lucioperca* and *Esox lucius*
(Starmach, in preparation).

## 18.54  ZONATION

It will be clear from the above account that both individual species
and communities as a whole show marked changes on passing down the
river.  Such zonation is evident not only in the main sequence of the
spring streams of the Dunajec, but also in its other tributaries.
Wayslik (1971) distinguished three zones of microphytes in the spring
tributaries (Table 18.4), based on community similarities calculated
according to the method of Czekanowski and Kulczyński (Pawłowski,
1959).  Their limits were indistinct and species composition changed
gradually.

In the first zone *Chamaesiphon polonicus* was particularly abundant.
*Homoeothrix varians* (probably identical with *H. janthina*: Starmach,
1966) and *Hydrurus foetidus* also formed rich aggregations.  These algae
were accompanied by numerous diatoms, forming the association Diatometo
hiemalis-Meridionetum with the characteristic species *Diatoma hiemale*,
together with its variety *mesodon,* and *Meridion circulare*.

In the second zone a community of intermediate character occurred.

*Table 18.4* Zonal distribution of algal communities in the upper Dunajec, showing the most abundant species. In the case of groups other than diatoms these are the species which form macroscopic conglomerations, covering more than 10% of bottom area. The diatoms are those species with more than 90 cells in an area of 20 mm$^2$ (modified from Wasylik, 1971).

| species | zone of Wasylik | 1 | 2 | 3 |
|---|---|---|---|---|
| | altitude (m) | 1500–1000 | 1000–750 | 750–600 |
| *Chamaesiphon fuscus* | | + | | |
| *Meridion circulare* | | + | | |
| *Diatoma hiemale* | | + | + | |
| *D. hiemale* var. *mesodon* | | + | + | |
| *Chamaesiphon polonicus* | | + | + | |
| *Phormidium favosum* | | + | + | |
| *Hydrurus foetidus* | | + | + | |
| *Ulothrix zonata* | | + | + | |
| *Homoeothrix* cf. *varians* | | + | + | + |
| *Ceratoneis arcus* | | + | + | + |
| *Cymbella ventricosa* | | + | + | + |
| *Synedra ulna* | | + | + | + |
| *Cocconeis placentula* | | + | | + |
| *Gomphonema intricatum* var. *pumilum* | | | + | |
| *Achnanthes pyrenaica* | | | + | + |
| *Diatoma vulgare* | | | + | + |
| *D. vulgare* var. *ehrenbergii* | | | + | + |
| *D. vulgare* var. *productum* | | | + | |
| *Cymbella helvetica* | | | + | + |
| *C. affinis* | | | | + |
| *Achnanthes minutissima* | | | | + |
| *Navicula cryptocephala* | | | | + |
| *N. cryptocephala* var. *intermedia* | | | | + |
| *N. gracilis* | | | | + |
| *N. radiosa* | | | | + |
| *Stigeoclonium tenue* | | | | + |
| *Cladophora glomerata* | | | | + |

*Hydrurus foetidus, Homoeothrix* cf. *varians* and *Chamaesiphon polonicus* were still quite abundant, but less so than in the first zone. *Phormidium favosum* and *Ulothrix zonata* were relatively frequent. The diatoms formed the sub-association vulgaretosum of the association Diatometo-hiemale-Meridionetum, in which, in addition to *Diatoma hiemale* and var. *mesodon, Diatoma vulgare* with varieties *ehrenbergii* and *productum* were numerous.

In the third zone *Cladophora glomerata* and *Stigeoclonium tenue* were frequent. The diatoms formed the association Diatometo vulgaris-Cymbelletum with the characteristic species: *Diatoma vulgare, D. vulgare* var. *ehrenbergii, Cymbella affinis, C. helvetica, Navicula gracilis.* This community resembles that in the middle section of the Dunajec, where *Cladophora glomerata* together with diatoms *Diatoma vulgare* (including var. *ehrenbergii* and var. *productum), Cymbella affinis, C. ventricosa, Navicula gracilis, N. radiosa, N. viridula, Nitzschia dissipata* and *Rhoicosphenia curvata* were the dominant species (Chudyba, 1965).

Three general zones of microphytes and invertebrates were also recognized in the streams of the High Tatras by Kawecka *et al.* (1971). These authors distinguished the main zones of microphytes on the basis of the most numerous species in limited sections of streams (Kawecka, 1971), and of invertebrates by means of so-called domination indices (Kownacki, 1971). The most numerous microphytes and invertebrates of the highest values of domination index determined the zones (Table 18.5). The three zones were called (I) the high montane streams, (II) streams of the wooded hills and (III) streams and rivers of the foot of the Tatra Mts. Two sub-zones determined mainly by the dominant chironomids were additionally distinguished in both zones I and II. Zone I is missing entirely in the West Tatras (Kownacki, pers. comm.).

The zones distinguished by Kownacka (1971) and Kownacki (1971) depended heavily on representation of various species of chironomid. This was a consequence of the method used for statistical analysis of the data, which depends on the most numerous and frequent species in the samples collected. Chironomids usually exceed 50% of specimens in samples from Tatra streams.

The classification of rivers and streams using chironomids as indicator species is perhaps well grounded if one takes into account their numerical predominance in aquatic habitats. So far, however, the ecologies of individual species are poorly understood. Such a classification is also made difficult by their small size, poor morphological differentiation, and the taxonomic problems associated larvae and with imaginal forms. For instance, *Orthocladius rivicola* and *O. thienemanni* were both given as characteristic for zone III because their larvae are indistinguishable. *Cricotopus* group *algarum* comprises at least nine species which are not distinguishable at the larval stage (Dratnal *et al.*, 1979).

Another attempt to distinguish general zones of distribution along the whole river was made by Sowa (1975a) exclusively on the basis of mayflies. The author analysed their distribution in the main aquatic systems of the Polish Northern Carpathians. He distinguished five zones in running waters by grouping stations of a similar species composition within any one stream, and then applying the percentage coefficient of similarity of Jaccard. In addition to these five zones the author distinguished zone 0, spring outflows where mayflies do

Table 18.5 Zonal distribution of benthic communities in streams of the High Tatras. (After Kawecka et al., 1971)

| zone | sub-zone | upper limit (m) | dominant taxa algae | dominant taxa macroinvertebrates |
|---|---|---|---|---|
| I high-mountain streams | 1 | 2100 | crust of blue-green algae | *Diamesa steinbocki* |
| | 2 | 2000 | *Chamaesiphon polonicus* | *Diamesa gr. latitarsis* Simuliidae (*Prosimulium* sp.) |
| II streams of wooded hills | 3 | 1550 | *Hydrurus foetidus* *Homoeothrix janthina* *Diatoma hiemale* | *Eukiefferiella minor* *Parorthocladius nudipennis* *Baetis alpinus* *Rhithrogena loyolaea* |
| | 4 | 1500 | *D. hiemale* var. *mesodon* | *Parorthocladius nudipennis* *Orthocladius rivicola* *Baetis alpinus* *Rhithrogena loyolaea* |
| III streams and rivers at foot of the Tatras | 5 | 1000 | *Ulothrix zonata* *Diatoma vulgare* *D. vulgare* var. *ehrenbergii* *Cymbella affinis* *Synedra ulna* | *Orthocladius rivicola + O. thienemanii* *Cricotopus gr. algarum* Simuliidae |

*Table 18.6*  Altitudinal distribution of Trichoptera in upper Dunajec (B. Szczęsny, unpublished).

| species | \makebox presence at various altitudes (m) | | | | | | | | | | | | | | |
|---|---|---|---|---|---|---|---|---|---|---|---|---|---|---|---|
|  | 1680 | 1600 | 1500 | 1400 | 1300 | 1200 | 1100 | 1000 | 900 | 800 | 700 | 600 | 500 | 400 | 300 |
| *Drusus monticola* | + | + | + | + | + |  |  |  |  |  |  |  |  |  |  |
| *Rhyacophila philopotamoides* | + | + |  | + | + |  |  |  |  |  |  |  |  |  |  |
| *R. glareosa* | + | + | + | + | + |  | + | + |  |  |  |  |  |  |  |
| *Allogamus uncatus* | + |  | + | + | + | + | + | + |  |  |  |  |  |  |  |
| *Acrophylax zerberus* | + | + | + | + | + | + | + | + | + |  |  |  |  |  |  |
| *Drusus discolor* | + | + |  | + | + |  | + | + | + | + |  |  |  |  |  |
| *Chaetopteryx polonica* |  | + | + | + | + |  |  | + |  |  |  |  |  |  |  |
| *Lithax niger* |  |  |  | + | + |  |  |  |  |  |  |  |  |  |  |
| *Apatania carpathica* |  |  |  | + | + |  | + |  |  |  |  |  |  |  |  |
| *Psilopteryx psorosa psorosa* |  |  |  | + | + | + | + | + |  |  |  |  |  |  |  |
| *Rhyacophila tristis* |  |  |  | + |  |  |  | + | + | + | + |  |  |  |  |
| *Drusus carpathicus* |  |  |  |  | + |  |  |  |  |  |  |  |  |  |  |
| *Melampophylax nepos* |  |  |  |  | + | + | + | + |  |  |  |  |  |  |  |
| *Acrophylax vernalis* |  |  |  |  | + |  | + | + | + |  |  |  |  |  |  |
| *Rhyacophila vulgaris* |  |  |  |  | + |  | + | + | + |  |  |  |  |  |  |
| *Halesus rubricollis* |  |  |  |  | + |  | + | + | + |  |  |  |  |  |  |
| *Rhyacophila polonica* |  |  |  |  | + |  | + | + | + |  |  |  |  |  |  |
| *R. fasciata* |  |  |  |  | + |  | + |  | + |  |  |  |  |  |  |
| *Allogamus starmachi* |  |  |  |  |  | + | + |  |  |  |  |  |  |  |  |
| *Drusus annulatus* |  |  |  |  |  |  | + | + |  |  |  |  |  |  |  |
| *Rhyacophila obliterata* |  |  |  |  |  |  | + | + |  | + |  |  |  |  |  |
| *Drusus biguttatus* |  |  |  |  |  |  | + | + | + | + |  |  |  |  |  |
| *Potamophylax cingulatus* |  |  |  |  |  |  | + | + | + | + |  |  |  |  |  |
| *Allogamus auricollis* |  |  |  |  |  |  | + | + | + | + | + |  |  |  |  |
| *Ecclisopteryx madida* |  |  |  |  |  |  |  | + | + | + |  |  |  |  |  |
| *Glossosoma conformis* |  |  |  |  |  |  |  | + | + | + |  |  |  |  |  |
| *Potamophylax latipennis* |  |  |  |  |  |  |  |  | + | + | + |  |  |  |  |
| *Rhyacophila mocsaryi* |  |  |  |  |  |  |  |  | + | + | + |  |  |  |  |
| *Ecclisopteryx dalecarlica* |  |  |  |  |  |  |  |  | + | + | + |  |  |  |  |
| *Glossosoma intermedia* |  |  |  |  |  |  |  |  | + | + | + | + |  |  |  |
| *Halesus interpunctatus* |  |  |  |  |  |  |  |  | + | + |  |  |  |  |  |
| *Annitella obscurata* |  |  |  |  |  |  |  |  | + | + |  |  |  |  |  |
| *Silo pallipes* |  |  |  |  |  |  |  |  | + |  | + |  |  |  |  |
| *Sericostoma sp.* |  |  |  |  |  |  |  |  | + | + | + | + |  |  |  |
| *Rhyacophila nubila* |  |  |  |  |  |  |  |  | + | + | + | + |  | + | + |
| *Polycentropus flavomaculatus* |  |  |  |  |  |  |  |  | + | + | + | + |  | + | + |
| *Hydropsyche pellucidula* |  |  |  |  |  |  |  |  | + | + | + | + |  | + | + |
| *Glossosoma boltoni* |  |  |  |  |  |  |  |  |  | + |  | + |  |  |  |
| *Psychomyia pusilla* |  |  |  |  |  |  |  |  |  | + | + | + |  | + | + |
| *Hydropsyche bulbifera* |  |  |  |  |  |  |  |  |  | + |  |  |  |  |  |
| *Chaetopteryx fusca* |  |  |  |  |  |  |  |  |  | + |  |  |  |  |  |
| *Annitella thuringica* |  |  |  |  |  |  |  |  |  | + |  |  |  |  |  |
| *Goera pilosa* |  |  |  |  |  |  |  |  |  | + |  |  |  |  |  |
| *Lasiocephala basalis* |  |  |  |  |  |  |  |  |  | + |  | + |  |  |  |
| *Agapetus delicatulus* |  |  |  |  |  |  |  |  |  |  |  | + |  |  |  |
| *Brachycentrus subnubilus* |  |  |  |  |  |  |  |  |  |  |  | + |  |  |  |
| *Cheumatopsyche lepida* |  |  |  |  |  |  |  |  |  |  |  | + |  | + |  |
| *Hydroptila forcipata* |  |  |  |  |  |  |  |  |  |  |  | + |  | + | + |
| *Oligoplectrum maculatum* |  |  |  |  |  |  |  |  |  |  |  | + |  | + | + |

*Table 18.7* Altitudinal distribution of Ephemeroptera in upper Dunajec. (From Sowa (1980) and M. Olechowska (unpublished).)

| species | \multicolumn{15}{c}{presence at various altitudes (m)} | | | | | | | | | | | | | | |
|---|---|---|---|---|---|---|---|---|---|---|---|---|---|---|---|
| | 1680 | 1600 | 1500 | 1400 | 1300 | 1200 | 1100 | 1000 | 900 | 800 | 700 | 600 | 500 | 400 | 300 |
| Rhithrogena loyolaea | + | + | + | + | + | + | + | + | | | | | | | |
| Baetis alpinus | + | + | + | + | + | + | + | + | + | + | + | + | + | | |
| Ameletus inopinatus | | | | | | + | + | + | + | | | | | | |
| Rhithrogena iridina | | | | | | | + | + | + | | | | | | |
| R. hybrida | | | | | | | + | + | + | + | + | | | | |
| Ecdyonurus venosus | | | | | | | + | + | + | + | + | + | + | + | + |
| Baetis vernus | | | | | | | | + | + | + | + | + | + | + | + |
| B. muticus | | | | | | | | + | + | + | + | + | + | + | + |
| Rhithrogena ferruginea | | | | | | | | + | + | + | + | + | + | + | + |
| Baetis melanonyx | | | | | | | | | + | + | + | | | | |
| Ephemerella krieghoffi | | | | | | | | | + | + | + | + | + | + | + |
| Habroleptoides modesta | | | | | | | | | | + | + | + | + | + | + |
| Baetis rhodani | | | | | | | | | | + | + | + | + | + | + |
| Rhithrogena hercynia | | | | | | | | | | + | + | + | + | + | + |
| Epeorus sylvicola | | | | | | | | | | + | + | + | + | + | + |
| Baetis sinaicus | | | | | | | | | | + | + | + | + | + | + |
| Caenis beskidensis | | | | | | | | | | + | + | + | + | + | + |
| Ecdyonurus lateralis | | | | | | | | | | | + | + | + | + | + |
| Ephemerella major | | | | | | | | | | | + | + | + | + | + |
| Baetis scambus | | | | | | | | | | | + | + | + | + | + |
| B. lutheri | | | | | | | | | | | + | + | + | + | + |
| Siphlonurus lacustris | | | | | | | | | | | | + | + | + | + |
| Ecdyonurus dispar | | | | | | | | | | | | + | + | + | + |
| Centroptilum luteolum | | | | | | | | | | | | + | + | + | + |
| Ephemera danica | | | | | | | | | | | | + | + | + | + |
| Habrophlebia fusca | | | | | | | | | | | | + | + | + | + |
| Baetis fuscatus | | | | | | | | | | | | + | + | + | + |
| Ephemerella ignita | | | | | | | | | | | | + | + | + | + |
| Oligoneuriella rhenana | | | | | | | | | | | | + | + | + | + |
| Rhithrogena semicolorata | | | | | | | | | | | | + | + | + | + |
| R. germanica | | | | | | | | | | | | + | + | + | + |
| R. diaphana | | | | | | | | | | | | | + | + | + |
| Ecdyonurus insignis | | | | | | | | | | | | | + | + | + |
| E. torrentis | | | | | | | | | | | | | + | + | + |
| Heptagenia sulphurea | | | | | | | | | | | | | + | + | + |
| Habrophlebia lauta | | | | | | | | | | | | | + | + | + |
| Baetis vardarensis | | | | | | | | | | | | | + | + | + |
| Centroptilum pennulatum | | | | | | | | | | | | | + | + | + |
| Procloeon bifidum | | | | | | | | | | | | | + | + | + |
| Cloeon cognatum | | | | | | | | | | | | | + | + | + |
| Potamanthus luteus | | | | | | | | | | | | | | + | + |
| Caenis horaria | | | | | | | | | | | | | | + | + |

not occur at all.  The upper limits of the vertical range of zones in
the Tatra tributaries of the Dunajec are shown in Table 18.6.

*Table 18.8*  Upper limit of zones populated with determined communities of Ephemeroptera (see text) in Tatra streams (after Sowa (1980)
and Olechowska (1982) ).

| zone | High Tatras | West Tatras |
|------|-------------|-------------|
| 0 | 2100 | 1700 |
| 1 | 1450 | 1680 |
| 2 | 1050 | 1150 |
| 3 | 680 | 700 |
| 4 | - | 560 |

Some of the zones determined by Kawecka *et al.* (1971) and Sowa (1980)
coincide.  Zones I-II of the former authors coincide with zones 0-I of
the latter.  The group best suited to indicate the character of the
environmental factor will be  one whose representatives are narrowly
adapted to the particular factor.  At the same time the scale of adap-
tation of the whole group to this factor ought to be as wide as
possible.  An example of the different character of distribution of
two groups of invertebrates important in the Dunajec is provided by a
comparison of caddis-flies and mayflies (Tables 18.6 and 18.7).
Among caddis-flies is a regular exchange of species with decrease in
altitude.  With mayflies this exchange is only slight and instead the
number of species increases with decrease in altitude.

   Without going into the problem as to which factors determine the
distribution of Trichoptera and Ephemeroptera down the river, it is
evident that with decreasing altitude and stream gradient, the sections
inhabited by the particular species or groups of species become longer.
This needs to be borne in mind when choosing sample sites.


18.55   SEASONAL CHANGES

The development of plant and animal communities takes place throughout
the year in the Dunajec and most of its tributaries.  Ice rarely covers
the whole surface of a river or stream and the water is frozen to the
bottom only near the banks.  Although data on algae in the Dunajec in
winter are lacking, changes have been followed over a whole year in
streams of the High Tatras (Kawecka, 1980).  In general, it is possible

to distinguish the spring period (May, June) when communities are subjected to heavy destruction caused by the rise in water resulting from
melting snow.  In summer (July, August) there occurs the natural re-
construction of communities, although hampered by frequent rain.  In
autumn (September, October) and winter (mid-November to mid-May in the
Tatras) algal communities change relatively little.  The greatest diversity of species is recorded in spring and summer and the least in
winter.  Most diatoms show a tendency to increase in abundance in
winter and spring; green algae develop well in summer and blue-green
algae all the year though with a tendency to increase in biomass in
autumn; *Hydrurus foetidus* is conspicuous in winter and spring. Wasylik
(1971) reported that in the Czarny Dunajec the maximum algal crop occurs
usually in September, October and November, and the minimum one in May
or June.  In the course of studies carried out from spring to winter,
Chudyba (1965) did not notice distinct differences in the occurrence
of algae in the middle and lower sections of the Dunajec.

A full year's quantitative studies on invertebrates in the Pieniny
section of the Dunajec (Dratnal *et al.*, 1979) showed that the greatest
density occurs in June (Fig. 18.4).  This results partly from the mass
development of *Nais* (Fig. 18.5) associated with the growth of an algal-
layer on stones.  The small size of the bodies of *Nais*, together with
its method of reproduction by fragmentation, enables the species to
live even in very rapid currents.

Emergent insects are characterized by two maxima, in January-March
and in June-September.  The period of decrease in April and May occurs

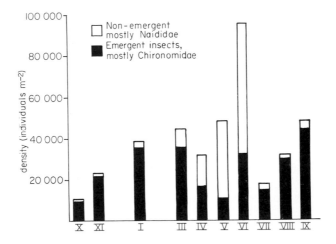

*Fig. 18.4* Seasonal changes in density of invertebrates in the
Dunajec at stations 165 - 170 km upstream from the mouth.

after the emergence of winter generations and before the increase in
numbers of the young larvae of the summer generations. The beginning
of the second insect maximum coincides with the June maximum number

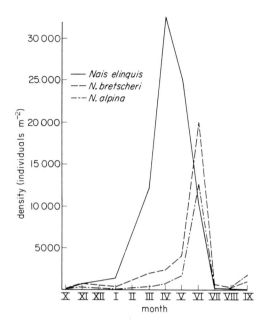

Fig. 18.5 Seasonal changes in density of three *Nais* species in
the Dunajec at stations 165 - 170 km upstream from the mouth.

for *Nais*; hence the greatest number of invertebrates occurs in June in this section of the river. Summer floods reduce pronouncedly the number of invertebrates, especially the worms. This decrease in number of worms depends to a considerable extent on the destruction of the algal layer. Under extreme conditions it may approach 100%.

The seasonal success of species within the genus *Nais* (Fig. 18.5) is probably connected with conditions in the river after the ice cover has melted. *Nais elinguis*, a species preferring habitats rich in the organic matter, first dominates; it is then succeeded by *N. bretscheri* and *N. alpina*.

## 18.56  PARTICULAR HABITATS

The most peculiar aquatic communities are those in the sections of Tatra streams which become dry or freeze at the end of autumn or in winter. Here there are rich aggregations of blue-green algae, which appear to be very resistant to such environmental conditions. *Chamaesiphon polonicus* is the dominant among these species. Its cell envelopes perhaps provide some protection from the effects of drying (Kann, 1978). Among the bottom fauna *Melampophylax nepos, Allogamus starmachi* and the chironomid *Diamesa starmachi* live exclusively in this habitat. *Allogamus starmachi* is endemic to the Tatra. All three species survive the period of freezing in the form of eggs laid in late autumn, which begin to hatch only during the spring flow of waters from snow-melt. It seems likely that development from eggs is possible only after freezing, but no experimental studies have been made.

## 18.6  PLANTS AND ANIMALS OF THE LOWER
## (REGULATED) RIVER

The section of the Dunajec from the present dams to its entry into the Vistula was regulated before the Second World War and as a consequence its natural ecological conditions were changed. Further changes occurred during the war due to the damming and in the post-war period due to discharges from plants of the nitrogen industry in Tarnów through the Biała Tarnowska river.

In the algal and animal communities inhabiting this section of the river there is a decrease in number of species combined with mass development of a few of them. This is manifest in the abundant development of *Sphaerotilus natans*, especially in the mouth part of the river (Chudyba, 1965). The algal community lacks golden-browns and reds (Table 18.5). Among the green algae, *Cladophora glomerata*, which is very abundant in the middle river, disappears and *Stigeoclonium tenue* is abundant instead. This species is very resistant to organic (Palmer, 1969) and heavy metal pollution (Harding & Whitton, 1976). In comparison with the middle river, the number of species of diatom

decreases and *Diatoma vulgare* disappears. *Nitzschia palea*, another
species known to be resistant to many types of pollution, also increases
(Kawecka, 1981). *Cyclotella meneghiniana, Melosira granulata* and
*Fragilaria crotonensis* are abundant. The reservoirs provide an inoculum for all three species although the outflow water does lie below the
thermocline in summer (18.36); these species can apparently continue to
grow well in the plankton of the main river (Bucka, 1965).

Among the bottom fauna the most obvious example of a decrease is in
the Plecoptera, which are completely absent; the groups next more
seriously affected are the Ephemeroptera, Trichoptera and even the
Diptera. Ephemeroptera and Trichoptera are eliminated not so much by
changes in the environment caused by flow regulation as by pollution.
In the polluted section a mollusc (*Ancylus fluviatilis*), a bloodsucker
(*Erpobdella octoculata*), worms (*Nais elinguis*, Tubificidae) and midges
(*Cricotopus* spp., *Microcricotopus bicolor, Limnochironomus* ex. gr.
*nervosus*) develop in such masses that there is an almost three-fold
increase in macroinvertebrate density as compared with upstream sites.

Below the dam reservoirs the following species of fish prevail:
*Alburnus alburnus, Leuciscus leuciscus, L. cephalus, Barbus barbus,
Rutilus rutilus, Esox lucius, Perca fluviatilis.* The lowest part of
the Dunajec was examined extensively by the Fisheries Division of the
Institute of Applied Zoology, University of Agriculture, Kraków, during
1973-1976. Special attention was paid to the influence of domestic and
chemical effluents from Tarnów. The study included cage experiments,
aquarium biotests with rainbow trout and electro-fishing. The results
revealed that this section of the river is drastically affected by
these effluents. The number of fish species, amounting to about 20
upstream of these effluents was drastically reduced to 6 in a zone
6 - 20 km downstream of them. Much of the zone was in fact totally
fishless and most fish that were caught were unsuitable for consumption
because of their abnormal "chemical" odour (A. Żysak, pers. comm.).

## 18.7  CONCLUDING COMMENTS

It will be evident that there are  many gaps in our present knowledge
of the hydrobiology of the Dunajec. In fact, many of the results so
far obtained can be considered only as preliminary. Nevertheless they
do indicate the value of studies on this river. The extensive stretches
of stream and main river which are in more or less natural condition
make the studies of much more than local interest. Unfortunately,
there are also sections where the pronouncedly destructive activity of
man (regulation, pollution) is marked. Polish hydrobiologists are
under an obligation to study the river biocoenosis in all its aspects
to determine changes connected with the gradual eutrophication of the
environment, and to participate actively in the protection of the
river.

The authors are very grateful to Professor K. Starmach and Professor
S. Wróbel for critical comment on the manuscript. Thanks are due also
to Professor A. Łysak and Professor J. Starmach for aid with the section on fish and to Dr B. A. Whitton for valuable comments.

## REFERENCES

Bieniarz K. & Epler P. (1972)  Ichtiofauna niektórych rzek Polski Połud-
    niowej. *Acta Hydrobiol*. 14, 419-444.  (English abstract and summary.
    Ichthyofauna of certain rivers in Southern Poland.)
Brański J. (1975)  Assessment of the Vistula catchment area denudation
    on the basis of bed load transport measurements. *Prace Inst. Gosp.
    Wodn*. 6, 5-58.
Bucka H. (1965)  The phytoplankton of the Roznów and Czchów reservoirs.
    Limnological investigations, in the Tatra Mountains and Dunajec River
    basin. *Kom. Zagospod. Ziem Górskich PAN* 11, 235-63.
Chudyba D. (1965)  Benthic algae in the river Dunajec.  Limnological
    investigations in the Tatra Mountains and Dunajec River Basin. *Kom.
    Zagospod. Ziem Górskich PAN* 11, 153-59.
Cummins K.W. (1974)  Structure and function of stream ecosystems.
    *Bioscience* 24, 631-41.
Dratnal E., Sowa R. & Szczęsny B. (1979)  Zgrupowania bezkregowców
    bentosowych Dunajca na odcinku Harklowa - Sromowce Niżne.  *Ochrona
    Przyrody* 42, 183-215.  (English summary.  Benthic invertebrate
    communities in the Dunajec River between Harklowa and Sromowce Nizne.)
Dratnal E. & Szczesny B. (1965)  Benthic fauna of the Dunajec river.
    Limnological investigations in the Tatra Mts and Dunajec River Basin.
    Kom. Zagospod. *Ziem Gorskich PAN* 11, 161-214.
Figuła K. (1956)  Monografia górnego Dunajca. *Pr. i Stud. Kom. Gosp.
    Wodnej* 1, 327-57. (Monograph of the upper part of the Dunajec River.)
Gołek J. (1961)  Termika rzek Polskich. *Prace Inst. Hydrol.-Meteorol*. 62,
    5-79.  (Ru. and Fr. summaries.  Thermal features of Polish rivers)
Harding J.P.C. & Whitton B.A. (1976)  Resistance to zinc of *Stigeo-
    clonium tenue* in the field and the laboratory. *Br. phycol. J*. 11,
    417-26.
Hess M. (1965)  Piętra klimatyczne w Polskich Karpatach Zachodnich.
    Prace Naukowe Uniwersytetu Jagiellońskiego, CXV, Prace Geograficzne
    Seszyt 11, *Prace Instytutu Geograficznego, Zeszyt* 33, 5-267.
    (English summary.  Vertical climatic zones in the Polish Western
    Carpathians.)
Informacje G.U.S. (1979)  Zasoby wodne i gospodarka wodno-ścienkowa
    Wisly i jej dorzecza w 1978 r. Warszawa.  (Water resources and
    management of the Upper Wistula basin ....  This information is
    available only at the Central Statistical Office.)
Kamler E. (1967)  Distribution of Plecoptera and Ephemeroptera in
    relation to altitude above mean sea level and current speed in Moun-
    tain waters. *Pol. Arch. Hydrobiol*. 14(27), 2: 29-42.

Kann E. (1978) Systematik und Ökologie der Algen österreichischen Bergbächen. *Arch. Hydrobiol., Suppl.* 53, 405-643.

Kawecka B. (1971) Zonal distribution of alga communities in streams of the Polish High Tatra Mountains. *Acta Hydrobiol.* 13, 393-414.

Kawecka B. (1980) Sessile algae in European mountain streams. 1. The ecological characteristics of communities. *Acta Hydrobiol.* 22, 361-420.

Kawecka B. (1981) Sessile algae in European mountain streams. 2. Taxonomy and autecology. *Acta Hydrobiol.* 23, 17-46.

Kawecka B., Kownacka M. & Kownacki A. (1971) General characteristics of the biocenosis in streams of the Polish High Tatras. *Acta Hydrobiol.* 13, 465-76.

Kołder W. (1967) Rybactwo w dorzeczu Dunajca. *Gospodarka Rybna* 12, 18-19. (Fisheries of the Dunajec basin.)

Kownacka M. (1971) The bottom fauna of the stream Sucha Woda (High Tatra Mountains) in the annual cycle. *Acta Hydrobiol.* 13, 425-38.

Kownacka M. & Kownacki A. (1965) The bottom fauna of the river Białka and of its Tatra tributaries the Rybi Potok and Potok Roztoka. *Kom. Zagospod. Ziem Górskich PAN* 11, 130-52.

Kownacka M. & Kownacki A. (1968) Wplyw pokrywy lodowej na faunę denną potoków tatrzańskich. *Acta Hydrobiol.* 10, 90-102. (English abstract and summary. The influence of ice cover on bottom fauna in Tatra streams.)

Kownacki A. (1971) Taxocens of Chironomidae in streams of the Polish High Tatra Mountains. *Acta Hydrobiol.* 13, 439-64.

Kownacki A. (1977) Biocenosis of a high mountain stream under the influence of tourism. 4. The bottom fauna of the stream Rybi Potok of the High Tatra Mountains. *Acta Hydrobiol.* 19, 293-312.

Luther H. & Rzóska J. (1971) Project AQUA. *A source Book of Inland Waters Proposed for Conservation.* IBP Handbook No. 21. 239 pp. Blackwell, Oxford.

Nowicki M. (1880) *Ryby i Wody Galicji pod wzgledem rybactwa krajowego, Kraków (u Korneckiego)*, 96 pp. (Fishes and Waters of Galicia, Poland.)

Nowicki M. (1882) Przeglad fauny i rozsiedlenia ryb w wodach Galicji. *Gazeta Rolnicz, Kraków* 39, 465-66. (List of fauna and fish distribution in the waters of Galicia.)

Nowicki M. (1883) Przeglad rozsiedlenia ryb w Wodach Galicji według dorzeczy krain rybnych. (Mapa) Wien. (Review of fish distribution in Galicia water according to fish zones.)

Olechowska M. (1982) Zonation of mayflies (Ephemeroptera) in several streams of the Tatra Mountains and the Podhale region. *Acta Hydrobiol.* 24.

Oleksynowa K. & Komornicki T. (1965) The chemical composition of water in the Polish Tatra Mountains and the problems of its variation in time. *Kom. Zagospod. Ziem Górskich PAN* 11, 91-111.

Palmer C.M. (1969) A composite rating of algae tolerating organic pollution. *J. Phycol.* 5(1), 78-82.

Pasternak K. (1968) Charakterystyka podłoza zlewni rzeki Dunajec. *Acta Hydrobiol.* 10, 299-317. (Abstract and summary in English. Characteristics of the substratum of the River Dunajec catchment basin.)

Pawłowski B. (1959) Skład i budowa zbiorowisk roślinnych oraz metody
    ich badania. In: Szafer W. (Ed.) *Szata roslinna Polski* 1. pp.
    229-74. (Composition and structure of plant communities and methods
    of investigation.)

Romer E. (1949) Regiony klimatyczne Polski. *Pr. Wroc. Tow. Nauk. B.*
    16, 452-72. (Climatic regions of Poland.)

Solewski W. (1963) Lipień (*Thymallus thymallus* L.) potoku Rogoźnik.
    *Acta Hydrobiol.* 5, 229-43. (English summary. The grayling -
    *Thymallus thymallus* L. - in the Rogoźnik stream.)

Solewski W. (1964) Pstrag potokowy (*Salmo trutta morpha fario* L.)
    niektórych rzek karpackich Polski. *Acta Hydrobiol.* 6, 227-53.
    (English summary. *Salmo trutta* f. *fario* L. in some Polish Carpathian
    rivers.)

Solewski W. (1965) Rybostan potoku Białka Tatrzańska ze szczególnym
    uwzględnieniem charakterystyki pstraga potokowego (*Salmo trutta
    morpha fario*). *Acta Hydrobiol.* 7, 197-224. (The ichthyofauna of
    the Białka Tatrzańska stream with particular reference to the brown
    trout (*Salmo trutta* f. *fario* L.)

Sowa R. (1975a) Ecology and biogeography of mayflies (Ephemeroptera)
    of running waters in the Polish part of the Carpathians. 1. Distri-
    bution and quantitative analysis. *Acta Hydrobiol.* 17, 223-97

Sowa R. (1975b) Ecology and biogeography of mayflies (Ephemeroptera)
    of running waters in the Polish part of the Carpathians. 2. Life
    cycles. *Acta Hydrobiol.* 17, 319-53.

Sowa R. (1980) La zoogeografie, l'ecologie et la protection des
    Ephemeroptères en Pologne et leur utilisation en tant qu'indicateurs
    de la puretée des eaux courantes. In: Flannagan J. F. & Marshall K.E.
    (Eds) *Advances in Ephemeroptera Biology*, pp. 141-54. Freshwater
    Inst., Winnipeg, Canada. Plenum Press, New York & London.

Starmach J. (in preparation) Stan ichtiofauny górnego dorzecza Dunajca
    w obecnych warunkach środowiska. (State of the ichthyofauna of the
    upper part of the Dunajec basin under present environmental condi-
    tions.)

Starmach K. (1966) *Cyanophyta-Sinice, Glaucophyta-Glaukofity*, 807 pp.
    Pańtswowe Wydawnictwo Naukowe, Warszawa. (Flora Słodkowodna
    Polski.)

Wasylik K. (1971) Zbiorowiska glonów Czarnego Dunajca i niektórych
    jego doplywów. *Fragm. Flor. et Geobot.* 15, 257-354. (English
    summary. Algal communities in the Czarny Dunajec River (Southern
    Poland) and in some of its effluents.)

Wojtas F. (1964) Widelnice (Plecoptera) Tatr i Podhala, pp. 1-29.
    Uniwersytet Łodzki. (German summary. Stoneflies (Plecoptera) of
    the Tatra Mountains and Podhale region.)

Wróbel S. (1965) The bottom deposits in dam reservoirs at Roznów and
    Czchow. *Komitet Zagospod. Ziem Górskich PAN* 11, 289-94.

Ziemońska Z. (1965) Hydrographic characteristics of the Dunajec River
    basin. Along the Dunajec River. In: Starmach K. (Ed.) *XVI Limnolo-
    gorum Conventus in Polonia MCMLXV*, pp. 5-10. Polish Academy of
    Science Hydrobiological Committee, Kraków.

# 19 · Llobregat

N. F. PRAT, M. A. PUIG, G. GONZALEZ, M.F. TORT & M. ESTRADA

19.1   Introduction: Geography of the catchment          528

19.2   General data                                      532

19.3   Water quality                                     534

       19.31   Temperature                               534
       19.32   Chemistry                                 534
       19.33   Pollution problems                        538

19.4   The Llobregat and its life                        538

       19.41   Introduction                              538
       19.42   Study methods and taxonomic composition   542
       19.43   Longitudinal distribution                 543
       19.44   Seasonal variation                        545
       19.45   Changes in relative abundance along the   546
                  river
       19.46   Zonation of communities                   547

References                                               550

## 19.1   INTRODUCTION:   GEOGRAPHY OF THE CATCHMENT

The Llobregat, a typical Mediterranean river, runs across Catalonia in
N-E. Spain (Fig. 19.1).  The river flows south-east from its source, a
spectacular spring in the Cadi Mountains of the Pre-Pyrenees, to the
Mediterranean sea.  The main channel is 145 km long and drops 1360 m
from origin to mouth (19.2).  Its slope diminishes very steeply in the
first few kilometres.

The drainage basin of the Llobregat has an area of 4948 km$^2$ and comprises several climatic and geological zones.  The hydrological regime
of the river reflects the Mediterranean climate of its basin.  Characteristic features are the variation in discharge (19.2 and Table 19.1)
with frequent floods in spring and autumn.  The recent construction of
a dam at La Baells has allowed some degree of flow regulation.

Geographically the Llobregat can be divided into zones of different
physical and geological characteristics (Fig. 19.3).

(i)   The head streams are located in the Cadi Mountains and traverse
deep valleys carved by the erosion processes in the calcareous blocks.

(ii)   The largest zone of the river basin occupies part of the
Catalonian central plain, between the Cadi Mountains and the junction
of the Cardoner and Llobregat.  The head streams of the Anoia watershed
are also situated in this area.  Detritic materials cemented by carbonates and layers of gypsum and salts form the river bed in this plain.

(iii)   After the confluence with the Cardoner, the main river crosses
the pre-littoral chain near the Montserrat Mountains.  The river runs
through a canyon excavated on carbonate conglomerates.  Further to the
west the  Anoia also traverses the pre-littoral chain.  The Llobregat
and Anoia join in the pre-littoral plain, just in the vicinity of the
littoral chain.  The pre-littoral plain is recent in geological time

Fig. 19.1 Location of the Llobregat
catchment.

Fig. 19.2 Longitudinal profile of the Llobregat.

(Miocene) and contains detritic materials and outcrops of gypsum and limestone.

(iv)  The littoral chain is composed of granite and paleozoic meta-morphic shales, but these materials are often covered locally by cal-careous soils.  After crossing the littoral chain the Llobregat flows into the Mediterranean through a typical delta plain.

The Llobregat receives many tributaries (Fig. 19.4).  From the left side, the Marles and the Gabarresa, which also arise in the Cadi Moun-tains, join the main river in the Catalonian central plain.  The heavily polluted waters of Riera de Rubi reach the Llobregat in the left lower part of its course.  From the right, the Cardoner meets the Llobregat in the Catalonian central plain and the Anoia in the limits between pre-littoral plain and the littoral chain.  The Cardoner, which rises also in the Cadi Mountains, crosses several salt deposits and carries waters of very high salinity.  The Anoia is a typical stream of limestone country with strong organic and industrial pollution in its lower reaches.

Along the course of Llobregat, numerous weirs retain water for urban and industrial use and irrigation.  As a result, in summer only a small volume of water reaches the Mediterranean through the river channel. The principal factors influencing the hydrological, physical and chemi-cal features of the water are flow modification by abstraction, agri-cultural, lumbering and mining activities and urban and industrial wastes.  The river may be regarded as strongly polluted in the lower stretch and this is even more so in some tributaries, where human settlements and industries are more abundant.

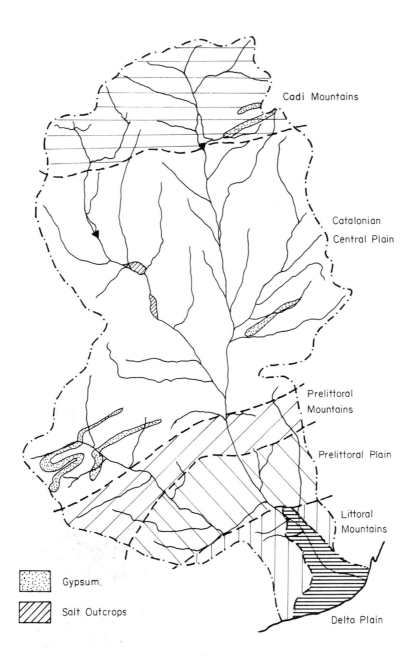

*Fig. 19.3* Geographical units in the Llobregat basin. (For details of geology see 19.2)

*Fig. 19.4* Llobregat and main tributaries, showing location of sampling stations.  The principal towns are indicated in hatched poly-
gons whose areas are proportional to the human population.
A, Sant Ponç reservoir;  B, La Baells reservoir.

## 19.2   GENERAL DATA

DIMENSIONS
   Area of catchment:                 4938 km$^2$
   Length of river                     145 km
   Highest point in catchment         2531 m
   Source of river                    1360 m

CLIMATE
   (Data from Panareda, 1979)
   The river runs across three climatic zones from spring to mouth.
   (i)   The Pre-Pyrenean zone, with high rainfall in spring and relatively
abundant rainfall in summer.  There are differences between the eastern
and western basin areas with mean annual precipitation of 905 and 703 mm
respectively.
   (ii)  Central zone, including part of the central plain and the mountains
of pre-littoral chain and the pre-littoral zone, with dry summers and abundant rainfall in spring and autumn. Rainfall is abundant in the top of some
mountains (800 mm yr$^{-1}$), but scarce in valleys (550 mm yr$^{-1}$).  The mean
temperatures are low to very low in winter.
   (iii)  Littoral zone, with low annual rainfall (550 mm yr$^{-1}$) most of
which occurs in spring and autumn.  Moderate temperatures.

GEOLOGIC
   Related to geographical zones (Riba, 1979)
   (i)   Cadi mountains.  Mesozoic calcareous rocks with deep canyons.
   (ii)  Central plain.  Eocene and Oligocene detritic materials, with
   calcareous cement and gypsum and saline outcrops.
   (iii)  Pre-littoral chain.  Calcareous conglomerates of Eocene origin.
   In the central part of this chain these form the very characteristic
   Montserrar Mountains.  Very altered metamorphic materials of Palaeozoic
   origin, often covered by Quaternary materials are also present.
   (iv)  Pre-littoral plain. Detritic materials of Miocene origin with
   some layers of gypsum and limestone.
   (v)   Littoral chain.  Granite and Palaeozoic shales covered by Mesozoic
   limestones.
   (iv)  Delta plain. Quaternary sediments.

POPULATION
   About 782000 people inhabit the watershed, but if the city of Barcelona
(located near the river mouth) is considered, this number rises to 25 x 10$^6$.
The mean density in the drainage area (Barcelona excluded) is 116.25 km$^2$
(Linati *et al.*, 1977).

LAND USE
   In the higher parts forestry and cattle rising activities are important.
There are also some industries (textile, papermills) and mines.  Alteration
to waters is limited in extension.
   In the intermediate zone (central plain and pre-littoral chain) agriculture and livestock are important, together with textile industries.
Mining (sodium and potassium salts) has a strong influence on water
quality.  Many weirs provide water to textile industries for electricity
power generation, and to villages for drinking water supply (Manresa).
In this zone Anoia is grossly polluted by industrial activities, especially
wastes from paper-mills.

In the lower zone many industries with toxic wastes are situated in Riera de Rubi. Untreated wastes are discharged also in the Llobregat. Abstraction of water for irrigation and urban water supply can leave the end part of the river channel with a very reduced flow during the dry period of the year.

PHYSICAL FEATURES OF RIVER

Flow at selected points (See also Table 19.1)

| village | sampling station no. | range of monthly mean discharge $(m^3\ s^{-1})$ | annual discharge $(Hm^3)$ | maximum recorded $(m^3\ s^{-1})$ |
|---------|---------------------|-----------------------------------------------|---------------------------|----------------------------------|
| Castellbell | 101 | 13.4 - 26.8 | 576 | 1502 |
| Martorell | 91 | 14.6 - 29.4 | 680 | 2200 |
| Prat del Llobregat | 106 | 7.2 - 68 | 707 | 3800 |

Seasonal changes in flow

Minimum and maximum recorded discharge fluctuate between 2 and 3800 $m^3\ s^{-1}$, according to fluctuations of the Mediterranean climate. Storms in spring and fall may cause catastrophic floods.

Snowfall is only important in the higher parts of the basin and its melt does not represent an important input to the discharge of the river. On the other hand the head of the watershed (Cadi Mountains) is a karstic zone and acts as a regulator of flow. A further regulation is provided by La Baells reservoir.

On-stream reservoirs and major weirs

Reservoirs La Baells (Llobregat) 125 Hm$^3$ (1976)
Sant Pono (Cardoner 247 Hm$^3$ (1957)

Many weirs along the river, very frequent in central plain.

Abstraction

350000 $m^3\ d^{-1}$ abstracted for Barcelona

Substratum (related also to geographical zones)

(i) Cadi Mountains. Abundant rocks of various sizes and sheets of rock. Sedimentation zones scarce.

(ii) Central plain. Pebbles, with cobbles and in some zones, boulders. Riffles and pools developed. Weirs are frequent and accumulate large amount of silt.

(iii) Pre-littoral chain. The Anoia and the Llobregat flow through canyons. The substratum is formed by boulders and cobbles, but pools are also common.

(iv) Pre-littoral plain. Sand and pebbles. Sand and gravel are extracted for building purposes, creating artificial ponds near the river banks. These ponds are in some cases filled with urban or industrial waste.

(v) Delta plain. Lotic areas are scarce. Cobbles, pebbles and sand. Silted bottoms in large pools are frequent. Sand and gravel extraction has been carried out intensively. In the final part the river runs as a channel.

Temperature

Thermal constancy near the sources and regulated temperature below La Baells reservoir. For explanation, see Fig. 19.5.

## 19.3  WATER QUALITY

### 19.31  TEMPERATURE

Reservoirs play an important role in temperature regulation of the
rivers along which they are sited; they lower the river temperature in
summer and increase it in winter,  especially when the water flows from
the hypolimnion (cf Ward, 1976a, b; Short & Ward, 1980).  Such an effect
is evident in the middle and lower stretches of the Llobregat due to the
presence of the La Baells dam (Fig. 19.4).  In the study summarized in
Fig. 19.5, differences in the water temperature between stations upstream
and downstream of the reservoir (stations 53 and 60) were evident at all
sampling periods.  Except in summer, there was very little change in the
temperature during passage of the river downstream of La Baells.  In
summer, at the tail of the reservoir the temperature was high (23 $^{\circ}$C)
due to the reduction of flow caused by water abstraction.  The temperature was lower (15 $^{\circ}$C) below the dam,  but rose steadily as the water
passed downstream because of the effect of the high air temperatures
and low flow.  The maximum temperature downstream of the dam was recorded in autumn (19 $^{\circ}$C), coinciding with the vertical mixing of the
reservoir water.

*Fig. 19.5* Temperature profiles along the river in four sampling
periods.  The numbers on the abcissa correspond to the stations on
Fig. 19.4.

### 19.32  CHEMISTRY

Due to the composition of the geological substratum, the concentrations of carbonates and bicarbonates are very high in the whole

Llobregat basin, and sulphate is also high in some tributaries.  Alkalinity is always higher than 2 meq $l^{-1}$ and exceeds 4 meq $l^{-1}$ in some limestone bed streams of the Anoia basin.  As a result of mining activities
in salt beds (Queralt, 1974), the water is loaded heavily with sodium
and potassium chlorides in middle and lower parts of the Cardoner and
Llobregat.

A summary of surveys of water quality in the Llobregat (*La contaminación de los cauces públicos*, Ed. Laia, 1977) includes chapters by
Linati *et al*. (1977), who studied sources of pollution in the surface
waters and Vilaro (1977) who dealt with groundwater problems.  Episodic
but dangerous incidents of heavy metal contamination were reported by
Guardiola *et al*. (1977).

In order to provide a more comprehensive framework for biological
and ecological studies, the Department d'Ecologia of the Universitat de
Barcelona has carried out an extensive survey on the Llobregat basin.
Part of this work is described by Prat *et al*. (1982).  In this survey
samples were taken on four occasions (September 1979, December 1979,
April 1980, August 1980) for physicochemical analysis of the water and
to study the benthic fauna at 75 stations from the Llobregat drainage
area (Fig. 19.4).  The analytical methods employed were those given in
Margalef *et al*. (1976).  Certain geographical variables were determined
for all sampling stations.

From the 27 geographical and physicochemical variables considered,
19 were selected as the most representative to perform a multivariate
statistical analysis:  altitude, slope, distance from origin, drainage
area, conductivity, temperature, dry matter, alkalinity, dissolved
oxygen, chloride, silicate, sulphate, nitrite, nitrate, phosphate,
calcium, magnesium, sodium, potassium.  The technique chosen was principal component analysis.  (This and other related methods have been
applied only rarely to rivers - see, for example, Angelier *et al*. (1978)
- but can be very useful in summarizing information and pointing out
trends of variation contained in large data sets.)  The analysis was
based on the correlation matrix among the selected variables, without
previous logarithmic transformation of the data and following the
methods described in Estrada (1975) and Margalef *et al*. (1976).  248
samples were included.

The two first principal components explain,  respectively, 30 and
17% of the variance of the original data (Fig. 19.6).  As can be seen
in this figure, the first component is correlated positively with
altitude and slope and negatively with distance from the origin,
conductivity, chloride, sodium, potassium and magnesium.  Thus the
major source of variability of the original data is related to the
increase in conductivity due to the ion content that increases along
the river (Figs. 19.7, 19.9, 19.10).  (The distribution of the mean
values of the score of the first component for all the samples taken
at each point, reflects this situation and presents decreasing values
from the upper to the lower parts of the basin (Fig. 19.7).  A strong

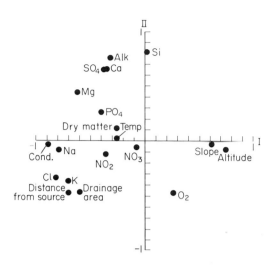

*Fig. 19.6* Position of the extremes of the vectors corresponding to the 19 variables in the space of the first two axes of the principal component analysis.

drop in the scores occurs at stations 68 of the Llobregat and 42 of the Cardoner, corresponding to the passage across salt deposits and the quantitatively important input to the river of chloride, sodium and potassium (Figs. 19.9, 19.10).   Between sampling points 102 and 101 there is a rise in conductivity and the score for the first component drops in the Llobregat as a result of the entry of the Cardoner which carries the higher salt concentration (Fig. 19.9).

The strongly negative scores of the first component in the Anoia and Riera de Rubi (Fig. 19.7) are not related to mining activities, but reflect elevated ionic concentrations due to high solubility of the rocks in their upper basin areas and to heavy industrial and urban pollution in their lower stretches, where chloride concentrations may reach 1 g $l^{-1}$.

The second principal component (Fig. 19.6) shows positive correlations with alkalinity, silicate, sulphate, calcium and magnesium, and negative correlations with distance from the source, drainage area and oxygen concentration.

The mean scores of the second component at each sampling point (Fig. 19.8) appear to discriminate between the Anoia basin and the other basins.  The values are negative with a decreasing trend down the Llobregat course and positive in the Anoia and the polluted Riera de Rubi.  The source of variation expressed by the second component is

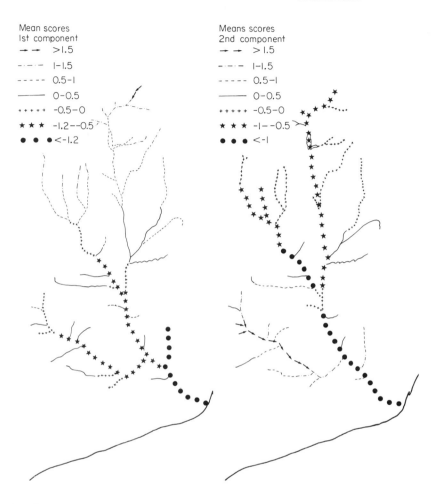

Mean scores
1st component
→  →   >1.5
—·—·—   1-1.5
— — — —   0.5-1
———————   0-0.5
+ + + +   -0.5-0
★ ★ ★   -1.2--0.5
● ● ●  <-1.2

Means scores
2nd component
→  →   >1.5
—·—·—   1-1.5
— — — —   0.5-1
———————   0-0.5
+ + + +   -0.5-0
★ ★ ★   -1--0.5
● ● ●  <-1

*Fig. 19.7*  Mean scores of the first principal component in the Llobregat catchment.  Note the range along the river.

*Fig. 19.8*  Mean scores of the second axis principal component in the Llobregat basin.  Note the differences between the Anoia and Riera de Rubi (positive values) and the other river basins (negative values).

due to the concurrence, in the Anoia basin, of high alkalinity and elevated concentrations of calcium, magnesium and sulphate of geological origin, with high silicate and phosphate concentrations due to the inflow of urban and industrial wastes.  The Riera de Rubi has lower alkalinity and concentrations of sulphate, calcium and magnesium, but more severe pollution problems.

In the middle and lower stretches of the Llobregat and Cardoner, silicate concentrations are sometimes reduced by phytoplankton and periphyton growth.  Alkalinity, sulphate, calcium and magnesium do not vary in a significant way along these rivers and oxygen depletion is rare ; in some cases oversaturation may exist during daylight hours due to the activity of the primary producers.  The negative scores of the Llobregat and Cardoner for the second component reflect these features.


## 19.33  POLLUTION PROBLEMS

The Llobregat basin is densely populated (19.2).  There are virtually no sewage treatment plants and the run-off from villages and industries goes directly to the river channel, which thus receives numerous sources of pollution.  In the lower parts of the Llobregat and specially in the Anoia basin there is intense organic pollution leading to high BOD and COD values which have increased progressively in recent years (Canto *et al.*, 1975).

Monthly data collected by river authorities (D.G.O.H., 1978) at stations coinciding with our sampling points show the increase in BOD and COD on passing down the river (Table 19.1).  Variability in the values was related principally to river discharge (Canto *et al.*, 1975) and as a result, marked fluctuations in BOD and COD occurred during the year.  The maximum values were recorded in summer, when the river flow was minimum.

The saline load due to substratum composition and mining activities has been already considered (19.32).  Cases of pollution by heavy metals such as chromium have occasionally forced the water supply to one of the drinking water treatment plants for Barcelona to be discontinued (Guardiola *et al.*, 1977).

As a result of organic and other type of contamination, plant nutrients such as phosphate show also a progressive increase along the Llobregat course (Fig. 19.11).  This trend persists throughout the year but is specially evident in summer, coinciding with the lower water flow.


## 19.4  THE LLOBREGAT AND ITS LIFE

## 19.41  INTRODUCTION

Very few biological studies have been carried out on any Spanish river (Meynell, 1973 ; Prat *et al.*, 1979 ; Prat, 1981 ; Viedma & Jalon, 1980). In the case of the Llobregat, there are the publications of Oliver & Bernis (1967), Puig *et al.* (1981) and Tomas & Sabater (in press).

The data on submerged macrophytes and periphyton growths on stones

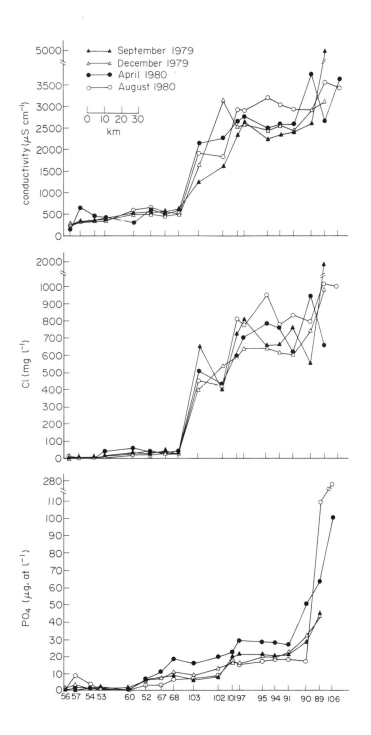

*Figs. 19.9, 19.10, 19.11* Profiles of conductivity, chloride and phosphate along the Llobregat.

540    PRAT *ET AL.*

A

C

B

*Fig. 19.12* Llobregat.  A) Station 53: immediately upstream of La Baells reservoir showing Roman Bridge.  B) Station 102: Pont de Vilomira, above the confluence of the Llobregat with the Cardener. C) Station 91: Martorell, downstream of the entry of the Anoia.

*Table 19.1* Geographical features, discharge, COD (chemical oxygen demand) and BOD (biological oxygen demand) along the Llobregat. Data from D.G.O.H. (1978), with mean, standard deviation (in some cases) and range for the three variables.

| sampling station | nearest village | alti-tude (m) | catch-ment (km$^2$) | distance from springs (km) | discharge annual mean (m$^3$ s$^{-1}$) | COD annual mean (mg l$^{-1}$) | BOD annual mean (mg l$^{-1}$) |
|---|---|---|---|---|---|---|---|
| 53 | Figols | 620 | 352 | 22.3 | 10.1 ± 6.8 ( 5.4 – 12.5) | 5.32 (1.8 – 16) | 2.46 (0.5 – 5.3) |
| 68 | Balsareny | 229 | 1093 | 61 | 14.4 ± 4.7 ( 5.3 – 22.7) | 3.37 (2.6 – 4.6) | 2.87 (0.5 – 6.2) |
| 102 | Vilomara | 176 | 1885 | 87 | 16.3 ± 5.5 ( 5.9 – 25.8) | 4.51 (3 – 7.8) | 3.55 (1.4 – 6.1) |
| 97 | Castellbell | 155 | 3293 | 99 | 25 ± 10.9 ( 1.3 – 38.5) | 5.03 (3 – 9.3) | 3.91 (1 – 7.2) |
| 91 | Martorell | 44 | 4561 | 131 | 43.6 ± 59 ( 3.9 – 176) | 32.12 (6.8 – 276) | 20.58 (4.4 – 163) |
| 89 | Sant Joan Despi | 17 | 4848 | 149.4 | 46.3 ± 60.4 ( 8.5 – 180) | 11.85 (7.9 – 23) | 9.12 (5.6 – 23.7) |

are sparse.  Some general information on the algae is given in Margalef
(1951), while Tomas and Sabater have surveyed the diatom flora.  In
the zone above the La Baells reservoir (see 19.36) the most abundant
species are *Achnanthes minutissima, Cymbella minuta, Diatoma hiemale*
var. *mesodon* and, in winter, *Meridion circulare*.  In a zone from stations 60 to 94 the above species are gradually replaced by species of
*Navicula* and *Gomphonema*.  The last zone of the Llobregat, from station
91 downstream, only pollution-tolerant species are found.  These include
*Nitzschia frustulum, N. palea, Navicula pellulosa, N. atomus, N.
permitis* and various centric diatoms.  In the lower zone of the watershed where chloride concentrations are high, there is a proliferation
of *Cladophora* (see Margalef, 1951).

*Cladophora* is very abundant on stones from station 57 down to the
lower river.  After station 68 which coincides with the high salt inputs (Fig. 19.10) *Potamogeton pectinatus* appears in addition to the
abundant *Cladophora*.  Between stations 68 and 102 clumps of this species
are only infrequent, but below the entrance of the Cardoner it can cover
up to 50% of the river bed at some stations. In the Cardoner, in areas
of scarce flow, *P. pectinatus* forms a dense carpet covering all the
bottom of the river.  Animals associated with *Cladophora* and *P.
pectinatus* may be very abundant and some macroinvertebrates are found
only on these plants.

Although fishing is a popular sporting activity, the fish fauna has
not been surveyed extensively in the Llobregat.  As in other Spanish
rivers, fish diversity is low:  only four species can be found with
some abundance.  Rainbow trout (*Salmo gairdneri*), which has been introduced from fish-farms, is abundant in inpolluted stretches of the upper
courses.  *Barbus* spp. and especially *Leuciscus cephalus* are frequent in
the middle and lower river; carp (*Cyprinus carpio*) can be found in the
lower river where pollution is not so strong.

The more detailed information on macroinvertebrates reported below
is based on an extensive survey carried out by Departament d'Ecologia
of the Universitat de Barcelona between September 1979 and August 1980,
the same period as that covered by the physicochemical survey.


19.42   STUDY METHODS AND TAXONOMIC COMPOSITION

Samples for qualitative study were collected during the four sampling
periods at all the stations in Fig. 19.4.  The sampling strategy was
based on the stratification guide-lines suggested by Resh (1979).  All
attached material was scraped from stones taken, where possible, from
areas subject to fast or moderate current speeds in the central part of
the stream.  The organisms (mostly larval and pupal insect stages)
retained by a net of 250 μm mesh size were fixed with 10% formalin and
transported to the laboratory for identification and enumeration under
a stereoscopic microscope (with a magnification of 10x).  This account
is based on all the material obtained from the 18 Llobregat stations,

with the exception of the chironomids, which have been studied only in samples from the first sampling period (September 1979). More detailed studies can be found in Prat *et al.* (1983), Gonzalez *et al.* (in press), Millet & Prat (in press) and Prat (in press).

More than 120 species or groups of species have been found in these 18 stations. As with for other rivers (Illies & Botosaneanu, 1963; Hynes, 1970), insects form the major part of the fauna, with 7 species of Plectoptera, 15 Ephemeroptera, 14 Trichoptera, 9 Coleoptera and 65 Diptera. From the Diptera, 45 were Chironomidae, 2 Blepharoceridae and 10 Simuliidae. Among the other groups, the most common were the Turbellaria (1), Hirudinea (2), Mollusca (6) and Ostracoda. However, while some species were very frequent, others could be found only in one or very few sampling points.

## 19.43 LONGITUDINAL DISTRIBUTION

Fig. 19.13 shows the longitudinal distribution of the most important macroinvertebrates along the Llobregat course. As can be seen, several species showed a very wide distribution along the river. This group included the chironomids *Rheocricotopus chalybeatus*, *Cricotopus* gr. *tremulus* and *Polypedilum* gr. *convictum*, the caddis *Hydropsyche exocellata* and *Hydroptila* sp ; the mayfly *Caenis moesta* and the molluscs *Physa acuta* and *Ancylus fluviatilis*. On the other hand, some rheophilic and stenothermic species were found only at the most upstream station (56); these comprised the blepharococerid *Liponeura cinerascens*, the chironomid *Diamesa* sp., the stoneflies *Isoperla* sp. and *Protonemura risi spinulosa*, the caddis *Tinodes dives* and *Micrasema morosum* and the flatworm *Polycelis felina*.

Many species were found only between stations 68 (280 m altitude) and 57 (840 m) or between stations 57 and 53 (500 m), upstream of La Baells reservoir. These included mayflies (*Baetis rhodani, Rhithrogena diaphana*), stoneflies (*Dinocras cephalotes*) and the blackflies *Simulium monticola* and *S. rheophilum*.

No stonefly larvae were found downstream of station 60, which can be attributed to their cold stenothermy (Berthélemy, 1966); they were very abundant at stations 56 and 54, but absent from the intermediate station 57, probably due to pollution by wastes of nearby town and a papermill. The Simuliidae were also restricted to the upper part of the river (Fig. 19.13).

Some species, such as the mayfly *Baetis meridionalis* and caddisflies *Hydropsyche contubernalis* and *Psychomyia pusilla*, and the chironomids *Cardiocladus fuscus* and *Orthocladius* sp. 2 were found exclusively in intermediate zones of the river. Their distribution downstream was probably limited by pollution.

Only very few species were present in the heavily contaminated lower reaches of the river. Among them were the chironomids *Cricotopus* gr.

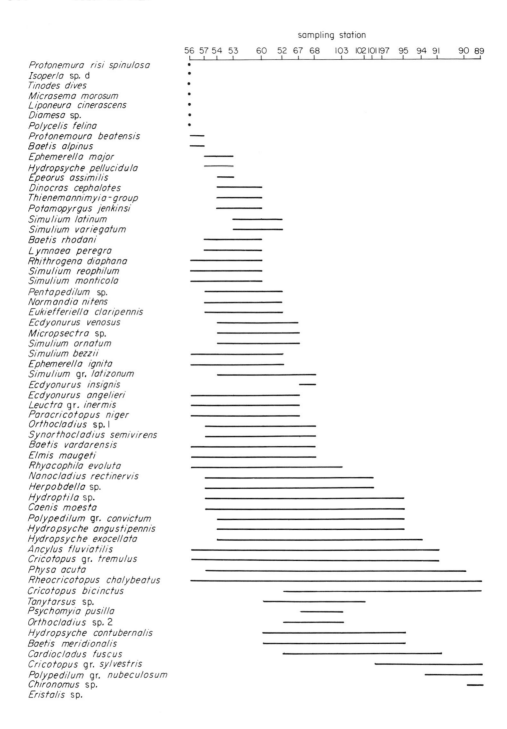

*Fig. 19.13* Longitudinal distribution of macroinvertebrates along the Llobregat. Numbers correspond to stations in Fig. 19.4.

*sylvestris, Polypedilum* gr. *nubeculosum* and *Chironomus* sp. and another Diptera, *Eristalis* sp.  Tubificids were also abundant here.


## 19.44  SEASONAL VARIATION

The relative importance (based on total number of individuals collected in each station) of the main macroinvertebrate groups was relatively
constant for the different sampling periods (Fig. 19.14).  The highest
temporal differences were shown by the chironomids, which were very
abundant in spring in polluted zones.  Blackflies, which were present in

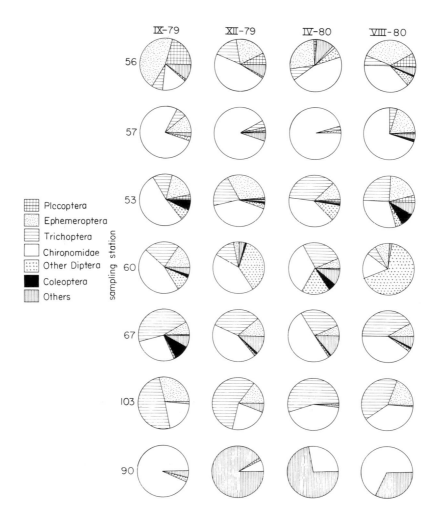

*Fig. 19.14* Percentage composition (based on number of animals) of
the macroinvertebrate fauna at seven stations of the Llobregat during
four different sampling periods.  Only the most important groups are
figured.

masses in both summer and winter at station 60, just below the La Baells
reservoir, were scarce in autumn and spring at the same site.  Plankton
richness and low temperatures seem to be the two key factors bringing
about this abundance; the decrease in numbers could be related to the
life cycles of the species present.  Among the caddisflies, *Psychomyia
pusilla* proliferated on stones in spring and summer and was scarce in
autumn and winter, as had been reported previously in the river Ter
(Prat, 1981).  *Hydroptila* larvae were also more abundant in spring and
summer, the behaviour in this case being due to its association with
*Cladophora*.


19.45   CHANGES IN RELATIVE ABUNDANCE ALONG THE RIVER

The relative abundance of many species (based on percentage of all
individuals taken at each station in September 1979) showed interesting
changes along the Llobregat.  The genus *Baetis* presented a gradual
pattern of species replacement (Fig. 19.15) from the upper to the lower
parts of the river course.  The sequence was: *B. alpinus, B. rhodani,
B. vardarensis* and *B. meridionalis* and is consistent with known ecological characteristics of these species.  For example, *B. alpinus* has
been found in Andorra above 1000 m (Puig, 1981) and *B. meridionalis* is
common in the lower stretches of many rivers (Müller-Liebenau, 1969).
Its absence from the most downstream sampling sites on the Llobregat can
be attributed to pollution.  Among the caddis, *Rhyacophila evoluta*
could not be found downstream of station 68 and was more abundant in
the upper river.  Conversely *Hydropsyche exocellata* was common in the

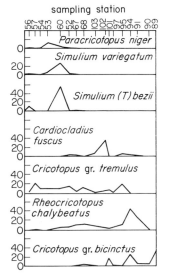

*Fig. 19.15* Distribution of the most common species of Ephemeroptera,
Plecoptera and Trichoptera along the Llobregat.

*Fig. 19.16* Distribution of the most common Diptera along the
Llobregat.

middle river, especially between stations 67 and 101, where it was the
most abundant invertebrate.

In the case of midges (Chironomidae), the lack of detailed data for
three of the sampling periods and the difficulties of species determination (only possible by examination of mature pupae with male geni-
talia) made it difficult to interpretate their abundance patterns.  In
general, they prefer polluted zones (compare, for instance, their
numbers at the polluted station 57 with those at neighbouring stations
in Fig. 19.14).  However examples were found in each section of river
where particular species reached their greatest abundance.  *Paracricotopus niger* was most abundant upstream La Baells reservoir (Fig. 19.14),
*Cardiocladus fuscus* larvae in the central part of the river and
*Cricotopus sylvestris* and *Polypedilum* gr. *nubeculosum* in lower polluted
stretch.  The abundance of some midge species but related to their substratum preferences *Rheocricotopus chalybeatus* was especially common
on stones with *Cladophora* or on *Potamogeton* clumps, and its abundance
increased in the middle and lower river (Fig. 19.16).  Classification
difficulties made it impossible to distinguish patterns of distribution
within the *Cricotopus* gr. *tremulus* species complex, which involves two
or three species (*Cricotopus annulator, C. triannulatus* and another
unidentified species).  *C. bicinctus* is an example of species where
pollution was the key factor;  as can be seen in Fig. 19.16, it was
abundant both in the lower Llobregat and also in the polluted station
57 in the upper river.


## 19.46  ZONATION OF COMMUNITIES

The establishment of macroinvertebrate communities along the course of
rivers has been subject of much discussion.  River zonation has been
considered as a succession of more or less well defined discrete units
(Illies & Botosaneanu, 1963) or as the result of continuous change in
faunistic composition throughout the river (Hynes, 1970; Vannote *et al.*,
1980).  Sharp faunistic changes can be found as the result of sudden
variations in water quality due, for example, to pollution sources and
in rivers less affected by contamination a gradual replacement of
species can be found without sharp changes (Hawkes, 1975).

Analysis of common species between neighbouring stations following
the method of Illies & Botosaneanu (1963) has been applied to the data
of September 1979 (Fig. 19.17).  If the longitudinal distribution of
species and their relative importance is considered in the Llobregat,
five faunistic units can be distinguished (Table 19.2).

The first zone includes only most upstream station (56), which has a
species composition very similar to that of Pyrenaean streams (see
for example Prat *et al.*, 1979).  In this area water is very cold,
current speeds are high and dissolved oxygen is near saturation.

The limit between the second and third zone is set by the La Baells

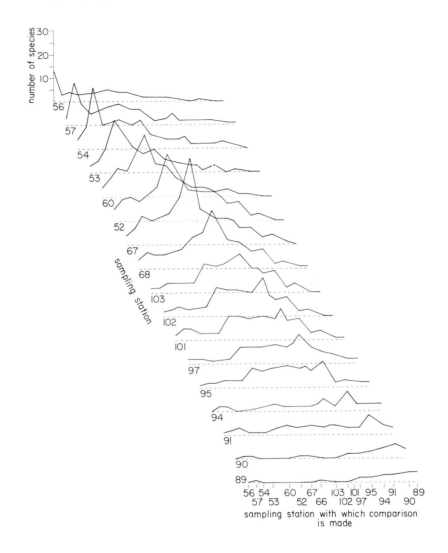

*Fig. 19.17* Number of species at each sampling station on the Llobregat in September 1979 present also at other stations at the same time. Each graph corresponds to the station indicated in the left side of the abcissa.

reservoir. Upstream of the dam (the second zone) the Llobregat is a typical mountain stream, with cool water, a relatively steep slope, large rocks in the river bed and a poorly developed riffle-pool structure. The faunistical similarity among the three stations of the second zone (57, 54, 53) is somewhat confused because of local pollution (station 57) or water abstraction (station 53). Station 60, downstream of La Baells reservoir, is influenced strongly by the hydrological regime of reservoir and the fauna is intermediate between second

Table 19.2  Zonation along the Llobregat: sampling stations, physicochemical characteristics (with mean and range) and important indicator species.

| | FIRST ZONE (epirhithron) | SECOND ZONE (upper metarhithron) polluted | SECOND ZONE (upper metarhithron) clean | BELOW RESERVOIR | THIRD ZONE (lower metarhithron) | FOURTH ZONE (hyporhithron) | FIFTH ZONE (potamon) |
|---|---|---|---|---|---|---|---|
| sampling station | 56 | 57 | 54, 53 | 60 | 52, 67, 68 | 103, 102, 101, 97, 95 | 94, 91, 90, 89 |
| slope (range) (%) | 16 | 8.6 | 1.11 - 1.4 | 1.1 | 0.46 - 0.88 | 0.25 - 0.52 | 0.12 - 0.35 |
| temperature (°C) | 8.5 (7 - 10) | 10.8 (4 - 17) | 10.5 (1.5 - 23.5) | 12.7 (8.5 - 19) | 14.2 (6.5 - 23) | 16.5 (7.5 - 28.5) | 17.6 (8.2 - 28.5) |
| conductivity ($\mu$S cm$^{-1}$) | 228 (159 - 270) | 418 (330 - 667) | 396 (329 - 474) | 493 (323 - 605) | 566 (498 - 672) | 2369 (1250 - 3200) | 3082 (2400 - 5000) |
| phosphorus ($\mu$g - at l$^{-1}$) | 0.67 (0.01 - 1.45) | 3.14 (0.8 - 9.7) | 1.8 (0.58 - 4.2) | 1.87 (0.8 - 1.76) | 8.69 (3 - 19.2) | 17.95 (7 - 30.6) | 40.5 (19.1 - 112.6) |
| oxygen (mg l$^{-1}$) | 9.31 (8.61 - 10.8) | 11.5 (8.4 - 16.4) | 9.0 (4.1 - 12.0) | 10.8 (6.41 - 8.5) | 9.1 (6.5 - 12.9) | 9.3 (<0.1 - 16.4) | 8.1 (2.5 - 21.0) |
| INDICATOR SPECIES exclusive | *Protonemura risi*<br>*Tinodes dives*<br>*Micrasema morosum*<br>*Liponeura cinerascens*<br>*Diamesa* spp.<br>*Polycelis felina* | | *Epeorus assimilis*<br>*Paracricotopus niger* | *Simulium bezzii*<br>*Simulium ornatum* | ***Ecdyonurus insignis*** | *Baetis meridionalis*<br>*Hydropsyche contubernalis*<br>*Cardiocladius fuscus*<br>*Psychomyia pusilla* | *Cricotopus* gr. *sylvestris*<br>*Polypedilum* gr. *nubeculosum*<br>*Chironomus* spp.<br>*Eristalis* spp. |
| more abundant in this zone | *Baetis alpinus*<br>*Protonemura beatensis* | *Cricotopus bicinctus* | *Baetis rhodani*<br>*Rhithrogena diaphana*<br>*Dinocras cephalotes*<br>*Hydropsyche pellucidula*<br>*Ephemerella major* | *Baetis vardarensis* | *Hydropsyche exocellata* | | |

and third zone.  As has been indicated previously the high abundance of blackflies is a very characteristic feature here at some seasons.

The third zone extends between stations 52 and 68, where many species have their downstream limit due to sharp changes in water quality and geographical characteristics.  High faunistic variety characterises this zone; the maximum number of species per sample was recorded at station 67.

Dominance of *Hydropsyche exocellata* is the most striking feature of the fourth river zone, between stations 103 and 97.  High salinity and conductivity, warm summer temperatures and the succession of weirs and riffle-pool areas are the most characteristic physicochemical features. The presence of abundant filter-feeding species on stones is associated with the development of plankton behind weirs and pools due to nutrient inputs of sewage from villages and farms.

Downstream of station 95 (fifth zone) the benthic fauna is very poor.  The river increases in width and depth and true lotic zones are infrequent.  The concentration of salts is very high and there is a high contribution of urban and industrial pollution.  Dissolved oxygen concentrations are low in some periods of the year.  Only a very few species resist these conditions, some of which are abundant but some of which are very scarce.

*Acknowledgements*

We thank Prof. R. Margalef, J.D. Ros and R. Julià for their helpful criticisms and for reviewing the text, Anna M. Domingo for the drawings and V. Rull, X. Millet, C. Marrasé, T. Vegas, Nuria Prat, T. Cassasayas and J. Carbonell for sampling assistance.  Financial support has been provided by Universitat de Barcelona and "Servei del Medi Ambient, Diputació Provincial de Barcelona".

REFERENCES

Angelier E., Bordes J.M., Luchetta J.C. & Rochard M. (1978)  Analyse statistique des paramètres physico-chimiques de la rivière Lot. *Annls Limnol.* 14, 29-58.
Berthélemy C. (1966)  Recherches écologiques et biogéographiques sur les Plécoptères et Coléoptères d'eau courante (Hydraena et Elminthidae) des Pyrénées. *Annls Limnol.* 2, 227-458.
Canto J., Guardiola J. & Salvatella N. (1975)  Evolución de la polución del agua en el rio Llobregat. *Agua* 91, 15-24.
D.G.O.H. (1978)  *Análisis de la calidad de aguas. 1978-1977.* Publicaciones del M.O.P.U., Centro de Estudios Hidrográficos.  No. 129. Ministerio de Obras Públicas, Madrid.

Estrada M. (1975)  Statistical considerations of some limnological
    parameters in Spain reservoirs. *Verh. int. Verein. theor. angew.
    Limnol.* 19, 1849-59.
Guardiola J., Salvatella N. & Canto J. (1977)  Polución del agua por
    Cromo en el rio Llobregat.  In: *La contaminacion de cauces públicos*,
    251 pp., pp. 117-137.  Laia, Barcelona.
Gonzalez G., Puig M.A., Millet J. & Prat N. (in press)  Patterns of
    macroinvertebrate distribution in the Llobregat river basin (NE
    Spain). *Verh. int. Verein. theor. angew. Limnol.* 22
Hawkes H.A. (1975)  River zonation and classification.  In: Whitton
    B.A. (Ed.) *River Ecology*, 725 pp., pp. 313-374.  Blackwell, Oxford.
Hynes H.B.N. (1970)  *The Ecology of Running Waters*, 555 pp.  Liverpool
    University Press, England; University of Toronto Press, Canada.
Illies J. & Botosaneanu L. (1963)  Problèmes et méthodes de la classi-
    fication et de la zonation écologique des eaux courantes, consi-
    derées surtout du point de vue faunistique. *Mitt. int. verein.
    theor. angew. Limnol.* 12, 57 pp.
Linati J.A., Aparicio I., Guardiola J. & Verges J.C. (1977)  El problema
    de la contaminación en la cuenco del rio Llobregat.  In: *La
    contaminación de los Cauces Publicos*, 251 pp., pp. 75-92.  Laia,
    Barcelona.
Margalef R. (1951)  Regiones limnológicas de Catalunya y ensayo de
    sistematización de las asociaciones de algas. *Collectanea Botánica*
    3, 43-67.
Margalef R., Planas D., Armengol J., Vidal A., Prat N., Guiset A., Toja
    J. & Estrada M. (1976) *Limnologia de los Embalses Espanolcs*. Publi-
    caciones de M.O.P.U. Centro de Estudios Hidrográficos.  No. 123,
    454 pp.  Ministerio de Obras Públicas, Madrid.
Meynell P.J. (1973)  A hydrobiological survey of a small Spanish river
    grossly polluted by oil refinery and petrochemical wastes. *Freshwat.
    Biol.* 3, 503-20.

Millet J. & Prat N. (in press)  Las comunidades de macroinvertebrados
    a lo largo del rio Llobregat.  Actas del Segundo Congreso Espanol
    de Limnologia.  Murcia, April 1983.  (Spanish = Macroinvertebrate
    communities along the River Llobregat.)
Müller-Liebenau I. (1969)  Revision de europäischen Arten der Gattung
    *Baetis* Leech (Insecta, Ephemeroptera). *Gewässer Abwässer* 48/49,
    1-214.
Oliver B. & Bernis J. (1967)  Contribución al estudio biológico del
    rio Llobregat y sus afluentes. *Doc. Inv. Hidrológica* 2, 27-71.
Panareda J.M. (1979)  El clima y les aigües.  In: *Geografia Fisica
    dels Paisos Catalans*, pp. 69-102.  Ketres, Barcelona.
Prat N. (1981)  The influence of reservoir discharge on benthic fauna
    in River Ter (NE Spain).  In: Moretti G.P. (Ed.) *Proceedings of
    the 3rd International Symposium on Trichoptera*, pp. 295-303.  Series
    Entomologicae Vol. 20.  W. Junk, The Hague, The Netherlands.

Prat N. (in press)  Los rios.  In: *Curso Fundamentos de Limnologia*.
    Murcia, April 1983.  Publicaciones de la Universidad de Murcia.
Prat N., Bautista M.I., Gonzalez G. & Puig M.A. (1979)  Els cursos
    d'aigua.  In: Folch R. (Ed.) *El Patrimoni Natural d'Andorra*, pp.
    261-309.  Ketres, Barcelona.

Prat N., Gonzalez G. & Puig M.A. (1983)  Predicció i control de la
    qualitat de les aigües dels rius Besós i Llobregat.  II. El pobla-
    ment faunistíc i la seva relació amb la qualitat de les aigües.
    Monografies, 9.  164 pp.  Servei del Medi Ambient.  Diputació de
    Barcelona.  (Catalan = Prediction and control of water quality in
    the Rivers Besós and Llobregat.  II. Faunistic composition in rela-
    tion to water quality)
Prat N., Gonzalez G., Puig M.A. & Millet J. (in press)  Chironomid
    longitudinal distribution and macroinvertebrate diversity along the
    river Llobregat (NE Spain).  Memoirs of the American Entomological
    Society.
Prat N., Puig M.A., Gonzalez G. & Tort M.J. (1982)  Predicció i control
    de la qualitat de les aigües dels rius Besós i Llobregat.  I. Carac-
    terístiques fisicio-químiques.  Monografies 6, 206 pp.  Servei del
    Medi Ambient.  Diputació Provincial de Barcelona.  (Catalan = Pre-
    diction and control of water quality in the Rivers Besós and
    Llobregat.  I. Physico-chemical features)
Puig M.A. (1981)  Contribució al coneixement dels Plecópters i Efemer-
    ópters d'Andorra.  *Bull. Ins. Cat. Hist. Nat.* (45 Sec. Zool. 3).
    (Cat. = Contribution to knowledge of Plecoptera and Ephemeoptera ..)
Puig M.A., Bautista I., Tort M.J. & Prat N. (1981).  Les larves de
    trichoptères de la rivière Llobregat (Catalogne, Espagne).  Distri-
    bution longitudinale et relation avec la qualité de l'eau.  In:
    Moretti G.P. (Ed.) *Proceedings of the 3rd International Symposium
    on Trichoptera*, pp. 305-311.  Series Entomologicae, vol. 20.  Junk,
    The Hague, The Netherlands.
Queralt A. (1974)  La contaminación de las aguas.  *C.A.U.* 25, 83-110.
Resh V.H. (1979)  Sampling variability and life history features: basic
    considerations in the design of aquatic insect studies.  *J. Fish.
    Res. Bd Canada* 36, 290-311.
Riba O. (1979)  El relleu dels Paisos Catalans.  In: *Geografia Fisica
    dels Paisos Catalans*, pp. 7-66.  Ketres, Barcelona.
Short R.A. & Ward J.V. (1980)  Leaf litter processing in a regulated
    rocky mountain stream.  *Can. J. Fish. aquat. Sci.* 37, 130-7.
Tomàs X. & Sabater S. (in press)  The diatom associations of the
    Llobregat River and their relations to water quality.  *Verh. int.
    Verein. theor. angew. Limnol.* 22.
Viedma M.G. de & Jalon D.G. de (1980)  Descriptions of four larvae of
    *Rhyacophila (Parahyacophila)* from the Lozoya river, central Spain,
    and key to the species of the Iberian peninsula (Trichoptera:
    Rhyacophilidae).  *Aquatic Insects* 2, 1-12.
Vilaro F. (1977)  La contaminación de las aguas subterráneas.  In:
    *La contaminación de los cauces públicos*, 251 pp., pp. 139-159.
    Laia, Barcelona.
Ward J.W. (1976a)  Effects of thermal constancy and seasonal tempera-
    ture displacement on community structure of stream invertebrates.
    *ERDA Symposium Series*, pp. 302-7.
Ward J.W. (1976b)  Effect of flow patterns below large dams on stream
    benthos.  A review.  In: Osborn J.F. & Allman C.H. (Eds) *Instream
    flows*, vol. II, pp. 235-53.  *American Fisheries Society.*

# 20 · Rönneå

## A. ALMESTRAND & A. ALMESTRAND

20.1 Introduction                                          554

20.2 General data                                          556

20.3 Physiography and hydrology                            558

20.4 Ringsjön and its tributaries                         559

20.5 The main river                                        562

    20.51 Changes on passing downstream           562
    20.52 Usage of the water and water quality     565

20.6 The largest tributaries                              568

    20.61 Ybbarpsån                                568
    20.62 Bäljaneå-Perstorpsbäcken                 568
    20.63 Pinnån                                   569
    20.64 Rössjöholmsån-Kägleån                    570

20.7 Concluding comments                                  572

References                                                574

## 20.1 INTRODUCTION

Sweden, a long, narrow country with north-south extension, exhibits many
types of river, due to large differences in climate, morphology and
geology.  There are slow-moving lowland rivers almost hidden in abundant
vegetation and mighty, rushing Norrland rivers (Sw. älv) with huge
hydroelectric plants.

This chapter describes a river in Scania (Sw. Skåne) which runs
mainly through farming land, but whose tributaries drain forest-covered
districts.  The hydrology of this and other Scanian rivers has been much
changed by drainage and the lowering of lake water-levels; farmers
strove to improve and extend their fields in this way long ago.

Skåne is the southern-most province of Sweden, with 1/40 (11000 km$^2$)
of the country, but about 1/8 (1.2 x 10$^6$) of the population.  It is the
most densely populated province; in Malmöhus county there are about 150
inhabitants per km$^2$ and in Kristianstad county, 43 per km$^2$.  The
climate is mild due to the long coast-line.  Precipitation varies from
800 mm per year in the north to <500 mm in the south-west.  Skåne is
poor in lakes except in the north-east; the biggest is Ivösjön (54.2
km$^2$, 50 m depth).  The rivers follow in the western part the line of
the landscape, from S-E. to N-W.  Typical examples are Rönneå, Vegeå
and Kävlingeån.

Northern Skåne is a monotonous Archaean rock plain rich in mires and
sloping gradually towards the S-W.  The Archaean rock ridges (Sw. åsar)
south of the plain have come up as horsts and present a striking
feature, although they rise to only about 200 m above sea-level.  They
belong to the system of diagonal parallel axes which can be drawn over
Skåne from N-W. to S-E.  Some axes constitute depressions, others ridges,
but all are reflected in both bedrock and surface morphology.  The
typical plains of Skåne are in the N-W., N-E. and south; thanks to these
Skåne has become "the granary of Sweden".

The geology of Skåne differs from the rest of the country (Lundegårdh
*et al.*, 1964).  A coloured map is like a mosaic.  More than half is
connected geologically with Middle Europe via Denmark.  Along the divi-
ding lines mentioned above, movements in the earth's crust occurred over
a long period and probably did not end until the Tertiary.  In the
depressions "young" rocks (sandstones, slates, limestones) from the
Cambrian to Cretaceous layers have been preserved; in other places
erosion has sometimes reached down to Archaean rocks.  The final surface
forms were built up during the Quaternary period.  The most important
moraine soil in the Rönneå district derives from the N-E. ice (Sw.
nordostisen).  Ice river deposits gave rise to eskers, gravel ridges
and plateaus.

Less than 50% of Skåne is cultivated and this is continuing to decrease.
Farming is not profitable and cattle-breeding in particular is on the
decline.  Meadows where cows once grazed are now planted with spruce.
Skåne is however highly industrialized, with food production predominant.
Industry employs c. 35% of the working population.

*Fig. 20.1* Survey of Rönneå catchment area.  Inset: Map of Sweden; Rönneå river system is dotted;
S, Stockholm; G, Göteborg; M, Malmö.

## 20.2  GENERAL DATA

DIMENSIONS

| | |
|---|---|
| Area of catchment | 1890 km$^2$ (including 66.5 km$^2$ of lake) |
| Ybbarpsån | 89 km$^2$ |
| Bäljaneå | 228 km$^2$ |
| Pinnån | 206 km$^2$ |
| Rössjöholmsån & Kälgeån | 270 km$^2$ |
| Highest point in catchment | 160 m |
| Source | c. 160 m: streams draining the ridge Linderödsåsen |
| Length: total | 120 km |
| downstream of Ringsjön | 83 km |
| Drop in height below Ringsjön | 54 m = 0.54 m km$^{-1}$ |

CLIMATE

| | |
|---|---|
| Precipitation | 700 mm (approx.) |
| Evaporation (mean) | 350 mm |
| Temperature (mean annual) | 7.8 °C |

GEOLOGY

| | |
|---|---|
| (i) Northern side of river | Archaean |
| Southern side of river | younger |
| close to Ringsjön | Silurian |
| Ängelholm | Rhaet-Lias |

(ii) In upper part, within the triangle Ringsjön, Lippan and Hässleholm, about fifty basalt domes some 50 m high - probably remnants of Tertiary volcanoes.

(iii) soils: Moraines with and without clay, sea clay, sand and gravel.

LAND USE

Agricultural fields 570 km$^2$; grazing 45 km$^2$; forest 1200 km$^2$.

POPULATION

Farming 13%; industry 40%; trade 17%; transport 6%; public services 24%.

PHYSICAL FEATURES OF RIVER

<u>Discharge</u>   (Data obtained before 1970; $m^3 s^{-1}$)

| | at Ringsjön | 2 km upstream of Klippan | Klippan, 2 km downstream of Bäljaneå | at Skälderviken | Tributary Bäljaneå at Klippan |
|---|---|---|---|---|---|
| maximum flood | 36 | 77 | 137 | 200 | 41 |
| normal flood | 18 | 48 | 75 | 100 | 21 |
| mean flood | 4.0 | 10.1 | 13.7 | 21.0 | 3.3 |
| normal low flood | 0.9 | 2.2 | 2.5 | 4.5 | 0.25 |
| minimum low flood | 0.4 | 0.16 | 0.75 | 2.0 | 0.07 |

total discharge into bay at Skälderviken $6 \times 10^8 m^3 yr^{-1}$
$$10\text{-}11 \ l \ s^{-1} \ km^{-2}$$

SWEDISH WORDS

Examples of terms frequently encountered in descriptions of Swedish rivers:

| | |
|---|---|
| å | river or small river |
| älv | large river |
| bäck | stream (beck) |
| sjö | lake |
| ås | ridge |
| slätt | plain |
| mosse | bog with peat |
| samhälle | small town |
| mölla | mill |

<u>Word endings</u>

A terminal 'n' after the name of a river means 'the'.  The present chapter mostly prefixes Rönneå with 'the', but adds the terminal 'n' for other rivers and omits 'the'.

## 20.3  PHYSIOGRAPHY AND HYDROLOGY

The Rönneå has the second largest catchment in Skåne.  (The largest
river, Helgeån, has its source in another province, Småland.)  The
Rönneå has its sources in the area east of (lake) Ringsjön and thus
drains the central part of the province.  The small source streams run
mainly through mixed forests, and pollution is negligible, since the
region is thinly populated.  At Hörby the river leaves the Archaean
ridge, passes through cultivated land and receives sewage effluent.
The stream is now called Hörbyån.

The main river begins below Ringsjön.  The bed is about 5 m wide
here, and the river runs in a N-W direction, at first through hilly
forest country, then through an open landscape covered with fields and
meadows.  It runs parallel to the Archaean rock horst Söderåsen.   The
N-E. side of the ridge is steep and the small pure streams flow very
rapidly - sometimes in narrow ravines - down to the Rönneå.  Of the
western tributaries, Bälganeå, rising in the southern-most part of the
ridge, is the largest.  Unlike the others it moves rather slowly and
drains agricultural land.  The greatest number and largest tributaries
however, come from the north-east.  These also have the most water.

In its downstream reaches the Rönneå becomes more and more markedly
a lowland river.  It empties into the sandy bay of Skälderviken 5 km
north of Ängelholm.  The influence of high sea water levels is obser-
vable up to Tranarps Bridge, about 28 km from the mouth.  The Rönneå
runs mainly through soils of moraine and glacial river gravel.  Upstream the soils are clayey, and downstream, clay-poor.  The hilly
landscape of the glacial riverain drifts consists of small ridges and
cavities sometimes filled by lakes.  It is often very beautiful and
excellent recreation ground.  The Counties therefore want these areas
to be protected against damage by housing schemes and deforestation.
In the downstream part of the main valley, up to Ljungbyhed, a glacial
sea bay has deposited varved clay and sand.  From Ängelholm to the
mouth the river passes through a post-glacial sand area.  The Ängelholm plain, built up of the glacial sea sediments, is very flat.  It
rises slowly from the sandy dune coast up to 40 m above sea-level at
Ljungbyhed (c. 35 km).  The Rönneå has often cut out 10 - 20 m deep
ravines in the plain.

The rapid changes between frost and mild weather in winter result
in frequent changes in type of precipitation.  The amount of discharge
therefore varies, although it remains generally high throughout this
season.  Usually there is sufficient snow to give rise to an obvious
contribution from snow-melt in March-April.  In summer flows are
usually low.  If rainfall is insignificant in autumn, the low flows
continue, and their total can extend for a very long time, e.g. until
December in 1975.  On the other hand, the rainfall can be very high
in autumn and winter and then there is extensive flooding, especially
where the side of the river is flat.  Flooded fields can cause the
farmers serious problems.  On a stretch of about 20 km between Höja

and Bäljaneå, as much as 750 ha have been inundated. As a consequence
of these floods alluvial clay is sedimented in the valley. The continuous regulation of discharge at power plants does, however, lessen
some of the effects of the floods.

For the main river the following discharges are characteristic:

| Level | below Ringsjön ($m^3$ $s^{-1}$) | Skälder- viken ($m^3$ $s^{-1}$) |
|---|---|---|
| maximum flood | 36 | 200 |
| normal flood | 18 | 100 |
| mean | 4.0 | 21.0 |
| normal low | 0.9 | 4.5 |
| minimum low | 0.4 | 2.0 |

## 20.4  RINGSJÖN AND ITS TRIBUTARIES

The Ringsjön is by far the largest lake within the catchment of the
Rönneå and will be described in some detail. It consists of two units,
E. Ringsjön with Sätoftasjön and W. Ringsjön (Ryding, 1983b).

| | area ($km^2$) | max. depth (m) | vol. ($x10^6$ $m^3$) | mean depth (m) | alti- tude (m) | drain- age ($km^2$) |
|---|---|---|---|---|---|---|
| E. Ringsjön | 20.8 | 16 | 104 | 5 | 53.7 | 325 |
| Sätoftasjön | 4.5 | 18 | 10.4 | 2.3 | 53.7 | 105 |
| W. Ringsjön | 15.4 | 6 | 52 | 3.4 | 53.7 | 347 |

History. The land around the Ringsjön became populated early. Archaeo-
logical investigations show that an extensive settlement occurred along
its shores and on swampy islands and also around the prehistoric lakes
Agerödsmosse and Rönneholms mosse to the west (Bunte, 1976; Eskeröd,
1976). The abundance of fish, birds and good hunting grounds first
attracted the nomads; subsequently there came the megalithic farmers.
When the Ringsjön was lowered in the 1880s, a number of Stone Age sites
around the lake were laid bare. The Rönneå was, in prehistoric time,
navigable for most of its course. The falls at Djupadal, Forsmöllan
and Klippan were probably no hindrance to seamen wishing to reach the
Ringsjön.

Geology and soils. Ringsjön is on the border between Archaean gneiss,
Rhaet-lias sediments (mainly sandstones) and Cambrosilurian shales.
The surrounding soils are to the north composed of Archaean rock moraine,
to the south of clayey moraine. Clays occur in many places at the
shore line. The agricultural ground round Ringsjön is extensive (60 -
70%) except north of the lake, where only 30 - 40% of the land consists
of fields.

Tributaries. Three larger tributaries empty into Ringsjön. Hörbyån, considered to be the source of the Rönneå, flows into E. Ringsjön, Ludvigsborgsån and Höörsån into Sätoftasjön. All have their sources in and flow through mixed woods and arable land. The water is therefore more or less polluted when it reaches Ringsjön. Until a few years ago Höörsån received the sewage effluent of the urban district of Höör (c. 5000 population equivalents). Hörbyån still receives sewage effluent from the town of Hörby. Even after biological and chemical purification of this effluent the bacterial content of the river is still so high that bathing in the Ringsjön is sometimes prohibited near the entry of Hörbyån.

Water quality. The quality of Ringsjön water has been studied more or less continuously since 1940 and obvious changes have taken place during this period. The maximum pH has increased from 8 to above 9 in summer and both alkalinity and conductivity are generally higher. The surface water is very often supersaturated with oxygen in summer, while in the hypolimnion the oxygen content is reduced. When the lake is covered in ice, the water at the bottom can be deficient in oxygen. The situation with respect to dissolved oxygen is better in the shallow W. Ringsjön than in the other parts of the lake. In 1978 the average total P concentration was about 0.15 mg $l^{-1}$ for the whole lake, thus indicating a clear eutrophic-hypereutrophic state. Phosphorus seems to be supplied also from the sediments. The P budget is now the subject of a comprehensive study, with the aim to elucidate the eutrophication process and to propose methods for reducing blooms (Ryding, 1983a).

Vegetation. Ringsjön provides a suitable environment for fringing vegetation. It is relatively shallow, the shores slope gradually and, especially with W. Ringsjön, the sediments are calcareous. Grazing still occurs around the lake and consequently the shore vegetation is nowhere intact. The level of Ringsjön was lowered in 1883 by 1.5 m, reducing the area of the lake by 3.4 $km^2$. New reed beds have developed, but the area overgrown is still small in relation to the total area of the lake. Phragmites australis is the most abundant component (Fig. 20.2). The floating leaf vegetation is insignificant, whereas the isoetid vegetation forms dense mats. This is more or less true for all three parts of the lake. In the 1950s the larger part of the lake bottom in W. Ringsjön was covered with submerged plants, the most abundant of which were charophytes, Myriophyllum alterniflorum, Naias flexilis, Ceratophyllum demersum and Elodea canadensis (Almestrand & Lundh, 1951). Nowadays, the submerged macrophytes have diminished, probably as a result of reduced illumination owing to the much increased phytoplankton crops. A notable finding as late as 1950 was a locality for Isoetes lacustris (Lundh, 1951b), a species belonging normally to non-humic oligotrophic lakes and thus perhaps an indicator of a former more oligotrophic condition in Ringsjön.

Plankton. The earliest detailed investigation of the Ringsjön plankton was made in the 1940s (Lundh, 1951a). In April 1949 Asterionella formosa was dominant and Melosira granulata frequent in E. Ringsjön;

*Fig. 20.2* W. Ringsjön at the outflow of Rönneå, showing the broad, dense reeds of *Phragmites* along the shore-line. The background shows a famous beech wood on the southern shore. May 1980.

in September *Ceratium hirundinella* was dominant and *Aphanizomenon flos-aquae* and *Melosira granulata var. angustissima* very frequent. In W. Ringsjön *Dinobryon sociale* dominated in April 1949, while *Asterionella, Melosira granulata* and *Synedra acus var. angustissima* were frequent. In September *Fragilaria crotonensis* was the dominant and *Melosira granulata var. angustissima* the next most frequent. Ringsjön could thus, by this time, be considered a more or less typical lowland lake when compared to other Scanian lakes (Lundh, 1951b).

Some sparse data from older literature (Trybom, 1893) may indicate that Ringsjön contained more oligotrophic species before the lowering of the water level in 1883. Lundh-Almestrand (1954) concluded that the relatively eutrophic species, *Melosira granulata* had increased since the end of the nineteenth century. It is, however, hardly likely that Ringsjön has ever been an extremely oligotrophic lake because of the geological conditions and its position in an ancient agricultural area. Changes in the species composition and abundance of the plankton have occurred very rapidly since the 1950s. During the middle part of the sixties large crops of *Aphanizomenon flos-aquae* were noted. Other blue-green algae species have appeared in tremendous quantities (water bloom) since the end of the 1970s, especially *Microcystis* species (Cronberg, 1983).

The accelerating eutrophication in recent decades is due especially to the increasing load of more or less purified sewage effluent and rapid increases in pollution from farm fertilizers. 75% of N and P comes from farming (Ryding, 1983a). Problems are also created by

recreational settlements, camping, boating and waterski-ing.  Ringsjön
was one of the Scanian lakes that was "colonized" by summer guests
particularly early, already by the turn of the century.  Everything so
far indicates that excess phosphorus is the key factor and the sewage
treatment plants of Höör and Hörby include chemical treatment to reduce
the phosphorus content.  Nevertheless, it cannot be said that the lake
has recovered yet - rather the reverse.

It is of great importance to try to stop the eutrophication process
as far as technology and economy will allow.  The detrimental changes
in Ringsjön influence not only aesthetic and recreational quality but
also fishing and grazing:  cattle have died after drinking lake water.
Of still greater importance is the fact that the lake serves as a
source of water for a large region including the towns of Lund, Eslöv,
Landskrona and Helsingborg.  The water is treated by microstraining,
chemical precipitation and filtration.


## 20.5  THE MAIN RIVER

### 20.51  CHANGES ON PASSING DOWNSTREAM

The total fall from Ringsjön to the sea is 53.7 m, 70% of which is
concentrated in a few rapids.  Much of the river has, therefore, slow
current speeds and in parts an almost lenitic character.  The main
Rönneå comes from W. Ringsjön, where a dam regulates the water level
of the lake.  The river then flows through a flat landscape between
two large bogs, in one of which peat-cutting is still going on; part
is used also as a site for the deposition of sludge from Eslöv treatment plant.

At Hasslebro summerhouses have been built along the riverside,
making access to the water impossible for anyone but the owner.  A
recent law now prohibits any more private houses so close to the
water's edge.  From Hasslebro the river passes over to a lotic section
(Stockamöllan) and the current speeds are high.  Where the river forms
rapids, even small ones, water power has been used in the past for
mills (Sw. möllor).  Most now no longer operate, but remnants are
still preserved.  The old mills have become parking places for nature
lovers.  At Stockamöllan a well-known truck industry has replaced the
old mill.

The next part of the valley offers a particularly beautiful natural
landscape.  The slopes are covered with forests of various types.  The
valley is narrow but sometimes there is a strip of swampy meadow along
the bank which is flooded in rainy seasons.  Canoeists love this quiet
landscape without houses and other tourists, where there is an opportunity to observe the rich undisturbed wild life.

Upstream of the mill-pond at Djupadal, the river Snällerödsbäcken

enters the Rönneå.  It drains, among other waters, the small lake Dagstorpssjön, which is surrounded by forests of beech and planted spruce
in a hilly terrain.  The landscape is appreciated as recreation ground
and it is possible to swim in the lake, which is said to have the
"purest water" in Skåne.  The plankton, which can be classed as meso-
trophic, does not seem to have changed much during the last thirty
years.

Apart from the vicinity of the outflow of Ringsjön, the aquatic
macrophytic vegetation downstream to Djupadal is relatively sparse.
This may be due partly to the high turbidity of the water during the
growing season as a result of phytoplankton originating from Ringsjön.
In the mill-pond at Djupadal, however, there are stands of *Glyceria
maxima*, and downstream from the bridge high stands of *Phalaris arundinacea* and *Phragmites australis*.  On stones at the bottom of the river
there are abundant growths of *Fontinalis antipyretica*, *Hildenbrandia*,
*Lemanea* and *Cladophora*.  The volcanic tufa found at this site is
perhaps responsible for the rich vegetation.

Leaving Djupadal, where the bed is very narrow, the river changes
direction towards the north-west.  It is fringed by marshy meadows and
surrounded by deciduous forests.  The valley is deep and the river
moves slowly in a meandering course.  At Ö. Forestad the river touches
a big esker, Färingtoftaåsen, and receives from the east the stream
Klingstorpabäcken (Färingtoftaån).  This arises in a lake-rich area of
gravel ridges and cavities.  The largest lake is Store Damm, which like
the other lakes in this area has peaty shores.  These lakes are known
for their rich bird life.  Aquatic birds, such as the whooper, both
rest and breed here.  The surface is partly covered with water lilies
mixed with *Equisetum fluviatile* and other species.  From the west
comes Bälganeå, which receives the sewage effluent of Röstånga
municipality, and also drains a big rural district.  The water of
this river has the highest conductivity (500 - 550 µS cm$^{-1}$) of the
whole Ronneå catchment.

Further downstream, at Riseberga, the Rönneå becomes wider and the
landscape flattens.  The riverside meadows broaden, becoming extensive
flooded areas during the rainy seasons.  At the old inn Spången the
river is 25 m wide and filled completely with *Glyceria maxima*
and *Scirpus lacustris*.     Within Ljungbyhed the current is rapid, but
later slows down and at the castle Herrevadskloster the river and its
surroundings are said to make a "pastoral idyll".  Here the river,
fringed by alders, winds along through grazing grounds with scattered
oaks, giving shelter for the cattle.  At the castle, once a monastery
in the Middle Ages, Ybbarpsån flows into the Rönneå.  The monks used
to use its water power.

A couple of kilometres downstream the river is transformed into a
long and narrow lake, the reservoir of Forsmöllan, which is owned by
Sydkraft AB and used for hydro-electricity.  The fall is 5 m.  The
reservoir and the rapids are popular places for excursions and anglers.

Between Forsmöllan and the next fall, situated about 2 km downstream at
Klippan, the current is very fast and the river falls 8 m.  In the
municipality of Klippan is the oldest industry of Skåne, the paper-mill
Klippans Finpappersbruk AB.

The Rönneå is then dammed again, forming a long and narrow lake;
this is the Stackarp reservoir (Fig. 20.3), which also has a hydro-
electric plant (6 m fall).  The reservoir is partly surrounded by marshy
ground overgrown with *Glyceria maxima*.  The shores are intensively
grazed.  Submerged macrophytes include *Elodea canadensis*, *Potamogeton
crispus* and filamentous algae (especially *Vaucheria*, *Microspora*,
*Spirogyra* and *Oedogonium*).  The water is strongly turbid with clay
particles and during summer also phytoplankton.

*Fig. 20.3*  Stackarp reservoir downstream from Klippan paper-mill.
The clayey shores are trampled heavily by cattle.  Saprobic vegetation can sometimes be observed here.  August 1979.


From Stackarp the Rönneå flows through a fertile open agricultural
district towards Ängelholm, sometimes like a broad ditch between fields.
Near Stackarp the river is bordered with alders.  The marshy land
adjacent is mostly used for grazing and is floristically rich.  It is
flooded at high water levels.  The river winds through Ängelholm in a
wide and deep bed fringed with a park-like vegetation.  Along the
remaining stretch down to the mouth there are forests of alder, beech
and pine.  The bed of the river is more or less overgrown with reeds,
which provide a favourable habitat for birds.  At its mouth the Rönneå
divides into two branches forming islands.  Brackish water can be
pushed upstream several kilometres.

## 20.52   USAGE OF THE WATER AND WATER QUALITY

The Rönneå passes through districts with various cultural traditions and its water has been exploited for different purposes.  Since time immemorial the river has surely been used for fishing, transport and water power.  Today, however, its principal importance lies in its utilization as a receiving body for polluted water.  A tragic fate!

Only two densely populated areas, Ljungbyhed (military airport) and Ängelholm, put treated sewage effluent directly into the river.  These include about 26000 people, since small densely populated areas in the vicinity are included.  The town of Ängelholm has, among other industries, a tannery which discharges chrome-rich waste water.

The reservoir at Stackarp also receives waste effluent, in this case from Klippans Pappersbruk, which produces annually about 30000 tonnes of paper.  Sufficient dilution of the waste water is made possible because of a law stipulating that Ringsjön is regulated to secure a minimum discharge of 2.5 $m^3$ $s^{-1}$ at Forsmöllan.  In addition, the factory has had to reduce fibre discharge by providing the paper machines with fibre regainers.  Finally the waste water is treated by sedimentation prior to its release to the Rönneå.

The water of the main river is influenced also by the quality of the tributaries.  Table 20.1 presents some results for 1980, based on monthly samples.  In general differences between minimum and maximum were extremely large in 1980, because of the high precipitation in June and October-December; total precipitation was, in spite of a dry spring, 40% higher than average.  Therefore, the analyses provide a good indication of seasonal variations in water quality.

Nowhere in the main river has the water anything like a natural character.  The water discharged from Ringsjön is already greatly affected by eutrophication processes in the lake.  During May-October the water leaving Ringsjön is extremely turbid due to the phytoplankton, and with pH values over 9.  The turbidity (up to 17 FTU-formazin turbidity units) is visible down to the mouth.  It is not aesthetically attractive.  At the time of year when the plankton is not so dense, there is an increase in turbidity in the lower river due to erosion and surface run-off from fields.  Thus, the lower stretch of the Rönneå is turbid throughout the year.  The same holds true for water colour.  The Ringsjön water is generally weakly coloured (equivalent $c$. 40 - 60 mg $l^{-1}$ Pt; in 1980, 15 - 80 mg $l^{-1}$), but the colour increases more and more downstream on account of the supply of humic water from the tributaries, especially Baljaneå and Pinnån (Table 20.1). In the rainy autumn of 1980 the water was intensively brown at the mouth (200 mg $l^{-1}$ Pt) upstream of Ängelholm.  The Rönneå water coming from Ringsjön has relatively high concentrations of total phosphorus (up to 0.3 mg $l^{-1}$ P) and total combined nitrogen (4.4 mg $l^{-1}$ N).  While phosphorus usually decreases slightly downstream, nitrogen increases. Thanks to the present effective purification of the sewage effluents,

Table 20.1 Chemistry of water in Rönneå and tributaries: data show the range found during monthly samplings during 1980. Analyses from tributaries refer to a station just above the entry to the main river.

| site | | pH | conductivity µS cm⁻¹ | alkalinity meq l⁻¹ | turbidity FTU | colour (as mg l⁻¹ Pt) | O₂ mg l⁻¹ | BOD₇ mg l⁻¹ | NH₄-N µg l⁻¹ | NO₃-N µg l⁻¹ | total-N µg l⁻¹ | total-P µg l⁻¹ |
|---|---|---|---|---|---|---|---|---|---|---|---|---|
| **Main river** | | | | | | | | | | | | |
| Rönneå, below W. Ringsjön | min. | 7.40 | 261 | 0.26 | 1.8 | 15 | 8.10 | 2.4 | 6 | 1 | 710 | 76 |
| | max. | 9.05 | 384 | 1.76 | 22 | 80 | 20.90 | 9.0 | 1110 | 1400 | 4400 | 294 |
| Rönneå, above Skälderviken | min. | 7.15 | 235 | 0.57 | 3.8 | 50 | 7.15 | 2.8 | 126 | 1200 | 3200 | 55 |
| | max. | 7.65 | 371 | 1.02 | 17 | 160 | 13.10 | 7.9 | 1590 | 3000 | 6500 | 210 |
| **Tributaries** | | | | | | | | | | | | |
| Bålganeå | min. | 7.30 | 365 | 0.43 | 2.8 | 40 | 8.15 | 2.2 | 29 | 1600 | 2800 | 37 |
| | max. | 7.65 | 469 | 2.29 | 4.6 | 60 | 11.90 | 4.6 | 1050 | 3500 | 7500 | 106 |
| Ybbarpsån | min. | 6.60 | 171 | 0.15 | 2.5 | 70 | 7.40 | 1.6 | 34 | 160 | 1400 | 32 |
| | max. | 7.50 | 344 | 0.81 | 9.5 | 180 | 11.70 | 6.7 | 470 | 1200 | 2600 | 140 |
| Bäljaneå | min. | 6.85 | 140 | 0.22 | 2.7 | 50 | 6.90 | 2.4 | 343 | 590 | 1900 | 44 |
| | max. | 7.35 | 329 | 1.00 | 18 | 400 | 13.50 | 9.5 | 3175 | 1600 | 5700 | 486 |
| Pinnån | min. | 6.70 | 140 | 0.18 | 3.0 | 60 | 7.50 | 3.6 | 485 | 1200 | 2700 | 80 |
| | max. | 7.15 | 266 | 0.30 | 7.6 | 250 | 13.15 | 9.4 | 2280 | 2790 | 6100 | 254 |
| Rössjöholmsån-Kägleån | min. | 7.10 | 186 | 0.52 | 2.0 | 50 | 9.10 | 1.4 | 22 | 1200 | 1900 | 47 |
| | max. | 8.10 | 239 | 0.89 | 7.0 | 200 | 13.75 | 3.1 | 416 | 2000 | 3200 | 101 |

there are seldom problems due to reduced oxygen concentrations.  The
waste water and surface run-off from Ängelholm have led to increased
heavy metal contents in the sediments, particularly in chromium and
lead.

As mentioned in 20.4, Ringsjön is used as a source of water for
several densely populated areas.   Klippans Pappersbruk takes in river
water for industrial usage.  Rönneå water is also important for the
irrigation of fields, which is increasing, and for watering cattle.
There are three large hydro-electric power plants on the main river,
as described above.  Most of the numerous water mills are no longer
functioning.

Ringsjön is vital for the fishing trade in the province, while
amateur angling is rewarding and contributes one/third of the total
catch.  The conditions for summer anglers have worsened in recent
years because of the enormous plankton crops.  Jigging for fish in
winter is more profitable.  Valuable species in Ringsjön include
eel, pike, perch and bream.  Since the eel cannot migrate up the Rönneå,
a yearly input is needed.  The one-time lucrative fishing for whitefish has lessened appreciably, probably another effect of the eutrophication.  At present there is serious discussion about the possibility of improving the water quality by selective fishing for bream.
A decrease in the present bream population ought to result in an
increase in the zooplankton density and hence decrease in the phytoplankton.  Economically unimportant fish have been used as fish food for
large-scale basin-culture of rainbow trout in the neighbourhood.  This
was situated close to the Höör waste water treatment plant, which provided heat energy to warm the basins (now used for algal cultures).  The
saprobic conditions which occurred at one time in Hörbyån and Höörsån
have disappeared once improvements in waste treatment started in 1975
and 1978, respectively.  The environment for fish will probably improve
in the long run.  Electric fishing in 1970-1972 for trout revealed
only a small population.  E. and W. Ringsjön have been popular bathing
places for a long time.  However, at present the water-bloom acts as a
deterrent.  In addition, bathing along the Hörby shore of E. Ringsjön
is often prohibited.

In the main river Rönneå, perch and pike can be caught, and, near
the mouth, also trout.  The migration of salmon is prevented at
Stackarp.  Around the upper reaches of the Rönneå there are a great
number of put-and-take waters.  In recent years the Rönneå has become
the favourite canoe-river in Skåne and the number of canoes is
increasing more and more.  A positive consequence of the enormous
interest in recreational activities in and near the river is that the
authorities have sharpened their watchfulness.  The river is subject
to continuous checking of public activities, and any deterioration
in water quality should be registered rapidly.  The countryside of
Skåne can be damaged easily if visited by too many people, such as by
trampling, discarding litter or disturbing of bird life.

20.6   THE LARGEST TRIBUTARIES

## 20.61   YBBARPSÅN

This river drains a series of shallow dammed-up, pond-like lakes in a
dead ice landscape south of Perstorp.  The lakes serve as regulating
reservoirs for Perstorp  AB, a chemical industry and one of the most
important in the world for production of laminates, plastics, formal-
dehyde and polyalcohols.  Ybbarpsån is the receiving body for the
waste effluent, which flows into the river two kilometres downstream
from the factories.  The Ybbarpsån soon widens to a long and narrow
reservoir, Storarydsdammen, built in 1960 to enhance the self-purification of the water.  About 7 km south-west of the factories the river
meets the 5 m deep E. Sorrödssjön, from which a canal cut by monks in
the Middle Ages carries the water to the very shallow pond-like W.
Sorrödssjön.  The Ybbarpsån flows into Rönneå at the castle Herrevadskloster.

Perstorp AB started just over a hundred years ago and by the end of
the 1940s had become a large industry.  The original product was acetic
acid made by dry distillation of beech wood.  Owing to rapid increase
in production, the industry became an environmental problem.  It caused
the first great environmental debate in Skåne (Lillieroth, 1949).  The
unfavourable effects on both water and air were discussed in the press
and journals.  In the 1940s the authorities still gave priority to
economic interests.  Although the water of the river and the lakes was
terribly polluted and directly poisonous due to phenol, it took some
decades to get the water restored.  The industry was at last able to
reduce the discharge of phenol by internal measures, and in 1971 an
effective treatment plant was built.  Only one year after the start of
this plant, it was possible to observe green algae in the water rather
than sewage-fungus, and a "normal" plankton developed in the two
eutrophic lakes.  The earlier good fishing conditions have partially
been restored and the lakes are again attractive to birds.  Many of
the substances produced are difficult to break down biologically and
often are also toxic.  Therefore, there is still a risk of disturbance
in the active sludge process and reduced efficiency in treatment.  In
such circumstances, purification is completed in the river and an
obvious oxygen decrease can occur, primarily in Storarydsdammen.

## 20.62   BÄLJANEÅ-PERSTORPSBÄCKEN

Bäljaneå, the outlet from the lake Bälingesjön, flows into the Rönneå
some kilometres west of the municipality of Klippan, and by that time
has received two smaller tributaries, Perstorpsbäcken and Smålarps-
bäcken.  Bälingesjön is an example of the results of the numerous
lake-lowering projects carried out in Sweden at the end of the nineteenth and the beginning of the twentieth century.  Some of these
projects were done without permission from the authorities.  The
lowering was not successful.  The ground gained was rich in boulders

partly covered with bog ore and unsuitable for the farmers.  The only
benefit was a football ground on a sandy shore.  The important lasting
result of the lowering appears to be a slight eutrophication, with
increased reed beds and plankton development.  However, a recent
comparison of plankton composition with that in the 1940s, when
detailed studies were made (Lillieroth, 1950) shows that the eutrophication process has remained moderate.  The lake played a valuable role
in increasing swimming possibilities for the people living there, far
from the coast, during the car-free part of this century.  At present
it serves as a place for swimming instruction, boating and fishing.
Perch, pike and pike-perch can be caught.

The Bäljaneå runs for most of its course through bog-rich forested
districts, including large spruce plantations.  Cultivated land is only
sparse.  The river water is therefore very brown, especially in rainy
periods.  Current speeds are very high in some places and small rapids
are formed.  In the eastern part of Klippan the bottom is stony and
shaded by alder and willow; the fern *Osmunda regalis* occurs here.  The
shaded stones are overgrown with aquatic mosses (e.g. *Fontinalis
antipyretica*) and liverworts.

Via Perstorpsbäcken, the Bäljaneå is the receiving body of water for
the municipality of Perstorp, and at Klippan it receives sewage
effluent following biological purification.  The waste water from the
tannery at Klippan carries elevated chromium, leading to chromium
accumulation in the sediments.  The purification of the polluted water
in Klippan is unsatisfactory, leading at times to marked decreases in
dissolved oxygen in the river and, at low dilution, also to very high
nitrogen and phosphorus contents.

Current speeds are less in the lowermost part of the river near its
entry to the Rönneå and there is a rich vegetation of, for example,
*Scirpus lacustris*, *Sparganium* and *Nuphar luteum*.  Salmon migration
occurs to a certain extent in the Bäljaneå.


## 20.63  PINNÅN

Pinnån, the most attractive tributary, is the outlet of Åsljungasjön
and flows via Hjälmsjön down to the Rönneå.  Åsljungasjön has been
lowered twice, both times without success.  Hjälmsjön is a well-
frequented lake, with many large rowing and canoe competitions in
summer.  The river Pinnån in its uppermost course goes through vast
forests with cultivated fields only close to the river.  Downstream
of Örkelljunga it flows through large glacial, fluvial deposits
including ridges, plateaus and terraces - a suitable recreation area.
At Stidsvig the river is dammed up to form a small lake, Kopparmöllesjön
(see below).  The Pinnån then comes to a fertile agricultural region,
where it winds between borders of alder and ash.  There are several
rapids, previously utilized for milling and nowadays for hydro-electric
power.  At Skvattemölla the water is used for fish ponds.  The largest

waterfall is at Stora Mölla (power plant).  At times of high flow the
excess water cascades over the rocks in the ravine and the spray glit-
ters like a rainbow.  This waterfall is unique in Skåne.

The course of Pinnån through grazing land down to the Rönneå is tran-
quil.  From the high valley slope at Stora Mölla there is a wonderful
panorama of the mouth area of Pinnån.  At spring high water there is an
awe-inspiring stretch of lake caused by the extensive flooding of the
two rivers over fields and grazing area.  The downstream part of Pinnån
is idyllic scenery, reminiscent of an English "park landscape".  It is a
pity, therefore,  that the river is exposed to industrial waste effluent
at Stidsvig from a large factory producing glue, fats, edible gelatin
and bone meal (Extraco AB).  The effluent has a high content of organic
matter, nitrogen and phosphorus.  It has led to a complicated river protection problem for a long time, since low flows in Pinnån can extend
for long periods.  Before 1966 the effluent was untreated, resulting in
a saprobic vegetation (*Sphaerotilus natans* and *Carchesium polypinum*) and
anaerobic conditions in Kopparmöllesjön and the river downstream, with
accompanying fish mortality.  The problem was compounded by discharges
of cooling water.  Thick layers of black anaerobic sediment accumulated
in the lake and in the river downstream large quantities of *Vaucheria*
grew.  Since 1966 the effluent has been treated in an activated sludge
plant and this has partially improved the state of the river.  The
growth of saprobic organisms has decreased and oxygen conditions improved.  Nevertheless, during summer the oxygen concentrations can still
become critical, because of the nitrification of ammonium (4 - 6 mg $l^{-1}$
N).  Fish kills have still occurred.  After installing equipment for
reduction of nitrogen discharge in 1982, conditions in the river have
stabilized and the future development looks promising.

Since 1967 Kopparmöllesjön has been invaded by an extremely dense
vegetation of *Ceratophyllum submersum, Potamogeton obtusifolius* and
*Callitriche*.  Downstream from the lake the river is filled with *Callitriche* and filamentous algae (*Vaucheria, Microspora, Tribonema, Oedogonium*).  A luxuriant growth of algae, often thickly covered with small
diatoms and mosses (*Fontinalis, Drepanocladus*), is characteristic of
Pinnån down to its mouth.  Salmon migrate up to Stora Mölla, where
the fall causes a natural hindrance for further movement upstream.


20.64  RÖSSJÖHOLMSÅN-KÄGLEÅN

Rössjöholmsån has the largest drainage basin of the tributaries of the
Rönneå.  The source area consists of the south side of the ridge
Hallandsåsen, including the lakes Västersjön and Rössjön, situated
along a fault which once gave rise to the horst.  Rössjön (3.6 km$^2$) is
the eastern lake.  Västersjön is more extensive, but not as deep.  The
shores are mostly minerogenic, often rich in boulders.  The littoral
vegetion is scant and that characteristic of oligotrophic *Lobelia*
lakes.  In Västersjön there are more or less dense beds of reeds in
shallow parts, with *Scirpus lacustris* the dominant.  These lakes occur

in a district which is very sparsely populated and have so far not been subject to serious pollution.  The water has a relatively low colour (20 - 40 mg l$^{-1}$ Pt) and Secchi disc transparency is about 2.5 m in Västersjön and 3.5 m in Rössjön; alkalinity is below 0.1 mmol, so the lakes may be susceptible to increasing acidification in the immediate future.

Västersjön is strongly exploited for recreational activities (Fig. 20.4).  Boating is especially disturbing and too many summerhouses have been built in city-like aggregates.  The idyllic bathing-places where the "natives" could sit on the grass are gone and the risk of serious pollution has become real.

*Fig. 20.4*  Västersjön on a summer Sunday: sailing, waterski-ing and sometimes car-washing!  August 1978.

While Västersjön in summer time echoes with the sound of motor engines, Rössjön is still quiet, since all boat traffic is prohibited. This lake has for a long time been protected by the owner of the Rössjöholm castle, and is now a "protected area for birds".  Apart from the castle the shores are quite free from settlement.  They are well-frequented by nature lovers, especially in spring and autumn.  The northern shores of Rössjön are more lobated than those of Västersjön and offer suitable picnic sites.  The rivulets from the ridge sometimes flow in impressive ravines, which served as effective hiding places for guerrillas during the Swedish-Danish war in the seventeenth century.

Västersjön and Rössjön can be classified as oligotrophic-mesotrophic lakes and total phosphorus does not exceed 0.015 mg l$^{-1}$ P.  The dominant phytoplankton species are *Gomphosphaeria naegeliana, Tabellaria*

and *Asterionella*. The blue-green alga brings about a plankton turbidity
of long duration. In Västersjön one can catch pike, perch, trout, bream,
vandace and eel. A fish pond system south of Rössjön, is used for
breeding pike spawn and to some extent for growing carp. The water discharged to the outlet from the two lakes is only slightly humic and on
the whole of good quality. The supply of water from the bogs south of
the lakes makes the river more brown and iron-rich, which is unfavourable for fish production (rainbow trout) in the large fish ponds downstream (Munka Ljungby). The low water level in summer is also detrimental.

Rössjöholmsån is at present the source of water supply for Ängelholm
community. The intake is situated in the lower part of the river on
the plain. The drinking water is taken from artificial ground water.
The varying quality of the river water (at times high contents of iron,
suspended matter and nitrate) creates problems in the filtration plant.
Ängelholm would therefore like to take water directly from Rössjön, at
about 12 m depth. The lake is already regulated with an amplitude of
1.4 m, which is rather wide with regard to the shore configuration.
The planned regulation would lead to an elevation of the present minimum
water level. This reduction ought to mean a positive ecological effect.
The proposed regulation would also affect the discharge variation in
the river. It seems that it might be advantageous to the fish ponds
in Munka Ljungby in extremely dry periods, since the release from
Rössjön will be limited under such conditions. According to the
present stipulation zero release is possible. Permission for the
abstraction of water from Rössjön has recently been approved by the
authorities. Rössjöholmsån is also essential as a reproduction area
for the salmon and trout populations in the water system of Rönneå.

During rainy seasons the lower part of Rössjöholmsån overflows as
is the case in the main river and in Pinnån. About 2 km from the mouth
the river is supplied with water from Kägleån, which changes the water
quality of Rössjöholmsån considerably. It is a lowland river with high
concentrations of nitrogen and phosphorus.


20.7   CONCLUDING COMMENTS

The River Rönneå has become "a problem child", according to an author in
Skånes Natur (Holm, 1976). Unfortunately a similar fate has happened
to other rivers in Skåne. However, it has now become apparent to the
authorities that drastic preservation measures must be taken. Areas
which are still fairly intact will be protected. Often the problem
is one of "wear" caused by too many people moving around within a
limited area. The Västersjön-Rössjön district is one such example
which requires protection.

Pollution connected with densely populated areas seemed at one time
to pose very complicated problems, but these have now been attacked

*Fig. 20.5* Rönneå. *Upper* Billinge, 10 km from W. Ringsjön, where
the shallow river is fringed by intensively grazed meadows and
canoeing usually starts. *Lower* South of Djupadal, 17 km from W.
Ringsjön, where the river meanders through marshy meadows. (Scandia
Photopress AB)

with success by constructing highly effective waste treatment plants. Diffuse pollution, however, still remains to be tackled. The crucial situation for water protection is during long dry periods, when the effects of eutrophication are particularly obvious on many of the rivers and lakes of the Rönneå system. Water levels must be increased. Steps required include the reduction of the present rate of abstraction for drinking water, regulation to retain more water in rainy periods and the transfer of water from other drainage basins.

There is reason to set great hopes on the "Bolmen project". This is the plan by nine communities in Skåne to take "hygienically not objectionable" water from Bolmen, a large pure lake in Småland. A tunnel from Bolmen, more than half-finished, is to connect with the catchments of Ringsjön and Vombsjön. This would make it possible to increase the quantity of water in Ringsjön and also to raise low flows in the Rönneå by direct input of Bolmen water. Whether this project will be brought to fruition seems to be a question of "profitability".

## REFERENCES

### 1.    Published

Almestrand A. & Lundh A. (1951)  Studies on the vegetation and hydro-
  chemistry of Scanian lakes.  I-II.  *Bot. Not., Suppl.* 2, 3.
Bunte C. (1976)  Dokument från det förgångna.  *Skånes Natur* 63, 27-40.
  (Documents from the past.)
Cronberg G. (1983)  Förändringar i Ringsjöarnas växtplankton under 1900-
  talet.  CODEN LUNBDS/(NBLI-3053)/1-57 (1983).  ISSN 0348-0798.
  (Changes in  the phytoplankton of Ringsjön during the twentieth
  century.)
Eskerod A. (1976)  Glimtar från Ringsjön.  *Skånes Natur* 63, 7-16.
  (Glimpses from Ringsjön.)
Holm H. (1976)  Naturvårdssyn på Ringsjöbygd.  *Skånes Natur* 63, 117-26.
  (Aspects on environment protection in the countryside of Ringsjön.)
(Kristianstads och Malmöhus) (1979)  *Vägen till naturen i Skåne.*  ISBN
  91-38-04171-5.  Utg. av länsstyrelserna i Kristianstads och
  Malmöhus län.  (*The Way to Nature in Skåne.*  Published by Counties of
  Kiristianstad and Malmöhus.)
Lillieroth S. (1949)  Om ogynnsamma följder av sjösänkning och vatten-
  förorening i nordvästra Skåne.  *Skånes Natur* 36, 6-48.  (Unfavourable
  consequences of the lowering of lakes and water pollution in N-W.
  Skåne.)
Lillieroth S. (1950)  Über Folgen kulturbedingter Wasserstandsenkungen
  für Macrophyten- und Planktongemeinschaften in seichten Seen des
  südschwedischen Oligotrophiegebietes. *Acta Limnologica* 3.  (English
  summary.)
Lundegårdh P.H., Lundqvist J. & Lindström M. (1964)  *Berg och jord i
  Sverige.*  Stockholm. (*Rock and Soil in Sweden.*)
Lundh A. (1951a)  Some aspects of the higher aquatic vegetation in the
  lake Ringsjön in Scania.  *Bot. Not.* 1951, 21-31.

Lundh A. (1951b)  Studies on the vegetation and hydrochemistry of
    Scanian lakes.  III.  Distribution of macrophytes and some algal
    groups.  *Bot. Not. Suppl.* 3, 1.
Lundh-Almestrand A. (1954)  *Melosira islandica* and *M. granulata* in the
    Scanian lake Ringsjön.  *Svensk bot. Tidskr.* 48, 491-95.
Statens Naturvårdsverk (1980)  Undersökningar i Rönneåns avrinnings-
    område 1977.  PM 1336.  (Investigations in the catchment of the
    Rönneå 1977.  Published by National Swedish Environmental Protection
    Board.)
Trybom F. (1893)  Ringsjön i Malmöhus län, dess naturförhållanden och
    fiske.  *Medd. Kongl. Landtbruksstyr. N:4.* Stockholm.  (Ringsjön in
    Malmöhus county, its nature and fisheries.)


2.  Duplicated reports


Almestrand A. (1976)  EXTRACO AB.  Redogörelse för undersökning av sedi-
    mentförhållandena i Kopparmöllesjön jämte biologiska undersökningar
    i Pinnån 1976.  (EXTRACO AB.  An investigation of the sediments in
    Kopparmöllesjön and biological studies in Pinnån in 1976.)
Almestrand A. (1977)  ÄNGELHOLMS KOMMUN.  Limnologiska förhållanden
    inom Västersjön-Rössjön-Rössjöholmsåns avrinningsområde under
    tidsperioden 1946-1977.  (Limnological conditions in the catchment
    of Västersjön-Rössjön-Rössjöholmsån during 1946-1977.)
Berzins B. & Lundqvist I. (1973)  Limnologisk undersökning av Ybbarps-
    ån 1972.  (A limnological study in Ybbarpsån in 1972.)
Björk S. & Lettevall U. (1968)  BOLMEN-LAGAN-RINGSJÖN.
Franzén E. & Olofsson S. (1971)  Vattendragsutredning för Rönneå.
    Etapp. 1.  (General planning for the Rönneå.)
Länsstyrelsen i Kristianstads län och Malmöhus lan (1975)  Naturvårds-
    plan Skåne.  (Kristianstad and Malmöhus Counties 1975.  Environmental
    protection planning in Skåne.)
Länsstyrelsen i Kristianstads län och Malmöhus lan (1976)  Rönneån.
    Förslag till samordnat kontrollprogram.  (Kristianstad and Malmöhus
    Counties, 1976.  The Rönneå Scheme for a co-ordinated control
    programme.)
Mårtensson B., Sandberg R., Andersson Th. & Alkne H. (1979)  Ringsjön
    som naturresurs.  (Ringsjön as a natural resource.)
Mellanskånes Planeringskommitté (1980)  Vattenundersökningar i Ringsjön
    1975-1979.  (The Planning Committee of Central Skåne, 1980.  Investi-
    gations on Ringsjön water 1975-1979.)
Nilsson B.M. (1974)  Fritidsfiske inventering för Malmöhus län 1973.
    (Inventory of amateur angling in Malmöhus County 1973.)
Ryding S.-O. (1983a)  Redovisning och utvärdering av limnologiska under-
    sökningar.  I: Ringsjöområdet, ekosystem i förändring.  Mellanskånes
    Planeringskommitté.  (Account and evaluation of limnological investi-
    gations.  In: Ringsjön, a changing ecosystem.  The Planning Committee
    of Central Skåne.)
Ryding S.-O. (1983b)  Vattenkvalitet och ämnestransport.  Ringsjön och
    dess tillflöden 1975-1980.  I: Ringsjoområdet, ekosystem i förändring.
    Bilaga I.  Mellanskånes Planeringskommitté.  (Water quality and

transport of nutrients in Ringsjön and its tributaries 1975-1980.
In: Ringsjön, an ecosystem in change.  App. I. The Planning Committee
of Central Skåne.)

# Appendix A: Guide to Further Literature on European Rivers

The following bibliography is intended to help the reader start to find
the way into the literature; it is very far from comprehensive, especially
in the case of the eastern and southern countries.  Articles are not
included if they are listed by Whitton (1975) or at the end of chapters
in the present volume.

### GENERAL PAPERS BASED LARGELY ON EUROPEAN STUDIES OR WITH EXTENSIVE BIBLIOGRAPHIES

| | | |
|---|---|---|
| Armitage *et al.* | 1983 | quality score based on macroinvertebrates |
| Armitage | 1980 | regulation in the U.K. |
| Backhaus | 1973 | algae as indicators |
| Breitag & von Tümpling | 1982 | D.D.R. biological methods |
| Brooker | 1981 | impoundment on fisheries, with extensive bibliography |
| Dawson & Haslam | 1983 | shading and marginal vegetation |
| Ellington & Burke | 1981 | environmental policy |
| Haslam | in press | macrophytes in E.E.C. rivers |
| Haslam & Wolseley | 1981 | assessment and management of river vegetation |
| Hellawell | 1978 | surveillance methods |
| Logan & Brooker | 1983 | macroinvertebrates (mainly U.K.) |
| Milner *et al.* | 1981 | extensive bibliography includes papers on fish not listed here |
| Müller | 1974 | drift |
| Newbold *et al.* | 1983 | semi-popular account of conservation and river engineering |
| Schmitz & Illies | 1980 | biological methods |
| Schmitz & Schuman | 1982 | temperature and salmonid zone |
| Sládeček | 1973 | saprobic system |
| Tarak | 1981 | law and pollution control |

### AUSTRIA

| | | | |
|---|---|---|---|
| Ritrodat-Lunz study area | Bretschko | 1981 | zoobenthos |
| various montane | Kann | 1978a | benthic algae |
| various montane | Kann | 1978b | benthic algae |
| Danube (Donau) | Kann | 1983 | benthic algae |
| Finstertaler Bach, Gurgler Ache | Kawecka | 1974b | organic pollution and algae |

| | | | |
|---|---|---|---|
| Pitburger Bach | Kownacka & Margreiter | 1978 | benthic macro-invertebrates |
| Limmat | Schanz & Betschart | 1979 | algal bioassay |

## BELGIUM

| | | | |
|---|---|---|---|
| Vesdre | Houba & Remacle | 1983 | heavy metals |
| Helle, Roer Superieure | Leclercq | 1977 | chemistry, vegetation |
| near Boom | Maes | 1982 | organic C |
| Meuse | Mouvet & Bourg | 1983 | heavy metals |
| Flavion | Stroot | 1984 | Trichoptera ecology in trout river |
| Lieve | Willem & Frans | 1972 | general survey |
| Samson | Maquet | 1983 | |
| Ely | Esho & Benson-Evans | 1983 | sewage fungus |

## BULGARIA

| | | | |
|---|---|---|---|
| Maljovica (Rila) | Kawecka | 1974a | benthic algae |
| Rila Mts various | Kawecka | 1981 | benthic algae |

## CZECHOSLOVAKIA

| | | | |
|---|---|---|---|
| Danube | Ertl *et al.* | 1972 | benthic algae |
| Danube | Holčík & Bastl | 1977 | fish yield |
| Schwarza | Kubicek | 1968 | drift |
| Vltava | Sládeček & Miskovsky | 1976 | bacterial periphyton |

## DENMARK

| | | | |
|---|---|---|---|
| various | Andersen *et al.* | 1984 | modification of Trent Index |
| Fjederholt Å, Granslev Å | Dawson *et al.* | 1982 | macrophytes on chemistry |
| Gudenå | Ernst & Nielsen | 1983 | grayling biology |
| Odense Kanal | Fjerdingstad | 1977 | benthic bacteria |
| Skravad | Hvitred-Jacobson | 1982 | sewer overflows on dissolved oxygen |
| Rold Kilde spring | Iversen & Jessen | 1977 | *Gammarus pulex* biology |
| various | Iversen & Mortensen | 1978 | assessment of pollution by Danish and English methods |
| three streams | Iversen | 1975 | law of leaf break-down |
| various streams | Kern-Hansen & Dawson | 1978 | macrophyte biomass |
| four streams | Kern-Hansen | 1978 | *Gammarus* |
| Ravnkilde spring | Lindegaard *et al.* | 1975 | macroinvertebrates |

| | | | |
|---|---|---|---|
| Sønderup Å | Madsen *et al.* | 1977 | Ephemeroptera |
| | Ministry of the Environment, Denmark | 1978 | environmental |
| Bisballe Baek | Mortensen | 1981 | energy flow in population |
| S-W., various | Sand-Jensen & Rasmussen | 1978 | acidic stream macrophytes |
| Himmerland springs | Thorup & Lindegaard | 1977 | macroinvertebrates |

ENGLAND

| | | | |
|---|---|---|---|
| E. & W. Allen | Abel & Green | 1981 | zinc on invertebrates |
| Nent | Armitage | 1980 | zinc, invertebrates |
| Trent | Aston & Milner | 1980 | power stations on Asellus |
| (Bristol) Avon | Aykulu | 1978 | phytoplankton |
| (Bristol) Avon | Aykulu | 1982 | epipelic algae |
| Wensum | Baker *et al.* | 1978 | macrophytes |
| Wensum | Baker & Lumbley | 1980 | general survey |
| Frome, Tadnoll Brook | Baldock *et al.* | 1983 | chalk stream protozoa |
| N. Tyne | Boon | 1978 | current speed and Trichoptera |
| N. Tyne | Brennan *et al.* | 1978 | particulate material |
| N. Tyne | Brennan *et al.* | 1978 | particulate matter, chironomids |
| Holme | Brown *et al.* | 1982 | chemistry |
| Hayle | Brown | 1977 | Cu, Zn, invertebrates |
| Holme | Brown | 1980 | *Hydrobia* and toxic discharges |
| Derwent (N.-E. England) | Burrows & Whitton | 1983 | heavy metals, invertebrates |
| Bere | Casey | 1981 | macrophytes and discharge |
| Eden (Cumbria) | Crisp & Cubby | 1978 | fish |
| Piddle | Dawson *et al.* | 1978 | plant succession |
| Bere | Dawson | 1976 | leaf litter in chalk streams |
| Bere | Dawson | 1980 | transport of plant material |
| Duddon | Elliott & Tullett | 1977 | larval drift |
| Ruding | Extence | 1981 | drought on invertebrates |
| Irwell | Eyres *et al.* | 1978 | oligochaete ecology |
| Hull | Goulder | 1981 | phytoplankton |
| Moat Brook | Gower & Buckland | 1978 | sewage effluent on *Chironomus riparius* |
| Kennet, Lambourn | Ham *et al.* | 1981/ 1982 | macrophytes |
| various | Hanbury *et al.* | 1981b | herbicide effects |
| (Dorset) Stour | Hansford | 1978 | *Simulium austeri* |
| various | Hargreaves *et al.* | 1975 | acidic: chemistry, plants |
| Nene | Harper & Davies | 1977 | macrophytes |
| Swale | Holmes & Whitton | 1977 | macrophytes |

| | | | |
|---|---|---|---|
| Wear | Holmes & Whitton | 1977b | macrophyte survey |
| N. Tyne | Holmes & Whitton | 1981 | macrophyte survey |
| various | Holmes N. T. H. | 1983 | macrophyte communities |
| Twilfin Beck | Jones | 1978 | epilithic algae |
| Wye | Jones | 1984 | phytoplankton |
| Thames | Lack *et al.* | 1978 | diatoms |
| Thames | Lack *et al.* | 1978 | phytoplankton |
| Frome | Ladle *et al.* | 1977 | Simuliidae |
| various | Learner *et al.* | 1978 | Naididae biology |
| Coln | Mackey *et al.* | 1982 | general survey |
| Thames, Kennet | Mackey | 1976a,b 1977a,b | chironomids |
| Thames | Kackey | 1979 | chironomids and fish |
| (Dorset) Stour | Mann | 1978 | perch biology |
| (Dorset) Stour | Mann | 1979 | coarse fish biology |
| borehole water | Marker & Casey | 1982 | benthic algae |
| Bere, Frome | Marker & Gunn | 1977 | benthic algae |
| Bere, Frome | Marker | 1976 | benthic algae |
| Wandle | McCrow | 1974 | invertebrates |
| (Bristol) Avon | McGill *et al.* | 1979 | chironomid exuviae in surveillance |
| Wye | Merry *et al.* | 1981 | macrophytes |
| Frome | Mills *et al.* | 1983 | stone loach biology |
| Duddon | Minshall & Minshall | 1978 | invertebrates |
| (Bristol) Avon | Moore | 1976 | seasonal succession of algae |
| Highland Water (stream) | Moore | 1977a | seasonal succession of algae |
| Wellow Brook | Moore | 1977b | seasonal succession of algae |
| Wylye | Moore | 1978 | seasonal succession of algae |
| Wey | Moss | 1977 | diatom biology |
| various | Murphy *et al.* | 1981a | berbicide |
| Frome | Pinder | 1980 | chironomid distribution |
| (Buckinghamshire) Ouse | Pratten | 1980 | crayfish biology |
| Wandle | Price & Price | 1983 | macrophytes |
| Lowther | Saunders & Eaton | 1976 | phytobenthos measurement |
| Pennine, various | Say & Whitton | 1981 | zinc on plants |
| Bere | Thommen & Westlake | 1981 | macrophyte distribution |
| Medway,(Sussex) Ouse | Townsend *et al.* | 1983 | invertebrate community structure |
| Team | Wehr *et al.* | 1981 | zinc pollution |
| Frome | Westlake & Dawson | 1982 | weed-cutting |
| Tadnoll Brook (Dorset) | Wetton | 1980 | organic detritus |
| Chew | Wilson | 1977 | chironomid exuviae |
| Coquet | Wise | 1976 | Ephemeroptera distribution |
| Lambourn | Wright *et al.* | 1982 | macrophytes |
| Lambourn | Wright *et al.* | 1983 | macroinvertebrates |

## FINLAND

| | | | |
|---|---|---|---|
| central region | Elorantas Kunnas | 1979 | benthic algae |
| Kalajoki, Pyhäjoki, Siikajoki | Kainua & Valtonen | 1980 | lamprey |

## FRANCE

| | | | |
|---|---|---|---|
| Haut-Rhône | Amoros & Mathieu | 1984 | cyclopoid ecology |
| Haut-Rhône | Amoros | 1980 | Cladocera, Copepoda |
| Estaragne | Backhaus | 1976 | benthic algae |
| Ain | Bournaud & Keck | 1980 | benthic macroinverte-brates |
| Haut-Rhône | Bournaud *et al.* | 1978 | macroinvertebrate colonization |
| Haut-Rhône | Bournaud *et al.* | 1982 | Hydropsychidae |
| Aston | Chauvet | 1983 | reduced flow in mountain stream |
| Seine | Couté | 1979 | planktonic blue-greens |
| Somme | Descy | 1976 | benthic algae |
| Haut-Rhône | Dessaix & Roux | 1980 | 2ndy production of gammarids |
| Estibère | Elliott & Tullett | 1977 | larval drift |
| Haut-Rhône | Fontaine | 1982 | Ephemeroptera |
| Riou Mort | Giani | 1983 | heavy metals on fauna |
| Haut-Rhône | Gibert *et al.* | 1977 | macroinvertebrates |
| Haut-Rhône | Ginet | 1982 | amphopods of interstitial water |
| Verdon | Gregoire & Champeau | 1981 | Ephemeroptera |
| Moselle | Hinzelin & Lectard | 1978 | yeasts |
| Haut-Rhône | Juget *et al.* | 1976 | hydrology, macro-invertebrates |
| Ardière | Khalaf & Tachet | 1977 | macroinvertebrate colonization |
| Ardière | Khalaf & Tachet | 1980 | macroinvertebrate colonization |
| Loire | Lécureuil *et al.* | 1983 | *Hydropsyche* biology |
| Haut-Rhône | Merle | 1980 | bacteria |
| S. Juva streams | Nelva | 1979 | invertebrate community structure |
| general | Neveu | 1983 | zoobenthos review |
| Haut-Rhône | Perrin & Roux | 1978 | benthic fauna |
| Haut-Rhône | Richardot-Coulet *et al.* | 1983 | benthic macro-invertebrates |
| Lunain | Rofes | 1975 | influence of chalk |
| Haut-Rhône | Roux *et al.* | 1976 | benthic macro-invertebrates |
| Haut-Rhône | Roux | 1982 | introduction to project |
| Haut-Rhône | Roux | 1983 | review, esp., inverte-brates |
| Massif Central, various | Say & Whitton | 1982a | heavy metals, algae |

| | | | |
|---|---|---|---|
| Pyrenees, various | Say & Whitton | 1982b | heavy metals, algae |
| Haut-Rhône | Seyed-Reihani *et al.* | 1982a | interstitial macro-invertebrates |
| Haut-Rhône | Seyed-Reihani *et al.* | 1982b | interstitial macro-invertebrates |
| general | Vernaux | 1983 | quality assessment |

## GERMANY (B.R.D.)

| | | | |
|---|---|---|---|
| Rheinisches Schiefergebirge streams | Caspers | 1975 | invertebrate pro-ductivity |
| Elbe freshwater estuary | Caspers | 1981 | nitrogen cycle |
| Ruhr system | Dietz | 1973 | heavy metals in macrophytes |
| general | Federal Ministry of the Interior | 1977 | management |
| S. Lower Saxony various | Grube | 1975 | macrophtes |
| Ruhr | Hartmann | 1983 | diatoms and water chemistry |
| Breitenbach | Illies | 1975 | invertebrate pro-ductivity |
| Oberpfalzer Wald | Kohler & Zeltner | 1974 | macrophytes |
| Friedberger Au various | Kohler *et al.* | 1974 | macrophytes |
| various | Kohler | 1978 | macrophytes as indicators |
| Bavaria, various | Kohler | 1981 | macrophyte- |
| Hase | Koste | 1976 | rotifers |
| various lowland | Kohmeyer & Krause | 1975 | shade and macrophytes |
| Munchen region | Kutscher & Kohler | 1976 | macrophytes |
| various | Lorris | 1979 | macrophytes as indicators |
| Danube | Maas & Kohler | 1983 | macrophytes |
| eastern Hesse | Marxsen | 1980 a,b | suspended bacteria |
| Nagold, Alb | Monschau-Dudenhausen | 1980 | macrophytes as indicators |
| upper Fulda catch-ment stream | Pieper *et al.* | 1982 | *Gammarus* and chemistry |
| Elbe, Stor, Trave, Schwentine | Rheinheimer | 1977 a,b | bacteria |
| Ibach, Schwarzen-bachle | Rosset *et al.* | 1982 | leaf decomposition |
| Schierenseebrooks | Statzner | 1981 | hydraulic stress on benthic macro-invertebrates |
| Moosach, Regen, artificial | Steinberg & Melzer | 1983 | C physiology |

| Eder | Steubing *et al.* | 1983 | chemistry, algae |
| Niedersachsen, various | Weder-Oldecop | 1971 | macrophytes |
| Niedersachsen | Wiegleb | 1979a | macrophytes and chemistry |
| Niedersachsen | Wiegleb | 1979b | macrophyte communities |
| N.-W. and S. | Wiegleb | 1981a | discriminant analysis: water, macrophytes |
| Niedersachsen | Wiegleb | 1981b | macrophyte communities |
| Netzbach (Saarland) | Zaiss | 1981 | methane oxidation |
| Silbergragen, Kernwasser | Ziemann | 1975 | chemistry |

## GERMANY (D.D.R.)

| Saale | Braune | 1972 | algal crops |
| various | Jorga & Weise | 1979 | macrophytes as indicators |
| Havel | Klose | 1968 | plankton |
| Saale | Krausch | 1976 | macrophytes |
| various | Pietsch | 1972 | macrophytes as indicators |
| various | Stolz & Weise | 1976 | macrophytes and N |
| Zschopau | Werner & Weise | 1982 | macrophytes and P |

## GREENLAND

| Narssaq streams | Jóhansson | 1980 | benthic algae |
| Ûntartoq | Pedersen | 1977 | hot springs |

## HUNGARY

| Danube | Dvilhally-Tamás | 1975 | primary production |
| Backony Mts, various | Erzsebét | 1973 | benthic algae |
| Daugava | Kumsare | 1972 | phytoplankton |
| Tisza | Uherkovich | 1966 | plankton |

## ICELAND

| south-west | Castenholz | 1976 | hot springs |
| Laxá | Jóhansson | 1979 | dimensions and animals |
| Kráká, Laxá | Lindegaard | 1979 | macroinvertebrates |
| by R. Grendalsa | Sandbeck & Ward | 1982 | decomposition in hot springs |
| south | Sperling | 1976 | hot springs in winter |

## IRELAND

| Slaney catchment | An Foras Forbartha | 1981a | management plan |

| | | | |
|---|---|---|---|
| Barrow catchment | An Foras Forbartha | 1981a | management plan |
| general | An Foras Forbartha | 1983 | detailed bibliography |
| general | An Foras Forbartha | 1983 | includes comprehensive bibliography |
| Liffey catchment | An Foras Forbartha | 1985 | management plan |
| Lough Leane catchment | Casey et al. | 1981 | P |
| general | Clabby | 1981 | biological monitoring |
| general | McCumiskey | 1981 | review of resources |
| general | Toner & Lennox | 1980 | nitrate |

## ITALY

| | | | |
|---|---|---|---|
| Montegrotto Terme | Andreoli & Tascio | 1975 | hot spring |
| Arno | Balloni et al. | 1979 | algae and eutrophication |
| Lambro | De Simone & Damiani Damiani | 1983 | PCBs, pesticides, mercury |
| rivers of Piacenza district | Ghetti et al. | 1982 | biological monitoring |
| Tevere (Tiber) | IRSA | 1978 | pollution survey |
| Ticino | Marchetti & De Paolis | 1983 | water quality |
| tributaries of Lago Maggiore | Mosello et al. | 1982 | water quality |
| Adige | Regione Veneto | 1977 | pollution |
| stream in Boschi di Carrega Park | Zullini & Ricci | 1980 | rotifers |

## NETHERLANDS

| | | | |
|---|---|---|---|
| Renkumse Beck | Bovee-Meijn | 1970 | invertebrates |
| Hierden stream | Higler & Repko | 1981 | pollution in animals |
| Meuse, Rhine | Sloof et al. | 1983 | monitoring for toxicity |

## NORTHERN IRELAND

| | | | |
|---|---|---|---|
| Lough Neagh tributaries | Smith et al. | 1982 | nitrate |
| Main | Stevens & Smith | 1978 | N, P |

## NORWAY

| | | | |
|---|---|---|---|
| various | Braekke | 1976 | acid precipitation effects |
| Sognsvannsbekken | Brettum | 1974 | diatom drift |
| Tovdel | Leivestad & Muniz | 1976 | low pH and fish |
| Brurskardsbekken | Tangen et al. | 1978 | periphyton, drift algae |
| various | Wright et al. | 1976 | acid precipitation effects |

## POLAND

| | | | |
|---|---|---|---|
| Oder | Boroweic & Tarnawski | 1982 | birds waste water on ciliate |
| Biała Przemsza | Czapik | 1982 | waste water on ciliate communities |
| Płoska | Czecuzuga et al. | | phytoplankton |
| Drwinka + tributaries | Dratnal & Dumnicka | 1982 | benthic invertebrates |
| Unieść, Skotawa | Korzeniewski et al. | 1982 | influence of trout culture on rivers |
| Tatra streams | Kownacka & Kownacki | 1972 | zoocoenoses |
| Grabia | Niesiolowski | 1980 | Similuliidae on plants |
| Waksmundzki, Pysznianski, Koscieliski streams | Olechowska | 1982 | mayfly zonation |
| Biała Przemsza catchment | Pasternak | 1973 | heavy metals |
| Pilica | Penczak et al. | 1976 | roach biology |
| Bzura, Ultrata | Penczak et al. | 1981 | fish population parameters |
| Bzura, Ultrata | Penczak et al. | 1984 | fish diet |
| Vistula (Wisła) | Skoczén | 1982 | lead |
| Utrata | Zalewski & Penczak | 1981 | fish community |

## PORTUGAL

| | | | |
|---|---|---|---|
| Azores, general | Johansson | 1976 | algae |

## RUMANIA

| | | | |
|---|---|---|---|
| Danube | Botnariuc & Bolder | 1975 | primary production |
| Nera | Botosaneanu & Negrea | 1976 | gastropod association |
| Fagaras various | Kawecka | 1981 | benthic algae |
| Jiu | Miháitá | 1970 | saprobic survey of invertebrates |
| Jiu | Miháitá | 1971 | saprobic survey of invertebrates |

## SCOTLAND

| | | | |
|---|---|---|---|
| Duchray Water | Harriman & Morrison | 1982 | acid precipitation invertebrates, fish |
| Tweed | Holmes & Whitton | 1975 a,b | macrophytes |
| streams near Aberfeldy | Smith et al. | 1983 | mine impact assessment |

## SPAIN

| | | | |
|---|---|---|---|
| Besós, Llobregat | Lucena *et al.* | 1982 | viruses |
| Llobregat | Prat *et al.* | 1983 | macroinvertebrates and water quality |
| Avenco | Sabater | 1983 | benthic algae |

## SWEDEN

| | | | |
|---|---|---|---|
| general | Ahl | 1980 | chemistry |
| Kaltisjokk | Backlund & Renburg | 1973 | drift algae |
| Lapland various | Bjarnborg | 1983 | acidification |
| Kaltisjokk, Ångerån | Harmanen | 1980 | Ephemeroptera, Plecoptera |
| various | Henricson & Müller | 1980 | regulation |
| N. Jämtland, various | Johansson | 1979a | benthic chlorophyll |
| N. W. Jämtland, various | Johansson | 1979b | benthic algal biomass |
| Björbäck, Bakvattsbäck | Johansson | 1980 | benthic algal biomass + productivity |
| Jämtland, various | Johansson | 1982a | benthic algae |
| Jämtland, various | Johansson | 1982b | benthic algae |
| Rickleån | Johnson | 1978 | moss and detritus |
| Rickleån | Karlström & Backlund | 1977 | organic matter and chlorophyll |
| Rickleån | Karlström | 1978 | detrital metabolism |
| Kebnekaise Mts, various | Kawecka | 1981 | benthic algae |
| various | Kohler | 1978 | vegetation |
| Stampen stream | Malmqvist | 1980 | habitat selection by lamprey |
| Stampen stream | Malmqvist | 1973 | lamprey |
| Abisko region | Mendl & Müller | 1978 | *Amphinemura* |
| Vetkeån, Tolångaån | Müller | 1978 | algal productivity |
| Ulmeälven | Nilsson | 1978 | hydro-electric plant on flora |
| south, various | Otto & Svensson | 1983 | brown waters and invertebrates |
| far north | Steffan | 1971 | glacier brook chironomids |
| Tjulån | Wotton | 1979 | blackfly distributions |

## SWITZERLAND

| | | | |
|---|---|---|---|
| Lützel | Bärlocher | 1983 | standing crop |
| Aare, Rhine | Bloesch | 1977 | benthic macroinvertebrates |
| artificial | Eichenberger *et al.* | 1981 | experiments on stream benthos |
| various | Gujer | 1976 | nitrification |
| Lussel, Lützel | Rosset *et al.* | 1982 | leaf decomposition |

Rhine + tributaries   Schanz & Juon        1983   periphyton bioassay
artificial            Wuhrmann &           1980   experimental studies
                        Eichenberger


## U.S.S.R

Volga                 Butorin *et al.*     1981   impact of reservoirs
Neris                 Knabikas *et al.*    1973   mutagenicity of water
Azerbaijan, Caucasus  Kownacka &           1972   zoocoenoses
  streams               Kownacki
Kola Peninsula        Kruglova             1973   zooplankton
Dnieper               Sivko *et al.*       1972   standing crop +
                                                    productivity


## WALES

Rheidol, Ystwyth      Brooker & Morris     1980   macroinvertebrates
Ely                   Burton *et al.*      1982   plasmids
Ely                   Esho & Benson-Evans  1983   primary productivity
Teifi                 Jenkins & Wade       1981   macroinvertebrates
Teifi                 Jenkins *et al.*     1984   macroinvertebrates
Clarach, Ystwyth      McLean & Jones       1975   algae and heavy metals
Elan, Tywi            Scullion             1983   impoundments on ecology
Usk                   Thirb & Benson-      1983   substrata for
                        Evans                       *Lemanea*
Emral Brook           Wilkinson & Jones    1977   dace biology


## YUGOSLAVIA

Velika Morawa         Jankovic             1977   chironomid communities
Drava                 Markič               1983   rotifiers
Pek                   Nedeljković          1978   benthic fauna
Crnojevića            Smith *et al.*       1978   P in various components
Savinja               Vrhovšek             1981   macrophytes

REFERENCES FOR APPENDIX A

Abel P.D. & Green D.W.J. (1981)  Ecological  and toxicological studies
    on invertebrate fauna of two rivers in the Northern Pennine Orefield.
    In: Say P.J. & Whitton B.A. (Eds) *Heavy Metals in Northern England:
    Environmental and Biological Aspects.*  198 pp., pp. 109-122.
    Department of Botany, University of Durham.
Ahl T. (1980)  Variability in ionic composition of Swedish lakes and
    rivers.  *Arch. Hydrobiol.* 89, 5-16.
Amoros C. (1980)  Structure et fonctionnement des écosystèmes du Haut-
    Rhône francaise.  15. Comparison des peuplementes Cladocères et
    Copepodes de deux anciens méandres d'âge différent.  *Acta Oecol.
    (Oecol. gen.)* 1, 193-208.
Amoros C. & Mathieu J. (1984)  Structure et fonctionnement des
    écosystèmes du Haut-Rhône français.  35. Relations entre les eaux
    interstitielles et les eaux superficielles; influence sur les
    peuplements de Copepodes Cyclopoëdes (Crustacées).  *Hydrobiologia*
    108, 273-280.
Anderson M.M., Riget F.F. & Sparholt H. (1984)  A modification of the
    Trent index for use in Denmark.  *Water Res.* 18, 145-151.
Andreoli C. & Rascio N. (1975)  The algal flora in the thermal baths
    of Montegrott Terme (Padua).  Its distribution over one-year period.
    *Int. Rev. ges. Hydrobiol.* 60, 857-871.
An Foras Forbartha (1981a)  *Draft Water Quality Management Plan for the
    River Slaney Catchment* (5 Vols).  WR/C60, An Foras Forbartha, Dublin.
An Foras Forbartha (1981b)  *Draft Water Quality Management Plan for the
    River Barrow Catchment* (5 Vols).  WR/C62, An Foras Forbartha, Dublin.
An Foras Forbartha (1982)  *Draft Water Quality Management Plan for the
    River Barrow Catchment* (5 Vols).  WR/C61, An Foras Forbartha, Dublin.
An Foras Forbartha (1983)  *A Review of Water Pollution in Ireland.  A
    Report to the Council.*  152 pp. Water Pollution Advisory Council,
    Dublin.
An Foras Forbartha (1984)  *Draft Water Quality Management Plan for the
    River Liffey catchment* (5 Vols).  (In preparation), An Foras
    Forbartha, Dublin.
Armitage P.D. (1980)  The effects of mine drainage and organic enrich-
    ment on benthos in the River Nent system, Northern Pennines.
    *Hydrobiologia* 74, 119-128.
Armitage P.D. (1980)  Stream regulation in Great Britain.  In: Ward
    J.V. & Stanford J.A. (Eds) *The Ecology of Regulated Streams.*
    pp. 165-181.  Plenum Press, N.Y.
Armitage P.D., Moss D., Wright J.F. & Furse M.T. (1983)  The perfor-
    mance of a new biological water quality score system based on macro-
    invertebrates over a wide range of unpolluted running-water sites.
    *Water Res.* 17, 333-347.

Aston R.J. & Milner A.G.P. (1980)  A comparison of populations of the isopod *Asellus aquaticus* above and below power stations in organically polluted reaches of the River Trent. *Freshwat. Biol.* 10, 1-14.

Aykulu G. (1978)  A quantitative study of the phytoplankton of the River Avon, Bristol. *Br. phycol. J.* 13, 91-102.

Aykulu G. (1982)  The epipelic algal flora of the River Avon. *Br. phycol. J.* 17, 27-38.

Backhaus D. (1976)  Beiträge zur Ökologie der benthischen Algen des Hochgebirges in den Pyrenäen. II. Cyanophyceen und übrige Algengruppes. *Int. Rev. ges. Hydrobiol.* 61, 471-516.

Backlund S. & Renberg I. (1973)  Driftens och kolonisationens utreckling under våren hos encelliga alger i Kaltisjokk. *Ökol. stat. Messawe Ber.* 26, 1-9.

Baker R., Driscol R.J. & Lambley P. (1978)  The River Wensum (1). *Trans. Norfolk & Norwich Naturalists' Soc.* 24 (4), 197-219.

Baker R. & Lambley P. (1980)  The River Wensum (2). *Trans. Norfolk & Norwich Naturalists' Soc.* 25, 65-81.

Baldock B.M., Baker J.H. & Sleigh M.A. (1983)  Abundance and productivity of protozoa in chalk streams. *Holarctic Ecology* 6, 238-246.

Balloni W., Vincenzini M. & Florenzano G. (1979)  Algae and eutrophication of Arno River. In: *Symposium on Eutrophication in Italy.* Roma 3-4 October 1978. pp. 309-328. (In Italian)

Barlocher F. (1983)  Seasonal variation of standing crop and digestibility of CPOM in a Swiss Jura stream. *Ecology* 64, 1266-1272.

Bjärnborg B. (1983)  Dilution and acidification effects during the spring flood of four Swedish mountain brooks. *Hydrobiologia* 101, 19-26.

Bloesch J. (1977)  Bodenfaunistische Untersuchungen in Aare and Rhein. *Schweiz. Z. Hydrol.* 39, 46-68.

Boon P.J. (1978)  The pre-impoundment distribution of certain Trichoptera larvae in the north Tyne river system (northern England), with particular reference to current speed. *Hydrobiologia* 57, 167-174.

Boroweic L. & Tarnawski D. (1982)  Przeloty i zimonwanit ptakow na Ordze pod Brzegiem. The passage and wintering of birds on the River Oder near Brzeg. *Acta Zool. Cracov* 26, 3-30. (In Polish, English summary)

Botnariuc N. & Boldor O. (1975)  Evolution de la production primaire dans les eaux de la zone inondable du Danube. *Verh. int. Verein. theor. angew. Limnol.* 19, 1823-1827.

Botosaneanu L. & Negrea A. (1976)  Sur quelques aspects biologiques d'une rivière des monts du Banat - Roumanie: la Nera (et surtout) sur une remarquable association de gastropodes. *Riv. Idrobiol.* 15 (3), 403-432. (In French, with Italian and English summaries)

Bournaud M., Chavanon G. & Tachet H. (1978)  Structure et fonctionnement des écosystèmes du Haut-Rhône français. 5. Colonisation par les macroinvertébrés de substrats artificiels suspendus en pleine eau ou posés sur le fond. *Verh. int. Verein. theor. angew. Limnol.* 20, 1485-1493.

Bournaud M. & Keck G. (1980)  Diversité spécifique et structure des peuplements de macro-invertébrés benthiques au long d'un cours d'eau: le Furans (Ain). *Acta Oecologica, Oecol. génér.* 1, 131-150.

Bournaud M., Tachet H. & Perrin J.F. (1982)  Structure et fonctionne-
    ment des écosystèmes du Haut-Rhône français.  32. Les Hydropsychi-
    dae (Trichoptera) du Haut-Rhône entre Généve et Lyon.  *Annls Limnol.*
    18, 61-80.

Bovée-Meijn N.P. (1970)  Een hydrobiologisch onderzoek van de Renkumse
    Beek.  *Overdruk uit De Levende Natuur* 73, 7-11.

Braekke F.H. (Ed.) (1976)  *Impact of Acid Precipitation on Forest and
    Freshwater Ecosystems.*  111 pp. Fragrapport FR 6/76.  Norges Land-
    bruksvitenskapelige forskningsrad.

Braune W. (1972)  Experimentelle Untersuchungen *in situ* zur Biomass-
    bildung von Mikroalgen und zur Entwicklung naturlicher Algen-
    Biozönosen im Fliessgewässer.  *Int. Rev. ges. Hydrobiol.* 57, 227-256.

Breitag G. & von Tumpling W. (1982)  *Ausgewählte Methoden der Wasser-
    suchung.  Band II.  Biologische, Mikrobiologische und Toxikologische
    Methoden.*  Gustav Fischer Verlag, Jena.

Brennan A., McLachlan A.J. & Wotton R.S. (1978)  Particulate material
    and midge larvae (Chironomidae: Diptera) in an upland river.
    *Hydrobiologia* 59, 67-73.

Bretschko G. (1981)  Vertical distribution of zoobenthos in an alpine
    brook of the RITRODAT-LUNZ study area.  *Verh. int. Verein. theor.
    angew. Limnol.* 21, 873-876.

Brettum P. (1974)  The relation between the new colonization and drift
    of periphytic diatoms in a small stream in Oslo, Norway.  *Norw. J.
    Bot.* 21, 277-284.

Brooker M.P. (1981)  The impact of impoundments of the downstream
    fisheries and general ecology of rivers.  In: Coaker T.H. (Ed.)
    *Advances in Applied Biology III.*  pp. 91-152.  Academic Press,
    London.

Brooker M.P. & Morris D.L. (1980)  A survey of the macro-invertebrate
    riffle fauna of the rivers Ystwyth and Rheidol, Wales.  *Freshwat.
    Biol.* 10, 459-474.

Brown B.E. (1977)  Effects of mine drainage on the River Hayle, Corn-
    wall.  A) Factors affecting concentrations of copper, zinc and iron
    in water, sediments and dominant invertebrate fauna.  *Hydrobiologia*
    52, 221-233.

Brown L. (1980)  The use of *Hydrobia jenkinsi* to detect intermittent
    toxic discharges to a river.  *Water Res.* 14, 941-947.

Brown L., Bellinger E.G. & Day J.P. (1982)  A case study of nutrients
    in the R. Holme, West Yorkshire, England.  *Environ. Pollut. B.* 3,
    81-100.

Burton N.F., Day M.J. & Bull A.J. (1982)  Distribution of bacterial
    plasmids in clean and polluted sites in a South Wales river.  *Appl.
    environ. Microbiol.* 44, 1026-1029.

Butorin N.V., Ziminova N.A., Romanenko V.I. & Ekzertsev E.A. (1981)
    Ecological changes in the river Volga due to creation of reservoirs.
    *Verh. int. Verein. theor. angew. Limnol.* 21, 877-879.

Casey H. (1977)  Origin and variation of nitrate nitrogen in the chalk
    springs, streams and rivers in Dorset and its utilisation by higher
    plants.  *Progr. Wat. Technol.* 8, 225-235.

Casey H. (1981)  Discharge and chemical changes in a chalk stream head-
    water affected by the outflow of a commercial watercress-bed.
    *Environ. Pollut. B.* 2, 383-385.

Casey H. & Farr I.S. (1982)  The influence of within-stream disturbance on dissolved nutrient levels during spates. *Hydrobiologia* 92, 447-462.

Casey T.J., O'Connor P.E. & Greene R.G. (1981)  A survey of phosphorus inputs to Lough Leane. *Irish J. environ. Sci.* 1, 21-34.

Caspers H. (1981)  Seasonal effects on the nitrogen cycle in the freshwater section of the Elbe estuary. *Verh. int. Verein. theor. angew. Limnol.* 21, 866-870.

Caspers N. (1975)  Productivity and trophic structure of some West German woodland brooklets. *Verh. int. Verein. theor. angew. Limnol.* 19, 1712-1716.

Castenholz R.W. (1976)  The effect of sulfide on the blue-green algae of hot springs. I. New Zealand and Iceland. *J. Phycol.* 12, 54-68.

Chauret E. (1983)  Influence d'une réduction de débit sur un torrent de montagne: l'Aston (Ariège). *Annls Limnol.* 19, 45-50.

Clabby K.J. (1981)  *The National Survey of Irish Rivers. A Review of Biological Monitoring.* 322 pp. An Foras Forbartha, Dublin.

Couté A. (1979)  Cyanophycées planctoniques du bassin de la Seine. *Bull. Mus. natn. Hist. nat., Paris, 4 ser.* 1, 267-283.

Crisp D.T. & Cubby P.R. (1978)  The populations of fish in tributaries of the River Eden on the Moor House National Nature Reserve, Northern England. *Hydrobiologia* 57, 85-93.

Czapik A. (1982)  The effect of waste water on ciliate communities in the Biala Przemsza River. *Acta Hydrobiol.* 24, 29-37.

Czeczuga B., Gradzki F. & Bobiatyńska-Ksok E. (1968)  Produkcja pier pierwotna na wybranym stowisku rzeki Ploski Cześć I. Prodikcja fitoplanktonu. Primary production in a chosen site of the River Ploska. Part I. Phytoplankton production. *Acta Hydrobiol.* 10, 85-94.

Dawson F.H. (1976)  Organic contribution of stream edge forest litter fall to the chalk stream ecosystem. *Oikos* 27, 13-18.

Dawson F.H., Castellano E. & Ladle M. (1978)  Concept of species succession in relation to river vegetation and management. *Verh. int. Verein. theor. angew. Limnol.* 20, 1429-1434.

Dawson F.H. & Haslam S.M. (1983)  The management of river vegetation with particular reference to shading effects of marginal vegetation. *Landscape Planning* 10, 147-169.

Dawson F.H., Kern-Hansen U. & Westlake D.F. (1982)  Water plants and the oxygen and temperature regimes of lowland streams. In: Symoens J.J., Hooper S.S. & Compère P. (Eds) *Studies on Aquatic Plants.* pp. 214-221. Royal Botanic Society of Belgium, Brussels.

Descy J.-P. (1976)  Estimation de la qualité des eaux de la Somme par l'étude des peuplements algaux benthiques. *Bull. fr. Pisciculture* 260, 143-147.

De Simone R. & Damiani V. (1983)  The Lambro River sediments: PCBs, pesticides and mercury. *Acqua e Aria* 9, 965-971. (In Italian)

Dessaix J. & Roux A.L. (1980)  Structure and dynamics of the secondary production of Gammarids in the main stream. *Crustaceana, Suppl.* 6, 242-252.

Dietz F. (1973)  The enrichment of heavy metals in submerged plants. In: Jenkins S.H. (Ed.) *Advances in Water Pollution Research.* 6th International Conference, Jerusalem, June 1972. pp. 53-60. Pergamon Press, Oxford.

Dratnal E. & Dunnicka E. (1982)  Compositjon and zonation of benthic invertebrate communities in some chemically stressed aquatic habits of Niepolomice Forest (South Poland).  *Acta Hydrobiol.* 24, 151-165.

Dvihally-Tamás S. (1975)  Primary production of the Hungarian Danube.  *Verh. int. Verein. theor. angew. Limnol.* 19, 1717-1722.

Edwards R.W. & Crisp D.T. (1982)  Ecological implications of river regulation in the United Kingdom.  In: Hey R.D., Bathurst J.C. & Thorne C.R. (Eds)  *Gravel-bed Rivers.*  pp. 843-865.  Wiley, Chichester.

Eichenberger E., Schlatter F., Weilenmann H. & Wuhrmann K. (1981)  Toxic and autrophicating effects of Co, Cu and Zn on algal benthic communities in rivers.  *Verh. int. Verein. theor. angew. Limnol.* 21, 1131-1134.

Ellington A. & Burke T. (1981)  *Europe: Environment. The European Communities' Environmental Policy.*  23 pp.  Ecobooks, 16 Strutton Ground, London S.W.1.  (Published with the assistance of the Commission of the European Communittis.)

Elliott J.M. & Tullett P.A. (1977)  The downstream drift of larvae of *Dixa* (Diptera: Dixidae) in two stony streams.  *Freshwat. Biol.* 7, 403-408.

Eloranta P. & Kumas S. (1979)  The growth and species communities of the attached algae in a river system in Central Finland.  *Arch. Hydrobiol.* 86, 27-44.

Ernst M.E. & Nielsen J. (1983)  Gådenåstallingens (*Thymallus thymallus* L.) gydebiologi.  *Meddelelser fra Ferskvandsfiskerilaboratoriet* 1983/1, 30 pp.  (The spawning biology of the grayling (*Thymallus thymallus* (L.)) from the river Gudenå)

Ertl M., Juris S. & Tomajka J. (1972)  Vorläufige Angaben über jahreszeitliche Veränderungen und die Vertikale Verteilung des Periphytons in mittleren Abschnitt der Donau.  *Arch. Hydrobiol., Suppl.* 44, 34-48.

Erzsebét K. (1973)  Az Északi- (Öreg-) Bakony területén vegzett algólogiai és hidrobiologiai kutatások rövid ismertetése.  *Veszprem Megyei Múzeumok Közleményei* 12, 143-163.  (Hungarian, with German and English summaries)

Esho R.T. & Benson-Evans K. (1983)  Studies on the sewage fungus complex in the River Ely, South Wales, U.K.  *Nova Hedwigia* 37, 519-534.

Extence C.A. (1981)  The effect of drought on benthic invertebrate communities in a lowland river.  *Hydrobiologia* 23, 217-224.

Eyres J.P., Williams N.V. & Pugh-Thomas M. (1978)  Ecological studies on Oligochaeta inhabiting depositing substrata in the Irwell, a polluted English river.  *Freshwat. Biol.* 8, 25-32.

Federal Ministry of the Interior, Germany (1977)  *Report on Water Management in the Federal Republic of Germany.*  47 pp.  (Published on the occasion of the United Nations' International Water Conference in March 1977 in Mar del Plata, Argentina)

Fjerdingstad E. (1977)  Effect of warm water discharges from power plants on the bacterial decomposition in the bottom mud of a stream.  *Arch. Hydrobiol. Beih.* 9, 223-237.

Fontaine J. (1982)  Le piégage lumineux, moyen d'approche de la faune entomologique d'un grand fleuve (Ephéméroptères, en particulier).  *Bull. mens. Soc. Linn. Lyon* 51, 81-89.

Ghetti P.F., Bernina F., Bonazzi G., Cunzola A. & Ravanetti U. (1982)
    *Biological Monitoring of the Piacenza District Rivers.*  29 pp.
    Pubb. Amm. Prov. Piacenza e P. F. Ambients. (In Italian)

Giani N. (1983) Le Riou Mort, affluent du Lot pollué par les metaux
    lourds. III. Etude faunistique générale. *Annls Limnol.* 19, 29-44.

Gibert J., Ginet R., Mathieu J., Reygrobellet J.L. & Seyed-Reihani A.
    (1977)  Structure et fonctionnement des écosystèmes du Haut-Rhône
    français.  IV. Le peuplement des eaux phréatiques; premiers
    résultats. *Annls Limnol.* 13, 83-97.

Ginet R. (1982)  Structure et fonctionnement des écosystèmes du Haut-
    Rhône francais.  Les Amphipodes des eaux interstitielles en amont
    de Lyon. *Polskie Archwm Hydrobiol.* 29, 231-237.

Goulder R. (1982)  A note on phytoplankton in the River Hull at Hemp-
    holme, North Humberside. *Naturalist, Hull* 106, 153-156.

Gower A.M. & Buckland P.J. (1978)  Water quality and the occurrence of
    *Chironomus riparius* Meigen (Diptera: Chironomidae) in a stream
    receiving sewage effluent. *Freshwat. Biol.* 8, 153-164.

Gregoir A. & Champeau A. (1981)  Les Ephéméroptères d'une rivière à
    debit regulé: le Verdon. *Verh. int. Verein. theor. angew. Limnol.*
    21, 857-865.

Grube H.-J. (1975)  Die Macrophytenvegetation der Fliessgewässer in
    Süd-Niedersachsen und ihre Beziehungen zur Gewässerverschmutzung.
    *Arch. Hydrobiol., Suppl.* 45, 376-475.

Gujer W. (1976)  Nitrifikation in Fliessgewässern - Fallstudie Glatt.
    *Schweiz. Z. Hydrol.* 38, 171-189.

Ham S.F., Wright J.F. & Berrie A.D. (1981)  Growth and recession of
    aquatic macrophytes in an unshaded section of the River Lambourn,
    England, from 1971 to 1976. *Freshwat. Biol.* 11, 381-390.

Ham S.F., Wright J.F. & Borie A.D. (1982)  The effect of cutting on
    the growth and recession of the freshwater macrophyte *Ranunculus
    penicillatus* (Dumort.) Bab. var. *calcareus* (R. W. Butcher) C.D.K.
    Cook. *J. environ. Mgmt* 15, 283-271.

Ham S.F., Cooling D.A., Hiley P.D., McLeish P.R., Scorgie H.R.A. &
    Berrie A.D. (1982)  Growth and recession of aquatic macrophytes on
    a shaded section of the River Lambourn, England, from 1971 to 1976.
    *Freshwat. Biol.* 11, 381-392.

Hansford R.G. (1978)  Life-history and distribution of *Simulium
    austeni* (Diptera: Simuliidae) in relation to phytoplankton in some
    English rivers. *Freshwat. Biol.* 8, 521-532.

Hargreaves J.W., Lloyd E.J.H. & Whitton B.A. (1975)  Chemistry and
    vegetation of highly acidic streams. *Freshwat. Biol.* 5, 563-576.

Harmanen M. (1980)  Der Einfluss saurer Gewässer auf den Bestand der
    Ephemeriden - und Plecopterenfauna. *Gewässer Abwässer* 66/67,
    130-136.

Harper D.M. & Davies C.E. (1977)  Aquatic plants of the River Nene.
    *J. Northamptonshire Natural History Society* 37, 142-147.

Harriman R. & Morrison B.R.S. (1982)  Ecology of streams draining
    forested and non-forested catchments in an area of central Scotland
    subject to acid precipitation. *Hydrobiologia* 88, 251-263.

Hartman D. (1983)  Beziehungen zwischen der Diatomeen-Flora und dem
    Wasserchemismus in Fliessgewässern des Sauerlandes. 1. Die Rhur.
    *Ber. naturhist. Ges. Hannover* 126, 91-135.

Haslam S.M. (in press) *River Plants of Western Europe. The Macro-phytic Vegetation of Watercourses of the European Economic Community.* Cambridge U.P., Cambridge.

Haslam S.M. & Wolseley P.A. (1981) *River Vegetation: Its Identification, Assessment and Management. A Field Guide to the Macro-phytic Vegetation of British Watercourses.* 154 pp. Cambridge U.P., Cambridge.

Henricson J. & Müller K. (1980) Stream regulation in Sweden with some examples from central Europe. In: Ward V. & Stanford J.A. (Eds) *The Ecology of Regulated Streams.* pp. 183-199. Plenum Press, N.Y.

Higler L.W.G. & Repko F.F. (1981) The effects of pollution in the drainage area of a Dutch lowland stream on fish and macro-invertebrates. *Verh. int. Verein. theor. angew. Limnol.* 21, 1077-1082.

Hinzelin F. & Lectard P. (1978) Les levures dans les eaux de la Moselle. *Hydrobiologia* 61, 209-224.

Holcík J. & Bastl I. (1977) Predicting fish yield in the Czechoslovakian section of the Danube river based on the hydrological regime. *Int. Rev. ges. Hydrobiol.* 62, 523-532.

Holmes N.T.H. (1983) Focus on Nature Conservation No. 4. *Typing British Rivers According to their Flora.* Nature Conservancy Council, Shrewsbury.

Holmes N.T.H. & Whitton B.A. (1975a) Macrophytes of the River Tweed. *Trans. bot. Soc. Edinb.* 42, 369-381.

Holmes N.T.H. & Whitton B.A. (1975b) Submerged macrophytes and angiosperms of the River Tweed and its tributaries. *Trans. bot. Soc. Edinb.* 42, 383-395.

Holmes N.T.H. & Whitton B.A. (1977a) Macrophytic vegetation of the River Swale, Yorkshire. *Freshwat. Biol.* 8, 545-558.

Holmes N.T.H. & Whitton B.A. (1977b) Macrophytes of the River Wear: 1966-1976. *Naturalist, Hull* 106, 53-73.

Holmes N.T.H. & Whitton B.A. (1981) Plants of the River Tyne system before the Kielder Water Scheme. *Naturalist, Hull* 106, 97-108.

Houba C. & Remacle J. (1982) Distribution of heavy metals in sediments of the River Vesdre (Belgium). *Environ. Technol. Lett.* 3, 237-240.

Huitved-Jacobsen T. (1982) The impact of combined sewer overflows on the dissolved oxygen concentration of a river. *Water Res.* 16, 1099-1105.

Illies J. (1975) A new attempt to estimate production in running waters (Schlitz studies on productivity, No. 12). *Verh. int. Verein. theor. angew. Limnol.* 19, 1705-1711.

IRSA (Istituto di Recerca sulle Acque, Consiglio Nazionale delle Ricerca) (1978) Indagine sulla 'inquinamento del fiume Tevere. *Quaderni IRSA* 27, 482 pp. (Researches on pollution of the River Tiber)

Iversen T.M. (1975) Disappearance of autumn shed beech leaves placed in bags in small streams. *Verh. int. Verein. theor. angew. Limnol.* 19, 1687-1692.

Iversen T.M. & Jessen J. (1977) Life-cycle, drift and production of *Gammarus pulex* L. (Amphipoda) in a Danish spring. *Freshwat. Biol.* 7, 287-300.

Iversen T.M. & Mortensen E. (1978)  Sammerligningen af danske
    bedømmelse af vandrecipeinters forureningsgrad og det engelske
    Trent Index. *Vand* 3, 65-67.  (Danish = A comparison between the
    method used in Denmark for assessment of the degree of pollution
    and the English Trent Index.)
Jankovic M.J. (1979)  Communities on chironomid larvae in the Velika
    Morawa river. *Hydrobiologia* 64, 167-173.
Jenkins R.A. & Wade K.R. (1981)  *A Survey of the Macroinvertebrate
    Fauna of the River Teifi*. Biological Report No. 13/81.  Welsh
    Water Authority Laboratory, Llanelli.
Jenkins R.A., Wade K.R. & Pugh E. (1984)  Macroinvertebrate - habitat
    relationships in the River Teifi catchment and the significance to
    conversion. *Freshwat. Biol.* 14, 23-42.
Johansson C. (1976)  Freshwater algal vegetation in the Azores. *Bot.
    Soc. Broteriana* 50 (2nd series), 117-141.
Johansson C. (1979a)  Chlorophyll content and the periphytic algal
    vegetation in six streams in Northern Jämtland, Sweden 1977.  *Medd.
    Växtbiologiska Institutionen, Uppsala* 1979/2,  1-53.
Johansson C. (1979b)  Biomass of the periphytic algal vegetation in
    some stony mountain streams in N.W. Jämtland, Sweden.  A preliminary
    report. *Medd. Växtbiologiska Institutionen, Uppsala* 1979/3, 1-16.
Johansson C. (1980)  Attached algal vegetation in some streams from
    the Narssaq area, South Greenland. *Acta Phytogeogr. Suec.* 68, 89-96.
Johansson C. (1982a) Attached algal vegetation in two stony streams in
    N.W. Jämtland, Sweden.  Ecology, volume and primary production.
    *Medd. Växtbiologiska Institutionen, Uppsala* 1980/1, 1-30.
Johansson C. (1982b) The ecological characteristics of 314 algal taxa
    found in Jämtland Streams, Sweden.  *Medd. Växtbiologiska Institu-
    tionen, Uppsala* 1982/2.
Jóhansson P.M. (1979)  The River Laxá ecosystem, Iceland. *Oikos* 32,
    306-309.
Johnston T. (1978)  Aquatic mosses and stream metabolism in a North
    Swedish river. *Verh. int. Verein. theor. angew. Limnol.* 20,
    1471-1477.
Jones F.H. (1984)  The dynamics of suspended algal populations in the
    lower Wye catchment. *Water Res.* 18, 25-36.
Jones J.G. (1978)  Spatial distribution in epilithic algae in a stony
    stream (Wilfin Beck) with particular reference to *Cocconeis
    placentula*. *Freshwat. Biol.* 8, 539-546.
Jorga W. & Weise G. (1979)  Zum Bioindikationswert submerser Makro-
    phyten und zur Rucklhaltung von Wasserinhaltsstoffen durch Unter-
    wasserpflanzen in langsam fliessenden Gewässern. *Acta hydrochim.
    hydrobiol.* 7, 43-76.
Jorga W., Heym W.-D. & Weise G. (1982)  Shading as a measure to
    prevent mass development of submersed macrophytes. *Int. Rev. ges.
    Hydrobiol.* 67, 271-281.
Juget J., Amoros C., Gamulin D., Reygrobellet J.L., Richardot M.,
    Richoux P. & Roux C. (1976)  Structure et fonctionnement des éco-
    systèmes du Haut-Rhône français.  II. Etude hydrologique et
    écologique de quelques bras morts.  Premiers resultats. *Bull.
    Ecol. fr.* 7, 479-492.

Kainua K. & Valtomen T. (1980)  Distribution and abundance in European
    river lamprey (*Lampetra fluviatilis*) larvae in three rivers running
    in to Bothnian Bay, Finland. *Can. J. Fish. aquatic. Sci.* 37,
    1960-1966.
Kann E. (1978a)  Systematik und Ökologie der algen österreichischer
    Bergbache. *Arch. Hydrobiol., Suppl.* 53, 405-643.
Kann E. (1978b)  Typification of Austrian streams concerning algae.
    *Verh. int. Verein. theor. angew. Limnol.* 20, 1523-1526.
Kann E. (1983)  Die benthischen Algen der Donau im Raum von Wien.
    *Arch. Hydrobiol., Suppl.* 68, 15-36.
Karlström U. & Backlund S. (1977)  Relationship between algal cell
    number, chlorophyll and fine particulate organic matter in a river
    in Northern Sweden. *Arch. Hydrobiol.* 80, 192-199.
Karlström U. (1978)  Role of the organic layer on stones in detrital
    metabolism in streams. *Verh. int. Verein. theor. angew. Limnol.*
    20, 1463-1470.
Kawecka B. (1974a) Vertical distribution of algae communities in
    Maljovica stream (Rila - Bulgaria). *Polskie Archwm Hydrobiol.* 21,
    211-228.
Kawecka B. (1974b) Effect of organic pollution on the development of
    diatom communities in the alpine streams Finstertaler Bach and
    Gurgler Ache (northern Tyrol, Austria). *Ber. nat.-med. Ver.*
    *Innsbruck* 61, 71-82.
Kawecka B. (1981)  Sessile algae in European mountain streams. 2.
    Taxonomy and autecology. *Acta Hydrobiol.* 23, 17-46.
Kawecka B. (1982)  Stream ecosystems in mountain grassland (West Car-
    pathians). *Acta Hydrobiol.* 24, 357-365.
Kern-Hansen U. (1973)  Drift of *Gammarus pulex* L. in relation to macro-
    phyte cutting in four small Danish lowland-streams. *Verh. int.*
    *Verein. theor. angew. Limnol.* 20, 1440-1445.
Kern-Hansen U. & Dawson F.H. (1978)  The standing crop of aquatic
    plants of lowland streams in Denmark and the inter-relationships
    of nutrients in plant, sediment and water. In: *5th International*
    *Symposium on Aquatic Weeds, September 1978,* pp. 143-150.  European
    Weed Research Society.
Khalaf G. & Tachet H. (1977)  La dynamique de colonisation des sub-
    strats artificiels par les macro-invertébrés d'un cours d'eau.
    *Annls Limnol.* 13, 169-190.
Khalaf G. & Tachet H. (1980)  Colonization of artificial substrata by
    macroinvertebrates in a stream and variations according to stone
    size. *Freshwat. Biol.* 10, 475-482.
Klasvik B. (1974) *Computerized Analysis of Stream Algae.*  100 pp.
    Växtekologiska Studier 5.
Klose H. (1968)  Untersuchungen uber den Indikationswert des Potamo-
    planktons. *Int. Rev. ges. Hydrobiol.* 53, 781-805.
Knabikas A., Lekevičius R. & Adomonyte L. (1983)  Evaluation of
    mutageneity of water in the Neris River. *Acta Hydrobiologica*
    *Lituanica* 4, 59-68.  (In Russian, English summary)
Kohler A. (1978) *Bericht uber die Forschungsreisen nach Südengland*
    *und Südschweden im Sommer 1978.*  Institut für Landeskultur und
    Pflanzenökologie, Univ. Hohenheim, Stuttgart.

Kohler A. (1981)  *Fliessgewässer in Bayern.*  18 pp.  Akademie für
    Naturschutz und Landschaftspflege, Tagungsbericht 5/81.  (ISSN 0173-
    7724)
Kohler A., Brinkmeier R. & Vollrath H. (1974)  Verbreitung und
    Indikatorwert der submersen Makrophyten in den Fliessgewässern der
    Friedberger Au.  *Ber. bayer. bot. Ges.* 45, 5-36.
Korzeniewski K., Banat Z. & Moczulska A. (1982)  Changes in water of
    the Unieść and Skotawa Rivers, caused by intensive trout culture.
    *Polskie Archwm Hydrobiol.* 29, 683-691.
Koste W. (1976)  Über die Rädertierbestände (Rotatoria) der oberen und
    mittleren Hase in den Jahren 1966-1969.  *Osnabrucker Naturw.*
    *Mitteilungen* 4, 191-263.
Kownacka M. & Margreiter G. (1978)  Die Zoobenthos-Gesellschaften des
    Piburger Baches (Ötztal, Tirol).  *Int. Rev. ges. Hydrobiol.* 63,
    213-232.
Krausch H.-D. (1976)  Die Makrophyten der mittlerer Saale und ihre
    Biomasse.  *Limnologica, Berl.* 10, 57-72.
Kristjansson J.K. & Alfredsson G.A. (1983)  Distribution of *Themus*
    spp. in Icelandic hot springs and a thermal gradient.  *Appl.*
    *environ. Microbiol.* 45, 1785-1789.
Kruglova A.N. (1983)  Zooplankton in small rivers of the Kola penin-
    sula.  *Gibrobiol. Zh. Kiev* 19, 56-58.  (In Russian, English summary)
Kumsare A. (1972)  Zur hydrobiologischen Rayonierung der Daugava.
    *Verh. int. Verein. theor. angew. Limnol.* 18, 751-755.
Lack T.J., Youngman R.E. & Collingwood R.W. (1978)  Observations on a
    spring bloom in the River Thames.  *Verh. int. Verein. theor. angew.*
    *Limnol.* 20, 1435-1439.
Ladle M., Bass J.A.B., Philpott F.R. & Jeffrey A. (1977)  Observations
    on the ecology of Simuliidae from the River Frome, Dorset.
    *Ecological Entomology* 2, 197-204.
Learner M.A., Lochhead G. & Hughes B.D. (1978)  A review of the biology
    of British Naididae (Oligochaeta) with emphasis on the lotic
    environment.  *Freshwat. Biol.* 8, 357-376.
Leclercq L. (1977)  Végétation et caractéristiques physico-chimiques
    de deux rivières de Haute-Ardenne (Belgique):  la Helle et la Roer
    Supérieure.  *Lejeunia* 88, 1-42.
Lécureuil J.Y., Choret M., Bournault M. & Tachet H. (1983)  Descrip-
    tion, répartition et cycle biologique de la larve d'*Hydropsyche*
    *bulgaromanorum* Malicky 1977 (Trichoptera) in the lower Loire.  *Annls*
    *Limnol.* 19, 17-24.
Leivestad H. & Muniz I.P. (1976)  Fish kill at low pH in a Norwegian
    river.  *Nature, Lond.* 259, 391-392.
Lindegaard C., Thorup J. & Bahn M. (1975)  The invertebrate fauna of
    the moss carpet in the Danish spring Raunkilde and its seasonal,
    vertical, and horizontal distribution.  *Arch. Hydrobiol.* 75,
    109-139.
Lindengaard C. (1979)  A survey of the macroinvertebrate fauna with
    special reference to Chironomidae (Diptera) in the rivers Laxá and
    Kráká, northern Iceland.  *Oikos* 32, 281-288.
Lohmeyer W. & Krause A. (1975)  Über die Auswirkungen des Gehölz-
    bewuchses an kleinen Wasserläufen des Münsterlandes auf die Vegeta-
    tion im Wasser und an den Böschungen im Hinblick auf die Unterhal-
    tung der Gewässer.  *Schriftenr. Vegetationskr.* 9, 1-105.

Logan P. & Brooker N.P. (1983) The macroinvertebrate faunas of riffles and pools. *Water Res.* 17, 263-270.

Loris B. (1979) Bio-Indikatoren: Wasserpflanzen uberwachen unsere Flüsse. *Bild der Wissenschaft* 1, 44-52.

Lucena F., Fiance C., Jofre J., Sancho J. & Schwartzbrod L. (1982) Viral pollution determination of superficial waters (river water and sea-water) from the urban area of Barcelona (Spain). *Water Res.* 16, 173-177.

Maas D. & Kohler A. (1983) Die Makrophytenbestände der Donau im Raum Tuttlingen. *Landschaft + Stadt* 15 (2), 49-60.

Mackey A.P. (1976-77) Quantitative studies in the Chironomidae (Diptera) of the Rivers Thames and Kennet. (1976a). I. The *Acorus* zone. *Arch. Hydrobiol.* 78, 240-267. (1976b). II. The Thames flint zone. *Arch. Hydrobiol.* 78, 310-318. (1977a). III. The *Nuphar* zone. *Arch. Hydrobiol.* 79, 62-102. (1977b). IV. Production. *Arch. Hydrobiol.* 80, 327-348.

Mackey A.P., Ham S.F., Cooling D.A. & Berrie A.D. (1982) An ecological survey of a limestone stream, the River Coln, Gloucestershire, England, in comparison with some chalk streams. *Arch. Hydrobiol., Suppl.* 64, 307-340.

Madsen B.L., Bengtsson J. & Butz I. (1977) Upstream movement of some Ephemeroptera species. *Arch. Hydrobiol.* 81, 119-127.

Maes L. (1982) Dissolved organic carbon and UV-absorption in a polluted lowland brook - pond system. Influence of water aeration by the Phallus technique. *Hydrobiologia* 89, 269-276.

Malmqvist B. (1980) Habitat selection of larval brook lampreys (*Lampetra planeri*, Blach) in a South Swedish stream. *Oecologia, Berl.* 45, 35-38.

Malmqvist B. (1983) Growth, dynamics, and distribution of a population of the lamprey *Lampetra planeri* in a South Swedish stream. *Holarctic Ecology* 6, 404-412.

Mann R.H.K. (1978) Observations on the biology of the perch, *Perca fluviatilis*, in the River Stour, Dorset. *Freshwat. Biol.* 8, 229-240.

Mann R.H.K. (1979) Aspects of the biology of coarse fish in the Dorset Stour. *Freshwater Biological Association, U.K., Annual Report* 47, 51-59.

Marchetti R. & De Paolis A. (1983) Water quality of the River Ticino. *Parco Ticino Notizie, Suppl.* 17-18, 32 pp.

Marker A.F.H. (1976) The benthic algae of some streams in southern England. I. Biomass of the epilithon in some streams. *J. Ecol.* 64, 343-358.

Marker A.F.H. & Gunn R.J.M. (1977) The benthic algae of some streams in southern England. III. Seasonal variations in chlorophyll a in the seston. *J. Ecol.* 65, 223-234.

Marker A.F.H. & Casey H. (1982) The population and production dynamics of benthic algae in an artificial recirculating hard-water stream. *Phil. Trans. R. Soc. Lond. B* 298, 265-308.

Markic M. (1983) The Rotatoria-Monogononta of the River Drava in Slovenia, Yugoslavia. *Hydrobiologia* 104, 229-330.

Marxsen J. (1980a, b)  Untersuchungen zue Ökologie der Bakterien in der
    fliessenden Welle von Bachen.  I. Chemismus, Primarproduktion, $CO_2$-
    Dunkelfixierung und Eintrag von partikularen organischen Material.
    *Arch. Hydrobiol. Suppl.* <u>4</u>, 461-533.  II. Die Zahl der Bakterien im
    Jahreslauf.  *Arch. Hydrobiol., Suppl.* <u>57</u>, 461-533.
Maquet B. (1983)  Caractéristiques chimiques et biologiques des eaux
    de la vallée du Samson.  *Annls Soc. r. Belg.* <u>113</u>, 3-18.
McCumiskey L.M. (1981)  A review of Ireland's water resources - rivers
    and lakes.  *Irish J. environ. Sci.* <u>1</u>, No. 2, 1-13.
McGill J.D., Wilson R.S. & Brake A.M. (1979)  The use of chironomid
    pupal exuviae in the surveillance of sewage pollution within a
    drainage system.  *Water Res.* <u>13</u>, 887-894.
McLean R.O. & Jones A.K. (1975)  Studies of tolerance to heavy metals
    in the flora of the rivers Ystwyth and Clarach, Wales.  *Freshwat.
    Biol.* <u>5</u>, 431-444.
Mendl H. & Müller K. (1978)  The colonization cycle of *Amphinemura
    standfussi* Ris (Ins.: Plecoptera) in the Abisku area.  *Hydrobio-
    logia* <u>60</u>, 109-111.
Merle G. (1980)  Analyses bactériologiques de l'eau du Rhône au
    voisinage de la centrale thermique du Bugey.  *Cah. Lab. Hydrobiol.
    Montereau* <u>10</u>, 63-76.
Merry D.G., Slater F.M. & Randerson P.F. (1981)  The riparian and
    aquatic vegetation of the River Wye.  *J. Biogeog.* <u>8</u>, 313-327.
Mihăiţă A. (1970)  Cercetaria saprobiologice în Bazinul Rîului Jiu.
    Annal Analele Universitatii Burcuresti.  *Biologie Animala* <u>19</u>,
    101-107.  (Rumanian, with Russian and French summaries)
Mihăiţă A. (1971)  *Cercetări Saprobiologice în Bazinul Rîului Jiu.*
    42 pp.  Rezumatul tezei de doctorat, Universitatea din Bucuresti.
Mills C.A., Welton J.S. & Rendle E.L. (1983)  The age, growth and
    reproduction of the stone loach *Noemacheilus barbatulus* (L.) in
    a Dorset chalk stream.  *Freshwat. Biol.* <u>13</u>, 283-292.
Milner N.H., Scullion J., Carling P.A. & Crisp D.T. (1981)  The effects
    of discharge on sediment dynamics and consequent effects on inverte-
    brates and salmonids in upland rivers.  In: Coaker T.H. (Ed.)
    *Advances in Applied Biology Vol. VI.*  pp. 154-220.  Academic Press,
    London.
Ministry of the Environment, Denmark (1978)  *Environmental Protection
    Act.*  49 pp.  National Agency of Environmental Protection, 29
    Strandgade, Copenhagen.
Monschau-Dudenhausen K. (1983)  Wasserpflanzen als Belastungs-Indika-
    toren in Fliessgewässern dergestellt am Beispiel der Schwarzwald-
    flüsse Nagold und Alb.  *Beih. Veröff. Naturschutz Landschaftspflege
    Bad.-Württ.* <u>28</u>, 1-118.
Moore J.W. (1976)  Seasonal succession of algae in rivers.  I. Examples
    from the Avon, a large slow-flowing river.  *J. Phycol.* <u>12</u>, 342-349.
Moore J.W. (1977a)  Seasonal succession of algae in rivers.  II.
    Examples from Highland Water, a small woodland stream.  *Arch.
    Hydrobiol.* <u>80</u>, 160-171.
Moore J.W. (1977b)  Some factors affecting algal densities in a
    eutrophic farmland stream.  *Oecologia, Berl.* <u>29</u>, 257-267.

Moore J.W. (1978)  Seasonal succession of algae in rivers.  III.
  Examples from the Wylye, a eutrophic farmland river.  *Arch. Hydro-
  biol*. 83, 367-376.
Mortensen E. (1981)  Energy flow in trout (*Salmo trutta* L.) population
  and the benthic invertebrate community of Bisballe baek, Denmark.
  *I Jornadas de Ichtyologia Iberica*.  Leon, 25-30 mayo 1981.
Mosello R., Guidetti L., Calderoni A., Cominazzini C., De Giuli E. &
  Battioli M.T. (1982)  *Water Quality of Tributary Streams of Lago
  Maggiore*.  Symposium A.I.O.L., 1-3 December 1980, Chiavar.  (In
  Italian)
Moss M.O. (1977)  The temporal and spatial distribution of the valves
  of the diatom *Rhoicosphenia curvata* in the River Wye.  *Freshwat.
  Biol*. 7, 573-578.
Mouvet C. & Bourg A.C.M. (1983)  Speciation (including adsorbed species)
  of copper, lead, nickel and zinc in the Meuse River.  Observed
  results compared to values calculated with a chemical equilibrium
  computer program.  *Water Res*. 17, 641-649.
Müller C. (1978)  On the productivity and chemical composition of some
  benthic algae in hard-water streams.  *Verh. int. Verein. theor.
  angew. Limnol*. 20, 1457-1462.
Müller K. (1974)  Stream drift as a chronobiological phenomenon in
  running water ecosystems.  *A. Rev. Ecol. Systematics* 5, 309-323.
Nedeljkovic R. (1978)  Einflus der Anschwemmungen des Pek-Flusses auf
  die Bodenfauna der Donau.  *Verh. int. Verein. theor. angew. Limnol*.
  20, 1531-1535.
Nelva A. (1979)  Structure écologique de trois rivières du Jura
  méridional d'après l'analyse multivariée de la macrofaune benthique.
  *Arch. Hydrobiol*. 87, 327-346.
Neveu A. (1983)  Le zoobenthos des eaux courantes/The zoobenthos of
  running waters.  In: Association Française Limnologie, *Traveaux
  Française en Limnologie*.  pp. 48-52.  XX^e Congres SIL, Lyon,
  August 1983.  (French and English)
Newbold C., Purseglove J. & Holmes N. (1983)  *Nature Conservation and
  River Engineering*.  36 pp.  Nature Conservancy Council, Shrewsbury.
Niesiolowski S. (1980)  Studies on the abundance, biomass and vertical
  distribution of larvae and pupae of black flies (Simuliidae, Diptera)
  on plants of the Grabia river, Poland.  *Hydrobiologia* 75, 149-156.
Olechowska M. (1982)  Zonation of mayflies (Ephemeroptera) in several
  streams of the Tatra Mts and the Podhale region.  *Acta Hydrobiol*.
  24, 63-71.
Otto C. & Svensson B.S. (1983)  Properties of acid brown water streams
  in South Sweden.  *Arch. Hydrobiol*. 99, 15-36.
Pasternak K. (1973)  Rozprzestrzenienie metali ciezkich w wodach
  plynacych w rejonie wystepowania naturalnych złóż oraz przemyslu
  cynku i olowiu.  The spreading of heavy metals in flowing waters
  in the region of occurrence of natural deposits and of the zinc
  and lead industry.  *Acta Hydrobiol*. 14, 145-166.
Pedersen P.M. (1977)  Blue-green algae from the thermal springs of
  Ûnartoq, Southwest Greenland.  *Bot. Tidsskrift* 71, 80-83.
Penczak T., Moliński M., Kusto E., Palusiak H., Panusz H. & Zalewski,
  M. (1976)  The ecology of roach (Rutilus rutilus L.) in the barbel
  region of the polluted Pilica River.  2. Dry weight, ash and
  contents of some elements.  *Ekol. pol*. 24, 623-638.

Penczak T., Zalewski M., Suszycka R. & Moliński M. (1981)  Estimates of density, biomass and growth of fish populations in two small lowland rivers. *Ekol. pol.* <u>29</u>, 233-255.

Penczak T., Kusto E., Krzyżanowska D., Moliński M. & Suszycka E. (1984)  Food consumption and energy transformations by fish populations in two small lowland rivers in Poland. *Hydrobiologia* <u>108</u>, 135-144.

Perrin J.F. & Roux A.L. (1978)  Structure et fonctionnement des écosystèmes du Haut-Rhône français.  6. La macrofaune benthique du fleuve. *Verh. int. Verein. theor. angew. Limnol.* <u>20</u>, 1494-1502.

Pieper H.-G., Meijering M.P.D. (1982)  *Gammarus* occurrence as an indication for stable conditions in Hessian woodland brooks and rivers. *Polskie Archwm Hydrobiol.* <u>29</u>, 283-288.

Pietsch W. (1974)  Ökologische Untersuchwung und Bewertung von Fliessgewässern mit Hilfe hoherer Wasserpflanzen - Ein Beitrag zur Belastung aquatischer Ökosysteme. *Mitt. Sekt. Geobot. Phytotax. Biol. Ges. DDR,* 13-29.

Pinder L.C.V. (1980)  Spatial distribution of Chironomidae in an English chalk stream.  In: Murray D.A. (1980).  *Chironomidae: Ecology, Systematics, Cytology and Physiology.*  pp. 153-161.  Pergamon, Oxford.

Prat N., Rug M.A. & Gonzalez M.A. (1983)  *Predicció i Control de la Qualitat de les Aigües dels Rius Besòs i Llobregat, II. El Poblament faunistic i la Seva Relació amb la Qualitat de les Aigües.*  185 pp.  Estudis monografies 9, Servei del Medi Ambient. Diputació de Barcelona.

Pratten D.J. (1980)  Growth in the crayfish *Austropotamobius pallipes* (Crustacea Astacidae). *Freshwat. Biol.* <u>10</u>, 401-412.

Price S.M. & Price J.H. (1983)  The River Wandle: studies on the distribution of aquatic plants. *London Naturalist* <u>62</u>, 26-58.

Regione Veneto (1977)  *Symposium on Pollution of the River Adige.* 15 January 1977.  Verona.  (In Italian)

Reid J.M., Cresser M.S. & MacLeod D.A. (1980)  Observations on the estimation of total organic carbon from u.v. absorbance for an unpolluted stream. *Water Res.* <u>14</u>, 525-529.

Reinheimer G. (1977a, b)  Mikrobiologische Untersuchungen in Flussen. I. Fluoreszenzmikroskopische Analyse der Bakterienflora einiger norddeutscher Flüsse. *Arch. Hydrobiol.* <u>81</u>, 106-118.  II. Die Bakterienbiomasse in einigen norddeutschen Flüssen. *Arch Hydrobiol.* <u>81</u>, 259-267.

Richardot-Coulet M., Richoux P. & Roux C. (1983)  Structure et fonctionnement des écosystèmes du Haut-Rhône francais.  29. Structure des peuplements de macroinvertébrés benthiques d'un ancien méandre. *Arch. Hydrobiol.* <u>96</u>, 363-383.

Rofes B. (1975)  Influence de la craie en poudre sur les fonds envasés en eau courante. *Verh. int. Verein. theor. angew. Limnol.* <u>19</u>, 1693-1704.

Rosset J., Bärlocher F. & Oertli J.J. (1982)  Decomposition of conifer needles and deciduous leaves in two Black Forest and two Swiss Jura streams. *Int. Rev. ges. Hydrobiol.* <u>67</u>, 695-712.

Roux A.L. (1976)  Structure et fonctionnement des écosystèmes du Haut-Rhône français. 1. Presentation de l'étude. *Bull. Ecol.* 7, 475-478.

Roux A.L. (1983)  Un hydrosysteme fluviatile: le Haute-Rhône et sa vallée - A fluvial hydrosystem: the French Upper Rhône and its valley. In: Association Française de Limnologie, *Travaux Français en Limnologie*, pp. 252-257. XX Congres SIL, August 1983 (French and English)

Roux A.L., Tachet H. & Neyron M. (1976)  Structure et fonctionnement des écosystèmes du Haut-Rhône français. III. Une technique simple et peu onéreuse pour l'etude des macro-invertébrés benthiques des grands fleuves. *Bull. Ecol.* 7, 493-496.

Russev B. (1972)  Über die Migration der Rheobionten in Fliessgewässern. *Verh. int. Verein. theor. angew. Limnol.* 18, 730-734.

Sabater S. (1983)  Distribucion espacio-temporal de las poblaciones de algal del arroyo de l'Avenco (Barcelona). *Actas del Primer Congreso de Limnologia* (1983). pp. 159-166.

Sandbeck K.A. & Ward D.M. (1982)  Temperature adaptations in the terminal processes of anaerobic decomposition of Yellowstone National Park and Icelandic Hot Spring microbial mats. *Appl. environ. Microbiol.* 44, 844-851.

Sand-Jensen K. & Rasmussen L. (1978)  Macrophytes and chemistry of acidic streams from lignite mining areas. *Bot. Tidsskrift* 72, 105-112.

Saunders M.J. & Eaton J.W. (1976)  A method for estimating the standing crop and nutrient content of the phytobenthos of stony rivers. *Arch. Hydrobiol.* 78, 86-101.

Say P.J. & Whitton B.A. (1980)  Changes in flora down a stream showing a zinc gradient. *Hydrobiologia* 76, 255-262.

Say P.J. & Whitton B.A. (1981)  Chemistry and plant ecology of zinc-rich streams in the Northern Pennines. In: Say P.J. & Whitton B.A. (Eds) *Heavy Metals in Northern England: Environmental and Biological Aspects*. 198 pp., pp. 55-64.

Say P.J. & Whitton B.A. (1982a) Chimie et ecologie de la vegetation de cours d'eau en France à fortes teneurs en zinc. Massif Central. *Annls Limnol.* 18, 3-18.

Say P.J. & Whitton B.A. (1982b) Chemistry and plant ecology of zinc-rich streams in France. 2. The Pyrenees. *Annls Limnol.* 18, 19-31.

Scharz F. & Betschart B. (1979)  The use of periphyton from Lake Zurich to estimate the algal growth potential in River Limmat Water. *Schweiz. Z. Hydrol.* 41, 141-149.

Schmitz W. & Illies J. (1980)  Die Verfahren der biologischen Beurteilung des Gütezustandes der Fliessgewässer. *Studien zum Gewässerschutz* 5.

Schmitz W. & Schuman G.O. (1982)  Die sommerlichen Wassertemperaturen der Äschenzone mitteleuropäischer Fliessgewässer. *Arch. Hydrobiol.* 95, 435-443.

Scullion J. (1983)  Effects of impoundments on downstream bed materials of two upland rivers in mid-Wales and some ecological implications of such effects. *Arch. Hydrobiol.* 96, 329-344.

Seyed-Reihani A., Gilbert J. & Ginet R. (1982a)  Structure et fonctionnement des écosystèmes du Haut-Rhône français.  23. Ecologie de
deux stations interstitielles; influence de la pluviosité sur leur
peuplement. *Pol. Archwm Hydrobiol*. 29, 501-511.

Seyed-Reihani A., Ginet R. & Reygrobellet J.L. (1982b)  Structure et
fonctionnement des écosystèmes du Haut-Rhône français.  30. Le
peuplement de trois stations interstitielles dans la plaine de
Miribel-Jonage (Vallée du Rhône en amont de Lyon) en relation avec
leur alimentation hydrogéologique. *Rev. fr. Sci. Eau* 1, 163-174.

Sivko T.N., Kovalevskaja R.Z., Mikheeva T.M. & Ostapenia A.P. (1972)
Biological transformation of organic matter in a river. *Verh. int.
Verein. theor. angew. Limnol*. 18, 756-760.

Skoczén I. (1982)  The effect of rain waters on lead levels in the
Vistula in the region of Cracow agglomeration. *Acta Hydrobiol*. 24,
95-107.

Sládeček V. & Miskovsky O. (1976)  Bacterial periphyton of the Vltava
river. *Hydrobiologia* 51, 181-187.

Sloor W., de Zwart D. & van de Kerkhoff J.F.J. (1983)  Monitoring the
rivers Rhine and Meuse in the Netherlands for toxicity. *Aquat.
Toxicol*. 4, 189-198.

Smith B.D., Lyle A.A. & Maitland P.S. (1983)  The ecology of running
waters near Aberfeldy, Scotland, in relation to a proposed barytes
mine: an impact assessment. *Environ. Pollut. (Ser. A)* 32, 269-306.

Smith C.S., Adams R.S., Schmitt M.R. & Adams S.S. (1978)  Phosphorus
in the water, sediment, and vegetation of the Crnojevica River,
Montenegro, Yugoslavia. *Verh. int. Verein. theor. angew. Limnol*.
20, 1536-1542.

Smith R.V., Stevens R.J., Foy R.H. & Gibson C.E. (1982)  Upward trend
in nitrate concentration in rivers discharging into Lough Neagh for
the period 1969-1979. *Water Res*. 16, 182-188.

Sperling J.A. (1975)  Algal ecology of southern Icelandic hot springs
in winter. *Ecology* 56, 183-190.

Stazner B. (1981)  The relation between "hydraulic stress" and micro-
distribution of benthic macroinvertebrates in a lowland running
water system, the Schierenseebrooks (North Germany). *Arch. Hydrobiol*. 91, 192-218.

Steffan A.W. (1971)  Chironomid (Diptera) biocoenoses in Scandinavian
glacier brooks. *Canadian Entomologist* 103, 478-486.

Steinberg C. & Melzer A. (1983)  Aufnahme, Transport und Abgabe von
Kohlenstoff durch submerse Makrophyten von Fliesswasserstandorten.
*Schweiz. Z. Hydrol*. 45, 333-344.

Steubing L., Fricke G. & Jehn H. (1983)  Veranderung des Algenspek-
trums der Eder im Verlauf von vier Jahrzehnten. *Arch. Hydrobiol*.
205-222.

Stevens R.J. & Smith R.V. (1978)  A comparison of discrete and intensive sampling for measuring the loads of nitrogen and phosphorus
in the River Main, County Antrim. *Water Res*. 12, 828-830.

Stolz L. & Weise G. (1976)  Einflus des Nährstoffgehaltes auf die
Stoffproduktion submerser Macrophyten in Fliessgewässern, dargestellt
an *Fontinalis antipyretica*. *Limnologica, Berl*. 10, 405-417.

Stroot P. (1984) Faunistique et répartition longitudinale des Tri-
    choptères dans une rivière salmonicole de basse montagne, en Bel-
    gique. *Hydrobiologia* 108, 245-258.
Tangen K., Müller C. & Brettum P. (1978) Periphytic and drifting
    microalgae in a tributary stream of Øvre Heimdalsvattnet. *Hol-
    arctic Ecol.* 1, 148-154.
Tarak P. (1981) Environmental legislation and pollution control
    regarding water resources. In: *Water for Human Consumption. Man
    and his Environment.* A selection of papers prepared for the IVth
    World Congress of the International Water Resources Association
    (IWRA). pp. 494-506. Tycooly International Publishing Ltd, Dublin.
Thirb H.H. & Benson-Evans K. (1983) Studies on the life cycle of
    *Lemanea* using natural artificial substrates. *Hydrobiologia* 98,
    110-124.
Thomas E.A. (1976) Der flutende Hahnesfuss (*R. fluitans*), ein neues
    limnologisches Problem am Rhine. *Z. Wasser-Energie-Luft* 10,
    230-234.
Thommen G.H. & Westlake D.F. (1981) Factors affecting the distribution
    of populations of *Apium nodiflorum* and *Nasturtium officinale* in
    small chalk streams. *Aquatic Botany* 11, 21-36.
Thorup J. & Lindegaard C. (1977) Studies on Danish springs. *Folia
    Limnol. Scand.* 17, 7-15.
Toner P.F. & Lennox L.J. (1980) Nitrate content of Irish river waters.
    *Irish J. environ. Sci.* 1, 75-76.
Townsend C.R., Hildrew A.G. & Francis J. (1983) Community structure
    in some southern English streams: the influence of physicochemical
    factors. *Freshwat. Biol.* 13, 521-544.
Uherkovich G. (1966) Übersicht über das Potamophytoplankton der
    Tisza (Theiss) in Ungarn. *Hydrobiologia* 28, 252-280.
Vernaux J. (1983) Les methodes biologiques pratiques d'appreciation
    de la qualité. Practical biological methods used in France for the
    assessment of the quality of running waters. In: Association
    Francaise de Limnologie, Travaux Français en Limnologie, pp. 178-191.
    XX Congres SIL, Lyon, August 1983. (French and English)
Vrhovsek D., Martinincic & Kralj M. (1981) Evaluation of the polluted
    river Savinja with the help of macrophytes. *Hydrobiologia* 80,
    97-110.
Weber-Oldecop D.W. (1971) Wasserpflanzengesellschaften im östlichen
    Niedersachsen (II.) *Int. Rev. ges. Hydrobiol.* 56, 79-122.
Wehr J.D., Say P.J. & Whitton B.A. (1981) Heavy metals in an indus-
    trially polluted river, the Team. In: Say P.J. & Whitton B.A. (Eds)
    *Heavy Metals in Northern England: Environmental and Biological
    Aspects.* 198 pp. pp. 99-109. Department of Botany, University of
    Durham.
Welton J.S. (1980) Dynamics of sediment and organic detritus in a
    small chalk stream. *Arch. Hydrobiol.* 90, 162-181.
Werner I. & Weise G. (1982) Biomass production of submersed macrophytes
    in a selected stretch of the River Zschopau (South GDR) with special
    regard to orthophosphate incorporation. *Int. Rev. ges. Hydrobiol.*
    67, 45-62.
Westlake D.F. & Dawson F.H. (1982) Thirty years of weed cutting on a
    chalk-stream. In: *Proceedings, European Weed Research Society 6th
    Symposium on Aquatic Weeds.* pp. 132-140.

Whitton B.A. (1975) *River Ecology.* 725 pp. Blackwell Scientific Publications, Oxford.

Wiegleb G. (1979a) Der Zusammenhang zwischen Gewässergute und Macrophytenvegetation in niedersachsicher Fliessgewässern. *Landschaft + Stadt* 11, 32-35.

Wiegleb G. (1979b) Vorläufige Übersicht über die Pflanzengesellschaften der niedersächsischen Fliessgewässer. *Natursch. u. Landschaftspflege Niedersachsen* 11, 85-116.

Wiegleb G. (1981a) Application of multiple discriminant analysis on the analysis of the correlation between macrophyte vegetation and water quality in running waters of Central Europe. *Hydrobiologia* 79, 90-100.

Wiegleb G. (1981b) Struktur, Verbreitung und Bewertung von Makrophytengesellschaft niedersächsischer Fliessgewässer. *Limnologica (Berlin)* 13, 427-448.

Wilkinson D.R. & Jones J.W. (1977) The fecundity of dace, *Leuciscus leuciscus* (L.) in Emral Brook, Clwyd, North Wales. *Freshwat. Biol.* 7, 135-145.

Willis M. (1983) A comparative survey of *Ancylus fluviatilis* (MÜLLER) populations in the Afon Crafnant, N. Wales, above and below an input of zinc from mine-waste. *Arch. Hydrobiol.* 98, 198-214.

Wilson R.S. (1977) Chironomid pupal exuviae in the River Chew. *Freshwat. Biol.* 7, 9-18.

Wise E.J. (1976) Studies on the Ephemeroptera of a Northumbrian river system. I. Serial dilution and relative abundance. *Freshwat. Biol.* 6, 363-372.

Wotton R.S. (1979) The influence of a lake on the distribution of blackfly species (Diptera: Simuliidae) along a river. *Oikos* 22, 368-372.

Wright J.F., Cameron A.C., Hiley P.D. & Berrie A.D. (1982) Seasonal changes in biomass of macrophytes on shaded and unshaded sections of the River Lambourn, England. *Freshwat. Biol.* 12, 271-283.

Wright J.F., Hiley P.D., Cameron A.C., Wigham M.E. & Berrie A.D. (1983) A quantitative study of the macroinvertebrate fauna of five biotopes in the River Lambourn, Berkshire, England. *Arch. Hydrobiol.* 96, 271-292.

Wright R.F., Dale T., Gjessing E.T., Hendrey G.R., Henriksen A., Johanhessen M. & Muniz I.P. (1976) Impact of acid precipitation on freshwater ecosystems in Norway. *Water Air Soil Pollut.* 6, 483-499.

Wurhmann K. & Eichenberger D. (1980) Künstliche Bäche als Hilfsmittel der experimentellen Fliessgewässer-Ökologie. *Vom Wasser* 54, 1-18.

Zaiss U. (1981) Natural ebullition of mine gas and its microbial oxidation in the Netzbach brook, Saarland. *Verh. int. Verein. theor. angew. Limnol.* 21, 1381-1385.

Zalewski M. & Penczak T. (1981) Characterization of the fish community of the Utrata River drainage-basin, and evaluation of the efficiency of catching methods. *Pol. Archwm Hydrobiol.* 28, 385-396.

Zullin A. & Ricci C. (1980) Bdelloid rotifers and nematodes in a small Italian stream. *Freshwat. Biol.* 10, 67-72.

# Appendix B: Authorities, Synonyms and English Names

Latin names, authorities, synonyms and English common names for selected vascular plants and fish. (The synonyms given are those used in some chapters or which were in original manuscripts from authors.)

VASCULAR PLANTS

PTERIDOPHYTES

*Isoetes*
*Osmunda regalis* L.

DICOTYLEDONS

| | |
|---|---|
| *Apium graveolens* L. | wild celery |
| *Apium nodiflorum* (L.) Lag. | fool's watercress |
| *Berula erecta* (Huds.) Coville | lesser water-parsnip |
| (= *Sium erectum* Huds.) | |
| *Callitriche* | water-starwort |
| *Ceratophyllum demersum* L. | hornwort |
| *Filipendula ulmaria* (L.) Maxim. | meadowsweet |
| *Hottonia palustris* L. | water violet |
| *Impatiens glandulifera* Royle | Himalayan balsam |
| *Limosella aquatica* L. | mudwort |
| *Littorella uniflora* (L.) Aschers. | shoreweed |
| *Lobelia dortmanna* L. | water Lobelia |
| *Mentha aquatica* L. | water mint |
| *Myriophyllum* | water milfoil |
| *Nuphar lutea* (L.) Sm. | yellow water lily |
| *Nuphar pumila* (Timm.) DC | least water lily |
| *Nymphaea alba* L. | white water lily |
| *Nymphoides peltata* (S.G. Gmel.) O. Kuntze | fringed water lily |
| *Oenanthe* | water dropwort |
| *Ranunculus fluitans* Lam., *R. penicillatus* | water buttercup |
| (Dumort.) Bab. | |
| *Rorippa amphibia* (L.) Besser | great yellow-cress |
| *Rorippa islandica* (Oder) Borbas | marsh yellow-cress |
| *Rorippa nasturtium-aquaticum* (L.) Hayek | watercress |
| (= *Nasturtium officinale* R. Br.) | |
| *Rorippa sylvestris* (L.) Besser | creeping yellow-cress |
| *Utricularia* | bladder-wort |
| *Veronica beccabunga* L. | brooklime |

MONOCOTYLEDONS

| | |
|---|---|
| *Acorus calamus* L. | sweet fly |
| *Alisma plantago-aquatica* | water-plantain |
| *Butomus umbellatus* L. | flowering-rush |
| *Catabrosa aquatica* (L.) Beauv. | whorl-grass |
| *Cladium mariscus* (L.) Pohl | great fen-sedge |
| *Elodea canadensis* Michx. | Canadian pondweed |
| *Elodea ernstiae* St John | |
| *Eleocharis palustris* (L.) Roem & Schult. | common spike-rush |
| *Glyceria maxima* (Hartm.) Holmberg | reed sweet-grass |
| *Hydrocharus morsus-ranae* L. | frog-bit |
| *Juncus* | rush |
| *Lemna gibba* L. | fat duckweed |
| *Lemna minor* L. | common duckweed |
| *Lemna trisulca* L. | ivy-leaved duckweed |
| *Phalaris arundinacea* L. | reed canary-grass |
| *Phragmites australis* (Cav.) Trin. ex. Steud. | common reed |
| *Potamogeton lucens* L. | shining pondweed |
| *Potamogeton pectinatus* L. | fennel pondweed |
| *Scirpus fluitans* L. | floating club-rush |
| *Scirpus lacustris* L. (= *Schoenoplectus lacustris* (L.) Palla) | common club-rush |
| *Scirpus tabernaemontani* (= *Schoenoplectus tabernaemontani* (C.C. Gmel.) Palla) | |
| *Sparganium emersum* (= *S. simplex* Huds.) | unbranched bur-reed |
| *Sparganium erectum* L. | branched bur-reed |
| *Sparganium minimum* Wallr. | least bur-reed |
| *Stratrotes aloides* L. | water-soldier |
| *Typha angustifolia* L. | lesser bulrush |
| *Typha latifolia* L. | bulrush, great reedmace |
| *Wolffia arrhiza* (L.) Wimm. | rootless duckweed |
| *Zannichellia palustris* | horned pondweed |

FISH

| | |
|---|---|
| *Abramis brama* (L.) | bream |
| *Acipenser naccarii* Bp. | |
| *Acipenser sturio* L. | sturgeon |
| *Alburnus alburnus* (L.) | bleak |
| *Alburnus bipunctatus* (Bloch) = *Alburnoides bipunctatus* | bystryanka |
| *Alosa alosa* (L.) | allis shad, shad, mayfish |
| *Alosa fallax* (Lacepede) | twaite shad, twait |
| *Anguilla anguilla* (L.) | eel |
| *Aspius aspius* (L.) | asp, rapfen |
| *Barbus barbus* (L.) | barbel |
| *Barbus meridionalis* Risso | southern barbel |
| *Carassius auratus* (L.) | goldfish |

| | |
|---|---|
| *Carassius carassius* (L.) | crucian carp |
| *Chrondrostoma nasus* (L.) | nase |
| *Chrondrostoma toxostoma* (Vallot) | soiffe |
| *Cobitis taenia* (L.) | spined loach |
| *Coregonus oxyrhynchus* (L.) | houting |
| *Cottus gobio* (L.) | bullhead |
| *Cottus poecilopus* Heck | alpine bullhead |
| *Ctenopharyngodon idella* (Val.) | grass carp |
| *Cyprinus carpio* (L.) | carp |
| *Esox lucius* (L.) | pike |
| *Gambusia affinis* (Baird & Girard) | mosquito-fish |
| *Gasterosteus aculeatus* (L.) | three-spined stickleback |
| *Gobio gobio* (L.) | gudgen |
| *Gymnocephalus cernua* (L.) | ruffe |
| *Huso huso* (L.) | beluga |
| *Ictalurus melas* (Rafinesque) | black bullhead |
| *Ictalurus nebulosus* (Lesueur) | brown bullhead |
| *Lampetra fluviatilis* (L.) | river lamprey, lampern |
| *Lampetra planeri* (Bloch) | brook lamprey |
| *Lepomis gibbosus* (L.) | pumpkinseed |
| *Leucaspius delineatus* (Heck) | |
| *Leuciscus cephalus* (L.) | chub |
| *Leuciscus idus* (L.) | ide |
| *Leuciscus leuciscus* (L.) | dace |
| *Leuciscus souffia* | |
| *Liza ramada* (Risso) | thin-lipped grey mullet |
| = *Chelon ramada* (Risso) | |
| *Liza saliens* (Risso) | |
| *Lota lota* (L.) | burbot |
| *Micropterus salmoides* (Lacep.) | large-mouthed bass |
| *Noemacheilus barbatulus* (L.) | stone loach |
| *Oncorhynchus tshawytscha* (Walbaum) | Chinook salmon |
| *Padogabius martensi* (Gthr) | |
| *Perca fluviatilis* (L.) | perch |
| *Petromyzon marinus* (L.) | sea lamprey |
| *Phoxinus phoxinus* (L.) | minnow |
| *Platichthys flesus* (L.) | flounder |
| *Pungitius pungitius* (L.) | ten-spined stickleback, nine-spined stickleback |
| *Rutilus pigus* (Lacep.) | Danubian roach |
| *Rutilus rubilia* (Bp.) | |
| *Rutilus rutilus* (L.) | roach |
| *Sabanejewia larvata* (De Fil.) | |
| *Salmo gairdneri* | rainbow trout |
| *Salmo salar* | salmon |
| *Salmo trutta* (L.) (*fario*) | brown trout, brook trout |
| *Salmo trutta* (L.) (*marmoratus*) | |
| *Salmo trutta* (L.) (*trutta*) | sea trout |
| *Salvelinus alpinus* (L.) | Arctic charr |
| *Salvelinus fontinalis* (Mitch.) | brook charr |
| *Scardinius erythrophthalmus* (L.) | rudd |
| *Silurus glanis* (L.) | wels |

| | |
|---|---|
| *Stizostedion lucioperca* (L.) | zander, pike-perch |
| *Tinca tinca* (L.) | tench |
| *Thymallus thymallus* (L.) | grayling |

# Organism Index

*Abies pectinata* 241
*Abramis brama* 73, 230, 512
*Abutilon theophrasti* 283
Acanthocephala 192, 424
*Acanthocyclops robustus* 418
*Acer campestre* 416
*Acerina cernus* 412
*Achnanthes* 99, 161, 374-5, 378,
  486
*Achnanthes affinis* 375
*Achnanthes exilis* 375
*Achnanthes flexella* 375
*Achnanthes lanceolata* 336
*Achnanthes minutissima* 375, 379,
  413, 542
*Achnanthes minutissima* var.
  *cryptocephala* 375, 379
*Achnanthes pyrenaica* 513
*Achnanthes saxonica* 375
*Acilius sulcatus* 196
*Acipenser naccarii* 423
*Acipenser sturio* 293, 294, 423
Acipenseridae 422
*Acorus calamus* 37, 489
*Acrocladium cuspidatum* 155
*Acroloxus lacustris* 291, 336
  420, 460
*Acrophylax vernalis* 516
*Acrophylax zerberus* 516
*Actinastrum hantzschii* 161, 414
*Actinocyclus normannii* 450
  452, 453
*Actinophrys sol* 416
*Adenovirus* 413-14
*Adiceltà reducta* 196
*Agabus biguttatus* 195, 196
*Agabus bipustulatus* 195
*Agabus nebulosus* 196

*Agabus sturmi* 196
*Agapetus delicatulus* 516
*Agapetus fuscipes* 381
*Agraylea sexmaculata* 420
*Agrion splendens* 197
*Agrostis stolonifera* 133
*Alburnoides bipunctatus* 512
*Alburnus alburnus* 73, 230, 424,
  512, 522
*Alburnus alburnus alborella* 412,
  421-2, 423-4
*Alburnus bipunctatus* 294
*Alisma plantago-aquatica* 489
allis shad 73
*Allium schoenoprasum* 66
*Allogamus auricollis* 246, 250,
  254-5, 258, 516
*Allogamus starmachi* 516, 521
*Allogamus uncatus* 516
*Allotrichia pallicornis* 420
*Alnus glutinosa* 416
*Alona* 450
*Alona rectangula* 417
*Alosa alosa* 73, 230, 293-4, 455
*Alosa fallax* 73, 455
*Alosa fallax nilotica* 421-3
*Amaranthus* 416
*Amblystegium* 133
*Amblystegium fluviatile* 64
*Amblystegium riparium* 132, 135,
  143
*Amblystegium tenax* 64
*Ameirus nebulosus* 230
*Ameletus inopinatus* 517
Amphipoda 289, 418-19
*Amphora ovalis* 375, 486
*Anabaena* 220
*Anabaena flos-aquae* 491

*Anacaena limbata*   491
Anatidae   425
*Ancylus*
*Ancylus fluviatilis*   68, 101, 139,
    289-91, 381, 458, 460, 522, 543
*Anguilla anguilla*   73, 104, 230,
    421-2
Anguillidae   104, 422
Anisoptera   419
*Ankistrodesmus*   61
*Ankistrodesmus falcatus*   62,
    219, 491
Annelida   162
*Annitella obscurata*   516
*Annitella thuringica*   516
*Anodonta*   359
*Anodonta anatina*   454, 463
*Anodonta cygnea*   290
*Anomoeoneis exilis*   375
*Apatania carpathica*   516
*Aphanizomenon flos-aquae*   220,
    561
*Aphanocapsa*   375
*Aphanochaete repens*   376
*Aphanothece*   375
*Aphelocheirus*   396
*Aphelocheirus aestivalis*   68,
    444
*Aphelocheirus montandoni*   381
*Apium inundatum*   380
*Apium nodiflorum*   188-9
*Arcynopteryz compacta*   245, 247,
    251, 254, 257-8
*Aredea purpurea*   425
Arpacticidae   417
Ascaridae   424
*Asellus*   39, 69, 137-9, 359,
    395-6
*Asellus aquaticus*   68, 101, 291,
    336, 380, 418, 460
asp   493
*Asplanchna brightwelli*   335
*Astacus*   359
*Astacus astacus*   445
*Astacus fluviatilis*   71
*Asterionella*   285, 561, 572
*Asterionella formosa*   161, 284-6,
    444, 490-1, 494, 560
*Asymphylodora tincae*   424
*Atherix ibis*   68
*Athripsodes albifrons*   196

*Athyrium filix-femina*   63
*Atrichum crispum*   64
*Attheya zachariasi*   286
*Aulodrilus pluriseta*   68, 101,
    197
*Austropotamobius*   395-6
*Austropotamobius pallipes*   71,
    166
*Austropotamobius pallipes ful-
    cisianus*   419

Bacillariophyta   510
Baetidae   359, 395, 419
*Baetis*
*Baetis alpinus*   244-5, 247-52,
    254-5, 257, 515, 517, 546,
    549
*Baetis fuscatus*   517
*Baetis lutheri*   517
*Baetis melanonyx*   517
*Baetis meridionalis*   543, 546,
    549
*Baetis muticus*   68, 517
*Baetis rhodani*   39, 68, 94, 99,
    101, 163, 194, 380, 395-6,
    517, 543, 546, 549
*Baetis sinaicus*   517
*Baetis vardarensis*   517, 546, 549
*Baetis vernus*   140, 194, 517
*Bangia atropurpurea*   12, 279-80,
    359
barbel   73
*Barbilophozia floerkei*   64
*Barbus*   542
*Barbus barbus*   73, 230, 420, 423,
    445, 512, 522
*Barbus barbus plebejus*   421-2
*Barbus meridionalis*   422, 512
*Batrachium*   37
*Batrachospermum*   279, 378-9, 486
*Batrachospermum atrum*   375
*Batrachospermum moniliforme*   375,
    379
*Beggiatoa*   39
*Beraea maurus*   196
*Bereodes minutus*   196
*Bidens*   283
*Bidens tripartita*   489
*Bithynia*   359
*Bithynia tentaculata*   289, 291,
    336, 460

Bivalvia   192, 291
bleak   73, 493
Blepharoceridae   543
*Blicca Björkna*   230
*Bosmina coregoni*   417
*Botrydium granulatum*   283
*Botryococcus braunii*   376
*Brachionus calcyformis*   416
*Brachycentrus subnubilus*   68, 196, 516
*Brachyonus calyciflorus*
*Brachythecium plumosus*
*Brachythecium rivulare*   64, 244, 487
*Brachythecium rutabulum*   64
*Brassica nigra*   283
*Bratracobdella paludosa*   417
bream   73, 166, 493
*Bryocamptus zschokkei*   250, 252
Bryozoa   289, 291
*Bryum pseudotriquetrum*   64
*Bulbochaete*   376, 379, 486
bullhead   73,  166, 493
burbot   493
*Butomus umbellatus*   37, 282, 395

*Caddis*   39
Caenidae   419
*Caenis*   396, 419
*Caenis bcskidcnsis*   517
*Caenis horaria*   196, 517
*Caenis luctuosa*   289
*Caenis moesta*   68, 381, 543
*Caenis rivulorum*   163, 196, 381
*Calliergon stramineum*   64
*Callitriche*   37-8, 189, 191, 380, 489, 570
*Callitriche hamulata*   64-5
*Callitriche intermedia*   158
*Callitriche platycarpa*   158
*Callitriche stagnalis*   453, 455
*Calopteryx splendens*   290, 445
*Calopteryx virgo*   299, 445
*Calothrix braunii*   375
*Calothrix parietina*   155, 158, 375
*Caltha palustris*   450
*Caltha palustris* var. *araneosa*   450
*Cambusia affinis holbrooki*   422
*Campylopus atrovirens*   64
*Candida*   414

*Candida albicans*   414
*Candida guilliermondi*   414
*Candida tropicalis*   414
*Capnia atra*   492
*Capnia pygmaea*   492
Caradriidae   425
*Carassius auratus*   421-2
*Carassius carassius*   422
*Carchesium polypinum*   416, 570
*Cardiocladus fuscus*   543, 547, 549
*Carex aquatilis*   488-9
*Carex gracilis*   489
*Carex rostrata*   380, 488-9
*Cariophyllaeus*   424
carp   73, 166
*Carteria*   62
Centrarchidae   422
*Centroptilum pennulatum*   517
*Centroptilum luteolum*   517
*Ceratium hirundinella*   284, 561
*Ceratodon purpureus*   487
*Ceratoneis arcus*   62, 244, 375, 486, 513
*Ceratophyllum demersum*   282, 395, 415, 488-9, 560, 570
Ceratopogonidae 359, 455
*Chaemaesiphon confervicola*   375
*Chaemaesiphon incrustans*   375
*Chaemaesiphon minutus*   375
*Chaemaesiphon polonicus*   375, 521
Chaetophorales   377
*Chaetopteryx fusca*   516
*Chaetopteryx polonica*   516
*Chaetopteryx villosa*   196
*Chaetosphaeridium globosum*   376, 378
*Chamaesiphon fuscus*
*Chamaesiphon incrustans*
*Chamaesiphon polonicus*   512-13, 515
*Chantransia chalybea* var. *leibleinii*   486
char   493
*Chara*   282
*Chara aspera*   279
*Chara contraria*   279
*Chara hispida*   279
*Chara tomentosa*   359
*Chenopodium*   283
*Cheumatopsyche lepida*   68-9, 381, 516

Chidoridae  417
*Chiloscyphus pallescens*  454
*Chiloscyphus polyanthus*  13-14, 16
*Chimarrha marginata*  288, 381
Chironomidae  70, 101, 135-9, 162,
    164-5, 191-2, 194, 198, 245-6,
    250, 254, 289, 359, 381, 395-6,
    420, 444, 452, 454-5, 459-60,
    463, 494, 511, 543, 547
Chironominae  420
*Chironomius* cf. *plumosus*  420
*Chironomus*  39, 395, 545, 549
*Chironomus* f.l. *fluviatilis*  463
*Chironomus* f.l. *reductus*  463
*Chironomus riparius*  420
*Chironomus thummi*  336, 427
*Chlamydomonas*  219, 491
*Chlanis dilata*  335
Chlorococcales  219, 223, 334
*Chloroperla breviata*  251, 253,
    257
*Chloroperla burmeisterii*  40
*Chloroperla torrentium*  380
*Chloroperla tripunctata*  68
Chlorophyceae  286
Chlorophyta  510
*Chondrostoma nasus*  17, 230,
    445, 512
*Chondrostoma soetta*  422
*Chondrostoma toxostoma*  230, 422
*Chroococcus*  375
*Chroomonas*  285
*Chrysococcus*  491
Chrysophyceae  286
Chrysophyta  510
*Chydorus sphaericus*  450
*Cicuta virosa*  489
Ciliophora  416
*Cinclidotus aquaticus*  13
*Cinclidotus danubicus*  13-14
*Cinclidotus fontinaloides*  13,
    64, 380
*Cinclidotus mucronatus*  13
*Cinclidotus nitricans*  13
cisco  493
Cladocera  164-5, 192, 417
*Cladophora*  12, 38, 99, 100,
    131-3, 135, 143, 157, 354,
    379, 393-5, 542, 546-7,
    563
*Cladophora crispata*

*Cladophora fracta*
*Cladophora glomerata*  12, 37,
    98-9, 135, 158, 279, 335,
    513-14, 521
*Cladophora okamurai*  454
*Cleon dipterum*  196, 289
*Climacium dendroides*  64
*Cloeon cognatum*  517
*Cloeon dipterum*  419
*Closterium*  486
*Closterium pronum*  491
Clupeidae  422
Cobitidae  104, 421-2
*Cobitis barbatula*  230
*Cobitis taenia bilineata*  421-3
*Cocconeis placentula*  63, 99, 336,
    375, 453, 513
*Coelambus confluens*  196
*Coelambus impressopunctatus*  196
*Coelastrum microporum*  219, 414
*Coelastrum reticulatum*  219, 225
*Coelosphaerium*  375, 378
*Coenagrion puella*  381
*Coleochaete*  377
*Coleochaete orbicularis*  376
*Coleochaete scutata*  376
Coleoptera  162, 191-2, 197-8,
    289, 420, 543
*Colymbetes fuscus*  195, 196
*Condona sarsi*  455
*Conocephalum conicum*  64
Copepoda  164-5, 192, 417
*Coregonus oxyrhynchus*  294, 455
*Coregonus pollan elegans*  361
*Corixa panzeri*  197
*Corixa punctata*  197
Corixidae  396
*Corynoneura lobata*  245
*Coscinodiscus lacustris*  286
*Cosmarium*  486
*Cosmarium undulatum*  486
Cottidae  104, 422
*Cottus gobio*  73, 104, 199, 230,
    421-2, 512
*Cottus poecilopus*  512
*Coxsackie*  413-14
*Crangonyx*  194
*Crassula aquatica*  489
*Cratoneuron commutatum*  244
*Crenobia alpina*  246

*Cricotopus*  68, 381, 459, 463, 522

*Cricotopus* gr. *algarum*  514-15

*Cricotopus annulator*  547

*Cricotopus bicinctus*  68, 547, 549

*Cricotopus* gr. *sylvestris*  420, 543 545, 547, 549

*Cricotopus* gr. *tremulus*  543, 547

*Cricotopus triannulatus*  547

*Cricotopus trifascia*  68

crucian carp  493

*Crucigenia tetrapedia*  491

Crustacea  101, 162-3, 291, 424

*Cryptochironomus* gr. *defectus* 420, 463

*Cryptococcus*  414

*Ctenopharyngodon idella*  294

*Cucubalus*  283

Cyanophyceae  286

*Cyclidium glaucoma*  416

Cyclopidae  417

*Cyclops vicinus*  421

Cyclostoma  421

*Cyclotella*  61, 219, 285-6, 375

*Cyclotella bodanica*  284

*Cyclotella comta*  284, 414, 491

*Cyclotella melosiroides*

*Cyclotella meneghiniana*  13, 62, 161, 286, 414, 522

*Cyclotella socialis*  284

*Cyclotella stelligera*  491

*Cymbella*  158, 280, 375, 486

*Cymbella affinis*  486, 513-15

*Cymbella helvetica*  513-14

*Cymbella minuta*  542

*Cymbella ventricosa*  244, 513-14

*Cynodon dactylon*  282-3

*Cyperus glomeratus*  416

Cyprinidae  104, 422

*Cyprinus carpio*  73, 230, 422, 542

*Cyrnus trimaculatus*  460

dace  73, 166

Daphnidae  417

*Datura stramonium*  283

Decapoda  419

*Dendrocoelum lacteum*  291, 336

*Denticula*  375

*Dermatocarpon fluviatile*  99

Desmidiales  376

*Desmosiphon maculans*  375, 377

*Dialytrichia mucronata*  13

*Diamesa*  543, 549

*Diamesa hamaticornis*  245, 249-50

*Diamesa insignipes*  197, 198

*Diamesa* gr. *latitarsis*  515

*Diamesa starmachi*  521

*Diamesa steinbocki*  515

Diamesinae  249-50, 252, 420

Diaptomidae  417

*Diatoma*  61, 99, 375

*Diatoma elongatum*  161, 284, 444, 486-7, 491, 494

*Diatoma hiemale*  244, 512-13, 515

*Diatoma hiemale* var. *mesodon* 512-13, 515, 542

*Diatoma vulgare*  62-3, 99, 285, 336, 414, 491, 513-15, 522

*Diatoma vulgare* var. *ehrenbergii* 513-14

*Diatoma vulgare* var. *productum* 513-14

*Dichelyma falcatum*  487

*Dichodontium pellucidum*  64

*Dichothrix*  377

*Dichothrix baueriana*  375

*Dichthrix orsiniana*  375

*Dicranella palustris*  64

*Dicranota*  68, 381

*Dicrotendipes* gr. *nervosus*  459, 463

*Dictyosphaerium*  219

*Dictyosphaerium pulchellum*  376, 491

*Didymosphenia geminata*  158, 375, 377-8, 486-7

*Dinobryon borgei*  491

*Dinobryon divergens*  491

*Dinobryon sertularia*  284

*Dibobryon sociale*  491, 561

*Dinocras cephalotes*  543, 549

*Diplectrona felix*  69, 381

*Diploneis ovalis*  375

*Diplophyllum albicans*  64

*Diplotaxis tenuifolia*  283

Diptera  101, 162, 192, 245, 255, 289, 359, 396, 420, 511, 522, 543, 545-6

*Draparnaldia*  376

*Draparnaldia glomerata*  279, 486

*Dreissena polymorpha*  291, 460

*Drepanocladus*  570
*Drepanocladus exannulatus*  64, 487
*Drepanocladus fluitans*  64
*Drepanocladus uncinatus*  64
*Drusus*  254-5, 258
*Drusus annulatus*  516
*Drusus biguttatus*  516
*Drusus carpathicus*  516
*Drusus discolor*  246, 251-2, 257, 419, 516
*Drusus monticola*  516
*Drusus rectus*  245, 250, 252, 257
Dryopidae  455
*Dryopteris filix-mas*  63
*Dugesia lugubris*  291
*Dugesia polychroa*  460
*Dugesia tigrina*  291, 336, 417, 460
Dytiscidae  136-7, 396, 420
*Dytiscus marginatus*

*Ecclisopteryx dalecarlica*  516
*Ecclisopteryx madida*  516
Ecdyonuridae  164, 359
*Ecdyonurus*  288
*Ecdyonurus aurantiacus*  419
*Ecdyonurus dispar*  68, 99, 517
*Ecdyonurus insignis*  517
*Ecdyonurus lateralis*  517
*Edcyonurus torrentis*  381, 517
*Ecdyonurus venosus*  381, 517
*Echinogammarus stammeri*  418
*Echinogammarus veneris*  418
*Echo*  413, 414
*Ecnomus tenellus*  460
eel  73, 166
*Egeria densa*  415
*Egretta garzetta*  425
*Eiseniella tetraedra*  68, 195, 197
*Elatine hexandra*  380
*Elatine hydropiper*  489
*Elatine triandra*  489
*Eleocharis acicularis*  488-9
*Eleocharis palustris*  489
*Eleocharis uniglumis*  489
*Elimis aenea*  381
*Elmis*  359
*Elmis aenea*  68, 163
*Elodea*  38

*Elodea canadensis*  37, 39, 64-5, 160, 560, 564
*Elodea ernstiae*  282
*Elodea nuttallii*  135, 282

Empididae  101, 245, 257
Enchtraeidae  68, 101, 192
*Endochironomus albipennis*  459
*Enteromorpha*  279
*Enteromorpha flexuosa*  160
*Enterovirus*  414
*Epeorus assimilis*  549
*Epeorus insignis*  549
*Epeorus sylvicola*  517
*Ephemera danica*  40, 165, 196, 361, 517
*Ephemera vulgata*  419
*Ephemerella*  71, 138-9, 395-6
*Ephemerella ignita*  69, 72, 99, 101, 138, 194, 517
*Ephemerella krieghoffi*  517
*Ephemerella major*  517, 549
Ephemeridae  419
Ephemeroptera  40, 70, 135, 137, 162-3, 191-2, 194, 198, 244-6, 288-9, 291, 359, 419, 444, 457, 464, 491, 494, 510-11, 517-18, 522, 543, 546
Ephemerellidae  419
*Ephydatia fluviatilis*  336
*Ephydra*  381
*Epithemia zebra* var. saxonica  375
*Equisetum arvense*  63
*Equisetum fluviatile*  158, 380, 488-9, 563
*Equisetum palustre*  158
*Equisetum variegatum*  359
*Erica tetralix*  370
*Eriocaulon aquaticum*  380
*Eristalis*  39, 545, 549
*Erpobdella*  137, 139-40
*Erpobdella octoculata*  101, 291, 460, 522
*Erpobdella testacea*  417
*Eryngium campestre*  283
Esocidae  42
*Esolus parallelepipedus*  68-9, 381
*Esox lucius*  73, 230, 421-3, 512, 522

*Eucyclops serrulatus*  418
*Eudorina*  219
*Eukiefferiella*  197, 381
*Eukiefferiella calvescens*  245
*Eukiefferiella clypeata*  68
*Eukiefferiella discoloripes*  68
*Eukiefferiella minor*  515
*Eukiefferiella verralli*  68
*Eunotia*  374-5, 378
*Eunotia arcus*  486
*Euphorbia seguieriana*  283
*Eurytemora affinis*  450
*Eusimulium aureum*  68

*Festuca eskia*  241
*Filina*  335
*Fissidens*  380
*Fissidens bryoides*  454
*Fissidens crassipes*  64
*Fissidens grandifrons*  281
*Fissidens taxifolius*  454
flounder  73, 166
*Fontinalis*
*Fontinalis antipyretica*  13-14,
    37, 63-4, 99, 123-4, 132-3,
    258, 281, 380, 488, 563, 569
*Fontinalis dalecarlica*  488
*Fontinalis squamosa*  63-4, 99,
    130, 132
*Fragilaria*  486
*Fragilaria capucina*  487
*Fragilaria construens*  13, 486
*Fragilaria crotonensis*  219,
    284-6, 490-1, 494, 522, 561
*Fragilaria vaucheriae*  375
*Fraxinus excelsior*  416
*Fredericella sultana*  291, 336
*Frustulia*  378
*Frustulia rhomboides*  385, 378
*Frustulia rhomboides* var.
    *saxonica*  375
*Frustulia undulata*  375
*Fulica atra*  425
*Fusarium aqueductum*  97

*Gadidae*  422
*Gallinula chloropus*  425
*Gambusia affinis holbrooki*  421
Gammaridae  359
*Gammarus*  69, 164, 200, 203,
    359, 395-6, 418

*Gammarus duebeni*  380
*Gammarus pulex*  39, 101, 135, 139,
    192, 194, 196, 291, 454
*Gammarus zaddachi*  454
Gasterosteidae  104, 422
*Gasterosteus aculeatus*  73, 104,
    199, 422
Gastropoda  191-2, 198, 289, 291,
    336
Gaviidae  425
*Gemellicystis neglecta*  491
*Gerris*  197
*Gerris najas*  381
*Gloeocapsa sanguinea*  375
*Gloeotilopsis planctonica*  285
*Glossiphonia*  396
*Glossiphonia complanata*  101, 289,
    291, 460
*Glossiphonia heteroclita*  460
*Glossosoma boltoni*  288, 516
*Glossosoma conformis*  68, 516
*Glossosoma intermedia*  516
*Glyceria aquatica*  359
*Glyceria fluitans*  37, 64-5, 415
*Glyceria maxima*  37, 563-4
*Glyceria plicata*  64
*Glyphotaelius pellucidus*  196
*Glyptotendipes*  459, 463
Gobidae  421
Gobiidae  422
*Gobio gobio*  17, 73, 104, 230,
    422, 512
*Goera pilosa*  516
*Gomphonema*  99, 280, 374, 377-8,
    486, 542
*Gomphonema acuminatum*  385, 378
*Gomphonema acuminatum* var.
    *coronata*  375
*Gomphonema angustatum*  336
*Gomphonema constrictum*  375
*Gomphonema intricatum* var.
    *pumilum*  513
*Gomphonema olivaceoides*  375
*Gomphonema olivaceum*  486
*Gomphonema parvulum*  375
*Gomphosphaeria lacustris*  491
*Gomphosphaeria naegeliana*  571
*Gongrosira fluminensis*  376-7
*Goniochloris pulchra*  285
*Gonium*  219
*Gonium pectorale*  161

*Gonium sociale*   161
*Graptodytes pictus*   196
grayling   73, 166, 493
*Grimmia agassizii*   155
*Grimmia alpicola* var. *rivularis*
    380
gudgeon   73, 166
*Guignotus fuscillus*   420
*Guignotus pusillus*   196
*Gymnocephalus cernua*   73
*Gymnodinium*   161
*Gyrinus*   396

*Habroleptoides modesta*   517
*Habrophlebia fusca*   517
*Habrophlebia lauta*   517
*Halesus*   196
*Halesus interpunctatus*   516
*Halesus rubricollis*   419-20, 516
*Halipida*   420
*Halipus ruficollis*   420
Helmidae   395
*Heleniella ornaticollis*   197
*Helobdella*   420
*Helobdella stagnalis*   101, 291,
    417
*Helophorus*   196
*Helophorus arvernicus*   445
Hemiptera   191-2, 198, 419
*Heptagenia*   289
*Heptagenia coerulans*   419
*Heptagenia fuscogrisea*   40
*Heptagenia sulphurea*   40, 68, 517
Heptageniidae   138, 419
*Herpobdella*   39
Heteroptera   289
*Hildenbrandia*   99, 563
*Hildenbrandia rivularis*   279
*Hippuris vulgaris*   489
Hirudinea   101, 192, 198, 289,
    291, 336, 359, 380, 395, 380,
    395, 417, 424, 543
*Holosteum umbellatum*   283
*Homoeothrix*   375
*Homoeothrix janthina*   512, 515
*Homoeothrix varians*   512-13
*Hormidium rivulare*   376
*Hottonia palustris*   282
*Hucho hucho*   512
*Huso huso*   422
*Hydra*   164, 460

Hydracarina   68, 101, 192, 250,
    381
Hydrachnellae   162
Hydridae   68
*Hydrobius fuscipes*   195, 196
*Hydrocharis morsus-ranae*   282
*Hydrococcus cesatii*   375
*Hydrococcus rivularis*   375
*Hydrodictyon reticulatum*   279
*Hydroporus incognitus*   196
*Hydroporus marginatus*   195, 196, 198
*Hydroporus memnonius*   196
*Hydroporus nigrita*   196
*Hydroporus pubescens*   195, 196
*Hydropsyche*   291, 336, 359, 395
*Hydropsyche angustipennis*   459
*Hydropsyche bulbifera*   516
*Hydropsyche contubernalis*   68-9,
    290, 459-60, 543, 549
*Hydropsyche dissimulata*   420
*Hydropsyche exocellata*   543, 546,
    549-50
*Hydropsyche instabilis*   69
*Hydropsyche pellucidula*   67-9,
    101, 516, 549
*Hydropsyche siltalai*   68-9, 163,
    381
Hydropsychidae   462
*Hydroptila*   101, 196, 381, 543,
    546
*Hydroptila forcipata*   516
*Hydroptila pellucida*   381
*Hydroptila sparsa*   420
Hydrozoa   289
*Hydrurus*   253-5
*Hydrurus foetidus*   244, 247, 249,
    486-7, 511-13, 515, 519
*Hygrohypnum molle*   244
*Hygrohypnum ochraceum*   64, 130,
    132, 488
*Hyocomium armoricum*   64

Ictaluridae   422
*Ictalurus melas*   421-2
*Ictalurus nebulosus*   294
ide   493
*Ilyocoris cimicoides*   419
*Impatiens glandulifera*   135
*Innifargus elegans*   418
Insecta   101, 291
*Inula salicina*   356

*Iris pseudacorus* 64
*Isatis tinctoria* 283
*Ischnura* 336
*Ischnura elegans* 419
*Isocladius sylvestris* 68
*Isoetes echinospora* 489
*Isoetes lacustris* 560
*Isoperla* 359, 543-4
*Isoperla grammatica* 380
*Isoperla viridinervis* 245, 247-9, 251-4, 257-8
Isopoda 289, 291, 418
*Ithytrichia* 196
*Ithytrichia lamellaris* 381

*Juncus bulbosus* f. *fluitans* 380
*Juncus filiformis* 488-9

*Kephyrion spirale* 491
*Koeleria pyramidata* 282

*Laccobius bipunctatus* 196
*Laccobius striatulus* 196
*Laccophilus hyalinus* 420
*Laccophilus minutus* 196, 420
Lamellibranchiata 289
*Lampetra fluviatilis* 73
*Lampetra planeri* 73, 104, 199, 230
*Lampetra planeri zanadreai* 421-2
lamprey 73, 166
*Lamproglena pulchella* 424
*Larsia* 197
*Lasiocephala basalis* 516
*Lemanea* 99, 563
*Lemanea fluviatilis* 12, 375, 377-8, 486-7
*Lemna gibba* 282
*Lemna minor* 64-5, 415
*Lemna polyrrhiza* 395
*Lemna trisulca* 37
*Lepocinclis ovum* 161
*Lepomis* 294
*Lepomis gibbosus* 230, 421-2
Leptophlebiidae 419
*Leucaspius delineatus* 294
*Leuciscus cephalus* 420, 445, 512, 522, 542
*Leuciscus cephalus cabeda* 17, 73, 104, 230, 411-12, 421-23
*Leuciscus leuciscus* 17, 73, 104, 230, 445, 512, 522

*Leuciscus souffia agassizi* 294
*Leuciscus souffia muticellus* 422
*Leuctra* 137, 359, 395-6
*Leuctra fusca* 68
*Leuctra hippopus* 380
*Leuctra inermis* 68, 380
*Leuctra lamellosa* 247
Leuctridae 250
*Limnaea* 396
*Limnaea peregra* 139
Limnephilidae 194
*Limnephilus lunatus* 194, 196, 289
*Limnephilus vittatus* 197
*Limnius volckmari* 68, 163, 381
*Limnochironomus* ex gr. *nervosus* 522
*Limnodrilus* 427
*Limnodrilus claparedeianus*
*Limnodrilus hoffmeisteri* 68, 101, 417-18, 454, 462
*Limnodrilus profundicola* 417-18
*Limnodrilus udekemianus* 417-18, 463
*Limnophyes* 197, 198
*Limosella aquatica* 283, 289
*Liponeura cinerascens* 543, 549
*Lithax niger* 516
*Lithoderma fontanum* 279
*Littorella uniflora* 64, 380
*Liza ramada* 422
*Liza saliens* 422
*Lobelia* 570
*Lobelia dortmanna* 380
*Lobomonas* 161
*Lota lota* 422
*Lucioperca lucioperca* 230, 512
Lumbriculidae 101
*Lumbriculus variegatus* 68, 101, 195, 197, 336, 380
*Lunularia cruciata* 64
*Lycopersicon esculentum* 283
*Lycopodium selago* 63
*Lymnaea* 137, 140, 395
*Lymnaea peregra* 68, 101, 289, 460
*Lyngbya* 375, 486
*Lyngbya aestuarii* 375
*Lyngbya epiphytica* 375
*Lyngbya kuetzingii* 375
*Lype reducta* 196

*Macropelopia* sp.   197
Malacostraca   192
*Mallomonas elongata*   491
*Mallomonas reginae*   491
*Marchantia polymorpha*   64, 454
*Marsilea quadrifolia*   282
*Marsupella emarginata*   64
Megaloptera   192, 198
*Melampophylax mucoreus*   196
*Melampophylax nepos*   516, 521
*Melosira*   286, 375
*Melosira granulata*   219, 414, 444,
   450, 453, 522, 560-1
*Melosira granulata* var. *angustis-
   sima*   286, 491, 561
*Melosira islandica*   286
*Melosira islandica* var. *helvetica*
   284-5, 491
*Melosira italica*   13, 161, 491
*Melosira varians*   13, 62, 219,
   279, 285
*Mentha aquatica*   64, 189, 191
*Meridion circulare*   62, 244, 375,
   486, 512-13, 542
*Meridion circulare* var. *con-
   strictum*   375
*Merismopedia*   375
*Merismopedia tenuissima*   491
*Mesodinium pulex*   416
*Mesovelia fucata*   419
*Metacnephia amphora*   180
Metriocnemini   198
*Micractinium pusillum*   219
*Micrasema morosum*   251, 543, 549
*Microchironomus tener*   463
*Michocoleus vaginatus*   453
*Microcricotopus bicolor*   522
*Microcricotopus rectinervis*   68
*Microcystis*   561
*Microcystis aeruginosa*   414
*Micronecta meridionalis*   419
*Micropsectra*   68, 197
*Micropsectra bidentata*   245
*Micropterus salmoides*   230, 294
   421-3
*Microspora*   12, 564, 570
*Microspora amoena*   376-8, 486-7
*Microspora tumidula*   376, 379
*Microspora wittrockii*   376
minnow   73, 166, 493
*Mixobolus*   424

*Mnium*   454
*Moina branchiata*   417
*Molinia*   370
*Molinietalia coeruleae*   370
Mollusca   101, 162-4, 291, 381,
   396, 420, 543
*Monorhaphidium gracilis*   62
*Morhaphidium tortile*   62
*Mougeotia*   220, 376, 378-9, 486
*Mugil*   230
Mugilidae   422
Muscicapidae   425
Muscidae   101
*Myosotis scorpioides*   64
*Myricaria germanica*   282
*Myriophyllum alterniflorum*   158,
   380, 488-9, 560
*Myriophyllum spicatum*   64, 158,
   160, 282
*Myriophyllum verticillatum*   64-5,
   488-9
*Mysis*   359

*Naias flexilis*   560
*Naias marina*   281
*Naias minor*   281
Naidae   511
Naididae   101, 163, 192, 417
*Nais*   164-5, 511, 519-21
*Nais alpina*   68, 246, 521
*Nais barbata*   101
*Nais bretscheri*   101, 460, 521
*Nais elinguis*   101, 450, 521-2
*Nais simplex*   460
*Nanocladius*   459
*Nardia compressa*   63-4, 99
*Nasturtium officinale*   188
*Natarsia* sp.
*Navicula*   99, 280, 375, 486, 542
*Navicula avenacea*   98, 336
*Navicula cryptocephala*   62, 513
*Navicula vyrptocephala* var.
   *intermedia*   513
*Navicula gracilis*   513-14
*Navicula gregaria*   375
*Navicula mutica*   513-14
*Navicula radiosa*   513-14
*Navicula viridula*   62, 514
Nematoda   192, 424
*Nemoura cinerea*   198
Nemouridae   136

*Nemurella picteti* 197
*Nepa cinerea* 419
*Neureclipsis bimaculata* 381
*Niphargus aquilex* 195, 196
*Nitella opaca* 160, 378
*Nitzschia* 61, 99, 375, 486
*Nitzschia acicularis* 285, 414, 487
*Nitzschia amphibia* 62
*Nitzschia atomus* 542
*Nitzschia capitellata* 336
*Nitzschia dissipata* 514
*Nitzschia frustulum* 542
*Nitzschia fruticosa* 285, 414
*Nitzschia palea* 62, 336, 375, 486, 542
*Nitzschia paleacea* 336
*Nitzschia pellulosa* 542
*Nitzschia permitis* 542
*Noemacheilus barbatulus* 73, 104
*Nostoc* 375, 378
*Nostoc* cf. *verrucosum* 486
*Noterus clavicornis* 420
*Notonecta glauca* 419
*Nuphar lutea* 37, 65, 380
*Nuphar luteum* 458, 569
*Nycticorax nycticorax* 425
*Nymphaea alba* 282
*Nymphoides peltata* 281, 283

*Ochromonas* 161
Odonata 192, 198, 289, 336, 419
*Odontocerum albicorne* 196, 420
*Oedogonium* 376-8, 486, 564, 570
*Oedogonium undulatum* 378
*Oenanthe* 388
*Oenanthe crocata* 380
*Oenanthe fluviatilis* 282
Oligochaeta 101, 192, 198, 245, 257, 289, 359, 417, 511
*Oligoneuriella rhenana* 288, 517
*Oligoplectrum maculatum* 516
*Oncorhynchus tshawytscha* 294
*Ononis spinosa* 283
*Oocystis crassa* 62
*Ophiogomphus serpentinus* 445
*Orchestia cavimana* 452, 455-6
*Orconectes limosus* 445
*Orectochilus villosus* 444
Orthocladiinae 68, 163, 246, 249-50, 252-3, 420

*Orthocladius* 197, 543
*Orthocladius rivicola* 245, 514-15
*Orthocladius thienemanni* 514-15
*Oscillatoria* 375, 486
*Oscillatoria* cf. *bornetii* 486
*Oscillatoria* f. *tenuis* 490-1
*Oscillatoria brevis* 487
*Oscillatoria limosa* 487
*Oscillatoria rubescens* 284-6, 414
*Osmerus eperlanus* 361
*Osmunda regalis* 569
Ostracoda 102, 543
*Oulimnius tuberculatus* 68-9, 381
*Oxethira* 196

*Pachyleuctra benlocchi* 251
*Padogobius martensi* 421-2
*Palaemonetes antennaris* 419
*Pandorina morum* 219
*Parachironomus* gr. *arcuatus* 459, 463
*Parachironomus* gr. *longiforceps* 463
*Paracricotopus niger* 547, 549
*Paroigolaimella bernensis* 416
*Parorthocladius nudipennis* 515
Parvocaricetea 370
Pedicinae 258
*Pellia epiphylla* 64
*Peloscolex ferox* 64, 197
*Peplis portula* 489
*Perca fluviatilis* 73, 104, 230 422
perch 73, 166, 493
Percidae 104, 422
*Pergamasus brevicornis* 381
*Perla bipunctata* 380
*Perla grandis* 247, 251
*Perlodes microcephala* 40
*Petromyzon marinus* 73, 230, 294, 422
Petromyzonidae 104, 422
*Peucedanum chabraei* 283
*Phaenopsectra* 197
*Phagocata vitta* 68, 380
Phalacrocoracidae 425
*Phalaris arundinacea* 64, 133, 489, 563
*Philonotis fontana* 64, 155, 488
*Philopotamus ludificatus* 419

*Philopotamus montanus*  381
*Phormidium*  99, 150, 375, 377
*Phormidium autumnale*  336, 486
*Phormidium favosum*  336, 513
*Phormidium retzii*  375, 377
*Phormidium tenue*  336
*Phormidium uncinatum*  487
*Phoxinus phoxinus*  17, 73, 104,
    230, 422, 512
*Phragmites*  282, 417, 419-20,
    427, 561
*Phragmites australis*  37, 380,
    415, 452, 458, 488-9, 560, 563
*Phyllitis scolopendrium*  63
*Physa*  396
*Physa acuta*  291, 543
*Physa fontinalis*  101, 197, 291,
    460
*Physocomitrium pyriforme*  64
pike  73
pikeperch  493
*Pinnularia*  375, 378, 486
*Pinnularia borealis*  375
*Pinnularia* cf. *brevicostata*  486
*Pinnularia viridis*  62, 486
Pisidiidae  463
*Pisidium*  69, 381, 452, 454-5
*Pisidium casertanum*  198
*Pisidium henslowanum*  463
*Pisidium milium*  197
*Pisidium personatum*  197, 198
*Planaria*  245
*Planaria torva*  68, 336
Planorbidae  420
*Planorbis*  101, 395-6
*Platichthys flesus*  73, 104
*Platichthys flesus luscus*  422
Platyhelminthes  291
*Plathypnidium riparioides*  13-14
*Plea minutissima*  419
Pleocoptera  70, 135, 137-8, 140,
    162-4, 192, 245, 288-9, 291,
    359, 427, 491, 494, 510-11,
    522, 543, 546
*Plectonema*  375
*Pleurocladia lacustris*  279
Pleuronectidae  104, 422
*Plumatella*  291
*Plumatella fungosa*  336
*Plumatella repens*  336
*Poa bulbosa*  282

Podicipedidae  425
Poeciliidae  422
*Pohlia albicans*  488
*Pohlia carnes*  64
*Pohlia proligera*  64
*Polio*  413-14
*Polyarthra*  335
*Polyarthra vulgaris major*  416
*Polycelis felina*  197, 246, 543,
    549
Polycentropidae  136-7, 140, 359,
*Polycentropus*  395
*Polycentropus flavomaculatus*  68,
    101, 381, 516
*Polycentropus kingi*  381
*Polygonum*  283
*Polygonum amphibium*  64
*Polygonum hydropiper*  64-5, 455
*Polypedilum*  197
*Polypedilum* gr. *convictum*  543
*Polypedilum cutellatum*  420
*Polypedilum nubeculosum*  68, 545,
    547, 549
*Polypedilus nubeculosum*  68
*Polytrichum commune*  64
*Populus alba*  416
*Populus nigra*  416
Porifera  289, 291
*Potamanthus luteus*  71, 517
*Potamogeton*  65, 393-6, 452, 488,
    547
*Potamogeton acutifolius*  282
*Potamogeton alpinus*  37, 158,
    489
*Potamogeton angustifolius*  282
*Potamogeton coloratus*  282
*Potamogeton compressus*  282
*Potamogeton crispus*  37, 39, 131
    133, 160, 282, 415, 564
*Potamogeton densus*  282
*Potamogeton gramineus*  282, 489
*Potamogeton lucens*  37, 64, 282,
    415
*Potamogeton natans*  131, 133, 158
*Potamogeton nodosus*  458
*Potamogeton obtusifolius*  570
*Potamogeton pectinatus*  37, 39,
    64, 131-3, 135, 158, 281-2,
    335, 388, 394, 543
*Potamogeton perfoliatus*  37-9,
    64, 160, 281-2, 415, 488-9

*Potamogeton polygonifolius*  64-5
*Potamogeton praelongus*  37
*Potamogeton pusillus*  37, 160
*Potamogeton salicifolius*  64
*Potamophylax cingulatus*  516
*Potamophylax latipennis*  516
*Potamopyrgus*  359, 395-6
*Potamopyrgus jenkinsi*  101, 455
*Potamothrix hammoniensis*  101, 463
*Potamothrix moldaviensis*  454
   462-3
*Potentilla rupestre*  66
*Potentilla supina*  283
*Procladius*  463
*Procladius choreus*  68
*Procloeon bifidum*  517
Prodiamesinae  420
*Prosimulium*  515
*Protoderma*  280
*Protonemura beatensis*  549
*Protonemura risi*        549
*Protonemura risi spinulosa*  543-4
*Protonemura vandeli*  246
Protozoa  424
*Psammoryctides albicola*  463
*Psammoryctides barbatus*  68, 197,
   463
*Psectrotanypus varius*  197, 198
*Pseudamnicola confusa*  452, 455
*Psilopteryx psorosa psorosa*  516
*Psychomyia pusilla*  381, 516, 543,
   546, 549
*Pteridium aquilinum*  63
*Pteromonas angulosa*  161
*Pungitius pungitius*  199
*Pyramimonas*  161


*Quercus pedunculata*  416


*Racomitrium aciculare*  64, 130-1
*Racomitrium aquaticum*  64
*Radix peregra*  219
*Ranatra linearis*  419
*Ranunculus*  65, 189-90, 380, 388,
   393-4, 396
*Ranunculus aquatilis*  158
*Ranunculus flammula*  64
*Ranunculus fluitans*  64-5, 130-1,
   133, 135, 157-8, 281-2, 297
*Ranunculus omiophyllus*  64-5
*Ranunculus peltatus*  64-5, 191,
   489

*Ranunculus penicillatus*  64-5
*Ranunculus penicillatus* var.
   *calcareus*  160, 191
*Ranunculus reptans*  488-9
*Ranunculus trichophyllus*  488
*Reo*  413
*Rheocricotopus*  68
*Rheocricotopus chalybeatus*  543,
   547
*Rheosmittia spinocornis*  420
*Rheotanytarsus*  68, 71-2, 197,
   291
*Rhithrogena*  288
*Rhithrogena diaphans*  517, 543,
   549
*Rhithrogena ferruginea*  517
*Rhithrogena germanica*  517
*Rhithrogena hercynia*  517
*Rhithrogena hybrida*  517
*Rhithrogena iridina*  517
*Rhithrogena loyolaea*  245, 248-9,
   251-2, 254-5, 257, 515, 517
*Rhithrogena semicolorata*  68-9,
   163, 381, 517
*Rhizoclonium riparium*  453
*Rhizomnium punctatem*  64
*Rhizosolenia*  494
*Rhizosolenia eriensis*  491
*Rhizosolenia longiseta*  286, 491
*Rhodochorton violaceum*  375, 377
*Rhododendron*  244
*Rhododendron ferrugineum*  241
*Rhodomonas lacustris*  161
*Rhodomonas minuta* var. *nono-
   planctica*  285
Rhodophyta  510
*Rhodoplax schinzii*  376-7
*Rhodotorula*  414
*Rhoicosphenia curvata*  62, 453,
   514
*Rhopalodia gibberula*  375
*Rhyacodrilus coccineus*  68
*Rhyacophila*  136-40, 254-5, 258,
   396
*Rhyacophila contracta*  257
*Rhyacophila dorsalis*  101
*Rhyacophila evoluta*  247, 251,
   257, 546
*Rhyacophila fasciata*  516
*Rhyacophila galaerosa*  419
*Rhyacophila glareosa*  516

*Rhyacophila intermedia*  247, 251-2, 419
*Rhyacophila kelnerae*  419
*Rhyacophila mocsaryi*  516
*Rhyacophila nubila*  516
*Rhyacophila obliterata*  516
*Rhyacophila philopotamoides*  516
*Rhyacophila polonica*  516
*Rhyacophila tristis*  419, 516
*Rhyacophila vulgaris*  419, 516
*Rhynchostegium*  133
*Rhynchostegium riparioides*  14, 64, 99, 132, 135, 143, 208
*Riccardia pinguis*  64
*Ricinus communis*  283
*Rivularia*  155, 158
*Rivularia haematites*  279
roach  73, 166, 493
*Rorippa nasturtium-aquaticum*  64, 188-91, 199
*Rorippa sylvestris*  395
*Rotaria neptunia*  335
*Rotaria rotaria*  335
rudd  73, 166, 493
ruffe  73, 493
*Ruppia*  488
*Rutilus pigus*  422
*Rutilus rubilio*  421-2, 422
*Rutilus rutilus*  73, 104, 230, 294, 423, 512, 522

*Sabanejewia larvata*  412, 422
*Saccharomyces cerevisiae*  414
*Saccharomyces uvarum*  414
*Sagittaria*  359
*Sagittaria sagittifolia*  133, 489
*Salix*  416, 452
*Salix alba*  416
*Salmo gairdneri*  73, 259-60, 294, 421-3, 542
*Salmo salar*  42, 73, 199, 230, 294, 351, 396, 493
*Salmo trutta*  17, 41-2, 104, 199, 259-61, 294, 351, 396
*Salmo trutta fario*  73, 230, 396, 421-3, 512
*Salmo trutta marmoratus*  422
*Salmo trutta trutta*  73
*Salmonella*  414
*Salmonella agona*  414
*Samonella brandeburg*  414

*Salmonella panama*  414
*Salmonella paratyphi*  414
*Salmonella typhimurium*  414
Salmonidae  104, 170, 422
*Salvelinus fontinalis*  421-2
*Salvinia*  415
*Salvinia natans*  281-2
*Saponaria officinalis*  416
Sarcodinea  416
*Scapania undulata*  13-14, 63-4, 99, 130-1, 133
*Scardinium erythrophthalmus*  73, 230, 421-3
*Scenedesmus*  61-3, 219, 334, 486, 491
*Scenedesmus crassus*  227
*Scenedesmus obliquus*  11, 486
*Scenedesmus quadricauda*  13, 414
Scheuchzerietea  370
*Schistidium alpicola*  488
*Schizochlamys gelatinosa*  376, 379
*Schizomeria leibleinii*  280
*Schizothrix*  375
*Schoenoplectus lacustris*  65, 488
*Scirpus*  452, 458
*Scirpus fluitans*  380
*Scirpus lacustris*  489, 563, 569-70
*Scirpus maritimus*  450
*Scirpus tabernaemontanii*  489
*Scirpus triqueter*  450
*Scytonema*  486
*Selenastrum*  219
*Selenastrum capricornutuum*  426
*Sericostoma*  69, 516
*Sericostoma personatum*  197
*Siderocapsa treubii*  280
*Sigara concinna*  197-8
*Sigara dorsalis*  419
*Sigara italica*  419
*Sigara lateralis*  197
*Sigara nigrolineata*  197
*Silene conica*  283
*Silo pallipes*  516
Siluridae  422
*Silurus glanis*  421-2
Simuliidae  137-8, 161, 245, 250, 257, 359, 515, 543
*Simulium*  165, 359, 395-6
*Simulium bezzii*  549

*Simulium monticola* 543
*Simulium nitidifrons* 68
*Simulium ornatum* 68, 549
*Simulium reptans* 68
*Simulium reptans* var. *galeratum* 68
*Simulium rheophilum* 543
*Simulium truncatum* 492
*Simulium variegatum* 68
*Siphlonurus lacustris* 381, 419, 517
*Sium erectum* 36
*Sium latifolium* 356
smelt 493
*Solanum nitibacatum* 283
*Solenostoma triste* 64, 99
*Sparganium* 488, 569
*Sparganium angustifolium* 489
*Sparganium emersum* 133
*Sparganium erectum* 37, 64, 133, 415
*Sparganium minimum* 282
*Sparganium ramosum* 489
*Sparganium simplex* 37, 39, 489
Sphaeriidae 395-6, 420
*Sphaerium* 69, 101, 359, 381
*Sphaerium corneum* 68, 291, 452, 454, 463
*Sphaerium rivicola* 454, 463
*Sphaerium solidum* 452, 454, 463
*Sphaerobotrys fluviatilis* 280
*Sphaerocystis schroeteri* 284
*Sphaerotilus* 39, 285
*Sphaerotilus natans* 98, 510, 521, 570
*Sphagnetalia magellanici* 370
*Sphagnum* 152
*Sphagnum auriculatum* 64
*Spirogyra* 376, 378-9, 486, 564
*Spongilla lacustris* 291, 336
*Stactobiella risi* 288
*Staurastrum* 486
*Staurastrum alternans* 486
*Staurastrum apiculatum* 486
*Stauroneis anceps* 375
*Stauroneis phoenicenteron* 375, 378
*Stempellina bausei* 197
*Stenophylax* 196
*Stenopterobia intermedia* 375
*Stephanodiscus* 61, 219

*Stephanodiscus astrea* var. *minutula* 286
*Stephanodiscus dubius* 286
*Stephanodiscus hantzschii* 13, 285-6, 334, 444, 491
*Stephanodiscus subsalsus* 286, 334
*Stephanodiscus tenuis* 286
*Sterna albifrons* 425
*Sterna hirundo* 425
sticklebacks 73, 166, 493
*Stigeoclonium* 375, 378-9, 486
*Stigeoclonium tenue* 98-9, 486, 513-14, 521
*Stigonema mamillosum* 375, 378-9, 486
stoneloach 73, 166
*Stratiotes aloides* 282
*Stylaria lacustris* 68, 101, 460
*Stylodrilus heringianus* 68, 195, 197, 380
*Subularia aquatica* 488-9
*Surirella* 99, 375
*Surirella ovata* 62, 336
*Synchaeta* 335, 416
*Synedra* 61, 161
*Synedra actinastroides* 286
*Synedra acus* 444, 491
*Synedra acus* var. *angustissima* 28, 284, 562
*Synedra delicatissima* 284, 286
*Synedra ulna* 62, 285, 336, 414, 486, 491, 513, 515
*Synedra ulna* var. *oxyrhynchus* 285
*Synorthocladius semivirens* 68
*Synura* 491
*Synurella ambulans* 418

Tabanidae 101
*Tabellaria* 571
*Tabellaria fenestrata* 63, 284-5, 375, 486, 491
*Tabellaria fenestrata* var. *asterionelloides* 286, 491, 494
*Tabellaria flocculosa* 63, 375, 378, 486, 491
Tanypodinae 420, 459
Tanytarsinae 249-50, 252, 420, 459

*Tanytarsus*  68
tench  73
*Teucrium scordium*  356
*Thalassiosira*  61
*Thalassiosira fluviatilis*  62
*Thalictrum minus*  283
*Thamnobryum alopecurum*  64
*Thelohania contejeani*  72
*Theodoxus*  359
*Thereruyzon tessulatum*  417
*Thermacyclops crassus*  418
*Theromyzon tessulatum*
*Thienemanniella vittata*  68
*Thienemannimyia*  68
*Thorea ramosissima*  23, 279, 336
Thymallidae  422
*Thymallus thymallus*  17, 73, 199,
    421-2, 512
*Tinca tinca*  73, 230, 421-3
*Tinodes dives*  543, 549
*Tinodes machlachlani*  381
Tipulidae  101, 162, 165
*Tobrilus helveticus*  416
*Tobrilus longus*  416
*Tolypella intricata*  279
*Tolypellopsis stelligera*  281
*Tolypothrix penicillata*  375, 377
*Torulopsis*  414
*Torulopsis glabrata*  414
*Trapa natans*  281-2, 415
Trematoda  424
*Tribonema*  570
*Trichodina*  424
*Trichodrilus macroporophorus*
    245, 250
Trichoptera  101, 135, 137, 140,
    162, 191-2, 194, 198, 288-9,
    336, 359, 395, 419-20, 444,
    457, 491-2, 494, 510-11, 518,
    522, 543, 546
Tricladida  192, 198, 289
*Trocheta bykowskii*  452, 455
*Tubifex*  39, 417
*Tubifex tubifex*  01, 195, 197,
    418, 462-3
Tubificidae  101, 137-9, 163,
    395-6, 417, 452, 454-5, 463,
    511, 522
Turbellaria  336, 417, 543
Turdinae  425
twaite shad  73

*Typha*  417, 420, 427, 458
*Typha angustifolia*  415
*Typha latifolia*  37, 415

*Ulmus campestris*  416
*Ulothrix*  158, 486
*Ulothrix tenerrima*  453
*Ulothrix zonata*  279, 377-8, 486-7,
    513, 515
*Unio crassus* var. *batavus*  445,
    454
*Unio pictorum*  290, 463
*Unio tumidus*  290, 463
Unionidae  452, 463
*Utricularia bremii*  282
*Utricularia neglecta*  282

*Valvata*  396
*Valvata puscinalis*  197
*Vaucheria*  12, 130-1, 336, 379,
    452-3, 486-7, 564, 570
*Vaucheria bursata*  454
*Vaucheria canalicularis*  454
*Vaucheris compacta*  452-3, 455
*Vaucheria frigida*  454
*Vaucheria sessilis*  158
*Vaucheria walzi*  487
Vaucheriaceae  282
*Veronica beccabunga*  64-5, 189
*Verrucaria*  377
*Verrucaria rivularis*  219

whitefish  493
*Wilhelmia equinum*  69
*Wormaldia*  381
*Wormaldia copiosa*  419
*Wormaldia occipitalis*  419

*Xenochironomus xenolabis*

*Zannichellia palustris*  65, 160,
    282
*Zygnema*  376-9, 486-7
Zygoptera  419

# Subject Index

Aare, R.  279, 275
abstraction: *see also General
  Data sections*, drinking water,
  irrigation
  Llobregat  529
  Lot  231
  Rössjön  572
  Shannon  352
  Suir  391, 399
  Tees  148
  Wye  78
acid waters
  *Fusarium aqueductum*  97
  influence on fish  482
  Meuse system  6
  Västersjön and Rössjön,
    possible risk  570
Acoose R.  374
actinomycete  306
Acts of Parliament (on river
  pollution)  120-121
Addit (stream)  32
Afon Hirnant (R.)  198
Afon Lwyd (R.)  98
Agerodsmösse  559
air pollution  482
aldrin  412
ahen: *see* invader(s)
alkalinity, total alkalinity:
  *see also various chemical
    summaries*
  Mersey system  127
  Meuse system  6, 8, 13-14, 32
  Winterbourne  186-187
  Wye, influence of season on,
    58
    model for  60, 62
allochthonous *v.* autochothonous
  macroinvertebrate food
    Chapter 9

Neckar organic materials, break-
  down of  339
allyl thiourea  302
Alsace  295
aluminium
  industrial plant  349
älv  554
Amblève, R. 8, 15
  chromium pollution  18
  fish community  18
AMES-Test  299
Ammer, R.  330
ammonia, ammonium: *see also various
  chemical summaries*
  Ebbw  90-92
  Meuse fish kills  445
  Neckar  331, 340-341
    toxicity to fish in  342
  Pinnån  570
anaerobic
  Neckar water  341
Ängelholm  Chapter 20
angler(s)
  economics  76-77
  Ebbw  109
  Gudenå system  26-27, 41
  Rhine  291-294
  Shannon  351
  Suir  387
  Wye  52, 76-77
Anoia, R.  529, 536
anthracite  349
Antwerpen  2
Arbeitsgemeinschaft der Wasser-
  werke Bodensee - Rhein  267
ardennais, type (water)  6, 8
Ardennes  8
Århus, County  42, 48
Arigna power station  350
arsenic

arsenic (*cont.*)
  Po  411, 412
artificial substrata, use of
  biomass measurements  13
  difficulty  279
åser  478
Athlone  349, 354
Atna, R.  475
autotrophic, autrophy: *see also*
  primary production
  character of Estaragne  258
Aursunden, L.  474
Avon  77
Avon & Dorset River Authority
  181, 202

Baden-Württemberg  Chapter 11
Bagnor  186, 187, 188, 191
Balder, R.  152
Bälingsejön (L.)  568
Bäljaneå (R.)  559, 569
Ballinamore and Ballyconnell
  Canal  350
Banagher  354
barbel: *see also Organism Index*
  zone in Meuse  14
Barcelona  532
Barnard Castle  171
Barrow, R.  387
barytes
  export from Foynes  349
  influence on Teesdale streams
    150, 160
  Base Flow Index  151, 152
Basel  Chapter 10
Bassaleg Weir  103
battery (manufacture) effluent
  Tame  140
bed load transport  291, 508
benthic algae: *see also Organism
  Index*
  Caragh  374-379
  Dunajec  512-516, 519, 521
  Ebbw  98-99
  Estaragne  244
  Glåma  485-487
  Meuse  12
  Neckar  335-336
  Rhine  279-280
  Tees  158, 160
  vertical zonation in High Tatra
    515

Bergsche Maas  438, 447
$\alpha$-BHC, $\gamma$-BHC  411, 412
Biala Tarnowska, R.  507, 508, 510
Biesbosch  445-456
Billingham Beck,  152
Bimmen  287
Bingfoss  479
bioassay:  *see also* AMES-Test, OPL
  test
  algal growth potential in Glåma
    483, 485
  Neckar  337-339
  Rhine  305-308
  *Selenastrum capricornutum* on Po
    water  426
biotic index: *see index* bryophyte(s),
  Chandler biotic siore, diatom(s),
  diversity index
  Po, comparison of methods in  427
  Verneaux & Tuffery method  17
birds: *see also Organism Index*
  impact of tourism  567
  Po  424-425
  Ringsjön  559
Birmingham
  reservoir supply  52
Bislich  281
Bjerringbro  40
Black Forest (Schwarzwald)  320, 321
'Black list'  267
Black stream  394
blue-green alga(e): *see also*
  eutrophic, phytoplankton
  Glåma  490-491
  Po  414
  Ringsjön  467
  Shannon lakes  356
  Västersjön and Rössjön, influence
    on turbidity in  571
boating, boat: *see also* canoe, motor-
    boat, shipping route
  Gudenå  26
  Po  407
  Shannon  351
  Suir  388
BOD (B.O.D., biological oxygen demand,
  biochemical oxygen demand)
  Gudenå system  44-46
  Llobregat  538
  Mersey system  121-122, 126
  Neckar  339
  Po  410

Rhine   295, 302, 305
   Shannon   354
Bodensee   267, 268, 269, 287, 297, 305
Bolmen, Bolmen project   574
borehole(s)
   Suir catchment   391
   Thames valley   180
Borrisnoe Mts   386
Boyle R.   353, 356
Brandstrup Baek (stream)   41
Braskereidfoss   479
bream: *see also Organism Index (Abramis brama)*
   influence on plankton relationships   567
Bregenz Bay   268
Breisach 277
bricks
   possible use for dredged sediments   327
Bridia Valley   367, 374
Broken Scar gauging station (Tees)   147, 151
Bruxelles   2, 5
bryophyte(s): *see also* heavy metals, macrophytes
   alkalinity 'tolerance'   13-14
   Caragh   380
   Glåma   487-488
   index   15
   Meuse system   13-15
   monitoring heavy metals   18-20
   monitoring radionuclides   20
   role in trapping suspending matter   458
   Tees   155, 158, 160
   Wye   63
   zonation on trunks in Biesboch   454
Builth Wells   74
bullhead: *see also Organism index (Cottus gobio)*
   Ebbw system   103
   Goyt   141
   Tees   166-169, 172
   Winterbourne survey   199-200

Cabon Coch reservoir   57
Cadi Mts   Chapter 19

cadmium: *see also various chemical summaries,* heavy metals
   Ebbw system   107
   Gudenå, below detection limit in   33
   Meuse sediments   442
      water   443
   Rhine sediments   443
      water   298-299
caesium
   radionuclide in Meuse   12
Caher   391
Caher Mt   366
Cahors   213
calefaction: *see* heat waste
Camlin, R.   354
canalization: *see also* shipping route
   Neckar   322-323, 333-334, 337, 342
canals
   Ballinamore and Ballyconnel Canal   350
   Bijlands Kanaal   438
   Canal Albert   5
   Grand Canal   350, 359
   Lough Allen Canal   350
   Pannerdens Kanaal   438
   Royal Canal   350, 359
Candover Brook (Hampshire)   181, 202
canoe, canoeing
   Gudenå system   26
   Rönneå   567, 573
   Shannon   351
Caragh Bridge   369, 372
Cardoner, R.   Chapter 19
Carl Beck   162, 167
Carn, R. 103
carnivore
   Estaragne   247, 248-249, 250, 251, 254, 255
   Dunajec   511
Carrick-on-Shannon   349, 3-4
Carrick-on-Suir   391
Cashel   394
Catalonia   Chapter 19
Cauldron Snout   167, 168, 169
Causses   210
chalk stream(s)   Chapter 7   163, 169
   Meuse catchment   Chapter 8
Chandler biotic score   99
Characeae - pond   279

Charleroi   10
Chepstow   52, 54
Chew Brook   138
Chitterne Brook   202
chloride: *see also various chemical
  summaries*
  Agreement   267
  Alsace, pollution of Rhine from
    267, 296
  Caragh, influence of sea winds
    on   374
  Ebbw Fawr, influence of flow on
    93
  Green Sandstone, high content in
    370
  Heilbronn salt industry   329, 334
  Llobregat, influence of salt beds
    on   535-536
  Nieuwe Merwede, influence on
    Rhine on   447
  Wye, model for   60, 62
chlorine
  Goyt, pollution in   128, 131, 133
  Rhine, total organic in   299-300
chloroform   299
chlorohydrocarbon 299
chlorophenol   299
chlorophyll *a*
  Lot   219
  Meuse   13
Chochołowski stream   507
cholera
  19th century Irwell   120
  19th century Ebbw   85
cholinesterase inhibitor   459, 461
Chooz nuclear power station   4,
  11-12
chromium: *see also various chemical
  summaries*, tannery
  Amblève   18
  bryophyte monitoring   124, 133
  Dukinfield sewage treatment plant
    effluent   129
  Mersey-Etherow, intermittent
    pollution   126
  Meuse sediments   443
  Rhine sediments   443
  Rönneå sediments   567
  Tame   129, 133, 140
  Warche   10, 18
Claerwen reservoir   52
Clodiagh R. 293, 394

Clonmel   Chapter 14
Clow Beck   152, 160
coalfields, coal mining
  Decazeville and Fumel   210
  Sambre, Meuse   8
  S. Wales   Chapter 4
cobalt: *see also various chemical
  summaries*
  Ebbw system   107
  Meuse, radionuclide in   12
cockle bed   396
COD (chemical oxygen demand)
  Llobregat   538
  Po   410
  Rhine   301-302
coke-works (R. Skerne)   154, 171
coliforms
  Po   413, 414
Como, L.   Chapter 15
condrusien, type (waters)   6, 8
Connemara
  influence of wind on electrolytes
    372
conservation
  *A Nature Conservation Review* (U.K.)
    52, 181
  Conservation of Nature Act
    (Denmark)   48
  fragility of high Pyrenees stream
    system   262
  stream and rivers   181
  Taubergiessen   277
  Tees   173
  Wye   52
copper: *see also various chemical
  summaries*, heavy metals
  Ebbw rocks   88
  Mersey-Etherow, intermittent
    pollution   126
  Meuse sediments   443
  Po   411, 412
  Rhine sediments   443
  Rhine water   299
  Sambre   10, 20
Cotherstone   165
cotton industry   119
Cow Green reservoir
Craig Goch reservoir   52, 77-79
cresol
  degradation by bacteria   306-307
Croft   157, 167
Cwm   88, 91, 94

Czarny Dunajec   Chapter 18
cyanide
    and fish kills in lower Meuse
        445
    coke-making effluents   90
    Ebbw system   90, 91
    Mersey system   123
    Skerne   154
Czchów reservoir   505, 507
Czekanowski coefficient   193

Dagstorpssjön (L.)
dam(s): *see also* Cow Green reser-
    voir, Czchów reservoir, hydro-
    electric plant La Baells reser-
    voir, Roznów reservoir
    Haringsvliet mouth   438
    Lot system   Chapter 8
    Neckar   326
    Rhine   277
Danish Lake District   26
Danube (Donau)   266, 336
Darlington   Chapter 6
DDE   412
Decazeville (industrial basin)
        210, 216
Delta Works   438
denitrification
    Neckar   340
Dent Bank gauging station   147,
    152, 167
Denton sewage treatment works   124
Derg, Lough   Chapter 12
destratification   79
detergents
    Mersey   123
    Po   413
detritivore
    Dunajec   511
    Estaragne   244, 249
    Tees   165
Deutsche Arbeitsgemeinschaft der
    Rheinwasserwerke   267
Deutsche Kommission zur Reinhal-
    tung des Rhein   295
diatoms: *see also various floristic
    lists*, benthic algae, phyto-
    plankton, silica
    diatom index   12, 14-16
    Glåma, annual cycle on   490
    spring growths in Tees   158
dichlorbenzol   299

dieldrin   458
diversity index
    Po macroinvertebrates   427
    value   95-97
diving bell   290
Djupadal   Chapter 20
Dowra   354
drainage, influence on discharge
    151-152
dredging
    Neckar sediments   326-327
    Rhine gravel   278
drift
    chironomid   249
    Estaragne invertebrates   252-253
        plants   244, 253
    *Isoperla viridinervis*   249
    Tees invertebrates   163-165
    Wye invertebrates   70-71
drinking water: *see also General
    Data sections*
    groundwater sources   33
    Gudenå catchment   33
    Llobregat, influence of chromium
        in   538
    Lot   231
    Meuse catchment   33
    Rhine   266
    Rössjöholmsån, problems of   572
Drish R.   386
drought
    Winterbourne 1976/77   181
Drumshanbo   350
Dubby Sike   165, 168
Dukinfield Sewage (treatment) Works
    124, 133
dystrophic: *see also* humic substances
    water in Glåma catchment   483

Eastham Locks   119
East Stake Mill Stream   163
Ebbw Fach, R.   Chapter 4
Ebbw Fawr, R.   Chapter 4
Ebbw Vale   Chapter 4
economic trend
    influence of downward, on Rhine
        295
eel(s): *see also Organism Index
    (Anguilla)*
    Ebbw, former abundance in   87
    Ebbw Fawr, (re-)colonization of
        103

eel(s): (cont.)
  importance in Rhine fishing   293
  Llynfi, trap on   74
  Ringsjön, stocking in   567
  Shannon, fishing on   349, 351,
    360
effluents: see also sewage
  steel-making works   90-91
  coke-oven   90
  coal washeries   90
Eggleshope Burn   152
Egglestone   157, 168
Elan, R.   54, 57
  growth rate of pike   72
  phytoplankton   63
  reservoirs in catchment   61
  restricted fauna in   69
Electricity Supply Board
  (Ireland)   367
embankment: see also flood,
  polder
  Mersey   117
Emmerich   301
Emscher, R.   268, 301
encrusting algae   377
Endrick, R.   69
endrin   412
ENEL   408
energy flow
  Estaragne   255-258
Enz, R.   Chapter 11
erosion
  Dunajec catchment   505
Eschach, R.   330
Eslöv   562
Estibère (torrent)   258-261
Etherow, R.   Chapter 5
Ennis   348
euphotic zone   221, 227
European (Economic) Community,
  Council of European Communities
  267, 391, 353
euryhaline species   457
eutrophic, eutrophication
  and diatom index   15
  Belgian rivers   10
  Glåma   483, 487
  influence on Hochrhein inverte-
    brates   288-289
  influence on Lower Rhine flora
    283
  influence of Po on coastal
    waters   429

  influence on Upper Rhine valley
    flora   279
  Lough Derg   356
  Ringsjön   561-562
  Suir   Chapter 14
exuviae, chironomid   290
Eyach, R.   330

fagnard, type (waters)   6, 8
Färingtoftaån esker   563
feeding: see also fish, grazing,
  macroinvertebrates, trout
  Estaragne invertebrates   254
Fenethcarb   459
Fennoscandia   470
ferric hydroxide
  influence on macroinvertebrates
    69
fertilizers
  impact on N and P   47
Fetsund   475
fibre
  pollution in Ringsjön   565
Fils, R.   Chapter 11
filtration plint
  County of Århus   47
Findhorn, R.   169
fish, fishery: see also Organism
  Index, barnel, bream, eel, invader,
  migration, salmon, trout
  acidity, influence of   482
  cage experiments   108, 397, 522
  disease   360, 425
  Gudenå fish farms   33
  kills in Meuse   445
    Neckar   335
    Pinnån   570
    Rhine   293
    Tees   171
  marking   201
  pass   42-43
  pollution, use in monitoring   17,
    108, 397, 522
  ponds for carp   572
    in Munka Ljungby   572
    for pike spawn   572
  radionuclides in   12
  sport   14, 41, 77
  stand crop   14, 17
  stocking in Mersey system   142
  Tees   169-171
  Wye, economic value in   76-77
flood(s), see also General Data

*sections*, drift
Caragh, influence on invertebrates 381
Dunajec   507
Glåma, catastrophies   481
Mersey   117
Netherlands   447
Po, influence on oligochaetes in 417
Rhine   278
Rönneå 558-559
Shannon   348, 349
  'Sint Elizabethsvloed' (1421) 447
Tees   151, 160, 163
Folla, R.   475
Forsmöllan   559
Forth, R.   169
Foynes   348, 349
France   Chapters 1, 8, 9
Frankisches Schild   321
Freshwater Biological Association (U.K.)   162
Frome, R. (Dorset)   180, 186
Frome, R. (Wye catchment)   59, 60
frost
  influence on *Rorippa nasturtium-aquaticum*   188
Fumel   210

Gabarresa, R. 529
Gamsheim   276
Garda, L.   Chapter 15
Gastric evacuation
  trout   261
Geer, R.
  sugar-beet waste   15
Gent   2
Gjelbaek (Denmark)   35
Glenvickee   365
Glossop Brook   121, 137, 139
Glossop Sewage Works   121, 126, 137
Goyt, R.   Chapter 5
Granslev Å (R.)   39
Grand Canal (Ireland)   350, 359
Gjern Å (stream)   39
Granslev Å (stream)   39
gravel
  beds in Caragh   379, 380
  dredging   278
  pits   277

substratum for *Ranunculus* 130, 190
grazing: *see also* feeding
  minor role of zooplankton in Neckar   335
Great Eggleshope Beck   154
Greta, R.   152
'Grey List'   267
Grisle: see: salmon
Groote Waard   447
ground water, groundwater
  addition of borehole water to Thames   180
  impact of ecology of chalk rivers   181
  influence on Winterbourne discharge   185
  nitrate-rich
  occurrence of *Niphargus aquilex* in
  source for *Chara* ponds   279
growth rates
  brown trout   104-105
  phytoplankton   225-227
Gudbrandsdalen   481
Gudenå Committee   43, 47
Gueule, R.   8
Gussage stream   181, 197
Gwyddon, R.   103, 105, 107

Hadsten   44, 46
Hadsten Lilleå (R.)
  B.O.D., oxygen, sewage   44-46
  macrophytes   39
Hald (stream)   32
Haringvliet   438
Harwood Beck   152, 157
Hasselbro   562
Haut-Lot, R.   Chapter 8
heat waste
  artificial aeration of cooling water for   300, 327
  influence on tench and bream in Shannon   351
  Rhine   300-301, 304
  use of algal cultures   567
  use for rainbow trout ponds   567
heavy metals: *see also various chemical summaries*
  and *Stigeoclonium*   98, 521
  *Cladophora*, sensitivity of   98

heavy metals (*cont.*)
Ebbw, pollution in  91
Ebbw Fawr, pollution in   107
Gudenå  33
Kilmastulla, pollution incident
    354
Meuse sediments  442-443
  system, monitoring in  18-20
Neckar sediments  330
Po, pollution in  412
release by oxidation of sul-
    phides  88
Rhine, pollution in  298-299,
    443
  sediments  298
Riou Mort, pollution in  216
Tees catchment  154, 148
Heidelberg  Chapter 11
Heilbronn  Chapter 11
helicopter
  use for sampling Rhine  301
helophyte  280
Helsingborg  562
herbivore(s)
  Estaragne  244, 247, 248, 254
Hereford  52
heptachlorepoxide  411, 412
heron
  importance in Po food web  425
Herrevadskloster castle  563
hexachlorcyclohexane  299
High Force (waterfall)  150, 172
High Peak  117
Hinter Rhein (R.)  268
Hirschhorn  324
Hjalmsjön (L.)  569
Hochrhein  Chapter 10
Høgronden  478
Höja  559
Hollands Diep  438
Holycross  Chapter 14
Höor (sewage) treatment plant  567
Höörsån (R.)  567
Horb  331
Hörby  558
Hörbyån (R.)  567
Hoyoux, R.  15
humic substances: *see also* dys-
    trophic, peat
  Baljaneå  565
  influence on fish ponds  572
  Pinnån  565

Tees  154
Hurworth  157, 170
hydrochloric acid  90
hydroelectric plant, hydroelectric
    power
  Aare  275
  Gudenå  26, 42, 43
  hot system  208, 213, 231, 232
  Neckar  Chapter 11
  Pinnån  569
  Po  407
  Rönneå  565, 567
hydro-power (not electric)
  Gudenå mills  26
  Herrevadskloster (Ybbarpsån)
    563
hypnospore  305

ice
  Dunajec catchment, behaviour
    518
  Glåma catchment, duration  480
  Shannon lakes, short duration in
    bays of  349
  Tatra streams, freezing in sec-
    tions of  521
  Teesdale streams  153
IJssel  272,  Chapter 16
Imperial Chemical Industries Ltd
    146
impoundment: *see also* dam, regula-
    tion, reservoir
  effect on Mn and Fe in water  69
  release of water on *Rheotany-
    tarsus* and *Ephemerella ignita*
    72
indicator species: *see also* exuviae
  chironomids  514
Industrial Revolution  118, 125,
    135, 141, 143, 170
Inny, R.  346, 353
intermittent
  flows and effects  Chapter 7
  heavy metal pollution  126
  sections of Tatra streams, biota
    of  521
  pools with *Tolypella intricata*  279
Internationale Arbeitsgemeinschaft
  der Wasswerkeim Rheineinzugsgebiet
    267
Internationale Kommission für den
  Schutz des Rheins  267

Internationale Kommission für die
  Hydrologie der Rheingebietes
  263, 272, 321, 325
invader(s) (and introductions)
  *Bangia* in Lough Derg 359
  blue-green algae in Ringsjön
    561
  changes in Kopparmölesjön 570
  *Elodea nuttallii* in Mersey 135
  *Impatiens glandulifera* in the
    19th century 135
  *Orconectes limosus* in Nether-
    lands 445
  Rhine, fish, 294
    flora 283
  *Thorea ramosissima* in Neckar
    336
Inver, R. 379
Irfon, R. 71
Irish Tourist Board 351, 388
Irlam Weir 117, 125
iron-works
  in Ebbw catchment 84, 87
irrigation water: *see also*
  *General Data sections*, abstrac-
  tion
  Lot valley 231
  Meuse catchment 2
  Rhine 266
Irwell, R. Chapter 5
Istituto di Recerca sulle
  Acque 408, 413
Ithon, R. 57, 62, 63
Ivösjön (L.) 554

Jagst, R. Chapter 11
jigging (for fish) 567
Jara Mts 268

Kaltisjokk (stream) 379
Karlsruhe 293, 299, 300, 301
Kaub 301
Kävlingeån (R.) 554
Kehl 281
Key, Lough 351
kick-sampling for macroinverte-
  brates 94, 191
Kielder Water 146
Killaloe Chapter 12
Kilmastulla R. 354
Kimber's Cottage 185, 202
Kingsley, Charles 114

Kleve-Bimmen 299
Klingstorpabäcken (stream) 563
Klippan 559, 569
Klippans Finpappersbruk AB 564,
  567
Klosterkanal 26, 38, 47-48
Klostermølle 26, 43
Koblenz 266, 268, 299, 301
Kocher, R. Chapter 11
Kolke
  Lower Rhine 283
Köln Chapter 10
Kopparmöllesjön (L.) 569, 570
Kościeliski stream 507
Kromme Rijn 441
Krościenko 502 .

La Baells reservoir 531, 534, 540,
  542, 546, 548
Lahn, R. 268
Lambourn, R. Chapter 7
Lambro, R. 408
Landesregierung (of) Baden-
  Württemberg 327
Landskrona 562
land use: *see also General Data*
  *sections*
  Caragh 370-371
  Lot 210
  Neckar 323-325
  Rhine 270
  Wye 54
Lanesboro(ugh) power station 350,
  351, 354
Langdon Befk 152
lead: *see also various chemical*
  *summaries*, heavy metals
  bryophyte monitoring 124
  co-precipitation with manganese
    326
  Dukinfield S.T.W. effluent 129
  Ebbw system 88, 104
  Neckar 326
  Po 411, 412
  Rhine water 299
  Rönneå sediments 567
  Tame 129
Lek, R. 271, 445
Leven, R. 152
Liège 2, 10, 17
life cycle
  Estaragne invertebrates 250-252

Limerick   Chapter 12
limestone: *see also General Data
  sections*, chalk
  Anoia   529
  Ebbw   88
  Dunajec catchment   504
  Lot (Causses)   210
  Meuse   8
  Neckar   321, 324
  Shannon, W. of Limerick   346
  Suir catchment   386
  Upper Teesdale   150, 158
  Wye   54
Little Brosna R.   346, 353
Llangynidr reservoir   85
Little Ouse, R.   181
Ljungbyhed   558, 563
Llangorse Lake   74
Llyngynidr reservoir   85
Llynfi, R.   74
Lodgegill Sike   153
Loop Head   349
Lorelei Cliff   268, 290
Lörrach   281
loss of species: see also eutro-
  phic, salmon
  Hochrhein macroinvertebrates
    288
  Lower Rhine macroinvertebrates
    291
  Rhine fish   293
  shad, due to overfishing   455
  Tolypella-vegetation   279
  Upper Rhine macroinvertebrates
    281-282
Lough Allen   Chapter 12
Lough Allen Canal   350
Lough Caragh   Chapter 13
Lough Derg   Chapter 12
Lough Key   351, 356, 357
Lough Neagh   361
Lough Rae   357
Ludwigshafen   271
Lugg, R.   59, 62, 63, 65, 66
Lund, 562
Lune, R.   147, 171
Luxembourg   10

Macgillycuddy Reeks (mts) 367
macroinvertebrates: *see also
  various chapter contents*, drift
  Dunajec, feeding groups in   511

Estaragne, feeding   254
  life cycles   250-252, 262
High Trata vertical zonation   515
Winterbourne, community analyses
    193-194
Wye, community analyses   67
macrophytes, macrophytic vegetation:
  *see also* invader
  Biesbosch   450-453
  biomass data   39
  Caragh   379-380
  changing distribution   135
  Dunajec, absence in   511
  Ebbw   98-99
  Glåma   488, 489
  Gudenå   37-39
  Mersey system   129-135, 143
  Po   415-416
  Rhine   280-283
  Ringsjön   560
  Shannon system   358, 359
  Suir   389, 393-395
  survey techniques   134
  Tees   155, 158-160
  Winterbourne   188-191
  Wye   63-66
Maggiore, L.   Chapter 15
Magnesian Limestone   154
Main, R.   339
Mainz   301
Maize Beck   153, 163, 164, 165, 166
Malmöhus County   554
management of rivers
  Denmark   43
  Dunajec   506
  England, North West Water
    Authority   116
  Gudenå   47-48
  Lot (hydraulic)   230-232
Manchester   125
Manchester Ship Canal   117, 118, 119,
    120, 128, 137, 138
manganese
  Meuse, radionuclide   12, 20
  Neckar, increased content in   326
Mannheim   271, 275, 325
Manning's (frictional) coefficient
    66
manure   45
Marles, R.   529
mathematical model
  thermal behaviour in Meuse   10, 11

Streeter & Phelps oxygen    44
  quality data for Wye    60
  oxygen in Gudenå    44-47
  salmon in Wye    75, 77
Mattergill Sike    168
Mattrup Å (stream)    38
Mehaigne, R.
  fish    14
  sugar-beet waste    15
Meldon Hill    153
mercury
  Po    411, 412
  Rhine water    299
methylmercury    411
Mersey and Irwell Joint Com-
    mittee    121
Mersey and Weaver River
  Authority    129
Merthyr Tydfil    87
Merwede, R.    445
metallurgical works
Meuse    442
Middleton-in-Teesdale
  167, 168, 170
Milano    Chapter 15
migration: see also salmon,
  trout
  Alosa alosa in Rhine    293
  Baetis alpinus upstream flight
    of    248
  Gudenå fish    42-43
  Isoperla viridinervis larvae
    upstream movement of    253
  Mersey salmon and sea trout
    141
  Po fish    421
  Tees trout, salmon    169,
    170-171, 172
  Wye trout    74
mill(s)
  Gudenå    43
  Rönneå    559, 563, 567
Millingen    271
Mingevatn, L.    475
Mittelgebirge    268, 272
Mjøsa, L.    481
modelling
  phytoplankton productivity
    221-230
Money Point power station    351
monitoring: see also bryophyte,
  diversity index, fish

programme for lower Rhine    457
Monmouth    Chapter 3
Monmouthshire    87
Monnow, R.    52, 72
Montserrat Mts    528
Moor House National Nature Reserve
  146, 150, 162
Mosel, R.    Chapter 10
mosquito(es)
  blood-sucking blackflies on
    Glåma    492
  plagues in Upper Rhine plain    273
Moss Burn    168
Mossø L.    26
motorboats
  Gudenå    26, 48
  speed regulations    48
Multeen R.    393, 394
multivariate techniques    12
Munka Ljungby
  fish ponds    572
Murr, R.    325
mussel bed    396

Nantyglo, Nant-y-Glo    84, 87
National Board for Science and
  Technology (Ireland)    364
National Water Council (U.K.)
  river classification scheme    126
navigation: see shipping routes
Neagh, Lough    361
Neckar, R.    268, 286, 303, Chapter 11
Neckarwestheim nuclear power plant
  329-329
Nederijn Neder-Rijn, R.    268,
  Chapter 16
Neste d'Awe (Mts)    240
névé    241
Newport    90, 99
nickel: see also various chemical
  summaries
  Ebbw rocks    88
  lower Meuse, lack of binding on
    sediments    442
  Po    411
  Rhine water    299
Niederlandische Rijnkommissie
  Waterleidingbedrivjven    267
Nieuwe Merwede    447, 449
nitrate: see also various chemical
  summaries
  influence on phytoplankton

nitrate (*cont.*)
    physiology  225
    Neckar  331
    reduction  340
    Rössjöholmsån, high concentra-
        tion in  572
    Wye  58-59
nitrification
    Rhine  301-304
    Pinnån  570
    inhibitor  301
nitrogen (combined):
    ammonia, nitrate
    Gudenå catchment  35-37
    nitrophilous species by Lower
        Rhine  283
    industry at Tarnów  521
nitrogen-fixers
    Caragh  374
Nore, R.  387
Norfolk Broads  351
Nørre Å (R.)  45
North Tyne
    *Hydropsyche* spp. in  69
Northumbrian River Authority  147
Northumbrian Water Authority
    Chapter 6
North West Water Authority
    Chapter 5
Nowy Sącz  Chapter 18
nitrogen (combined) *see also*
    ammonia, nitrate
    industry pollution at Tarnów
        508
Nowy Targ  Chapter 18
nuclear power stations
    Chooz  11-12
    Neckarwestheim  328
    Obrigheim  328
    Po  407
    Tihange I  10-12, 20

Obrigheim nuclear power plant
    328
odour(s)
    in fish  522
    in water  306
Odum method for $O_2$  411
Oehningen  299
oligotrophic
    Caragh system  Chapter 13
    Glåma  487

Kaltisjokk  379
*Lobelia* lakes  570
Lough Allen  356
oil
    cover on Rhine surface  293
    ether-extrable in Po  413
    -fueled power station  349
    law on discharge by ships  293
    pollution in Tees  171
    refinery  349
    transport on Po  407
OPL test
    Neckar  337-339
    Rhine  305-308
Oredon, Lac  240
Örkelljunga  569
organophosphorus insecticide  459
Oude Maas  457
Odde Rijn, R.  441, 445
Over Dinsdale  167
Owthe, R.  2, 17
Owbeg R.  384
Owenroe R.  374
Oxbow
    Lower Rhine  283, 291
    Upper Rhine  281
oxygen, dissolved (*see also*
    varous chemical summaries)
    Biesbosch  447
    Gudenå  34, 44-47
    Neckar  341-342
    Pinnån  570
    Po  410-413
    Rhine  295-296
    Suir  398
Øyeren, L.  Chapter 17

Pannerden  271
paper manufacturing and wastes
    121, 123, 128, 133, 138, 139, 141,
    142, 532, 564, 565
particle size distribution  217
Parc National de Pyrénées (Pyrenees
    National Park)  239, 241
pathogen: *see also* cholera, porcelain
    disease, typhoid, ulcerative
    dermal necrosis, virology, yeast
    influence of transfer scheme on  78
peat
    Ardennes  8
    bog near Nowy Targ  505
    Caragh catchment, soils in  370

commercial use  349
cover on Pennine hills  117
erosion  153
influence of drainage on  445
influence on benthic fauna  359
lower Rhine, original course of
  438
organic phosphate, probable
  source of
power station  350
Shannon catchment, soils in  348
Shannon, siltation in  354
Store Damm shores  563
Tees catchment  151-153 158, 163
Pennines  Chapters 5, 6
periphyton
  measurement of biomass  13
Perstorp, Perstorpsbacken  569
pesticide  299, 411, 458, 459
Perstorps AB  568
Pforzheim  325
phenol(s): *see also* cresol
  Ebbw  90, 91
  Po  411
  Ybbarpsån  568
  Mersey system  123
  Skerne  154
  taste in Rhine salmon  293
pheromone  78
Philippburg  276
phosphate, phosphorus: *see also
  various chemical summaries*
  algal crops in Glåma, influence
    of  483
  algal reserves  227
  effluent loadings  44
  emission standards  43-44
  Gudenå systems  35-37
  Lot, seasonal changes in  216
  Meuse system  7
  phytoplankton physiology,
    influence on  225, 227
  *Ranunculus fluitans*, influence
    on  281
  Wye  58
phytoplankton
  Biesbosch  449-450
  Elan  63
  Glåma  488, 490-491
  Gudenå  44
  Lot  Chapter 8
  Lough Caragh  382

lower Meuse  444
Lujg  63
Meuse  13
Mosel  290
Neckar  341, 342
Po  414-415
Rhine  283-287
Ringsjön  567
Rönneå  563
Rössjön  571-572
Tees  161
Västersjön  571-572
Wye  56, 61-62
Pieniny National Park  500
Piercebridge  167
Plochingen  Chapter 11
Plynlimon  52
polder(s)  278, Chapter 16
Pontelagoscuro  Chapter 15
porcelain disease (of crayfish)
  72
Portumna
  freshwater laboratory  354, 356
potash industry  295
power station: *see also* hydroelectric
  plant, nuclear power station, oil,
  peat
  Shannon  349, 350-351
primary production
  Lot  221-230
production (*see also* primary pro-
  duction)
  Estargne invertebrates  254-254
psychrophilic bacteria  304
pteridophytes: *see also Osmunda
  regalis, Salvinia*
  Wye  63
Pyrenees  Chapter 9

radionuclides, radioactive pollution
  11-12
  bryophytes  12, 20
  fish  12
Rae, Lough  358
raft
  on Dunajec  500
rain: *see also General Data sections*
  influence on electrolytes in Caragh
    372
  influence on peat in Tees catchment
    153
  influence on sodium in Wye

rain (*cont.*)
    catchment  60
Randers Fjord  26, 43
Rastat  277
Reading University  181
Recess  379
Red Sike
regulation: *see also General Data
    section*, dam, impoundment
    reservoir
    Dunajec, influence on biota in
        521-522
    influence of karst  533
    Meuse, influence on inverte-
        brates in 444-445, 457
    Tees  160, 164-165, 171
    Wye, influence on fisheries
        in  79
Rems, R.  325
Rena, R.  475
reservoirs: *see also General Data
    sections*
    Belgium  3, 5
    Caban Coch  57
    Claerwor  52
    Cow Green  Chapter 6
    Craig Goch  52, 78-79
    Czchów  505, 507
    Elan catchment  61
    Haut-Lot  213
    La Baells  531, 534, 540, 542,
        546, 548
    Llangynidr  85
    Roznów  505, 507
    Sant Pono  533
    Truyère  213, 231
Restrhein, 'Rest Rhine'  277, 280
Rhayader  74
Rhee, R.  181
Rheidol, R.  77
Rheinfelden  275
Rhine - Chemistry Agreement  267
Rhine Falls  268
Rhiwderin  93
Rien, L.  474
Riera de Rubi, R.  529, 536-537
Rijksinstituut voor Zuivering
    van Afvalwater  443
Ringsjön (L.)  Chapter 20
Riou Mort (R.)  208
Risca  87, 89
Riseberga  563

Rixton  119
Rondeslottet  478
Rönneholms mosse  559
Rossestown R.  386, 393
Rössjöhn (L.)  570
Rössjöholsån (R.)  570-572
Ross-on-Wye  54, 74
Röstånga  563
Rottweil  331
Rough Sike  163
Royal Canal  350, 359
Roznów reservoir  505, 507, 508
Ruhr, R.  303
Rushyford Beck  152
Ry Mølle  43

Saddleworth Moor  117
Salford  141
Sambre, R.  8
    bryophytes, absence in parts
        15
    copper pollution  10
    phytoplankton  13
    thermal profile  11
salmon
    Baljaneå, migration in  569
    Caragh, fame of  382
    E.E.C. directive (salmonid
        waters)  353, 391
    influence of weirs on dis-
        appearance  230
    population structure  77
    Randers Fjord  43
    Salmon Fisheries Report (1861)  87
    Shannon  349, 351, 360
    Suir  387, 396-398
    Tees  169-171
    Thames catchment, re-estab-
        lishment  199
    ulcerative dermal necrosis
        360
    Wye, overfishingin  52, 74-76
samlet  52
Sand Sike  154, 155
Sandomierska Basin  502
Sant Ponc reservoir  531
Saprobien system
    use in Denmark  39-40
Sarpsborg  492
Sarpsfossen (waterfall)  475
Sätoftasjön (L.)  559
Scandes (mts)  478, 483

Schaffhausen   275
Schwäbische Alb   321
Schwarzwald (Black Forest)   320, 321
sediments: *see also* heavy metals, sludge
  influence of high organic content in   336
  problems due to heavy metal enrichment   326
self-purification
  characterization   301
  influence of toxic substances on   301
  Gudenå   44-46
  Neckar   301-303
Semois, R.
  absence in part of   15
Sett, R.   128, 138
Severn, R.   74, 77, 151
Severn Estuary   74
Sewage (many references, including)
  treatment in Belgium   10
  treatment in Gudenå catchment   33, 43-46
  treatment in Luxembourg   10
  treatment in Mersey system Chapter 5
  treatment in Rhine   301
  untreated   45, 85, 125
  use of treatment plant heat   567
sewage-fungus: *see also Fusarium aqueductum, Sphaerotilus*
  decrease in Ybbsarpsån   568
  Dunajec   510, 521
  Ebbw   94, 97-98
  Gudenå   39
  Suir   393
  Tame   130
sewers, sewerage: *see also* sewage
  diversion of industrial wastes to   125
  Gudenå   33, 35
  inadequate, in Mersey catchment   125
  overflows   47, 90, 109, 125-126, 139, 141
  Wye   54
shading
  influence of man on reduction of   130
  lack of, high Pyrenees   Chapter 8
  macrophyte study sites on Winterbourne   189
  reduction of aquatic weeds by   48
  tolerance of *Mentha aquatica*   191
Shannonbridge power station   354
Shannon diversity index   95-97
Shannon New Town   349
Shannon Pot   346
shipping
  influence on flora   283
  law on discharge of waste oil   293
shipping route
  Gudenå   26
  Lot   208
  Meuse   20
  Neckar   322-323
  Rhine   266, 277, 278
  Shannon   350
silage   47
silica, silicate: *see also various chemical summaries*
  Lot system   214-216
  spring decrease   216
  Tees, probable influence of reservoir   160
  Wye, influence of flow in   59
  Wye, use by diatoms in   57
Silkeborg   26
silting
  upper Rhine   281
Silvermine Mts   349, 354
'Sint Elizabethsvloed' (of 1421)   447
Sirhowy, R.   Chapter 4
Site of Special Scientific Interest Wye   52
Skälderviken   559
Skanderborg   35
Skåne   Chapter 20
Skånes Natur (journal)   572
Skerne, R.   Chapter 6
Slapestone Sike
slides
  for sampling algae   280
sludge   283
Småland   574
snow: *see also General Data sections*
  Glåma catchment   478
  influence on Estaragne   Chapter 9

snow (*cont.*)
  melt in Tees catchment   151
    frequency on Caragh   367
Söderåsen   558
sodium: *see also various chemical
  summaries*
  pollution
  Winterbourne, possible influence
    of road salt on   186
  Wye catchment, influence of rain
    in   60
Solbergfoss   481
Sörrodssjön (L.)   569
South Winterbourne (stream, Dorset)   185
spawning
  loss of grounds in Rhine   292
Spearman rank correlation
  use with macroinvertebrate
    surveys   68
Spey, R.   169
Speyer   276
splash zone
  occurrence of *Schizomeris
    leibleinii*   280
Spørring Å   38
spring algal maximum
  Caragh   377
springs
  chalk   Chapter 7
  Gudenå catchment   26, 32
  Meuse catchment   6
Stackarp reservoir   564
Stidsvig (factory effluent)   570
steel-works
  Ebbw Vale history   85
  effluent   90-93
  influence of closure on *Navicula
    avenacea*   98
  influence of improved treatment
    100, 103
Stigsholm, L.   38
Stockamöllan   562
Stockport   117, 122, 128, 135, 137
Stone Age sites   26, 559
Stora Mölla   570
Storarydsdammen   568
Strasbourg   266, 276
stratification
  Cow Green reservoir   153
  Glåma estuary, salinity   494
  La Baells reservoir, summer   534

Lot, summer   216
Rhine, winter   297
Roznów reservoir   504
Stromtalpflanzen   282
Stuttgart   Chapter 11
  sewage treatment plant   329
Suck, R.   346, 353
sugar-beet waste   15
sulphuric acid   90
suspended matter, solids: *see also
  various chemical summaries*
  Ebbw system   91
  size distribution in Lot   217
Sweden   152, 379, Chapter 20
swimming
  E. Ringsjön, ban in   567
  Po, basin   407
  Rhine, evaluation of   308
  Ringsjön, popularity   563, 567
  Västersjön, present unsuitability
    571
Symond's Yat   63

Taff (Taf), R.   87
tailings pond
  overspill to Shannon   354
Tame, R.   Chapter 5
Tange
  fish-pass   42
  hydro-electric power station   26
Tange Å (R.)   45
Tange, Lake   27
tannery effluent
  Ängelholm (Rönneå)   565
  Belgium   8
  Tame   129, 140
Tarbert power station   349, 350-351
Tarnów   Chapter 18
Tatra Mts   Chapter 18
Taubergiessen (region)   227, 281
Tees Fishery Board   170
Teesside   171
telodrin   458
Ter, R.   546
territory
  invertebrates in Estargne   247
textile works, effluent   121, 128,
  139, 141, 532
Thames Valley   77
Thames Water Authority   185, 187,
  199, 204
thermocline: *see* stratification

thermophilic species
  angiosperms by Lower Rhine
    282-283
  bacteria   304
Thet, R.   181
thiodan   411
Thomand Bridge   354
Thurles Bridge   Chapter 14
tidal movements: *see also* Biesbosch
  reduction by Haringvliet dam   438
  Glåma estuary   475
Tihange I nuclear power station
  4, 11-12, 20
Tinnet springs   26
Toma, L.   268
Torino (Turin)   402
tourism: *see also* boating, canoe,
  motorboat, swimming, tourist
  Dunajec   500, 505
  Estaragne   241
  Gudenå   26, 48
  Lot   231
  Rönnea   567, 572
  Shannon   349, 351
  Suir   388
total alkalinity: *see* alkalinity
transfer (water): *see also* tunnel
Tyne-Tees   146, 147, 172
  various Welsh transfers   54,
    77-78
Tredegar   88
Trent Biotic Index   96-97, 99,
  136-137, 139
Trilum   27
Trothy, R.   60
trout: *see also Organism Index*,
  migration
  cage experiments   108, 522
  Ebbw Fawr, (re-)colonization of
    103
  Ebbw system, movements   105-107
  gastric evacuation   261
  growth rate   104-105, 258-261
  marking, marking programme
    105, 201
  ova, fate of   106
  paper mill, influence on   142
  ponds in Rossjoholmsån catch-
    ment   572
  productivity in Brandstrup Baek
    41
  Shannon lakes, fishing on   361

stocking   106
Tees system   166-172
Winterbourne survey   201-203
zinc, probable influence on,
  125, 142
zone in Meuse   14
Trout Beck   151, 152
Truyère, R.   Chapter 8
tufa   563
Tulla Rhine rectification   275
tunnel (for water transport)
  Bolmen project   574
  Tyne-Tees   146, 147
  Tweed, R.   169
typhoid
  in 19th century R. Irwell   120

ulcerative dermal necrosis (U.D.N.)
  360
Usk, R.   67
Usk Estuary   84
Usk River Authority   88, 108

Valencia Observatory   365, 367, 372,
  373
Vecht, R.   441, 445
Vegeå (R.)   554
Vejle, County   42, 43, 48
Verviers   17
Vesdre, R.   8
Vestbirk power station   43
Vestvatn, L.   475
Viborg, County   35, 42, 43, 48
Vilholt Mølle   43
Villingen-Schwenningen   330
Viroin, R.   8
virology
  Po   413
Visé   17
Visterflo, L.   475
Vistula (Wisła), R.   500, 502
Voelbaek   35
Vombsjön (L.)   574
Vorder Rhein (R.)   268
Vorarlberg Valley   288
Vorre   26

Waal, R.   268, Chapter 16
Warche, R.   8
  chromium pollution   18
Warrington   119, 128
Water Authorities (England & Wales)

644

Water Authorities (*cont.*)
  water quality criteria   126
Watercourse Act (Denmark)   48
waterfall: *see also* weir
  Dunajec   504
  Glåma   473
  Tees   150, 172
water resources: *see also General
  Data sections,* transfer, tunnel
  Shannon   351-353
  Suir   390-391
Waterston stream   180
water-table
  lowering, influence on riparian
    forest   275
  preservation along Rhine   277
  winter rain, influence on
    Winterbourne   185
Wear, R.   146, 147
Wear and Tees River Board   147
Weeds, aquatic
  cutting   48
  influence on oxygen in Wye
    65-66
  influence on silting   281
  resistance to water-flow   66
Weel (Wheel)   158, 168, 172
Weelhead Sike   168
weir
  influence on disappearance of
    salmon in Lot   230
  influence on flow of IJssel   458
  major weirs on Meuse   440
Welsh Water Authority   101
Weser, R.   283
West Beskid Mts   502
Whaley Bridge   133, 138, 141
winterbourne(s)   Chapter 7

Wittlaer   300
Woolston Weir   123
Wupper, R.   301
Wye, R.   Chapter 4, 151, 169
Wye Fisheries Association   75
Wylye, R.   181, 202

Xanthen (old river branch)   281

Yarm
Ybbarpsån, R.   563, 568
yeasts
  Po   414
Ystwyth, R.   77

Zakopane   505
zinc: *see also various chemical
  summaries*
  *Calothrix parietina,* tolerance of
    158
  *Cladophora,* sensitivity to   132,
    133
  Ebbw system   88, 107
  in paper treatment waster   123
  Mersey-Etherow   123, 132, 139-142
  Meuse   442
  monitoring by bryophytes   18,
    123-125
  Po   411, 412
  Rhine water   299
zonation
  High Tatra stream biota   515-518
zoogloea   97
zooplankton
  Lough Derg   356
  Neckar   334-335
  Po   416-418
  Ringsjön   567